FUNCTIONALIZATION OF SEMICONDUCTOR SURFACES

FUNCTIONALIZATION OF SEMICONDUCTOR SURFACES

Edited by

Franklin (Feng) Tao

University of Notre Dame, Notre Dame, Indiana

Steven L. Bernasek

Princeton University, Princeton, New Jersey

WILEY

A JOHN WILEY & SONS, INC., PUBLICATION

Published by John Wiley & Sons, Inc., Hoboken, New Jersey
Published simultaneously in Canada

Library of Congress Cataloging-in-Publication Data:

Functionalization of semiconductor surfaces / edited by Franklin (Feng) Tao, Steven L. Bernasek.
p. cm.
Includes index.
ISBN 978-0-470-56294-9 (hbk.)
1. Semiconductors–Surfaces. 2. Semiconductors–Materials. I. Tao, Franklin (Feng), 1971- II. Bernasek, S. L. (Steven L.)
QC611.6.S9F86 2012
541′.377–dc23

2011046737

Printed in the United States of America

ISBN: 9780470562949

10 9 8 7 6 5 4 3 2 1

CONTENTS

PREFACE

Functionalization of semiconductor surfaces through direct molecule attachment is an important approach to tailoring the chemical, physical, and electronic properties of semiconductor surfaces. Incorporating the functions of organic molecules into semiconductor-based materials and devices can serve various technological applications, as in the development of microelectronic computing, micro- and optoelectronic devices, microelectromechanical machines, three-dimensional memory chips, silicon-based nano- or biological sensors, and nanopatterned organic and biomaterial surfaces. Dry organic reactions in vacuum and wet organic chemistry in solution are two major categories of strategies for functionalization of these surfaces, which is the focus of this book. The growth of molecular multilayer architectures on the formed organic monolayers is described. The immobilization of biomolecules such as DNA on organic layers chemically attached to semiconductor surfaces is also introduced. The patterning of complex structures of organic layers and metallic nanoclusters on surfaces for application in sensing technologies is discussed. This book covers both advances in fundamental science and the latest achievements and applications in this rapidly growing field over the past decade.

Surface analytical techniques are used to characterize the organic functionalized interface. Chapter 2 briefly introduces the main surface analytic techniques used in this field. The functionalization of semiconductor surfaces involves the chemical binding of organic molecules on active sites of the semiconductor surface. The creation of a reactive site comprising one to several atoms is the prerequisite for the functionalization of semiconductor surfaces. Chapter 3 describes the surface structures of semiconductors and the methods used to prepare them for the attachment of organic molecules. Early studies of the chemical attachment of organic molecules on semiconductor surfaces focused on the mechanistic understanding of pericyclic reactions of the simplest unsaturated organic molecules, acetylene and ethylene. Chapter 4 describes these early studies of pericyclic reactions and other small molecules with a single functional group. Later, efforts were made to attach aromatic molecules, as these five- or six-membered aromatic molecules are the building blocks for polymers or other functional materials. Chapter 5 summarizes the chemical binding of small aromatic molecules and the reaction mechanisms for this functionalization.

Selectivity of products in the functionalization of semiconductor surfaces is an important issue, since a homogeneous organic layer on the semiconductor surface is required for high-performance molecular and semiconductor devices. However, most organic materials are actually bifunctional or multifunctional molecules.

Understanding the competition and selectivity of different functional groups on the semiconductor surfaces is fundamentally important. Chapter 6 focuses on the influence of functional groups in substituted aromatic molecules on the selection of a reaction channel. Polycyclic aromatic hydrocarbons are comprised of multiple fused benzene rings. They are promising materials for the development of new semiconductor devices using organic materials as the active layer. The chemical binding of these large aromatic systems is thus very important for the field of organic electronic devices and nanodevices. Chapter 7 summarizes the covalent binding of polycyclic aromatic hydrocarbon systems on semiconductor surfaces.

In addition to chemical binding through the formation of strong covalent bonds at the semiconductor–organic interface, organic molecules may transfer electrons to or accept electrons from semiconductor surfaces, resulting in dative bonding. This bonding mode results from the availability of electron-rich and electron-deficient sites on semiconductor surfaces. Chapter 8 describes studies of dative bonding of organic molecules on semiconductor surfaces.

Theoretical simulation has been a very important component in the developing understanding of organic functionalization of semiconductor surfaces. It is widely used to mechanistically understand the binding configuration of organic molecules, particularly multifunctional organic molecules through the point of view of kinetics and thermodynamics. Chapter 9 exemplifies the integration of this theoretical component into fundamental studies of mechanism in the field of functionalization of semiconductor surfaces.

Besides the identification of the structure of surfaces and adsorbates atom by atom in real space, scanning tunneling microscopy (STM) has another important function in breaking chemical bonds of an adsorbate to create a reactive site or radical that can then act as a precursor for a subsequent new reaction on the elemental semiconductor surface. This is a promising approach to modification and functionalization of semiconductor surfaces at the atomic level. This approach is clearly described in Chapter 10.

In parallel with the early studies of the reaction mechanisms of organic molecules on semiconductor surfaces in vacuum, studies of the functionalization of semiconductor surfaces through solution phase (wet) chemistry have been carried out. The formation of organic layers through solution chemistry is described in Chapter 11. The chemical stability of organic thin films formed in this manner is reviewed in Chapter 12. On the basis of our fundamental understanding of the functionalization of semiconductor surfaces with small organic molecules, the functionalization of semiconductors with larger, biologically relevant molecules has developed recently. Application of these systems in biosensing is developing as a very exciting field. The progress made in this area is reviewed in Chapter 13.

In summary, this book reviews many of the important research areas in the field of functionalization of semiconductor surfaces from the past two decades. These reviews are provided by leading researchers across this exciting field of surface and materials chemistry. We hope that this volume will prove to be useful to active researchers in this field, as well as students and research scientists new to the field of semiconductor surface functionalization.

We thank the contributors to this collection of reviews for the elegant research that makes up the subject of this book. We also thank them for providing the critical reviews and commentaries on the field that comprise the individual chapters here. Finally, we acknowledge the support of the Chemistry Division of the National Science Foundation that supported the work of our laboratory described here, the Chemistry Department of the National University of Singapore for ongoing support of collaborative work in this area, and the support from Department of Chemistry and Biochemistry of University of Notre Dame.

FRANKLIN (FENG) TAO

STEVEN L. BERNASEK

CONTRIBUTORS

Damien Aureau, Department of Materials Science and Engineering, University of Texas at Dallas, Richardson, TX, USA

Stacey F. Bent, Department of Chemical Engineering, Stanford University, Stanford, CA, USA

Steven L. Bernasek, Department of Chemistry, Princeton University, Princeton, NJ, USA

Ying Wei Cai, Department of Chemistry, Princeton University, Princeton, NJ, USA; Befar Chemical Group Co., Ltd, Binzhou, Shandong, China

Yves J. Chabal, Department of Materials Science and Engineering, University of Texas at Dallas, Richardson, TX, USA

Robert J. Hamers, Department of Chemistry, University of Wisconsin-Madison, Madison, WI, USA

Keli Han, State Key Laboratory of Molecular Reaction Dynamics, Dalian Institute of Chemical Physics, Chinese Academy of Sciences, Dalian, China

Md. Zakir Hossain, Department of Materials Science and Engineering, Northwestern University, Evanston, IL, USA; Graduate School of Engineering, Gunma University, Kiryu, Japan.

Erik Johansson, Department of Chemistry, California Institute of Technology, Pasadena, CA, USA

Maki Kawai, RIKEN (The Institute of Physical and Chemical Research), Wako, Saitama, Japan; Department of Advanced Materials Science, The University of Tokyo, Kashiwa, Chiba, Japan

Do Hwan Kim, Division of Science Education, Daegu University, Gyeongbuk, Republic of Korea

Sehun Kim, Molecular-Level Interface Research Center, Department of Chemistry, KAIST, Daejeon, Republic of Korea

Hangil Lee, Department of Chemistry, Sookmyung Women's University, Seoul, Republic of Korea

Nathan S. Lewis, Department of Chemistry, California Institute of Technology, Pasadena, CA, USA

Young Hwan Min, Molecular-Level Interface Research Center, Department of Chemistry, KAIST, Daejeon, Republic of Korea

Leslie E. O'Leary, Department of Chemistry, California Institute of Technology, Pasadena, CA, USA

Yongquan Qu, State Key Laboratory of Molecular Reaction Dynamics, Dalian Institute of Chemical Physics, Chinese Academy of Sciences, Dalian, China

Franklin (Feng) Tao, Department of Chemistry and Biochemistry, University of Notre Dame, Notre Dame, IN, USA

Andrew V. Teplyakov, Department of Chemistry and Biochemistry, University of Delaware, Newark, DE, USA

Mark E. Tuckerman, Department of Chemistry and Courant Institute of Mathematical Sciences, New York University, New York, NY, USA

Keith T. Wong, Department of Chemical Engineering, Stanford University, Stanford, CA, USA

Guo-Qin Xu, Department of Chemistry, National University of Singapore, Singapore

Kian Soon Yong, Institute of Materials Research and Engineering, Singapore

Yanli Zhang, Department of Chemistry and Courant Institute of Mathematical Sciences, New York University, New York, NY, USA

Yuan Zhu, Department of Chemistry and Biochemistry, University of Notre Dame, Notre Dame, IN, USA

CHAPTER 1

Introduction

FRANKLIN (FENG) TAO, YUAN ZHU, AND STEVEN L. BERNASEK

1.1 MOTIVATION FOR A BOOK ON FUNCTIONALIZATION OF SEMICONDUCTOR SURFACES

Microelectronics has grown into the heart of modern industries, driving almost all the technologies of today. Semiconductor materials play ubiquitous and irreplaceable roles in the development of microelectronic computing, micro- and optoelectronic devices, microelectromechanical machines, three-dimensional memory chips, and sensitive silicon-based nano- or biological sensors. Being the most technologically important material, silicon and its surface chemistry have received phenomenal attention in the past two decades. One important motivation for semiconductor surface chemistry is to fine-tune the electronic properties of device surfaces and interfaces for applications in several technologically important areas. Chemical attachment of molecules to the semiconductor surface enables the necessary control over electron transfer through the semiconductor–organic interface. It also allows control of the architecture of the organic overlayer by chemical modification of the functionalized silicon-based templates. It provides a versatile and reproducible way to tailor the electronic properties of semiconductor surfaces in a controllable manner.

Organic molecules are widely used in areas from plastics to semiconductors. Compared to the world of inorganic materials, organic materials exhibit unique chemical and physical properties and biocompatibility. In addition, the availability of an enormous number of organic materials with a large number of different functional groups offers opportunity for tuning physical and chemical properties that is absent for inorganic materials. A few examples are organic semiconducting polymer materials including organic electroluminescent and organic light emitting diodes. The advantage of organic materials has driven the interest in incorporation of functional organic materials, such as size and shape effects, absorption spectrum, flexibility, conductivity, chemical affinity, chirality, and molecular recognition into

Functionalization of Semiconductor Surfaces, First Edition.
Edited by Franklin (Feng) Tao and Steven L. Bernasek.

existing silicon-based devices and technologies. Dry organic reactions in vacuum and wet organic chemistry in solution on 2D templates are the two major approaches for functionalization of these surfaces.

Functionalization of semiconductor surfaces has also been driven by significant technological requirements in several areas, including micro- and nanoscale electromechanical devices and new nanopatterning techniques. By combining molecular surface modification and nanofabrication of semiconductor materials and surfaces, selective functionalization on nanopatches and formation of organic nanostructures become quite important for nanopatterning of organic materials for application in devices. The development of these heterogeneous structures requires mechanistic understanding of organic modification at the nano- and even atomic scale.

These applications in several areas have driven the enormous efforts in functionalization of semiconductor surfaces with organic materials and the subsequent immobilization of biospecies at the surface in the past two decades. Significant achievements have resulted from these efforts. Reaction mechanisms of many organic molecules have been studied at the molecular level. Numerous organic monolayers have been grown. Furthermore, organic multilayer architectures have been developed as well. Incorporation of functional biospecies such as DNA has been demonstrated and prototype biosensor devices have been made. In light of these achievements in the past two decades, a book summarizing this progress and pointing the direction for future work in this area would certainly be useful.

1.2 SURFACE SCIENCE AS THE FOUNDATION OF THE FUNCTIONALIZATION OF SEMICONDUCTOR SURFACES

1.2.1 Brief Description of the Development of Surface Science

Historically, surface science has been developed since the spontaneous spreading of oil on water was studied by Benjamin Franklin [1]. From the 1900s to 1950s, surface science studies focused on the properties of chemisorbed monolayers, adsorption isotherms, molecular adsorption and dissociation, and energy exchange [2]. As surface science became important for understanding production processes in industries such as pretreatment, activation, poisoning, and deactivation of catalysts in production, it has become one of the major areas of chemistry and physics.

In the 1950s, surface science experienced an explosive growth driven by the advance of vacuum (UHV) technology and the availability of solid-state device-based electronics with acceptable cost [3]. Thus, many efforts were made in the study of surface structure and chemistry since clean single-crystal surfaces could be prepared in UHV at that time. In the 1960s, the advance of surface analytical techniques resulted in a remarkable development of surface science. Many surface phenomena such as adsorption, bonding, oxidation, and catalysis were studied at the atomic and molecular level.

In the 1980s, the invention of various scanning probe microscopes greatly accelerated the development of surface science [4], giving rise to a second explosive growth of surface science. These probing techniques make it possible to study surfaces and interfaces at the atomic level. Particularly important, these techniques allow scientists to actually visualize surfaces at the atomic level and to identify geometric structure and electronic structure of surfaces at the highest resolution. This breakthrough radically changed the scientists' vision of the properties of materials, from average information at a large scale to local information at the atomic scale. Numerous surface phenomena were reexamined at the atomic level. For example, scanning tunneling microscopy provided an opportunity to visualize atoms on various surfaces of metals and semiconductors [5,6]. Atomic level information achieved with these techniques significantly aided in the identification of specific sites of catalytic reactions [7,8]. In addition, the breakthrough in surface analytical techniques expanded the territory of surface science to almost all areas of materials science, physics, chemistry, and mechanical and electronic engineering. More importantly, semiconductor and microelectronic industries have largely benefited from the advancement of surface science [9–13] since all the protocols for the fabrication of semiconductor devices and microelectronic components extensively involve surface science and vacuum technology.

In recent years, the development of biochemistry and biomolecular engineering has given surface science another opportunity [14,15]. Surface science studies of various bioprocesses and biofunctions performed in nature largely rely on an understanding of the complicated liquid–liquid, liquid–solid, and liquid–gas interfacial phenomena in these biosystems. For example, the functions of some biospecies largely depend on the self-assembly of specific biomolecules at interfaces in nature. The functions and behaviors of some biospecies can be mimicked on a 2D chip toward the development of biosensing technology, which extensively involves interfacial chemistry. The terms "biosurface" and "biointerface" have been widely used to describe these studies.

1.2.2 Importance of Surface Science

The term "surface science" often makes people instantly have a connection to various surface analytical techniques used in their research fields of chemistry, materials science, and physics. It is true that the development of surface science has significantly relied on the invention and advance of new analytical techniques capable of providing different information at surfaces and interfaces [1,16]. In fact, every aspect of our daily life and work involves surface science. Most of the production processes in chemical industries involve catalytic reactions performed at the interface *between* solid catalysts at high temperature *and* gaseous phases under high pressure or liquid reactants with high flow rate. New energy conversion processes extensively involve heterogeneous catalysis such as (1) evolution of H_2 and O_2 on the surfaces of cocatalysts in solar-driven water splitting [17–22] and (2) generation of electricity from oxidation of fuel molecules on the surface of electrodes (Pt or Pt-based alloy) in fuel cells [23–25]. Most issues in environmental science involve chemical process occurring on the surface of various materials such as minerals under ambient conditions [26–28]. For example, chemical conversion of greenhouse gases to fuel and conversion of toxic emissions are typically heterogeneous processes occurring on specific catalysts [29,30].

The surface chemistry of semiconductors is essentially the core of the field of functionalization of semiconductor surfaces. This is because all the processes to functionalize the inorganic surface with organic molecules must be performed as interfacial reactions. In fact, the functions and behaviors of organic layers/devices developed on semiconductor surfaces are truly determined by the surface structure and reactive site of the semiconductor, the reactivity and selectivity of the organic molecules, and the binding strength of semiconductor–organic linkages such as Si–X (X = C, O, N, S, . . .). Thus, the fundamental studies of surface science in this field are crucial, which is abundantly demonstrated in the following chapters.

1.2.3 Chemistry at the Interface of Two Phases

Typically, the interactions at two different phases can be categorized into noncovalent weak interactions and covalent binding. Corresponding to this categorization, strategies used in the design of new materials and devices can be categorized as (1) molecular self-assembly through weak noncovalent forces and (2) breaking of chemical bonds and the formation of new ones [10,31,32]. The macroscopic self-assembled structure formed on a substrate is typically held together by various weak noncovalent forces between adsorbed molecules within a self-assembled structure and between the adsorbed molecules and template (Fig. 1.1). In this case, the ordered supramolecular systems with new structures and properties form spontaneously from the original components. By using weak noncovalent binding including electrostatic interactions between static molecular charges, hydrogen bonding, van der Waals forces, π–π interactions, hydrophilic binding, and charge transfer interactions, many

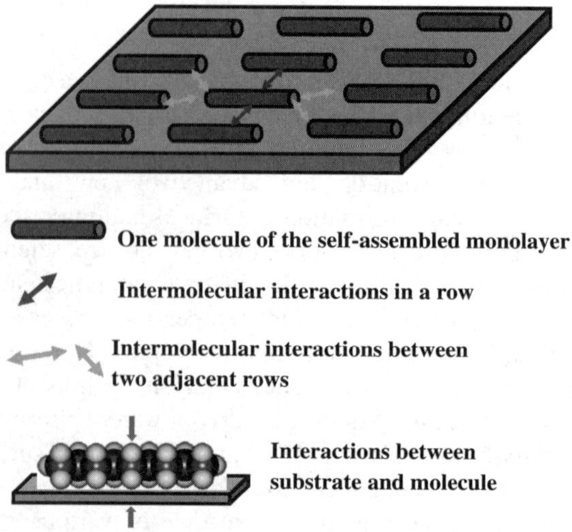

FIGURE 1.1 Schematic of a self-assembled monolayer on solid surfaces.

new self-assembled structures with various sizes, shapes, and functions have been produced [10,31,32].

In contrast to weak interactions in these systems, strong chemical bonding is commonly existent in many interfacial materials such as semiconductor surface materials and devices functionalized with organic molecules [10,31,33]. A large number of surface technologies rely on strong chemical binding at interfaces. For example, surface etching, chemisorption, and thin film growth strongly depend on the formation of chemical bonds at interfaces.

Other than the strong chemical binding and weak van der Waals interaction, chemical adsorption of molecules on metal surfaces in heterogeneous catalysis can be considered as the third type of interaction [2,16,34]. The strength of this type of interaction is between the weak van der Waals and the strong chemical binding (mostly covalent binding). Such binding with a medium strength is, in fact, necessary for heterogeneous catalysis since (1) binding of reactant molecules with certain strength results in a residence time for reactant molecules on the surface of the catalysts and the attainment of a certain coverage, and may aid in bond breaking in some cases, and (2) a strong binding will decrease molecular mobility on surfaces to some extent, which is necessary in producing intermediates or the final product molecules.

Regarding the functionalization of semiconductor surfaces for the preparation of new semiconductor devices, biosensors, molecular electronic devices, and nano-patterning templates, a strong and highly selective binding of organic molecules or biospecies is actually necessary. In most cases, the binding between the organic molecule and the semiconductor surface is covalent bonding instead of van der Waals forces.

1.2.4 Surface Science at the Nanoscale

Surface science has been studied at nanoscale well before the "nano" term was frequently used. Surface processes are performed at the nanoscale though the size of a surface could be as large as centimeter or more. In fact, the information volume along the surface normal is in the range of nanometers, since interaction at the interface is performed only in the surface region with a thickness of a few atomic layers, which is distinctly different from homogeneous process of organic reactions occurring in solution. In addition, STM has revealed that actually most samples are heterogeneous in lateral dimensions. Typically, a uniform surface feature is identified only at tens of nanometers. Thus, surface processes do occur at the nanoscale though the size of the material is macroscopic. For a crystallite with a size less than 100 nm such as 0D, 1D, 2D, and 3D nanomaterials, certainly the surface chemistry on these materials is already at the nanoscale. Overall, studies of chemistry on the surface at the nanoscale are important for understanding chemical and physical properties of solid surfaces. Thus, we term the surface chemistry on nanomaterials or nanoscale domain on the surface of materials with macroscopic size as *nanoscale surface science*.

For surfaces with different size at the nanoscale, there are size-dependent surface structural features. For example, as schematically shown in Fig. 1.2, fractions of

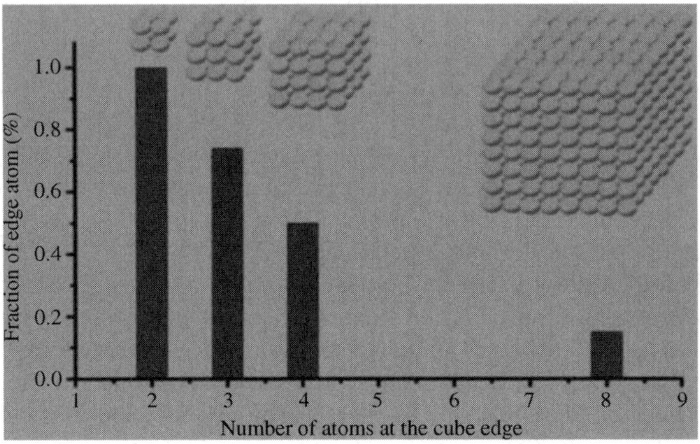

FIGURE 1.2 Fraction of atoms at edge and corner of nanoparticles with different size.

atoms at the edge of the surface increases with a decrease in size of the surface domain. This is also true for atoms at the metal–oxide interface (Fig. 1.3). More importantly, these size-dependent geometric structural factors can induce size-dependent electronic factors, surface chemistry, and functions of surfaces. The increased fraction of atoms on the surface results in large surface free energy. Chemical binding of organic molecules on these atoms at the edge of surface domains with low coordination numbers (Fig. 1.2) could be quite different from those at the center of surface domains. In addition, the packing of atoms on the surface and in surface region of nanomaterials could not follow the crystallographic periodicity of atomic packing of materials with a macroscopic size, which suggests different surface chemistry at the nanoscale in contrast to that on large domains and crystallites. Thus, size matters in surface chemistry of organic molecules on semiconductor surfaces.

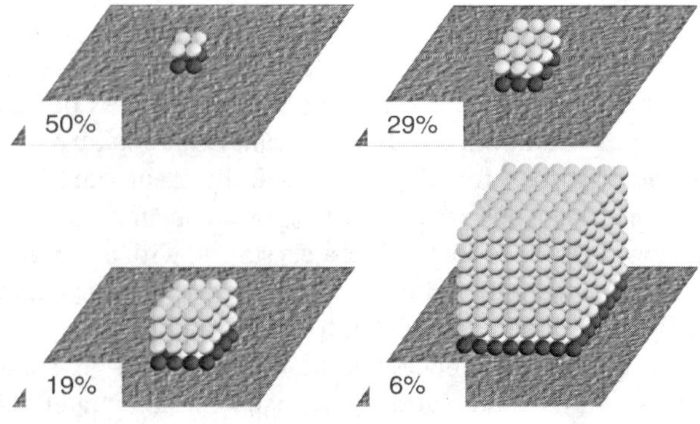

FIGURE 1.3 The size-dependent metal—oxide, per text interfacial area of catalysts. The atoms at the interface are highlighted in gray and the fractions of the interface atoms are shown at the corner of each model.

1.2.5 Surface Chemistry in the Functionalization of Semiconductor Surfaces

Chemical attachment of organic molecules to form organic thin films on different substrates is an important strategy for modification of chemical and physical properties of solid surfaces. Organic attachment is one of the main approaches to the functionalization of solid surfaces since the properties and functions of the attached organic layers are generally absent for inorganic substrates. More importantly, this organic modification and functionalization allows surface and interfacial properties to be tailored controllably since a myriad of organic molecules are available and the structure and property of organic materials can be systematically varied.

The surface and interfacial chemistry involved in the properties of semiconductor surfaces modified with organic molecules/biospecies includes surface structure, binding configuration, orientation of molecules, reaction mechanisms of organic molecules on those surfaces, and their connection to the function and behavior of the modified surfaces. Properties such as conductivity, surface polarity, friction, and biocompatibility can be modified and controlled by this functionalization.

Thus, all the aspects of functionalization of semiconductor surfaces indeed start from the fundamental surface chemistry of the semiconductor surface. From the point of view of information volume, it is at the nanoscale. In terms of reaction sites, most of the surfaces offer different reaction sites at the nanoscale. Thus, it is necessary to identify reaction details at the nanoscale. Overall, due to the nature of the heterogeneity of the functionalized surface, the understanding of surface chemistry in functionalization of semiconductor surface at the nanoscale is necessary.

1.3 ORGANIZATION OF THIS BOOK

The functionalization of semiconductor surfaces originated with fundamental studies of semiconductor surfaces at the atomic level for the successful development of semiconductor-based devices. This book covers (1) the early fundamental studies of semiconductor surface structure and the origin of surface reactive sites by using various vacuum-based surface analytical techniques, (2) creative and systematic studies of surface reactions of various organic molecules and the mechanistic understanding of reactions at semiconductor–organic interfaces at the atomic level, (3) chemical attachment of organic molecules and the formation of organic monolayers to template multilayer organic architectures on semiconductor surfaces, and (4) further functionalization of semiconductor surfaces by chemical reactions between biocompatible functional groups of organic layers and biospecies.

Characterization of the functionalized semiconductor surfaces at the molecular and atomic scales involves several techniques of spectroscopy and microscopy. The major surface science techniques will be briefly introduced in Chapter 2. Substrates used in these functionalization are typically Si(100), Si(111)-(7×7), Ge(100), and diamond(100) in the route of dry functionalization. Functionalization through wet chemistry uses hydrogenated or halogenated semiconductor surfaces (Si–H, Ge–H, Si–X, or Ge–X). Surface structure of these substrates and the origin of their reactive sites will be reviewed in Chapter 3.

The functionalization of semiconductor surfaces through dry chemistry and wet chemistry is the process that occurs at the organic molecule–semiconductor interface. Most of the chemical binding involved in these processes is strong covalent binding with a strength of 20–50 kcal/mol. The reaction mechanisms in the functionalization of these semiconductor surfaces are quite diverse because of the availability of reactive sites with different geometric and electronic structures and thus different reactivity toward organic molecules *and* definitely numerous organic materials with different functionalities. Significant efforts have been made in the understanding of these reaction mechanisms at the organic–silicon interface. Chapters 4, 5, 6, 7, 8 will review the main studies in terms of reaction mechanisms and summarize reaction mechanisms involved in most of the functionalization of semiconductor surfaces through dry chemistry. Chapter 9 reviews extensive theoretical studies of the mechanisms of organic functionalization of semiconductor surfaces. Focusing on the reaction of conjugated dienes on the semiconductor surface, insights into the reaction mechanisms and dynamics are provided.

As briefly introduced in Section 1.2.5, surface reactions are essentially performed at the nanoscale. The reaction at interfaces occurs on specific surface sites at the nanoscale. Characterization of these sites is an important component in mechanistic studies of reactions leading to the functionalization of semiconductor surfaces. One of the most important techniques to explore nanoscale surface chemistry is STM. Other than the basic function of imaging surface structure at the atomic level, STM has been used to create surface sites and further induce surface reaction of organic molecules for functionalization of semiconductor surfaces and formation of nanopatterns of organic molecules. In fact, tip-induced organic reaction can be considered as a separate strategy for functionalization of semiconductor surfaces. Chapter 10 will describe the function of STM in nanoscale surface chemistry toward functionalization of semiconductor surfaces.

Organic reactions on semiconductor surfaces performed in solution (wet chemistry) provide another important strategy for functionalization of semiconductor surfaces. These protocols and reaction mechanisms will be reviewed in Chapters 11 and 12.

Chapter 13 will summarize the applications of semiconductor surface tethered with organic molecules to the development of biosensing techniques. For example, growth of a multilayer thin film with a tunable thickness will possibly provide a flexible modification for the electronic properties of semiconductor-based devices, including electron transfer efficiency. In addition, multilayer architecture with outward facing functional groups, acting as a tether for a biospecies, is extremely important for designing biosensors. A change in physical properties such as tunneling current or fluorescence can be used to monitor the specific bioresponse. By identifying the change in physical signal induced by the binding of biospecies on the organic functionalized semiconductor surfaces, new diagnostic methods and biomedical sensing technologies can be developed.

The last chapter provides a perspective for the field of functionalization of semiconductor surfaces in the near future. It is possible to study the evolution of the surface chemistry of functionalized surfaces under reactive conditions such as in O_2 and humid environment at relatively high temperatures since *in situ* techniques such as ambient pressure XPS are available. Further research into the reaction mechanisms of immobilization of organic multilayer architectures and biospecies will be carried out. In addition, the incorporation of metal or semiconductor nanoparticles through organic reactions *between* the tethered functional groups of semiconductor surfaces *and* the end group of capping agents of nanoparticles is likely since such immobilization can bring unique physical and chemical properties into the field of functionalization of semiconductor surfaces.

REFERENCES

1. Franklin, B. *Philos. Trans. R. Soc. Lond.* **1774**, *64*, 445.
2. Somorjai, G. A.; Li, Y. *Introduction to Surface Chemistry and Catalysis*, 2nd ed. Wiley, Hoboken, NJ, **2010**.
3. Ertl, G.; Freund, H.-J. *Phys. Today* **1999**, *52* (1), 32.
4. Binnig, G.; Rohrer, H.; Gerber, C.; Weibel, E. *Phys. Rev. Lett.* **1982**, *49* (1), 57.
5. Besenbacher, F.; Lauritsen, J. V.; Linderoth, T. R.; Lægsgaard, E.; Vang, R. T.; Wendt, S. *Surf. Sci.* **2009**, *603* (10–12), 1315.
6. Vang, R. T.; Lauritsen, J. V.; Laegsgaard, E.; Besenbacher, F. *Chem. Soc. Rev.* **2008**, *37* (10), 2191.
7. Otero, R.; Rosei, F.; Besenbacher, F. *Annu. Rev. Phys. Chem.* **2006**, *57*, 497.
8. Lauritsen, J. V.; Besenbacher, F. *In Advances in Catalysis*, Vol. *50* Academic Press, New York, **2006**, p. 97.
9. Neergaard Waltenburg, H.; Yates, J. *T. Chem. Rev.* **1995**, *95* (5), 1589.
10. Tao, F.; Bernasek, S. L.; Xu, G.-Q. *Chem. Rev.* **2009**, *109* (9), 3991.
11. Stiévenard, D.; Legrand, B. *Prog. Surf. Sci.* **2006**, *81* (2–3), 112.
12. Filler, M. A.; Bent, S. F. *Prog. Surf. Sci.* **2003**, *73* (1–3), 1.
13. Wolkow, R. A. *Annu. Rev. Phys. Chem.* **1999**, *50* (1), 413.
14. Demers, L. M.; Ginger, D. S.; Park, S.-J.; Li, Z.; Chung, S.-W.; Mirkin, C. A. *Science* **2002** *296* (5574), 1836.
15. Taton, T. A.; Mirkin, C. A.; Letsinger, R. L. *Science* **2000** *289* (5485), 1757.
16. Ertl, G.; Knozinger, H.; Weitkamp, J. *Handbook of Heterogeneous Catalysis*. Wiley-VCH, **2008**.
17. Kamat, P. V. *J. Phys. Chem. Lett.* **2009**, *1* (2), 520.
18. Kamat, P. V. *J. Phys. Chem. C* **2008**, *112* (48), 18737.
19. Kamat, P. V. *J. Phys. Chem. C* **2007**, *111* (7), 2834.
20. Chakrapani, V.; Tvrdy, K.; Kamat, P. V. *J. Am. Chem. Soc.* **2010**, *132* (4), 1228.
21. Farrow, B.; Kamat, P. V. *J. Am. Chem. Soc.* **2009**, *131* (31), 11124.
22. Maeda, K.; Domen, K. *Chem. Mater.* **2009**, *22* (3), 612.

23. Adzic, R.; Zhang, J.; Sasaki, K.; Vukmirovic, M.; Shao, M.; Wang, J.; Nilekar, A.; Mavrikakis, M.; Valerio, J.; Uribe, F. *Top. Catal.* **2007**, *46* (3), 249.
24. Shao-Horn, Y.; Sheng, W.; Chen, S.; Ferreira, P.; Holby, E.; Morgan, D. *Top. Catal.* **2007**, *46* (3), 285.
25. Bing, Y. H.; Liu, H. S.; Zhang, L.; Ghosh, D.; Zhang, J. *J. Chem. Soc. Rev.* **2010**, *39* (6), 2184.
26. Centi, G.; Ciambelli, P.; Perathoner, S.; Russo, P. *Catal. Today* **2002**, *75* (1–4), 3.
27. Oberdorster, G.; Oberdorster, E.; Oberdorster, J. *Environ. Health Perspect.* **2005**, *113* (7), 823.
28. Weiss, W.; Ranke, W. *Prog. Surf. Sci.* **2002**, *70* (1–3), 1.
29. Liu, X. M.; Lu, G. Q.; Yan, Z. F.; Beltramini, J. *Ind. Eng. Chem. Res.* **2003**, *42* (25), 6518.
30. Bowker, M. *Chem. Soc. Rev.* **2008**, *37* (10), 2204.
31. Tao, F.; Bernasek, S. L. *Chem. Rev.* **2007**, *107* (5), 1408.
32. Love, J. C.; Estroff, L. A.; Kriebel, J. K.; Nuzzo, R. G.; Whitesides, G. M. *Chem. Rev.* **2005**, *105* (4), 1103.
33. Ma, Z.; Zaera, F. *Surf. Sci. Rep.* **2006**, *61* (5), 229.
34. Ertl, G. *Reactions at Solid Surface*.Wiley, **2009**.

CHAPTER 2

Surface Analytical Techniques

YING WEI CAI AND STEVEN L. BERNASEK

2.1 INTRODUCTION

A number of different surface analytical methods are used in the research described in this book. These methods provide structural, compositional, and molecular identity and reactivity information about the semiconductor surface before, during, and after functionalization. This chapter briefly summarizes the most commonly used of these techniques, providing a description of the operation of the method and an indication of the sort of information provided by the technique. There are a number of review papers and monographs [1] that describe these analytical tools in much more detail. This chapter is not meant to replace these references, but to provide a close at hand introduction to the techniques for researchers interested in the functionalization of semiconductor surfaces.

Compared to wet methods conducted in ambient environments, the dry methods for semiconductor surface modification and functionalization are performed in ultrahigh vacuum (UHV) that accommodates a wide spectrum of surface analytical techniques. With these techniques and under UHV conditions, the surface reaction process can be well controlled and characterized to study and explore the fundamentals of "*in situ*" modification and functionalization of semiconductor surfaces.

The development of UHV systems that can attain pressures from 10^{-7} to 10^{-11} Torr not only enables the preparation and maintenance of a clean surface for a reasonable period of time but also makes more accurate and surface-sensitive analytical techniques feasible. With the reduction of pressure, gas molecules behave independently and collide only with the chamber wall, as the mean free path of the molecules increases to a magnitude that is thousands of times larger than the dimension of the UHV chamber. Also, when the density of gas molecules is as low as $3 \times 10^{12}\ m^{-3}$ at 10^{-10} Torr, the rate of collision of gas molecules on the surface drops to about $10^{10}\ cm^{-2}\ s^{-1}$, keeping the sample surface uncontaminated for several

Functionalization of Semiconductor Surfaces, First Edition.
Edited by Franklin (Feng) Tao and Steven L. Bernasek.

hours. In addition, under this condition, ions, electrons, and photons can also travel freely without interaction with residual gas molecules, which is a critical condition for surface analytical techniques using these probes.

To achieve and maintain the UHV condition, a variety of pumping, sealing, and measuring technologies have been developed and implemented [2]. Vacuum pumps with different mechanisms work in different pressure ranges. Starting from atmospheric pressure, rotary mechanical pumps, sorption pumps, or turbomolecular pumps can attain pressure as low as 10^{-3}, 10^{-5}, or 10^{-7} Torr, respectively. Once the pressure is reduced to about 10^{-7} Torr, gas desorption, especially water vapor desorption, from the stainless steel chamber wall contributes significantly to the background pressure. At this stage, the chamber is baked at 150°C or higher to accelerate the desorption and pump away the gas. After cooling down, the background pressure can drop to about 10^{-10} Torr if pumped by turbomolecular pump, diffusion pump, or ion sputtering pump. The pressure is accurately measured by thermocouple gauge or ionization gauge for their respective pressure ranges. Thanks to the development of durable and high-temperature-tolerant sealing materials with extremely low vapor pressure, the UHV chamber can be baked and tightly sealed while the sample is transported, manipulated, and processed.

The above-mentioned critical vacuum components allow UHV chambers to accommodate a variety of scientific instruments. A mass spectrometer (MS) and an ion gun are generally included. In the MS, the gas molecule is ionized and fragmented, followed by examination of the fragment ions' mass-to-charge ratio. The distribution pattern of the charged fragments provides information about the gas composition in the UHV chamber and its working condition. Also, the MS is used as a detector for many analytical techniques. An ion gun ionizes and energizes inert gas molecules to bombard sample surfaces to remove contaminants for sample preparation.

Besides the MS and ion gun, a number of scientific instruments have been developed to characterize semiconductor surfaces, before and after modification. Although the principles and applications of these techniques vary widely, all share the characteristic of extreme surface sensitivity. A basic understanding of the frequently used surface analytical techniques will put the research described in this book into a clearer background and lead to an in-depth consideration of the development of research in related areas.

Based on the major applications of surface analytical technologies, the discussion in this chapter is organized to include surface structure, surface composition, electronic structure and vibrational properties, and kinetic and energetic probes.

2.2 SURFACE STRUCTURE

The structure of the semiconductor surface is the basis for further modification and functionalization, as it defines the properties, distribution, and spatial dimensions of reactive sites, which are the key to the understanding of the surface reaction mechanisms and processes. A number of technologies have been developed to detect and study the surface structure, among which low-energy electron diffraction (LEED), ion scattering methods (MEIS, ISS), and scanning tunneling microscopy (STM) are the most commonly used.

2.2.1 Low-Energy Electron Diffraction

The wave–particle duality suggests electrons diffract like water waves and photons. The diffraction of electrons was demonstrated experimentally by Davisson and Germer in 1927 [3]. However, only after its combination with UHV technologies did electron diffraction become a powerful tool to detect and study the structure of ordered surfaces.

The de Broglie relationship relates the electron's wavelength to its momentum, $\lambda = h/p$, where h is Planck's constant and p is the momentum. When the electron's energy is in the range of 20–200 eV, the range used in LEED, the wavelength of the electron varies from 0.866 to 2.74 Å. This wavelength is comparable to the dimension of the surface lattice, and the elastically backscattered electrons diffract and display unique patterns, revealing the structure of the surface. Electrons have a strong interaction with solid matter; thus, the mean free path of low-energy electrons is very short in solids (around 1 nm), making LEED highly surface sensitive.

In the experimental setup [4], a cathode filament held at a negative potential is electrically heated to emit electrons, which are then focused by electrostatic lenses, and accelerated onto the grounded metallic or semiconducting sample. If the electron-irradiated sample surface area is ordered, the elastically scattered electrons form a distinct diffraction pattern, which is displayed on a phosphor-coated screen. The inelastically scattered electrons are prevented from striking the phosphor display screen by a set of suppressor grids.

The LEED patterns and diffraction spot intensities provide information about the symmetry and atomic arrangements of superstructures, domains, and unit cells of the surface. Although the crystal structure of semiconductors and their surface periodicity are well understood today, LEED was the main technique decades ago used to investigate the semiconductor surface structures and atomic arrangements, based on the kinematic theory (single scattering), dynamical calculations, and crystallographic knowledge. LEED patterns and intensities can be calculated, and databases of observed LEED patterns have made the interpretation of LEED data more straightforward.

For organic modification and functionalization of semiconductor surfaces, LEED is normally used to verify and monitor the surface symmetry and structure during the sample preparation and modification, since the semiconductor crystal structure and surface reconstruction have been well studied and demonstrated. Moreover, the adlayers on modified semiconductor surfaces could also be characterized using this technique to determine the overlayer symmetry and order.

Even after the Si(111)-(7×7) reconstruction structure was well explained by the DAS (dimer–adatom–stacking fault) model [6], LEED still helped to experimentally improve the detailed knowledge about dimensions of the unit cell. For instance,

Webb and coworkers [7] studied the reconstructed surface with LEED and dynamical calculations based on the DAS model. By adjusting the cluster parameters, they simulated intensity variation in the diffraction features with changing incident electron energy, and thus further refined the dimensions of the surface unit cell.

2.2.2 Ion Scattering Methods

Ion scattering uses ions impinging on surfaces to study surface elemental composition, thin layer thickness, and atomic arrangement by analyzing the scattered ions. Compared to the scattering of electrons, the ions have comparable mass with surface atoms. Therefore, the elastically scattered ions will lose a significant amount of energy to surface atoms based on the conservation of energy and momentum. In addition, the de Broglie wavelength of noble gas ions is much less than the distance between surface atoms, avoiding obvious diffraction of scattered ions.

Also accommodated in the UHV chamber, ion scattering spectrometry consists of several main components, including an ion source, beam manipulators, sample manipulators, and detectors. The noble gas, such as helium or argon, is bombarded by electrons to produce positively charged ions in the ion source. The ions are then accelerated, focused, and directed to the sample surface, while the position and orientation of the sample with respect to the ion beam is adjusted by a sample manipulator. The scattered ions are gathered by the detectors, which convert the number of ions with specific energy into electrical signals [8].

When scattered by surface atoms with unique masses, the ions' kinetic energy carries identifying information about the scattering surface atoms. The energy and intensity of the scattered ions are analyzed by an electrostatic analyzer or are determined by the time of flight of the ion, while gradually changing the beam's incident angle. The elemental composition of the surface can be deduced from the mass and energy of the incident and elastically scattered ions. In addition, the intensity of scattered ions with a particular energy is proportional to the abundance of related surface atoms. Moreover, when the energy of scattered ions from a known element deviates from the expected value, the ion may have passed through a thin layer of matter and lost part of its energy. In this case, based on the relationship between the energy loss and the ion path length through the particular material, the depth of the scattering atoms and thickness of the layer on top of them can be directly calculated. In addition, if the ion–nucleus repulsion is taken into consideration, surface atoms will shadow the incident ions from colliding with some atoms along the incident direction. Thus, tracking the intensity variation of scattered ions as a function of incident angle reveals the atomic arrangement and lattice dimensions of the surface.

Before the Si(111)-(7×7) reconstruction structure was confirmed by scanning tunneling microscopy, the importance and complexity of this surface attracted intensive investigations, including ion scattering methods (ISS). Culbertson et al. [9] carefully determined the surface atomic displacement of about 0.4 Å by studying the variations in scattered ion peak intensity as a function of incident ion energy and sample orientation.

2.2.3 Scanning Tunneling Microscopy and Atomic Force Microscopy

Although the surface symmetry, composition, and even atomic arrangement can be carefully determined by experimental results from LEED and ISS, as discussed above, it is no doubt that the visualization of surfaces at atomic resolution provides more direct and accurate information. The invention of scanning tunneling microscopy enabled scientists to "see" surface atoms and earned the inventors Gerd Binnig and Heinrich Rohrer the Nobel Prize [10].

According to quantum mechanics, electrons have some probability to be present in classically forbidden energy regions. These tunneling effects are the basis of STM. When a metal tip is brought close to a surface, the vacuum gap between them results in an energy barrier, with electrons from the two materials transmitted from the one with the higher Fermi level to the other without reaching the vacuum energy level. Scanning tunneling microscopy works on this principle, measuring and controlling the tunneling current between the sample and the tip. The tunneling current is extremely sensitive to the local charge densities and the distance between the tip and the surface. After the tip scans around the surface like a finger touching and feeling it, the topography and the charge density distribution of the surface can be depicted using the information on the tip position and the measured tunneling current.

Atomic force microscopy (AFM) follows the concept of STM, but it is "feeling" the force applied to the tip by surfaces, rather than the tunneling current between them [11]. Thus, it can be applied on both conductors and insulators. The AFM probe is typically a silicon or silicon nitride cantilever with a sharp tip at the suspended end, and the tip's radius of curvature is on a nanometer scale. When the probe is scanned across surface in the "contact" mode, its deflection is monitored and recorded to map the surface at the atomic scale. Furthermore, in addition to the mechanical contact force and other short-range forces working in the contact mode, the tip is also influenced by some "long-range" forces, such as van der Waals forces, when it is away from the surface in "noncontact" mode. Therefore, when the cantilever vibrates around its resonance frequency close above the sample surface, the vibration is modified by this long-range force, and thus provides information about topography of the surface. In order to detect both the long- and the short-range force without trapping the tip by the surface, the contact and noncontact modes have been combined in the "tapping" mode. In the tapping mode, the cantilever vibrates around its resonance frequency with an amplitude ranging from 100 to 200 nm; thus, the tip contacts the surface intermittently.

The STM and AFM tips are all precisely controlled and manipulated by piezoelectric elements that operate at the atomic scale. The piezoelectric elements are, in turn, driven by stable high-voltage power supplies, which are monitored and controlled by computer. The computer also generates the surface image based on position and current information from the control circuit. Although the AFM image interpretation appears to be relatively straightforward, the STM image is a convolution of topography and electronic structure, and must be carefully considered and interpreted.

The imaging of the Si(111)-(7×7) surface stands among the most important contributions of the STM. Shortly after the invention of the STM, the Si(111)-(7×7) surface was imaged at atomic resolution [12], and the atomic arrangement and charge distribution among adatoms and rest atoms were demonstrated. This work then quickly led to the proposal of the well-accepted DAS model [13]. The atomic image of the Si(111)-(7×7) surface was also obtained with AFM, and the image resolution was further improved by finely controlling the amplitude of the cantilever vibration to enhance the short-range force sensitivity of the AFM [14].

2.3 SURFACE COMPOSITION, ELECTRONIC STRUCTURE, AND VIBRATIONAL PROPERTIES

The understanding of surface structures of semiconductors provides fundamental knowledge for their modification and functionalization, which in turn requires the analysis of surface composition, electronic structure, and vibrational properties. A variety of surface-sensitive analytical techniques are used to study the physical, chemical, and structural properties of modified surfaces and adlayers. Among these techniques, Auger electron spectroscopy (AES), photoelectron spectroscopy (PES; XPS, UPS), high-resolution electron energy loss spectroscopy (HREELS), and some synchrotron-based methods are the most widely used.

2.3.1 Auger Electron Spectroscopy

The Auger effect was first discovered in the 1920s by Lise Meitner and Pierre Auger, describing the electron emitting relaxation process that occurs after the formation of a core hole in an atom. The ejected electron is called an Auger electron. The Auger electron's kinetic energy approximately equals the energy difference between the initially excited state and the relaxed state, which are characteristics of the emitting element. The most prevalent Auger peaks observed in spectra are normally KLL or LMM types, where the letters represent the initial states of the core hole and the involved electrons in the relaxation.

AES uses an electron gun (generating electrons with an energy of several keV), an X-ray source, or ion gun incident on the surface to eject core level electrons, creating the core hole state. After core holes are created Auger electrons are emitted from the surface, and their kinetic energy is analyzed, typically by a concentric hemispherical or a cylindrical mirror electron energy analyzer. The electrons are amplified with an electron multiplier, and their kinetic energy spectrum collected and recorded.

If an electron beam is used for core hole excitation, the Auger peaks in AES spectra are superposed on a strong background of inelastically scattered electrons, appearing as sharp peaks on a rapidly varying slope. Thus derivative spectra are often used to eliminate the influence of this secondary electron background [16]. The characteristic AES peaks can identify the elemental composition of sample surfaces, and their values, compared to standard spectra, suggest the chemical states of the respective elements. In addition, with information about electron collision cross section and Auger relaxation probability, the abundance of the elements in the surface region can be determined. More commonly, similar information can be obtained by referring to external standards. Furthermore, when an AES study is conducted while the sample is sputtered by an ion beam at a known rate, the depth profile of elemental composition can be acquired.

The AES is normally used to check the semiconductor surface cleanliness, as the clean silicon surfaces show only the Si LVV peak around 92.6 eV without contaminant peaks caused by carbon (KLL at about 263.6 eV) or oxygen (KLL at 508.6 eV). Upon surface modification, the Auger peaks shift to indicate a change in chemical environment. More recently, AES was also proposed to characterize the thickness of graphene films based on AES peak intensities [17].

2.3.2 Photoelectron Spectroscopy

In addition to the Auger electrons, photoelectrons emitted from atoms in the surface region also carry information about their identity, abundance, and bonding status. Core level and valence level electrons can be excited to become photoelectrons by irradiation with X-ray and ultraviolet sources, respectively.

In X-ray photoelectron spectroscopy (XPS), the X-ray photons are normally emitted by electron-bombarded Al or Mg targets. The X-ray photons are then directed to the sample surface, ejecting photoelectrons from the surface atoms. The incident X-ray photons can be passed through a monochromator, improving the energy resolution of the photoelectron peaks. Since electrons interact with the matter strongly, the photoelectrons have a short mean free path [18] of several nanometers, making XPS surface sensitive. The photoelectrons ejected directly from the excited atoms travel trough vacuum to reach the analyzer. The analyzer is normally a concentric hemispherical analyzer with fine small apertures, followed by amplifiers and electron signal converters. The signal is monitored and recorded, indicating the core level electron binding energy (peak position) and intensity (peak area). Binding energy is obtained from a knowledge of the photon source energy and the measured electron kinetic energy ($B_E = h\nu - K_E$).

The core level binding energy, referred to as the Fermi level, not only identifies the elemental composition of the surface but also suggests the electronic environment because the electron density surrounding the nuclei partially screens the core level electrons. However, the experiments should be conducted with care to avoid any charging effect, which also causes the binding energy to shift. Owing to the energy width of the X-ray source, the XPS spectra normally require deconvolution to reveal the intrinsic peaks based on the knowledge of peak properties and composition information. There are several software packages available to fit and deconvolute XPS peaks [19]. After fitting and deconvolution, the ratio of peak area can be used to determine the composition and chemical evolution of sample surfaces. Furthermore, varying the photoelectron detection angle changes the origin depth of the detected electrons and makes XPS capable of providing depth profile information [20].

Ultraviolet photoelectron spectroscopy (UPS) is similar to XPS, but the photons from a helium discharge lamp of 21.2 (HeI) or 40.8 (HeII) eV are typically used instead of X-ray photons. Thus, the valence level electrons are excited to form photoelectrons, leaving the core levels unaffected. The UPS peaks for semiconductor surfaces represent surface states and molecular orbitals of the surface and adsorbed layers, which are directly involved in surface reactions. Hence, the spectral character and changes upon surface reactions and modifications directly indicate the evolution of involved surface sites and relevant molecular orbitals. In addition, the onset of the UPS spectra can be used to determine the work function of sample surfaces, which has important application for organic–semiconductor interface fabrication.

XPS and UPS are the most intensively employed methods to characterize modified and functionalized semiconductor surfaces. An example is given by the investigation of adsorption of acetonitrile on the Si(111)-(7×7) surface [21]. In this work, Xu and coworkers extensively applied XPS, UPS, HREELS (discussed in the following sections), and STM to experimentally study the molecular binding state of acetonitrile. In this study, XPS was used to observe the chemical shift of the core level electron binding energy to determine the respective atom bonding environment. The N atom is involved in the binding as its 1s binding energy shifts from 400.1 eV for physisorbed molecules down to 398.8 eV for chemisorption. This peak is attributed to the Si–N species. UPS monitors changes in surface states and the involvement of acetonitrile molecular orbitals upon adsorption and shows that the surface state at 0.3 and 0.7 eV below the Fermi level attenuates with the increase in acetonitrile coverage.

2.3.3 Inverse Photoemission Spectroscopy

Inverse photoemission spectroscopy (IPES) is a surface-sensitive spectroscopy that investigates the unoccupied surface states between the Fermi level and the vacuum level, which cannot be probed by PES. It employs an inverted photoelectric process. If energetic electrons are coupled with an unoccupied surface state or molecular orbital, relaxation to a lower unoccupied state can occur, emitting photons of characteristic energy. When the incident electron energy is scanned over a certain range, the energy level of the unoccupied state can be derived from the known incident electron energy and the emitted photon energy.

Besides an electron gun, a Geiger–Muller tube, which converts ionizing photons to current pulses, or a diffraction grating together with a two-dimensional position-sensitive detector, is used to quantify the energy and number of emitted photons. The assignment of these states requires careful experimental design and support from theoretical work [22].

After the common acceptance of the DAS model, much effort was put into the detailed experimental investigation of the distribution of charge and states on the Si(111)-(7×7) surface. Thus, IPES was employed to study the unoccupied states. In the study done by Nicholls and Reihl [23], the empty surface state, which is 0.5 eV above the Fermi level, can be identified with the IPES technique.

2.3.4 Vibrational Spectroscopy

For the organic–semiconductor interface, vibrational spectroscopy is widely used to study molecular structure and even orientation of adsorbates because it is versatile under many experimental conditions and conclusive for structural determination.

Among the vibrational spectroscopic methods, infrared (IR) spectroscopy and HREELS are most frequently used. However, these two techniques work on different principles with distinct selection rules and characters.

2.3.4.1 Infrared Spectroscopy IR spectroscopy is a conventional technique to excite and probe the characteristic molecular vibrations. Infrared photon energy is comparable to the molecular vibrational energy scale; thus, some of the IR photons with particular energy can be absorbed by molecules, making them vibrationally excited. The vibrational energy, which is determined by bond strength and the reduced mass of the vibrating atoms, provides the chemical bond's intrinsic character. Thus, IR spectroscopy is able to identify the chemical bonds and molecular structure of the adsorbed species. Especially, when polarized IR is used to selectively excite dipole moments in adsorbed molecules, their orientations can be deduced.

The continuous IR light is normally emitted from electrically heated inert solids. The continuous IR light is then dispersed in a monochromator, and respective absorption intensities for the adsorbed molecule are measured. More commonly, Fourier transform IR (FTIR) [24] is used instead of a dispersive spectrometer. In the FT-IR, broadband light from a Michelson interferometer is reflected from the surface and adsorbed layer. Upon interaction with the sample, some of the photons across the IR region of the incident spectrum are absorbed by the molecular overlayer, and photons reflected from the sample are detected with sensitive IR detectors. The data from the detector (intensity versus interferometer mirror displacement) is then converted to absorption spectra (intensity versus frequency) through Fourier transformation. The FTIR has the advantage of high speed, high resolution, and high throughput, resulting in high sensitivity.

The FTIR is a powerful tool to measure the molecular vibrations of adsorbates and determine their structures on semiconductor surfaces, especially when high resolution is required. For instance, the C–H stretching frequencies are substantially influenced by the rehybridization of the C atom, and can be monitored by IR. When the C atom is sp^3 rehybridized, the C–H stretching is mostly seen below 3000 cm^{-1}, and the C–H stretching with sp^2 rehybridization appears in the range from 3020 to 3100 cm^{-1}, while the sp rehybridized C–H bond can promote the vibration to around 3300 cm^{-1} [25].

2.3.4.2 High-Resolution Electron Energy Loss Spectroscopy HREELS uses a monochromatic low-energy electron beam of about 5 eV to probe adsorbed species on surfaces. When electrons impinge on solid surfaces, some of them are elastically scattered, as with low-energy electron diffraction, and some of the electrons lose energy to vibrationally excited surface species via impact, dipole, and intermediate negative ion resonance scattering mechanisms [26]. Similar to IR spectroscopy, the discrete energy losses correspond to the vibrational energy levels of adspecies, which in turn can be used to analyze their molecular and binding structures. Although the resolution is lower for HREELS (about 30 cm^{-1}) compared to FT-IR (as low as 2 cm^{-1}), HREELS is able to directly identify the vibrational features of substrate–adspecies bonds, which normally lie in the extremely low-energy range. In addition, careful comparison of the HREELS spectra collected in specular and off-specular directions, respectively, can shed light on the orientation of adspecies. Peak intensity contributed from the dipole scattering mechanism is sensitive to both adsorbate orientations and reflection angle, while the impact scattered electrons are not.

Several generations of HREELS instruments have been developed with increasing resolution and improved control systems. To achieve high resolution, two 127° sector electron energy monochromators are used in sequence to well define the electron beam emitted from a filament. After scattering from a sample surface, the electrons are analyzed also by two consecutive 127° cylindrical sector analyzers, and then counted by detectors [27].

The elastically scattered electrons create a sharp intense peak as an energy reference in HREELS spectra. The inelastic peaks appear on the sloping background of the elastic peak. A resolution of about 30 cm^{-1} or better can be routinely acquired. The interpretation and assignment of the HREELS peaks collected on modified semiconductor surfaces normally require assistance from theoretical work and well-studied IR or Raman spectra of related species.

The HREELS method has been extensively used in the study of organic modification of semiconductor surfaces. The HREELS technique can be used to detect the vibrations of substrate–adsorbate bonds (e.g., Si–O, Si–N, and Si–C) and to determine the structure and binding modes of adspecies. In addition, HREELS obtained in off-specular mode provides information about the geometry of the adsorbed molecules, as was found for adsorbed styrene molecules that adopt an orientation parallel to the surface on Si(111)-(7×7) [28].

2.3.5 Synchrotron-Based Methods

Synchrotron radiation, a tunable and highly monochromatic light source with high intensity and polarity, is key to a number of surface analytical techniques. Near-edge X-ray absorption fine structure spectroscopy (NEXAFS), energy scanned PES, and glancing incidence X-ray diffraction (GIXD) are frequently used to characterize the modified and functionalized semiconductor surface.

2.3.5.1 Near-Edge X-Ray Absorption Fine Structure Spectroscopy

When sample surfaces are irradiated with a tunable monochromatic X-ray beam from a synchrotron across a particular photon energy range, the sample absorbs photons of a specific energy, corresponding to core level electron excitation to the Fermi level, to unoccupied states or molecular orbitals of the surface–adsorbate complex. This excitation is followed by the Auger emission process. The electric current compensating the emitted Auger electrons from the grounded sample can be monitored to track the X-ray absorption [29]. Similar to XPS, NEXAFS presents peaks corresponding to core level electron binding energies. In addition to the main absorption peaks, X-ray absorption features near the transition arise from the excitation of core level electrons to unoccupied molecular orbitals, which are sensitive to the surrounding electronic and chemical environment of the surface adsorbed layer. These features serve as fingerprints for molecular and electronic structures. In addition, polarized X-rays from a synchrotron source can selectively excite electrons to certain unoccupied orbitals under particular selection rules, revealing their shape and orientation at the surface.

2.3.5.2 Energy Scanned PES This technique takes advantage of the ready tunability of light from a synchrotron source to enhance the information obtainable from photoelectron spectroscopy carried out with a fixed frequency photon source. A typical example is the use of variable photon energy to scan the kinetic energy of ejected photoelectrons originating from a specific binding energy transition. In this way, a vertical profile of the composition of a composite or layered sample can be obtained [30].

The ability to readily scan the photon energy enables a number of photoelectron-based methods that provide useful electronic structure information about the modified semiconductor surface. Some example methods [31] are as follows:

1. Photoemission yield spectroscopy, where all the photoelectrons emitted at a particular incident photon energy are collected as a function of photon energy.
2. Constant initial state spectroscopy, where the difference in energy between the incident photons and the photoelectron energy analyzer is kept fixed while scanning the incident photon energy. This scanning of the final state energies provides information about the unoccupied state density of states.
3. Constant final state spectroscopy, where the electron energy analyzer is kept at a fixed energy as the incident photon energy is scanned. This provides information about the occupied density of states of the sample.

2.3.5.3 Glancing Incidence X-Ray Diffraction X-ray Diffraction is a traditional crystal structure analytical methodology, which has contributed significantly to the determination of three-dimensional crystal and molecular structures. However, due to the ability of X-rays to penetrate deep into matter, conventional XRD technology is not sensitive to surfaces. To discriminate surface structural information from the bulk, a highly collimated intense X-ray beam from a synchrotron source irradiates the sample at a glancing incidence angle, which is typically smaller than the critical angle to enhance its surface sensitivity. While the incident beam and sample are fixed to maintain the incident angle, the scattered diffraction photons are detected at a continuously changing angle. The photon intensity versus detecting angle spectra provide information about crystal structures on surfaces and their atomic arrangements [32].

2.4 KINETIC AND ENERGETIC PROBES

Besides the compositional, electronic, and vibrational properties of the clean and modified semiconductor surfaces, the kinetic and energetic aspects of semiconductor surface reactions have attracted significant interest. The information related to reaction kinetics and energies provides fundamental scientific understanding and guides further modification and functionalization of semiconductor surfaces. Furthermore, after the modification and functionalization, the semiconductor surfaces are subject to a broad range of applications, which could require different surface properties such as stability on the one hand or reactivity on the other. Thermal programmed desorption (TPD) and molecular beam methods can help probe the reactions on surfaces from a kinetic and energetic aspect and provide information about the physical and chemical properties of the modified surface.

2.4.1 Thermal Programmed Desorption

Adsorbed molecules are bound in energy potential wells on the surface. Thermal energy can activate the adspecies in this potential well and cause it to desorb. The desorption temperature is indicative of the desorption energy, which is determined by the adsorption site, the binding configuration, and the interaction between adsorbates. By heating the sample while monitoring the desorbed fragments by mass spectroscopy, TPD can be used to study the surface coverage, thermal stability, and adsorption–desorption kinetics.

In the TPD experiment, the sample temperature is increased with a constant rate under computer control, which monitors the surface temperature and controls the heating power supply. Simultaneously, the desorbed species are tracked by mass spectrometry and the signal intensity of particular mass fragments is recorded versus temperature. Normally, the spectra acquired after different exposures to the sample are collected and compared.

The most straightforward information from the TPD spectra is the thermal stability of the adsorbed layers and the desorption temperature, which is directly linked to desorption and adsorption energies. The desorbed species also suggest the adsorption and desorption process and mechanisms. More importantly, the desorption peaks indicate different adsorption states with their corresponding desorption energies, differentiating physisorbed and chemisorbed species at different surface sites. The desorption peak area relates to the coverage of the species. With increased exposure, desorption peaks may change in position and shape, as well as peak area, providing information about desorption reaction order and other kinetic parameters [33].

For example, in the case of adsorption of acetonitrile on the Si(100) surface [34], the TPD spectra show first-order physisorption and two distinct chemisorption states, which desorb at 400 and 460 K, respectively. The desorption activation energies were calculated to be 24.6 and 29.8 kcal/mol, accordingly. On the basis of the spectral change with increasing surface exposure, the evolution of the adsorption states and a mechanism for the process was proposed. Combined with experimental results from other techniques, detailed adsorption structures and mechanisms can be concluded.

2.4.2 Molecular Beam Sources

A collimated molecular beam can be generated either by effusion of molecules from an oven or by high-pressure expansion through a skimmed nozzle source. These beam sources can be used as directed molecular dosers in the study of modified surfaces. One important application of these molecular beam sources is in the measurement of sticking coefficients of small molecules on the well-characterized surface. Because the adsorbed molecules do not contribute to the background pressure, the sticking coefficient can be determined on the basis of pressure variation in the experimental chamber when the sample surface is exposed to the collimated molecular beam. This method of King and Wells [35] is a widely used and accurate method for determining the probability of adsorption on a well-characterized modified surface. Measuring the sticking coefficient under a variety of conditions, such as differing coverages, temperatures, and preparation methods, leads to detailed information about adsorption kinetics and mechanisms.

2.5 CONCLUSIONS

A full spectrum of surface analytical techniques has been applied to study the modification and functionalization of semiconductor surfaces, ranging from structural analysis to reaction mechanism deduction. The accumulated knowledge of semiconductor surfaces and their modifications is based on these techniques. Also, the constant need for more accurate and more detailed understanding of the organic–semiconductor interface and its related applications motivates the development of surface analytical techniques. In order to obtain a complete picture of surface reactions, different surface analytical techniques are normally combined to conduct the analysis. The combination is determined by the properties of related surface reactions and the strengths of respective detection techniques. Although surface analytical techniques can provide direct evidence for properties and mechanisms of surface reactions and modification processes, theoretical work is typically needed in order to interpret the experimental data and make predictions on the basis of the results.

In the following chapters, we will constantly see the power and importance of these analytical techniques in their application to semiconductor surface modification and functionalization.

REFERENCES

1. (a) Ertl, G.; Küppers, J. *Low Energy Electrons and Surface Chemistry*. VCH, Weinheim, **1985**. (b) O'Connor, D. J.; Sexton, B. A.; Smart, R. S. C., Eds *Surface Analysis Methods in Materials Science*. Springer, Berlin, 2003. (c) Vickerman, J. C.; Gilmore, I., Eds *Surface Analysis: The Principal Techniques*. Wiley, West Sussex, UK, **2009**. (d) Riviere, J. C.; Myrha, S., Eds *Handbook of Surface and Interface Analysis*. Marcel Dekker, New York, 1998.
2. (a) Roth, A. *Vacuum Technology*. North Holland, Amsterdam, 1990. (b) O'Hanlon, J. F. *A User's Guide to Vacuum Technology*. Wiley, New York, 2003. (c) Hablanian, M. H. *High Vacuum Technology: A Practical Guide*. Marcel Dekker, New York, 1997.

3. Davisson, C.; Germer, L. H. *Phys. Rev.* **1927**, *30*, 705.
4. (a) Henzler, M. *Surf. Rev. Lett.* **1997**, *4*, 489. (b) Ertl, G.; Küppers, J. *Low Energy Electrons and Surface Chemistry*. VCH, Weinheim, 1985.
5. (a) Van Hove, M. A.; Weinberg, W. H.; Chan, C.-M., Eds. *Low Energy Electron Diffraction: Experiment, Theory and Surface Structure Determination*. Springer, Berlin, 1986. (b) Pendry, J. B. *Low Energy Electron Diffraction: The Theory and Its Application to the Determination of Surface Structure*. Academic Press, New York, 1974.
6. (a) Takayanagi, K. *Surf. Sci.* **1985**, *164*, 367. (b) Qian, G. X.; Chadi, D. J. *Phys. Rev. B* **1987**, *35*, 1288.
7. Tong, S. Y.; Huang, H.; Wei, C. M.; Packard, W. E.; Men, F. K.; Glander, G.; Webb, M. B. *J. Vac. Sci. Technol. A* **1988**, *6*, 615.
8. Rabalais, J. W. *Principles and Applications of Ion Scattering Spectrometry: Surface Chemical and Structural Analysis*. Wiley, New York, 2003.
9. Culbertson, R. J.; Feldman, L. C.; Silverman, P. J. *Phys. Rev. Lett.* **1980**, *45*, 2043.
10. Binnig, G.; Rohrer, H.; Gerber, C.; Weibel, E. *Phys. Rev. Lett.* **1982**, *49*, 57.
11. Eaton, P. J.; West, P. *Atomic Force Microscopy*. Oxford University Press, Oxford, 2010.
12. Binnig, G.; Rohrer, H.; Gerber, C.; Weibel, E. *Phys. Rev. Lett.* **1983**, *50*, 120.
13. Takayanagi, K. *Surf. Sci.* **1985**, *164*, 367.
14. (a) Giessibl, F. J.; Hernbacher, S.; Bielefeldt, H.; Mannhart, J. *Science* **2000**, *289*, 422. (b) Giessibl, F. J. *Science* **1995**, *267*, 68.
15. Auger, P. *C. R. Hebd. Seances* **1923**, *177*, 169.
16. (a) Briant, C. L.; Messmer, R. P. *Auger Electron Spectroscopy*. Academic Press, New York, 1988. (b) Palmberg, P. W. *Handbook of Auger Electron Spectroscopy*. Physical Electronics Industries, Edina, MN, 1972.
17. Xu, M.; Fujita, D.; Gao, J.; Hanagata, *N. ACS Nano* **2010**, *4*, 2937.
18. Powell, C. J.; Jablonski, A. *J. Phys. Chem. Ref. Data* **1999**, *28*, 19.
19. Walton, J.; Wincroft, P.; Fairley, N.; Carrick, A. *Peak Fitting with CasaXPS*. Acolyte Science, Cheshire, UK, 2010. Available at http://www.casaxps.com.
20. Briggs, D.; Seah, M. P. *Practical Surface Analysis: Auger and XPS*. Wiley, New York, 1990.
21. Tao, F.; Chen, X. F.; Wang, Z. H.; Xu, G. Q. *J. Phys. Chem. B* **2002**, *106*, 3890.
22. Smith, N. V. *Rep. Prog. Phys.* **1988**, *51*, 1227.
23. Nicholls, J. M.; Reihl, B. *Phys. Rev. B* **1987**, *36*, 807.
24. Buffeteau, T.; Desbet, B.; Turlet, J. M. *Appl. Spectrosc.* **1991**, *45*, 380.
25. Socrates, G. *Infrared and Raman Characteristic Group Frequencies: Tables and Charts*, 3rd ed.Wiley, Chichester, UK, 2001.
26. Ibach, H.; Mills, D. L. *Electron Energy Loss Spectroscopy and Surface Vibrations*. Academic Press, New York, 1982.
27. (a) Woodruff, D. P.; Delchar, T. A. *Modern Techniques of Surface Science*. Cambridge University Press, Cambridge, 1994. (b) Ertl, G.; Küppers, J. *Low Energy Electrons and Surface Chemistry*. VCH, Weinheim, 1985.
28. Tao, F.; Wang, Z. H.; Lai, Y. H.; Xu, G. Q. *J. Am. Chem. Soc.* **2003**, *125*, 6687.
29. Stohr, J. *NEXAFS Spectroscopy*. Springer, Berlin, 1992.
30. Haverkamp, R. G.; Marshall, A. T.; Cowie, B. C. C. *Surf. Interface Anal.* **2011**, *43*, 847.

31. IUPAC *Analytical Compendium,* Chapter 17. Available at http://old.iupac.org/analytical-compenium.

32. Robinson, I. K.; Tweet, D. J. *Rep. Prog. Phys.* **1992**, *55*, 599.

33. (a) Falconer, J. L.; Schwarz, J. A. *Catal. Rev. Sci. Eng.* **1983**, *25*, 141. (b) Miller, J. B.; Siddiqui, H. R.; Gates, S. M.; Russell, J. N.; Yates, J. T., Jr., ; Tully, J. C.; Cardillo, M. J. *J. Chem. Phys.* **1987**, *87*, 6725.

34. Tao, F.; Wang, Z. W.; Qiao, M. H.; Liu, Q.; Sim, W. S.; Xu, G. Q. *J. Chem. Phys.* **2001**, *115*, 8563.

35. King, D. A.; Wells, M. G. *Surf. Sci.* **1972**, *29*, 454.

CHAPTER 3

Structures of Semiconductor Surfaces and Origins of Surface Reactivity with Organic Molecules

YONGQUAN QU AND KELI HAN

In recent years, the interface of semiconductor materials has been a subject of intensive studies. Achieving rationally designed semiconductor surfaces with controllable configurations and special properties benefits the further applications of devices and lithography at the atomic level. Thus, various investigations have been performed to study the surface reactivity of organic molecules on semiconductor surfaces. Especially, single crystalline Si(100), Ge(100), Si(111), and diamond(100) are of particular interest. In this chapter, the surface structures and reactivity of the four semiconductor surfaces will be discussed.

3.1 INTRODUCTION

The chemistry of organic molecules on semiconductor surfaces has attracted much attention and has become a central issue in the field of surface chemistry [1–12]. Interaction between semiconductor surfaces and organic molecules is expected to be a crucial step toward applications in the areas of hybrid organic-semiconductor devices, biological devices, and nanolithography [1–12]. Many organic molecules, such as ketones, amino acids, nitrile compounds, and aromatic compounds, have been investigated to elucidate the mechanism of surface reactions between organic molecules and semiconductor surfaces. Investigations have shown that organic molecules containing more than one functional group cannot create a high degree of order at the semiconductor surface because of competitive reactions between different functional groups of the molecule and the semiconductor surface. Therefore, nonselective bonding at the semiconductor surfaces is observed. Fundamental understanding of

Functionalization of Semiconductor Surfaces, First Edition.
Edited by Franklin (Feng) Tao and Steven L. Bernasek.

the attachment chemistry of organic molecules at semiconductor surfaces and controllable surface modification are still a challenging topic in the functionalization of semiconductor surfaces.

Studying the interaction between molecules and semiconductor surfaces also elucidates the oxidation, reduction, doping and etching mechanisms of semiconductor materials [2, 11–13]. For example, the interactions between halogens and silicon are vital in understanding the semiconductor etching process, as well as the deposition process using chorine-containing precursor such as SiH_2Cl_2. Understanding the interaction between oxygen with semiconductor surfaces will unveil the mechanism of the oxidation process of semiconductor materials, and may further improve the performance of electronic devices. Understanding surface reactions between semiconductor surfaces and etchants will help optimize the etching conditions and improve etching selectivity of different semiconductor materials or different orientations of the same semiconductor materials. Furthermore, understanding the surface of semiconductor materials also will aid the investigation of epitaxial growth on semiconductor surfaces by chemical vapor deposition (CVD) methods. For example, the interaction of disilane (Si_2H_6) with Si(100) has been widely studied as a model system for understanding the chemical vapor deposition growth process.

Semiconductor surfaces that have garnered particular interest are Si(100), Ge(100), diamond(100), and Si(111). The reactivity of clean semiconductor surfaces and hydrogen- or halogen-terminated semiconductor surfaces differ because of their surface geometry and electronic structures. In order to explain and predict the selectivity of multifunctional organic molecules adsorbed on semiconductor surfaces and to control the oxidation, reduction, doping and etching processes in the current semiconductor industry, it is extremely important to understand the surface electronic structures and reactivity of semiconductors. The interactions of various organic molecules with Si(100), Ge(100), diamond(100) and Si(111) surfaces have been well reviewed [1–12]. This significant requirement has been stimulated by the development of technologies with atomic resolution such as scanning tunneling microscopy (STM), high-resolution electron energy loss spectroscopy (HREELS), synchrotron radiation (SR), and low energy electron diffraction (LEED). With the invention of the STM technology in the early 1980s [14, 15], it has been become possible to directly observe the surface structure and electronic properties of semiconductor materials at the atomic level. Herein, we summarize the structures and preparation methods of single crystal semiconductor surfaces including clean Si(100), Si(111)-(7×7), Ge(100), and diamond(100) for dry organic reactions in vacuum and hydrogen/halogen-terminated semiconductor surfaces for wet chemistry. The reactive sites and activity of these semiconductor surfaces for organic functionalization will also be discussed.

3.2 GEOMETRY, ELECTRONIC STRUCTURE, AND REACTIVITY OF CLEAN SEMICONDUCTOR SURFACES

Clean semiconductor surfaces are generally prepared by annealing substrates under ultrahigh vacuum (UHV) at high temperature, followed by a sputtering process with

Ar^+ ions to remove surface contamination such as adsorbed molecules, oxides and carbides. A slow cooling down process induces surface reconstruction of the semiconductor material. Surface structures are sensitive to vacuum. The cleanest silicon surface is obtained with around 1% defects if the initial background pressure is held within the range of 10^{-9}–10^{-11} Torr [1].

3.2.1 Si(100)-(2×1), Ge(100)-(2×1), and Diamond(100)-(2×1) Surfaces

Silicon is a group IV semiconductor with a bulk crystal structure known as the diamond structure, which consists of atoms with a coordination number of four in a tetrahedral geometry as shown in Fig. 3.1a [4]. Truncation of a bulk diamond structure along the (100) direction cuts through two Si–Si bonds for each surface silicon atom and leaves surface atoms with two covalent bonds to the bulk and two dangling bonds with one electron in each. Clearly, such a structure is thermodynamically unstable and the clean surface of Si(100) will undergo surface reconstruction to achieve a lower energy state. The earliest proposed surface reconstruction of Si(100)-(2×1) was a simple lateral translation, in which two adjacent surface silicon atoms pair together, forming a surface dimer as shown in Fig. 3.1b [16]. Although a number of alternative surface reconstruction models of Si(100) have been proposed [17, 18], theoretical calculations indicate that the most thermodynamically stable surface structure is the formation of dimers [19–21]. Lately, Si(100)-(2×1) reconstruction was finally confirmed on the basis of STM studies carried out by Hamers et al. [22–24]. Fig. 3.2 shows STM images of highly ordered surface silicon dimers for both filled and empty states. Filled and empty states were controlled by STM tip bias.

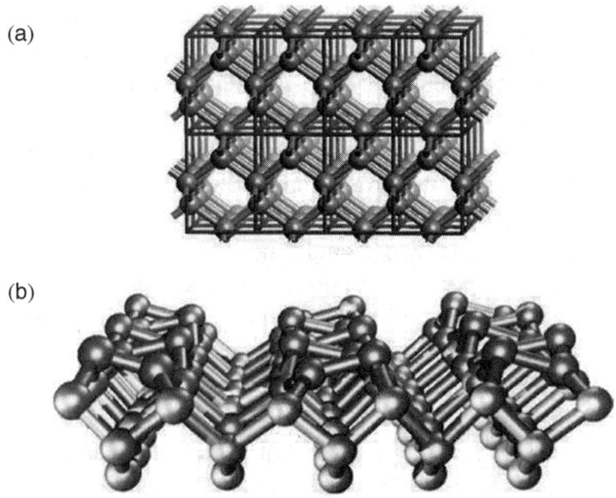

FIGURE 3.1 Surface structure of the clean Si(100) surface. (a) Bulk diamond structure of crystalline silicon. Each atom is tetrahedrally bonded to four neighboring atoms. (b) Surface structure of Si(100) surface after (2 × 1) reconstruction. The dimer rows are spaced by trenches. Reprinted with permission from Ref. 4. Copyright 2003 Elsevier Ltd.

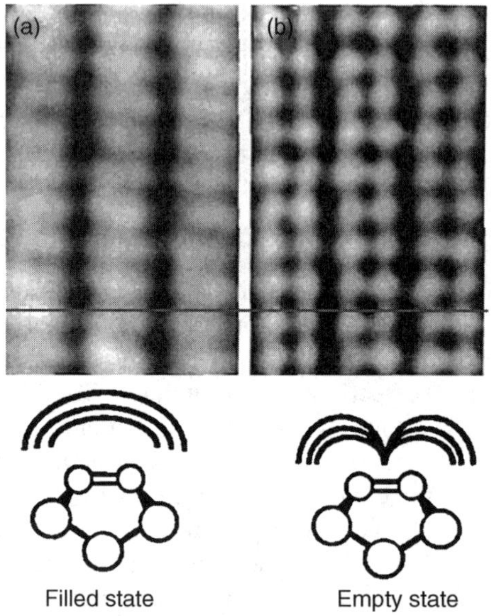

FIGURE 3.2 STM images of clean Si(001) surface probing the filled surface states and empty surface states: (a) $V_{sample} = -2.0$ V and (b) $V_{sample} = 2.3$ V. $I_{tunnel} = 0.2$ nA; dimensions = 23 Å × 77 Å. Reprinted with permission from Ref. 12. Copyright 1996 American Chemical Society.

Due to the solid-state electronic effects, the surface dimers tilt out of the surface plane. However, STM measurements at room temperature have observed nontilted surface dimers as the major surface structure as shown in Fig. 3.2 [22–24]. The tilted dimers are typically observed near edges or defects due to the break of intrinsic symmetry [22–24]. A model was proposed to explain this conflict between STM measurements and structural calculations. The surface dimers are tilted, but flipping back and forth between two tilt directions due to thermal excitation [22]. STM measurements confirm that silicon dimers are asymmetrically tilted at low temperature (120K) while measurements at room temperature display symmetric dimers [25]. Molecular dynamics simulations confirm this finding and demonstrate that silicon dimers dynamically tilt on the picosecond timescale at room temperature [26]. If the tilting of dimers is alternative between dimer rows, a $c(4\times2)$ structure can observed. Otherwise, a $p(2\times2)$ structure will be formed on the clean silicon surface. Previous LEED studies showed the $c(4\times2)$ structure, indicating an alternating arrangement of the surface dimers [27].

Fig. 3.3 shows the reactive site types of the reconstructed Si(100)-(2×1) surface. Each surface dimer contains a strong σ-bond and a weak π-bond, as shown in Fig. 3.3a. The reported values of this weak π-bond energy is approximately 2–8 kcal/mol [28–33], which is far less that the 64 kcal/mol of traditional alkenes [34]. This surface dimer with a strong σ-bond and a weak π-bond still can show similar chemical reactivity with alkenes from classic organic chemistry and lead to pericyclic reactions on Si(100)-(2×1) surfaces. Another character of the reconstructed

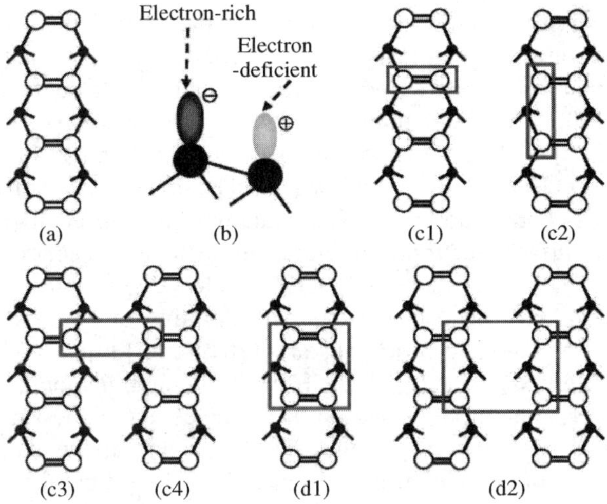

FIGURE 3.3 Surface structure and reactive sites of Si(100)-(2 × 1) reconstructed surfaces. (a) Top view of dimer and dimer row. (b) Electron transfer and distribution of buckled-down and buckled-up silicon atoms of a Si=Si dimer. (c1), (c2), (c3), and (c4) are disigma binding sites. (d1) and (d2) are tetrasigma binding sites. Reprinted with permission from Ref. 8. Copyright 2009 American Chemical Society.

Si(100)-(2×1) surface is thermodynamically tilted dimers. Electronic charge is transferred from a recessed silicon atom or "down atom" to the protruding, or "up atom," as shown in Fig. 3.3b, resulting in the electrophilic down silicon atom and the nucleophilic up silicon atom for a surface dimer. The charge separation between the nucleophilic and electrophilic dimer atoms adds zwitterionic character to the surface dimers [35], leading to the observation of nucleophilic–electrophilic reactions. Numerous experiments and calculations have indicated that surface reactions could happen across several rows or on several surface dimers of a row, which depends on the size of organic molecules and total energy after surface adsorption. Fig. 3.3c1–c4 shows the possible reactive sites for disigma binding between Si(100)-(2×1) surfaces and adsorbents on the same dimer (Fig. 3.3c1), two adjacent surface dimers of the same row (Fig. 3.3c2), or two surface dimers of two adjacent rows (Fig. 3c3–c4). The possible tetrasigma binding reactive sites are shown in Fig. 3.3d1 and d2.

Clean Ge(100) surfaces show surface (2×1) reconstruction, similar to that of the clean Si(100) surface. Thus, Ge(100)-(2×1) surfaces also show reactivity to many organic molecules due to their surface dimer characterization. Compared to Si(100)-(2×1) surfaces, Ge(100)-(2×1) surfaces exhibit statically tilted dimers at both low temperature and higher temperature [36]. Clean diamond(100) surfaces show similar surface reconstruction with a (2×1) geometry. However, the surface dimers of diamond(100) surfaces remain symmetric, leading to different chemistry activity compared to those of clean Si and Ge(100) surfaces [9, 37].

Reactivity of various molecules on semiconductor surfaces is contingent on the electronic structure and geometry of the semiconductor surfaces. Cycloaddition reactions are discussed as an example to illustrate the relationship between reactivity and surface structures. Experiments indicate that cycloaddition reactions happen on the Si(100)-(2×1) surface in a surprisingly facile pathway. In contrast, the reactivity on clean Ge(100)-(2×1) and diamond(100)-(2×1) surfaces are quite low. The reactivity of [2+2] cycloaddition of cyclopentene on Si(100)-(2×1) is ca. 10 times faster than that on Ge(100)-(2×1) surface [37]. The same reaction on diamond(100)-(2×1) surfaces is barely observed. These dramatically different results could be attributed to the surface electronic structures of the three semiconductor surfaces. In classic organic chemistry, the [2+2] C=C cycloaddition reaction is kinetically forbidden and strongly depends on the π–π^* splitting of the reactants [38]. The values of π–π^* surface-state band gaps are 110 kJ/mol [39], 140 kJ/mol [40], and 350 kJ/mol [41] for Si(100)-(2×1), Ge(100)-(2×1), and diamond(100)-(2×1) surfaces, respectively. Therefore, based on these values, the highest reactivity for the [2+2] cycloaddition reactions on the Si(100)-(2×1) surface can be predicted. Experiments also confirm this trend of reactivity. The difference of reactivity on these three semiconductor surfaces can be explained by the unique geometry of the surface dimers, which may play a critical role in controlling surface reactivity. The tilted dimers on Si(100)-(2×1) and Ge(100)-(2×1) surfaces show an asymmetric character, lowering the energy barrier for [2+2] cycloaddition reactions [4, 11]. *Ab initio* theoretical calculations also show [2+2] cycloaddition reactions on Si(100)-(2×1) and Ge(100)-(2×1) surfaces go through low barriers via such a tilted surface structure. The absence of a tilted dimer structure on diamond(100)-(2×1) surfaces is largely responsible for its low [2+2] cycloaddition reaction rate [37, 42, 43]. Additionally, molecules absorbed on Ge(100)-(2×1) surfaces can be desorbed at elevated temperatures [44]. In contrast, the absorbed molecules on Si(100)-(2×1) surfaces are much more stable. This could be attributed the Si–C bond, which is much stronger than the Ge–C bond. Theoretical calculations indicate that the Si–C bond is stronger than the Ge–C bond by 8.9 kcal/mol for 1,3-butadiene on Si(100)-(2×1) and Ge(100)-(2×1) surfaces [42].

[4+2] cycloaddition reactions are kinetically favorable for all semiconductor surfaces. Reactivity of [4+2] cycloaddition on semiconductor surfaces is similar to that of [2+2] cycloaddition. Notably, adsorption of a diene-like molecule on semiconductor surfaces shows the structure of [4+2] cycloadditions are much more stable than those of [2+2] cycloaddition. Considering the dissociated and formed bonds for both reactions, the stability of the products is mainly determined by the surface stabilization energy. The four-membered ring structures formed in [2+2] cycloadditions show much higher strain than six-membered ring structures from [4+2] cycloaddition reactions. However, adsorption of dienes on Si(100)-(2×1) and Ge(100)-(2×1) surfaces leads to nonselective surface binding due to the low energy barriers for both [2+2] and [4+2] cycloaddition [37, 42].

Another type of reaction is nucleophilic/electrophilic reaction of the Si(100)-(2×1) and Ge(100)-(2×1) surfaces due to the tilted surface dimers. The "down" surface atom (electron deficient) can react with a nucleophile to form a dative bond [45–49]. The "up" surface atom also can donate its electron density to a molecule containing an electron deficient center and form a dative bond [50]. Certain functional groups, including amino, phosphine, alcohol, carbonyl, and nitrile, can form dative bonds with the surface dimers [4]. The strength of the dative bonds is determined by the electronegativity of the atoms involved in the formation of the dative bond and the hybridization. Due to close values of electronegativity of Si(1.9) and Ge(2.01), experiments and calculations have demonstrated that the Si(100)-(2×1) and Ge(100)-(2×1) surfaces do not differ significantly in the formation of dative bonds for the same molecules [4].

3.2.2 Si(111)-(7×7) Surface

Reconstruction of Si(111) surfaces has been widely studied through various experimental techniques and theoretical calculations. The current accepted model is dimer-adatom-stacking (DAS) fault model proposed by Takayanagi et al. [51]. This structure is formed through a complicated layer-by-layer reconstruction on a base layer with 49 silicon atoms of the (111) surface, which reduces the number of 3-coördinated Si atoms from 49 to 19 in each unit cell. A top view of a 7×7 unit cell is schematically presented in Fig. 3.4a. Fig. 3.4b1 and b2 shows the unoccupied and occupied state STM images of a clean Si(111)-(7×7) surface, respectively. The 19 dangling bonds are located at 7 spatially inequivalent types of Si atoms, namely, corner and center adatoms on both faulted and unfaulted halves, faulted and unfaulted rest atoms, and the corner-hole atom. Upon reconstruction, each rest atom or corner atom has a formal charge of −1, while each adatom has electron occupancy of only 5/12, leading to a formal charge of approximately +7/12. Furthermore, each center adatom has two neighboring rest atoms, but only one for each corner adatom. The amount of charge transferred from a center adatom to the rest atom is roughly twice as much as that from a corner adatom. Consequently, the corner adatom has a higher electron density of occupied states than the center adatom, implying their different reactivities. The inherent differences in the electron

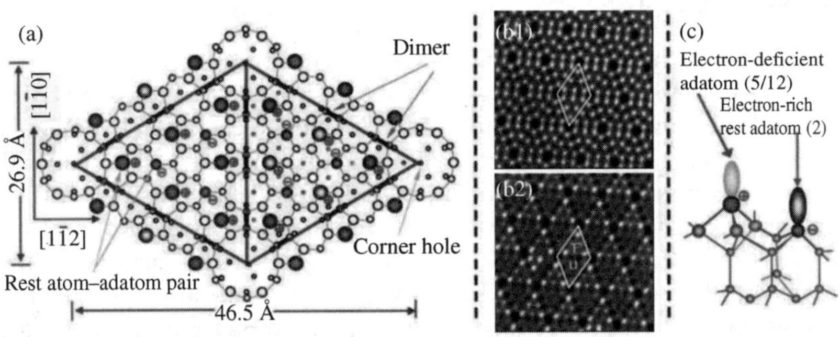

FIGURE 3.4 Surface structure and reactive sites of Si(111)-(7 × 7). (a) Top view of a reconstructed 7 × 7 unit cell. (b1) STM images of unoccupied states of a clean Si(111)-(7 × 7) surface. (b2) STM images of occupied states of a clean Si(111)-(7 × 7) surface. (c) Electronic transfer and distribution of the reactive site, a pair of adatoms, and its adjacent rest atom. Reprinted with permission from Ref. 8. Copyright 2009 American Chemical Society.

structure among these atoms are readily distinguishable in the STM images. It is evident that this surface provides a number of spatially and electronically inequivalent reactive sites. Because of the large difference in electron density among surface atoms containing dangling bonds, an adatom coupled with one adjacent rest atom can act as a strong dipole, making addition reactions of alkenes with Si(111)-(7×7) possible, similar to the polar additions of unsaturated organic molecules with Si(100) [6–8]. For convenience, the reaction involving one C=C bond is termed as [2+2]-like addition. Similarly, if a conjugated diene participates, it is called [4+2]-like addition.

3.3 GEOMETRY AND ELECTRONIC STRUCTURE OF H-TERMINATED SEMICONDUCTOR SURFACES

Hydrogenation of semiconductor surfaces changes the chemical and physical properties of semiconductors. Hydrogenation will remove surface dangling bonds of clean semiconductor surfaces, resulting in low reactivity. Hydrogenation will change surface states of semiconductor surfaces and, thus, change the physical properties of the semiconductor. For example, a clean diamond surface is an insulator, while H-terminated diamond is a p-type semiconductor [52]. There are many applications for H-terminated semiconductor surfaces including nanolithography and molecular electronic devices [1]. To obtain controllable H-terminated semiconductor surfaces will be a key issue for future applications. In this section, the preparation of H-terminated semiconductor surfaces and structures will be discussed. There are three general methods to create H-terminated semiconductor surfaces: (1) hydrogenation under UHV with hydrogen atoms; (2) hydrogenation by chemical methods; and (3) hydrogen plasmas.

3.3.1 Preparation and Structure of H-Terminated Semiconductor Surfaces Under UHV

This method is the most common method of preparation for H-terminated semiconductor surfaces. The interaction between clean semiconductor surfaces and hydrogen molecules is extremely weak because of the high bond energy of hydrogen. A hot tungsten filament is used to generate hydrogen atoms by thermal decomposition of hydrogen molecules under UHV conditions [53–55]. The method is a simple and a general method for creating H-terminated Si(100)-(2×1), Ge(100)-(2×1) and Si (111)-(7×7) surfaces. Adjustable experimental parameters that will affect the structures of the semiconductor surfaces include temperature of the tungsten filament, temperature of the semiconductor surface, distance between the filament and sample, and pressure of hydrogen in the chamber at the time of deposition [1, 8]. The surface structures and formation of hydrides on the H-terminated semiconductor surfaces can be determined by STM, RHEED, and HREELS. Complete hydrogenation can be achieved by the above method. Herein, H-terminated Si(100)-(2×1) surfaces will be discussed as a model system for study of the hydrogenation process.

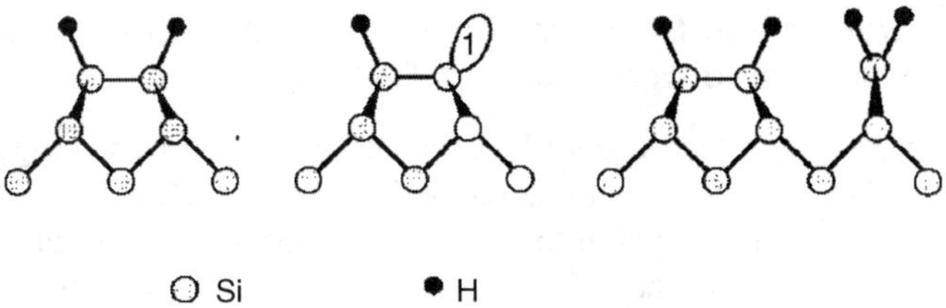

FIGURE 3.5 Structural model of hydrogen on Si(001), showing monohydride, hemihydride, and (3×1) structures. Reprinted with permission from Ref. 12. Copyright 1996 American Chemical Society.

STM studies indicate that there are three structures for absorbed hydrogen including monohydride, hemihydride, and ordered (3×1) structures as shown in Fig. 3.5 [56].

The common structure of H-Si(100) surfaces is the monohydride, in which hydrogen atoms bond to each end of the dimers. At lower coverage, it is possible to form hemihydride structures, in which only one hydrogen bonds to one atom of the dimers and leaves one dangling bond with one electron [56]. However, the hemihydride structure is less stable than the monohydride structure. This is easily understood after considering the energies of formation of each structure during absorption. Adsorption of two hydrogen atoms on same dimer silicon atoms only breaks one surface dimer, leading to the formation of monohydride. In contrast, the formation of hemihydride for two hydrogen atoms on the surface requires the breaking of two surface dimers. Therefore, it is thermodynamically favorable for H atoms to pair onto a single silicon dimer. All theoretical calculations support such a behavior. Boland performed an STM study that confirmed that there was a strong tendency for H atoms to bond onto the same dimer even at low coverage [54, 55, 57]. At coverage greater than one monolayer, it is possible to form an ordered (3×1) structure containing monohydride units with dihydride (SiH_2) groups as shown in Fig. 3.5 [56]. The high H exposure results in a disordered mixture of monohydride, dihydride, and trihydride [56].

3.3.2 Preparation and Structure of H-Terminated Semiconductor Surfaces in Solution

H-terminated semiconductor surfaces also can be prepared through wet chemistry methods. Clean silicon or germanium substrates with a layer of native oxide are immersed into dilute HF or NH_4F solution to remove native oxide and subsequently form H-terminated surfaces [58–60]. The surface smoothness can be identified by the sharpness of vibrational peaks of ν(Si-H) stretch at $\sim$2080 cm^{-1} or the ν(Ge-H) stretch at $\sim$2100 cm^{-1} [59]. Multiple hydrides, such as SiH, SiH_2, and SiH_3, can be formed by adjusting reaction conditions. Treatment of crystal silicon(100) wafers with dilute (1–2%) aqueous HF yields the dihydride SiH_2 surface with roughness on the nanometer scale [61]. Atomically flat surfaces with monohydride SiH termination can be achieved through treatment of commercial Si(111) orientation with degassed 40% aqueous NH_4F [62].

3.3.3 Preparation and Structure of H-Terminated Semiconductor Surfaces Through Hydrogen Plasma Treatment

The H-diamond(100) surface cannot be obtained by thermal cracking of hydrogen molecules and, therefore, is H-terminated via hydrogen plasma. Microwave heating of hydrogen gas at a relatively high pressure generates a hydrogen plasma that is in contact with the sample surfaces for hydrogenation [63]. Hydrogen plasma plays three roles during the process, including removal of surface roughness, etching of surface graphite-like carbon, and hydrogenation. The surface of H-terminated diamond(100) is stable in air for several weeks.

3.3.4 Reactivity of H-Terminated Semiconductor Surface Prepared Under UHV

For fully hydrogenated semiconductor surfaces, all atoms are fourfold coordinated. It is generally believed that the surface would be passive or inert with respect to other molecules because of the strong Si–H bonds. However, the reaction of disilane (Si_2H_6) has been observed on the monohydride Si(100) surfaces with a much slower reaction rate compared to that on the clean Si(100) surface [64]. This nonzero reactivity of monohydride Si(100) surface implies it is possible to grow epitaxial silicon by CVD process at low temperature because thermal desorption of H atom is not necessary. Additionally, exposure of monohydride Si(100) surfaces to H atoms also can cause desorption of H_2 [65, 66]. Generally, however, the fully hydrogenated semiconductor surfaces are nearly inert to surface chemical reaction. For future applications, it is important to generate activated H-terminated semiconductor surfaces. One approach is to create partially hydrogenated semiconductor surfaces, leaving some surface atoms free for desired surface functionalization.

3.3.5 Preparation and Structure of Partially H-Terminated Semiconductor Surfaces

Three general methods are used to partially release hydrogen atoms bonded on the semiconductor surfaces, including electron stimulated desorption (ESD), photon stimulated desorption (PSD), and thermal desorption. If selective removal of hydrogen atom from H-terminated semiconductor surfaces could be achieved, the surface chemical and electronic properties could be controlled.

ESD uses an electron beam to irradiate selective area of samples to induce desorption of hydrogen atoms. ESD has the advantages of a selective energy window of electrons and the angle of incident incoming electrons. Because the size of electron beam varies from 100 nm^2 to $\sim mm^2$ the desorption is an average over thousands of Si–H cleavage reactions. Using an STM tip as an electron beam source can achieve atomic scale manipulation. One can selectively remove single hydrogen atoms in a very selective area [1].

PSD can be carried out with lasers or synchrotron radiation. The synchrotron-based PSD has the advantage of a tunable energy window for selective excitation of chemical bonds. Laser-based PSD has the same advantages with selective wavelengths for surface desorption. By controlling the intensity and beam size, and using different patterns of multiple beams, PSD and ESD can create patterned surface active sites.

Thermal desorption, a nonlocalized technique, deals with the whole semiconductor surface. For H-terminated Si(100) surfaces, hydrogen diffusion along the silicon surface was observed at 300–350°C [67]. Annealing at 450–500°C results in desorption of surface hydrogen atoms [67]. The hydrogen desorption of the H-terminated diamond(100) surface happens at higher temperatures (>900°C) [1].

H-terminated surface structures can be studied by photoemission spectra. Figure 3.6 shows the results from a fully H-terminated Si(100) surface under high temperature annealing and synchrotron radiation [68]. The spectra of the thermally annealed surface shows a decreased intensity of Si–H bands (3.6, 4.2, 5.8, and 10.1 eV) with increasing the annealing temperature (Fig. 3.6a). The spectra fit well with a surface state of 1.0 eV, corresponding to the π-orbital of isolated silicon dimers. However, the spectra cannot be fit if the surface dangling bonds are considered. The results indicate that the thermal annealing causes formation of

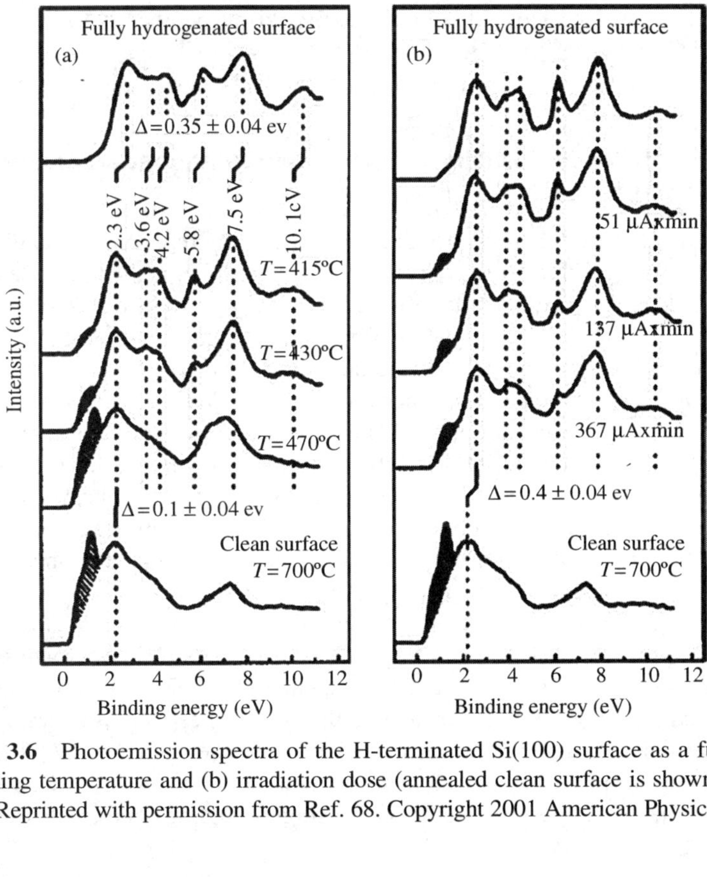

FIGURE 3.6 Photoemission spectra of the H-terminated Si(100) surface as a function of (a) annealing temperature and (b) irradiation dose (annealed clean surface is shown for comparison). Reprinted with permission from Ref. 68. Copyright 2001 American Physical Society.

surface dimers rather than surface dangling bonds. This phenomenon is explained by the fact that hydrogen diffusion leads to the pairing of dangling bonds, forming the more energetically favorable π-bonded silicon dimers during the cooling process. In contrast, synchrotron irradiation does not introduce heat and the photoemission spectra (Fig. 3.6b) shows a state at 0.7 eV that is only observed at the exposed area. This peak is explained by the existence of surface dangling bonds.

In contrast to the clean H-terminated silicon and germanium surfaces, H-terminated diamond(100) surfaces have different characteristics. Compared to the rapid oxidation of H-terminated silicon surfaces after exposure to ambient conditions, the H-terminated diamond(100) surfaces are stable in air for many weeks. H-terminated diamond surfaces have a negative electron affinity [69, 70]. C–H bonds terminating diamond surfaces create a surface dipole and cause the significant upwards band bending of the valence and conduction band, resulting in the conduction band shifting above the vacuum level. Therefore, the electrons will be spontaneously emitted when the H-terminated diamond surface is irradiated with light. In addition, diamond surfaces are converted into conducting p-type semiconductor surfaces upon hydrogenation because the surface bound hydrogen atoms play the role of a dopant by being directly responsible for the hole accumulation layer, forming shallow electron acceptors [71, 72]. The surface states of hydrogen on the H-terminated diamond(100) surface have also been monitored by photoemission spectra. C–H surface states are observed at 8.8, 11.4, and 16.9 eV, which disappear after hydrogen removal but can be recovered after hydrogenation [73].

3.3.6 Reactivity of Partially H-Terminated Semiconductor Surfaces Under Vacuum

For partially H-terminated semiconductor surfaces, the regenerated dangling bonds as the active sites exhibit reactivity with other atoms or molecules. Reactions of organic and inorganic molecules on such partially hydrogenated semiconductor surfaces have been studied. Single dangling bonds, created on fully H-terminated Si (100) surfaces by STM, can adsorb one incoming molecule. At this level, rationally designed surfaces with special properties can be obtained. Many large molecules such as norbornadiene, copper phthalocyanine (CuPc), and C60 have been studied on the partially H-terminated Si(100) surface through this method [74]. However, the binding of a single molecule to one dangling bond does not occur in the same way due to the existence of multifunctional groups in large molecules. For example, CuPc attached the silicon dangling bonds through the center copper atoms or through π-interaction with pyrrole groups [74].

Of particular interest is the local manipulation of hydrogen atoms and the subsequent promotion of surface reactivity at these sites for the growth of organic nanostructures through surface reaction mechanisms [75–77]. The growth of atomic wires of styrene on partially H-terminated Si(100) has been reported through radical generation and a hydrogen abstraction mechanism as shown in Fig. 3.7 [76]. A bias is applied to a specific hydrogen atom to generate a surface silicon dangling bond. The adsorption of a styrene molecule on this single dangling bond creates a radical on

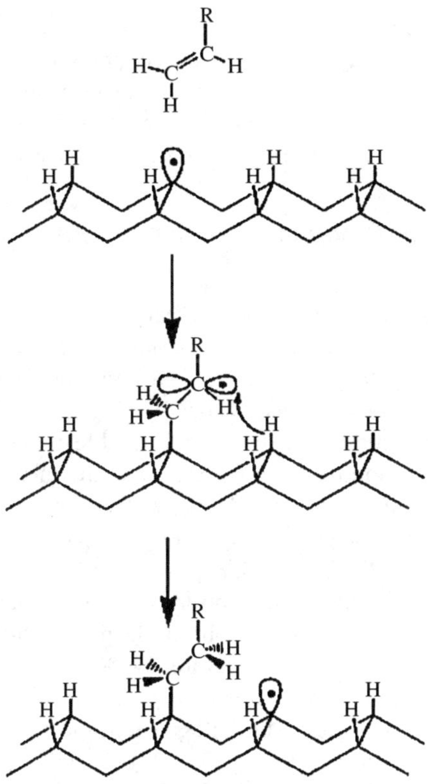

FIGURE 3.7 Proposed chain reaction mechanism for the growth of molecular nanostructures on partially H-terminated Si(100) surface. This involves the reaction of an alkene with a single dangling bond on an otherwise hydrogen-terminated Si(100) surface. Reprinted with permission from Ref. 76. Copyright 2000 Nature Publication Group.

the vinyl group. This vinyl group withdraws one hydrogen atom from an adjacent dimer terminated with hydrogen atoms, creating another dangling bond site on this dimer. This alternating creation of dangling bonds on a silicon dimer and a radical on a vinyl group leads to the formation of an atomic wire. The growth is terminated at a defect site on the dimer row.

3.4 GEOMETRY AND ELECTRONIC STRUCTURE OF HALOGEN-TERMINATED SEMICONDUCTOR SURFACES

Interactions between halogens and semiconductor surfaces are important to semiconductor etching processes as well as other surface functionalizations. In this section, the preparation and structure of halogen-terminated semiconductor surfaces are presented. The reactivity of halogen-terminated semiconductor surfaces via wet chemistry methods will be discussed in the next part. Similar to hydrogen, halogens require only one electron to form a closed-shell electronic structure on the surfaces. As a result, halogens can readily react with the semiconductor surfaces and terminate dangling bonds. The halogen-terminated semiconductor surfaces can be synthesized by exposing clean semiconductor surfaces to various halogen sources under UHV or by reacting them with H-terminated semiconductor surfaces.

3.4.1 Preparation of Halogen-Terminated Semiconductor Surfaces Under UHV

Exposure of clean semiconductor surfaces to halogen precursors (Cl_2, Br_2, etc.) leads to the formation of halogen-terminated semiconductor surfaces. One of the examples is the Cl-terminated Si(100)-(2×1) surface. At low coverage, a metastable bridge-bonded configuration can be observed as shown in Fig. 3.8b [78]. Annealing at 673K, the bridge-bonded configuration converts into stable monochloride configuration (Fig. 3.8c) [78]. At one monolayer of Cl coverage, the monochloride structure was observed on Si(100) surfaces [79, 80]. At higher Cl_2 exposure, dichloride structures are observed [81]. Br-terminated Si(100) surface shows a similar structure as that of Cl-terminated surfaces. $SiBr_2$ and $SiBr_3$ species are also observed [82]. When the temperature is higher than 850K, desorption of Br atoms lead to the formation of (3×1) structures, in which two SiBr monobromide dimers alternate with vacancies as shown in Fig. 3.8d. At higher temperature (>900°C), (3×1) structures are not observed due to the rapid desorption of $SiBr_x$.

In contrast to numerous studies on halogen-terminated Si(100)-(2×1) surfaces, the adsorption of halogen atoms on Ge(100)-(2×1) and diamond(100)-(2×1) surfaces have been studied less intensively. This is partially due to the less active surfaces of Ge(100) and diamond(100) compared to that of Si(100)-(2×1). Currently, experimental data available for the adsorption of iodine on the Ge(100)-(2×1) surface. STM investigation has shown the dominant GeI species with minor GeI_2 structure under low coverage [83, 84].

For Cl-terminated Si(111)-(7×7) surfaces, STM studies indicate the formation of Cl-terminated adatoms at low coverage [85, 86]. Upon further exposure to chlorine, Cl atoms insert into backbonds between the adatom and rest atom layer. Further exposure breaks one of the bonds between the chlorinated adatom and rest atom, producing a $SiCl_3$ group bonded directly above a rest atom layer. Saturated coverage leads to the formation of SiCl, $SiCl_2$, and $SiCl_3$ species [87]. Annealing a saturated

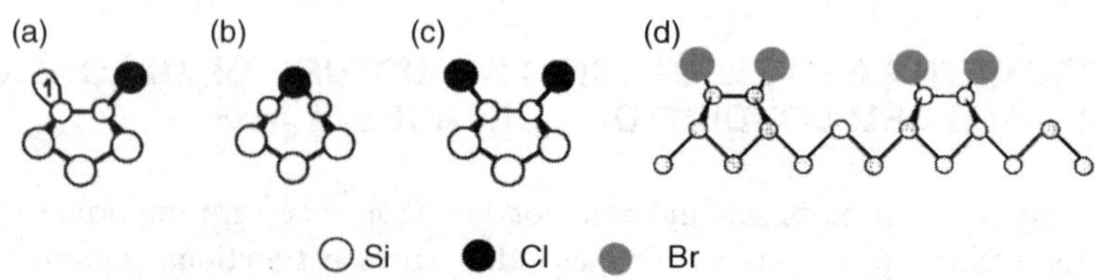

FIGURE 3.8 Structural model of halogens on Si(100), showing hemichloride (a), bridge-bonded structure (b), monochloride (c), and (3 × 1) structures (d). Reprinted with permission from Ref. 12. Copyright 1996 American Chemical Society.

surface at 675K causes surface reconstruction in which the rest atom layer is exposed, with all atoms terminated by Cl, and adatoms made up of higher $SiCl_x$ species agglomerated together into small islands [87].

3.4.2 Preparation of Halogen-Terminated Semiconductor Surfaces from H-Terminated Semiconductor Surfaces

The halogen-terminated semiconductor surface can be prepared by starting from H-terminated semiconductor surfaces. Two known methods involve treating H-terminated silicon surfaces with PCl_5 [88] at 80–100°C and *N*-bromosuccinimide [85] at 60°C using benzoyl peroxide as a radical initiator. Cl_2 or CCl_3Br can initiate the conversion from H-terminated silicon surfaces into halogen-terminated silicon surfaces [88–91]. A chlorinated germanium surface can be prepared by a reaction with gaseous HCl or 10% aqueous HCl on Ge(111) wafers pretreated with H_2O_2 and oxalic acid [92, 93].

3.5 REACTIVITY OF HYDROGEN- OR HALOGEN-TERMINATED SEMICONDUCTOR SURFACES IN SOLUTION

Surface modification of semiconductor surfaces under UHV conditions is costly under restricted reaction conditions. Therefore, various methods have been developed for semiconductor surface functionalization in solution starting from hydrogen- or halogen-terminated surfaces, such as radical-initiated hydrosilylation, thermal-driven hydrosilylation, photomediated hydrosilylation, metal complex-catalyzed hydrosilylation, electrochemical grafting, Grignard-related hydrosilylation, and mechanical carving with simultaneous hydrosilylation. In addition, chlorinated germanium surfaces can be functionalized by using ethyl Grignard reagents. These methods have all been well reviewed [2, 8].

3.5.1 Reactivity of Si and Ge Surfaces in Solution

The reaction mechanisms for hydrogen- or halogen-terminated Si surfaces are summarized in Fig. 3.9. For radical-initiated hydrosilylation, a chemical with an unsaturated bond reacts with surface Si–H or Ge–H bonds, forming an alkenyl or alkyl monolayer through the formation of Si–C or Ge–C covalent bonds [94, 95]. Generally, this type of reaction is processed by a radical initiator. Diacyl peroxide is often used to decompose and form an alkyl radical, which can abstract a hydrogen atom from the surface silicon and forms a surface silicon radical. This silicon radical can rapidly react with an unsaturated bond of an organic molecule and therefore form a radical on the carbon of the immobilized organic chain. The radical on carbon chain then takes a hydrogen atom from its adjacent Si–H bond and forms surface silicon radical. The reaction mechanism of such a reaction is illustrated in Fig. 3.9a. In the absence of radical initiator, hydrosilylation can be induced by a thermal or photo initiators. High temperature, generally 150–200°C, promotes the cleavage of surface

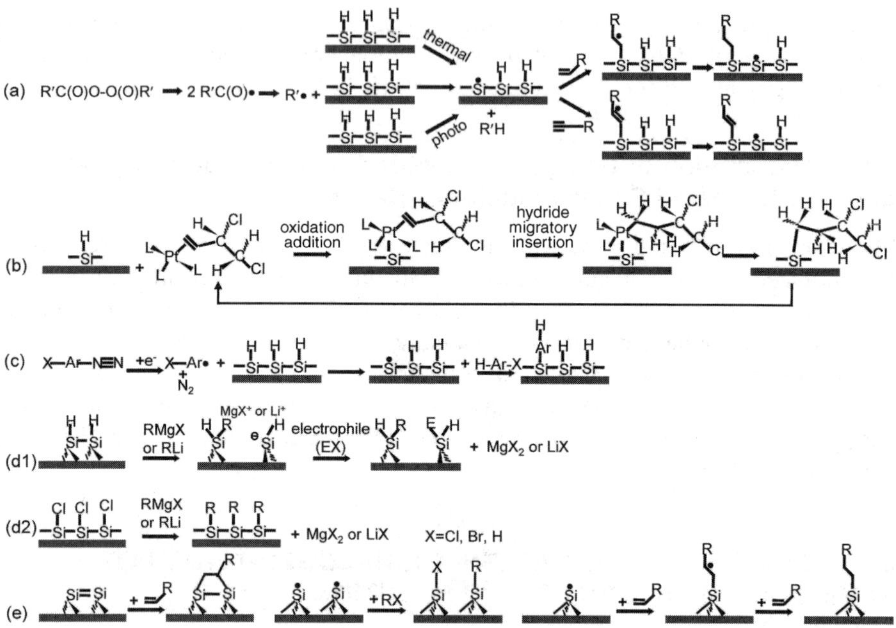

FIGURE 3.9 Reaction mechanisms for organic functionalization of hydrogenated and halogenated semiconductor surfaces in solution: (a) radical-initiated hydrosilylation, thermal-driven hydrosilylation, and photomediated hydrosilylation; (b) metal complex-catalyzed hydrosilylation; (c) electrochemical grafting; (d) Grignard-related hydrosilylation and halogenosilylation; and (e) mechanical carving and simultaneous hydrosilylation in solution. Reprinted with permission from Ref. 8. Copyright 2009 American Chemical Society.

Si–H bond to generate silicon radicals on silicon surfaces [94–101]. Such a radical on a silicon atom shows the same activity as that of hydrosilylation with the radical initiator. Photolysis is an alternative method to break Si–H bonds and generate surface silicon radicals for further reaction with unsaturated bonds of organic molecules [91, 102–106]. Hydrosilylation with the radical initiator happens homogenously in the solution without area selectivity. Thermally driven hydrosilylation may damage the semiconductor surfaces due to high temperature. In contrast, photo-driven hydrosilylation is promising in applications due to its selective exposure area and low temperature. By using masks with different sizes, shapes, and/or multiple beams with certain pattern and spot size, semiconductor surfaces can be selectively functionalized [91, 102, 103, 105–107].

The second method for surface hydrosilyation at liquid–solid interfaces is assisted by the presence of catalysts. Similar to the catalytic reactions between alkenes with silanes in classical organic chemistry, Pt(0) complexes can catalyze hydrosilylation at the liquid-solid interface between unsaturated organic molecules and a H-terminated silicon surface in solution as shown in Fig. 3.9b [108]. Notably, the Pt(0) complex can catalyze oxidation between the silicon surfaces and a trace amount of water or oxygen in solution. Lewis acid catalysts can also be used as an alternative route for surface hydrosilyation to avoid the possible surface oxidation [109–111]. Lewis acids have been successfully used to functionalize H-terminated Si(111) surfaces with various organic compounds at a relatively high temperature, such as 100°C.

Electrochemistry-based functionalization can avoid the problem of oxidation during the hydrosilyation process thanks to the electron-rich cathode, which draws nucleophilic attacks of water. Aryl radicals can be generated by applying a negative potential of 1 V in an electrolyte containing HF and bromobenzene diazonium salt. Then, aryl radicals can initiate surface modification by replacing hydrogen atom and forming Si–C covalent bonds on H-terminated silicon surfaces [112, 113].

The addition of organolithium and Grignard reagents to the silicon surface has been carried out on H-terminated porous silicon surfaces [114]. The reaction proceeds through a transmetalation reaction (Fig. 3.9d1) in which the formation of Si–C bonds results from the attack of the weak Si–Si bonds by a carbanion nucleophile. This approach was also applied to modify the flat silicon surface through a two-step halogenation/alkylation route [88]. Immersion of Cl-terminated silicon substrate into a solution composed of alkyllithium or Grignard at 80°C for various periods of time results in LiCl or MgXCl and an alkyl group bound to silicon substrate through Si–C covalent bond. It is also possible to directly form alkylated silicon surfaces on H-terminated substrate with butyl-, hexyl-, phenyl-, and 5-(*N*-pyrrolyl)pentyllithium reagents in THF at room temperature [115]. The reaction is similar to that of H-terminated porous silicon, by which cleavage of the relatively weak Si–Si bond leads to attachment of the organic molecule. Direct alkylation on H-terminated Si(111) was also observed with decylmagnesium bromide in diethyl ether for 16 h [105]. Since there are no Si–Si surface bonds on H-terminated Si(111), the reaction was suggested to proceed through cleavage of Si–H bonds.

The last pathway involves mechanical removal of the native oxide and instant chemical functionalization in solution. The removal of natural oxide creates dangling bonds that may exhibit reactivity similar to that of a Si=Si dimer formed under UHV conditions [116, 117]. The possible surface reactions are listed in Fig. 3.9e.

In contrast to rich surface chemistry on hydrogen- or halogen-terminated Si surfaces in solution, Ge surfaces are seldom studied. There are two reported approaches for functionalization of Ge surfaces. The first method is that of alkyl Grignard reagent reacting with the halogen-terminated Ge(111) surface, forming an alkylated Ge surface [118]. The other approach is hydrogermylation of the H-terminated Ge(100) surface by inserting unsaturated C–C bonds into the Ge–H bond [119].

3.5.2 Reactivity of Diamond Surfaces in Solution

The high stability of H-terminated diamond(100) surfaces makes it extremely difficult to functionalize. The first surface reaction on H-terminated diamond (100) was achieved through radical reactions of aryl diazonium salt, similar to the radical reactions on silicon surfaces [120, 121]. The pioneering work of alkenes

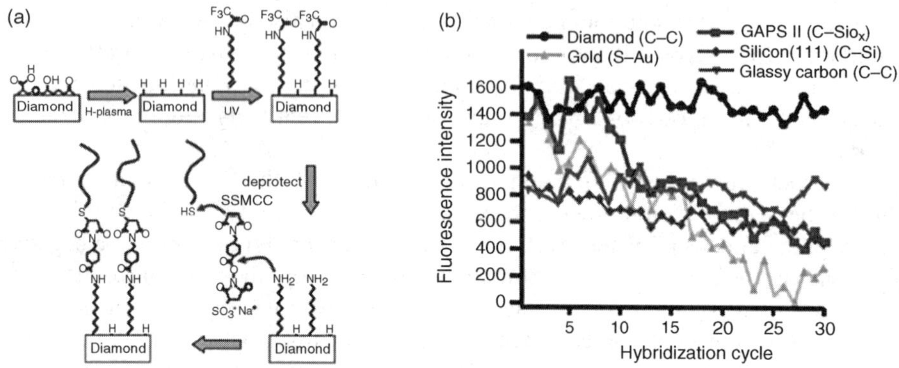

FIGURE 3.10 (a) Schematic showing the steps in covalent bonding of DNA to diamond. Reprinted with permission from Ref. 10. Copyright 2009 American Chemical Society. (b) Stability of DNA-modified surfaces in repeated cycles of hybridization and denaturation using 8.3 M urea. Reprinted with permission from Ref. 122. Copyright 2003 Nature Publishing Group.

grafting to the diamond surface was carried out via C=C group activated by a 254 nm light [122]. The original work used robust diamond surfaces for fabricating a DNA biosensor. Thc trifluoroacetamide-protected 10-aminodec-1-ene (TFAAD) was photografted on the H-terminated diamond(100) surface as shown in Fig. 3.10a. Compared to devices based on other surfaces, diamond and glassy carbon exhibit the best stability. The stability is not only attributed to the diamond but is also a consequence of the intrinsic stability of the interface C–C bond.

The photografting reaction of alkenes on H-terminated diamond is quite vague, because diamond (5.48 eV band gap, 226 nm) and most organic alkenes are almost completely transparent to the most commonly used light source (254 nm). For the photografting reactions of H-terminated silicon surface, light with energy larger than the band gap of silicon creates electron-hole pairs in silicon [105, 106]. It is critical to understand the behavior of a transparent substrate and a transparent alkene under light illumination. Experiments have discovered that surface activity of photografting is related to the functional groups of the alkene molecules [123, 124]. TFAAD grafting was much faster than that of *tert*-butyl oxycarbamate (tBoc) grafting and photografting of 1-dodecene was almost undetectable under the same conditions on diamond surfaces. Energy-level differences of diamond and bifunctional molecules are used to explain the different behavior of the surface photografting. Bifunctional alkenes have similar donor levels, but their acceptor levels are quite different as shown in Fig. 3.11a. Trifluoroacetamide group of TFAAD is a good electron acceptor and its affinity level is below the conduction band of diamond. Such an electronic structure favors electron transfer from the diamond surface to the organic molecule, leading to the highest reactivity. tBoc and 1-dodecene have higher affinity levels, and consequently, are less reactive. Fig. 3.11b shows the photoemission spectra of TFAAD and tBoc on the diamond(100) surface, which directly monitors electrons ejected from the valence band of diamond into acceptor levels of adjacent molecules.

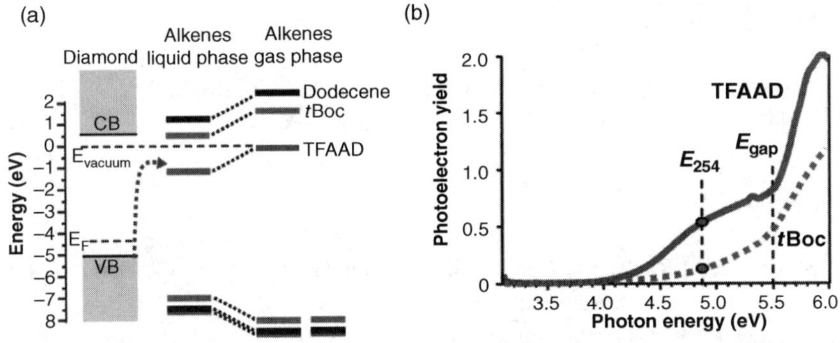

FIGURE 3.11 Photoemission initiation of grafting of alkenes to diamond: (a) energy-level diagrams for donor and acceptor levels; (b) direct measurements of photoelectron emission from diamond into TFAAD and tBoc. Reprinted with permission from Ref. 124. Copyright 2009 American Chemical Society.

For TFAAD, significant photoemission yields are observed at 3.9 eV, below the band gap of diamond. Actually, this value is close to the gap between the diamond valence band and the TFAAD affinity level. The difference in photoelectron yield at 254 nm between TFAAD and tBoc corresponds to the differences in reactivity.

3.6 SUMMARY

Significant efforts have been taken to understand the organic reactions on semiconductor surfaces, including experimental techniques and theoretical predictions, especially for Si(100), Ge(100), diamond(100), and Si(111) surfaces. The semiconductor surface geometry and electronic structure are extremely important in understanding surface reactivity with organic molecules and hence obtain controllable functionalization on semiconductor surfaces. The clean surface of Si(100), Ge(100), and diamond(100) undergo (2×1) reconstruction and form surface dimers to achieve a minimum energy state. Asymmetrical dimers for Si(100) and Ge(100) and symmetrical dimers for diamond(100) have different electronic structures and result in higher reactivity with organic molecules on Si(100) and Ge(100) surfaces. The pair of adatoms and rest atoms of Si(111)-(7×7) surface are similar to surface dimers of Si(100)-(2×1) surface, resulting in high surface reactivity toward surface organic reactions. H-terminated semiconductor surfaces always show weak reactivity due to their high stability. Partially H-terminated semiconductor surfaces with dangling bonds exhibit controllable surface modification on selective areas. Various methods including radical initiation, catalysts, and electrochemistry have been introduced to activate surface functionalization on hydrogen- or halogen-terminated semiconductor surfaces with organic molecules with unsaturated groups. Future investigations will focus on the interface of semiconductor surfaces with controllable properties at selected areas at the molecular level for wide applications of organic–inorganic hybrid devices.

ACKNOWLEDGMENTS

This work was supported by NKBRSF (No. 2007CB815202), NSFC (No. 20833008), and 863 (Nos. 2006AA01A119, 2009AA01A130).

REFERENCES

1. Mayne, A. J.; Riedel, D.; Comtet, G.; Dujardin, G. *Prog. Surf. Sci.* **2006**, *81*, 1.
2. Buriak, J. M. *Chem. Rev.* **2002**, *102*, 1271.
3. Wolkow, R. A. *Annu. Rev. Phys. Chem.* **1999**, *50*, 413.
4. Filler, M. A.; Bent, S. F. *Prog. Surf. Sci.* **2003**, *73*, 1.
5. Hamers, R. J.; Coulter, S. K.; Ellison, M. D.; Hovis, J. S.; Padowitz, D. F.; Schwartz, M. P.; Greenlief, C. M.; Russell, J. N. *Acc. Chem. Res.* **2000**, *33*, 617.
6. Leftwich, T. R.; Teplyakov, A. V. *Surf. Sci. Rep.* **2008**, *63*, 1.
7. Tao, F.; Xu, G. Q. *Acc. Chem. Res.* **2004**, *37*, 882.
8. Tao, F.; Bernasek, S. L.; Xu, G. Q. *Chem. Rev.* **2009**, *109*, 3991.
9. Ristein, J. *Surf. Sci.* **2006**, *600*, 3677.
10. Wang, X. Y.; Landis, E. C.; Franking, R.; Hamers, R. J. *Acc. Chem. Res.* **2010**, *43*, 1205, doi10.1021/ar100011f.
11. Hamers, R. J.; Wang, Y. J. *Chem. Rev.* **1996**, *96*, 1261.
12. Simpson, W. C.; Yarmoff, J. A. *Annu. Rev. Phys. Chem.* **1996**, *47*, 527.
13. Rodriguez-Reyes, J. C. F.; Teplyakov, A. V. *Chemistry* **2007**, *13*, 9164.
14. Binnig, G.; Rohrer, H. *Helve. Phys. Acta.* **1982**, *55*, 726.
15. Binnig, G.; Rohrer, H.; Gerber, C.; Weibel, E. *Phys. Rev. Lett.* **1983**, *50*, 120.
16. Schlier, R. E.; Farnsworth, H. E. *J. Chem. Phys.* **1959**, *30*, 917.
17. Phillips, J. C. *Surf. Sci.* **1973**, *40*, 459.
18. Poppendieck, T. D.; Gnoc, T. C.; Webb, M. B. *Surf. Sci.* **1974**, *43*, 647.
19. Appelbaum, J. A.; Baraff, G. A.; Hamann, D. R. *Phys. Rev. B* **1976**, *14*, 588.
20. Appelbaum, J. A.; Baraff, G. A.; Hamann, D. R. *Phys. Rev. Lett.* **1975**, *35*, 729.
21. Appelbaum, J. A.; Baraff, G. A.; Hamann, D. R.; Hagstrum, H. D.; Sakurai, T. *Surf. Sci.* **1978**, *70*, 654.
22. Hamers, R. J.; Tromp, R. M.; Demuth, J. E. *Phys. Rev. B* **1986**, *34*, 5343.
23. Tromp, R. M.; Hamers, R. J.; Demuth, J. E. *Phys. Rev. Lett.* **1985**, *55*, 1303.
24. Hamers, R. J.; Tromp, R. M.; Demuth, J. E. *Surf. Sci.* **1987**, *181*, 346–355.
25. Wolkow, R. A. *Phys. Rev. Lett.* **1992**, *68*, 2636.
26. Weakliem, P. C.; Carter, E. A. *J. Chem. Phys.* **1992**, *96*, 3240.
27. Tabata, T.; Aruga, T.; Murata, Y. *Surf. Sci.* **1986**, *179*, L63.
28. Wu, C. J.; Carter, E. A. *Chem. Phys. Lett.* **1991**, *185*, 172.
29. Develyn, M. P.; Yang, Y. M. L.; Sutcu, L. F. *J. Chem. Phys.* **1992**, *96*, 852.
30. Hofer, U.; Li, L. P.; Heinz, T. F. *Phys. Rev. B* **1992**, *45*, 9485.
31. Nachtigall, P.; Jordan, K. D.; Sosa, C. *J. Phys. Chem.* **1993**, *97*, 11666.
32. Flowers, M. C.; Jonathan, N. B. H.; Liu, Y.; Morris, A. *J. Chem. Phys.* **1993**, *99*, 7038.

33. Mui, C.; Bent, S. F.; Musgrave, C. B. *J. Phys. Chem. A* **2000**, *104*, 2457.
34. Carey, F. A.; Sundberg, R. J. *Advanced Organic Chemistry*. Plenum Publishing Corporation, **2001**.
35. Konecny, R.; Doren, D. J. *J. Chem. Phys.* **1997**, *106*, 2426.
36. Kubby, J. A.; Griffith, J. E.; Becker, R. S.; Vickers, J. S. *Phys. Rev. B* **1987**, *36*, 6079.
37. Hovis, J. S.; Coulter, S. K.; Hamers, R. J.; D'Evelyn, M. P.; Russell, J. N.; Butler, J. E. *J. Am. Chem. Soc.* **2000**, *122*, 732.
38. Woodward, R. B.; Hoffmann, R. *The Conservation of Orbital Symmetry*. Academic Press, New York, **1970**.
39. Cricenti, A.; Purdie, D.; Reihl, B. *Surf. Sci.* **1995**, *331*, 1033.
40. Wormeester, H.; Wentink, D. J.; Deboeij, P. L.; Wijers, C. M. J.; Vansilfhout, A. *Phys. Rev. B* **1993**, *47*, 12663.
41. Mercer, T. W.; Pehrsson, P. E. *Surf. Sci.* **1998**, *399*, L327.
42. Choi, C. H.; Gordon, M. S. *J. Am. Chem. Soc.* **1999**, *121*, 11311.
43. Liu, Q.; Hoffmann, R. *J. Am. Chem. Soc.* **1995**, *117*, 4082.
44. Lal, P.; Teplyakov, A. V.; Noah, Y.; Kong, M. J.; Wang, G. T.; Bent, S. F. *J. Chem. Phys.* **1999**, *110*, 10545.
45. Mui, C.; Han, J. H.; Wang, G. T.; Musgrave, C. B.; Bent, S. F. *J. Am. Chem. Soc.* **2002**, *124*, 4027.
46. Li, J.; Qu, Y. Q.; Han, K. L.; He, G. Z. *Surf. Sci.* **2005**, *586*, 45.
47. Qu, Y. Q.; Han, K. L. *J. Phys. Chem. B* **2004**, *108*, 8305.
48. Qu, Y. Q.; Li, J.; Han, K. L. *J. Phys. Chem. B* **2004**, *108*, 15103.
49. Qu, Y. Q.; Wang, Y.; Li, J.; Han, K. L. *Surf. Sci.* **2004**, *569*, 12.
50. Cao, X. P.; Hamers, R. J. *J. Phys. Chem. B* **2002**, *106*, 1840.
51. Takayanagi, K.; Tanishiro, Y.; Takahashi, M.; Takahashi, S. *J. Vac. Sci. Technol. A* **1985**, *3*, 1502.
52. Landstrass, M. I., Ravi, K. V. *Appl. Phys. Lett.* **1989**, *55*, 975.
53. Boland, J. J. *Adv. Phys.* **1993**, *42*, 129.
54. Boland, J. J. *Phys. Rev. Lett.* **1990**, *65*, 3325.
55. Boland, J. J. *Phys. Rev. Lett.* **1991**, *67*, 1539.
56. Sakurai, T.; Hagstrum, H. D. *Phys. Rev. B* **1976**, *14*, 1593.
57. Boland, J. J. *J. Phys. Chem.* **1991**, *95*, 1521.
58. Burrows, V. A.; Chabal, Y. J.; Higashi, G. S.; Raghavachari, K.; Christman, S. B. *Appl. Phys. Lett.* **1988**, *53*, 998.
59. Higashi, G. S.; Chabal, Y. J.; Trucks, G. W.; Raghavachari, K. *Appl. Phys. Lett.* **1990**, *56*, 656.
60. Chabal, Y. J. *Physica B* **1991**, *170*, 447.
61. Dumas, P.; Chabal, Y.; Jakob, P. *Surf. Sci.* **1992**, *269/270*, 867.
62. Higashi, G. S.; Becker, R. S.; Chabal, Y. J.; Becker, A. J. *Appl. Phys. Lett.* **1991**, *58*, 1656.
63. Thoms, B. D.; Owens, M. S.; Butler, J. E.; Spiro, C. *Appl. Phys. Lett.* **1994**, *65*, 2957.
64. Boland, J. J. *Phys. Rev. B* **1991**, *44*, 1383.
65. Naitoh, M.; Morioka, H.; Shoji, F. *Surf. Sci.* **1993**, *297*, 135.

66. Sinniah, K.; Sherman, M. G.; Lewis, L. B.; Weinberg, W. H.; Yates, J. T., Jr.,; Janda, K. C. *Phys. Rev. Lett.* **1989**, *62*, 567.
67. Boland, J. J. *J. Vac. Sci. Technol. A* **1992**, *10*, 2458.
68. Bobrov, K.; Comtet, G.; Dujardin, G.; Hellner, L. *Phys. Rev. Lett.* **2001**, *86*, 2633.
69. Diederich, L.; Kuttel, O.; Aebi, P.; Schlapbach, L. *Surf. Sci.* **1998**, *418*, 219.
70. Maier, F.; Ristein, J.; Ley, L. *Phys. Rev. B* **2001**, *64*, 165411.
71. Landstrass, M. I.; Ravi, K. V. *Appl. Phys. Lett.* **1989**, *55*, 1391.
72. Looi, H. J.; Pang, L. Y. S.; Molloy, A. B.; Jones, F.; Foord, J. S.; Jackman, R. B. *Diam. Relat. Mater.* **1998**, *7*, 550.
73. Bobrov, K.; Comtet, G.; Dujardin, G.; Hellner, L. *Surf. Sci.* **2001**, *482*, 437.
74. Hersam, M. C.; Guisinger, N. P.; Lyding, J. W. Silicon-Based Molecular Nanotechnology, *Nanotechnology II*, **2000**, p. 70.
75. Kato, K.; Kajiyama, H.; Heike, S.; Hashizume, T.; Uda, T. *Phys. Rev. Lett.* **2001**, *86*, 2842.
76. Lopinski, G. P.; Wayner, D. D. M.; Wolkow, R. A. *Nature* **2000**, *406*, 48.
77. Piva, P. G.; DiLabio, G. A.; Pitters, J. L.; Zikovsky, J.; Rezeq, M.; Dogel, S.; Hofer, W. A.; Wolkow, R. A. *Nature* **2005**, *435*, 658.
78. Cheng, C. C.; Gao, Q.; Choyke, W. J.; Yates, J. T., Jr., *Phys. Rev. B* **1992**, *46*, 810.
79. Szabo, Z.; Farral, P. D.; Engel, T. *Surf. Sci.* **1994**, *312*, 284.
80. Gao, Q.; Cheng, C. C.; Chen, P. J.; Choyke, W. J.; Yates, J. T., Jr., *J. Chem. Phys.* **1993**, *98*, 8308.
81. Bronikowski, M. J.; Hamers, R. J. *J. Vac. Sci. Technol. A* **1995**, *13*, 777.
82. Rioux, D.; Chander, M.; Li, Y. Z.; Weaver, J. H. *Phys. Rev. B* **1994**, *49*, 11071.
83. Gothelid, M.; Le Lay, G.; Wigren, C.; Bjorkqvist, M.; Karlsson, U. O. *Surf. Rev. Lett.* **1998**, *5*, 119.
84. Gothelid, M.; Le Lay, G.; Karlsson, U. O. *Surf. Sci.* **2004**, *556*, 203.
85. Boland, J. J.; Villarrubia, J. S. *Science* **1990**, *248*, 838.
86. Boland, J. J.; Villarrubia, J. S. *Phys. Rev. B: Condens. Matter* **1990**, *41*, 9865.
87. Schnell, R. D.; Rieger, D.; Bogen, A.; Wandelt, K.; Steinmann W. *Solid State Commun.* **1985**, *53*, 205.
88. Bansal, A.; Li, X. L.; Lauermann, I.; Lewis, N. S.; Yi, S. I.; Weinberg, W. H. *J. Am. Chem. Soc.* **1996**, *118*, 7225.
89. He, J.; Patitsas, S. N.; Preston, K. F.; Wolkow, R. A.; Wayner, D. D. M. *Chem. Phys. Lett.* **1998**, *286*, 508.
90. Zhu, X. Y.; Boiadjiev, V.; Mulder, J. A.; Hsung, R. P.; Major, R. C. *Langmuir* **2000**, *16*, 6766.
91. Terry, J.; Linford, M. R.; Wigren, C.; Cao, R. Y.; Pianetta, P.; Chidsey, C. E. D. *Appl. Phys. Lett.* **1997**, *71*, 1056.
92. Lu, Z. H. *Appl. Phys. Lett.* **1996**, *68*, 520.
93. Choi, K.; Buriak, J. M. *Langmuir* **2000**, *16*, 7737.
94. Linford, M. R.; Fenter, P.; Eisenberger, P. M.; Chidsey, C. E. D. *J. Am. Chem. Soc.* **1995**, *117*, 3145.
95. Chatgilialoglu, C. *Acc. Chem. Res.* **1992**, *25*, 188.
96. Sung, M. M.; Kluth, G. J.; Yauw, O. W.; Maboudian, R. *Langmuir* **1997**, *13*, 6164.
97. Sieval, A. B.; Vleeming, V.; Zuilhof, H.; Sudholter, E. J. R. *Langmuir* **1999**, *15*, 8288.

98. Sieval, A. B.; Linke, R.; Heij, G.; Meijer, G.; Zuilhof, H.; Sudholter, E. J. R. *Langmuir* **2001**, *17*, 7554.
99. Bateman, J. E.; Eagling, R. D.; Worrall, D. R.; Horrocks, B. R.; Houlton, A. *Angew. Chem., Int. Ed.* **1998**, *37*, 2683.
100. Boukherroub, R.; Morin, S.; Wayner, D. D. M.; Bensebaa, F.; Sproule, G. I.; Baribeau, J. M.; Lockwood, D. J. *Chem. Mater.* **2001**. *13*, 2002.
101. Bateman, J. E.; Eagling, R. D.; Horrocks, B. R.; Houlton, A. *J. Phys. Chem. B* **2000**, *104*, 5557.
102. Terry, J.; Mo, R.; Wigren, C.; Cao, R. Y.; Mount, G.; Pianetta, P.; Linford, M. R.; Chidsey, C. E. D. *Nucl. Instrum. Methods Phys. Res., Sect. B* **1997**, *133*, 94.
103. Terry, J.; Linford, M. R.; Wigren, C.; Cao, R. Y.; Pianetta, P.; Chidsey, C. E. D. *J. Appl. Phys.* **1999**, *85*, 213.
104. Cicero, R. L.; Linford, M. R.; Chidsey, C. E. D. *Langmuir* **2000**, *16*, 5688.
105. Stewart, M. P.; Buriak, J. M. *Angew. Chem., Int. Ed.* **1998**, *37*, 3257.
106. Stewart, M. P.; Buriak, J. M. *J. Am. Chem. Soc.* **2001**, *123*, 7821.
107. Imanishi, A.; Yamane, S.; Nakato, Y. *Langmuir* **2008**, *24*, 10755.
108. Lewis, L. N. *J. Am. Chem. Soc.* **1990**, *112*, 5998.
109. Boukherroub, R.; Morin, S.; Bensebaa, F.; Wayner, D. D. M. *Langmuir* **1999**, *15*, 3831.
110. Buriak, J. M.; Allen, M. J. *J. Am. Chem. Soc.* **1998**, *120*, 1339.
111. Buriak, J. M.; Stewart, M. P.; Geders, T. W.; Allen, M. J.; Choi, H. C.; Smith, J.; Raftery, D.; Canham, L. T. *J. Am. Chem. Soc.* **1999**, *121*, 11491.
112. deVilleneuve, C. H.; Pinson, J.; Bernard, M. C.; Allongue, P. *J. Phys. Chem. B* **1997**, *101*, 2415.
113. Hurley, P. T.; Ribbe, A. E.; Buriak, J. M. *J. Am. Chem. Soc.* **2003**, *125*, 11334.
114. Viellard, C.; Warntjes, M.; Ozanam, F.; Chazalviel, J.-N. *Proc. Electrochem. Soc.* **1996**, *95*, 250.
115. Kim, N. Y.; Laibinis, P. E. *J. Am. Chem. Soc.* **1999**, *121*, 7162.
116. Niederhauser, T. L.; Jiang, G. L.; Lua, Y. Y.; Dorff, M. J.; Woolley, A. T.; Asplund, M. C.; Berges, D. A.; Linford, M. R. *Langmuir* **2001**, *17*, 5889.
117. Niederhauser, T. L.; Lua, Y. Y.; Sun, Y.; Jiang, G. L.; Strossman, G. S.; Pianetta, P.; Linford, M. R. *Chem. Mater.* **2002**, *14*, 27.
118. He, J. L.; Lu, Z. H.; Mitchell, S. A.; Wayner, D. D. M. *J. Am. Chem. Soc.* **1998**, *120*, 2660.
119. Sharp, I. D.; Schoell, S. J.; Hoeb, M.; Brandt, M. S.; Stutzmann, M. *Appl. Phys. Lett.* **2008**, *92*, 223306.
120. Allongue, P.; Delamar, M.; Desbat, B.; Fagebaume, O.; Hitmi, R.; Pinson, J.; Save´ant, J. M. *J. Am. Chem. Soc.* **1997**, *119*, 201.
121. Kuo, T. C.; McCreery, R. L.; Swain, G. M. *Electrochem. Solid State Lett.* **1999**, *2*, 288–290.
122. Yang, W. S.; Auciello, O.; Butler, J. E.; Cai, W.; Carlisle, J. A.; Gerbi, J. E.; Gruen, D. M.; Knickerbocker, T.; Lasseter, T. L.; Russell, J. N., Jr., ; Smith, L. M.; Hamers, R. J. *Nat. Mater.* **2002**, *1*, 253.
123. Wang, X.; Colavita, P. E.; Metz, K. M.; Butler, J. E.; Hamers, R. J. *Langmuir* **2007**, *23*, 11623.
124. Wang, X.; Colavita, P. E.; Streifer, J. A.; Butler, J. E.; Hamers, R. J. *J. Phys. Chem. C* **2009**, *114*, 4067–4074.

CHAPTER 4

Pericyclic Reactions of Organic Molecules at Semiconductor Surfaces[1]

KEITH T. WONG AND STACEY F. BENT

4.1 INTRODUCTION

Pericyclic reactions comprise an important class of the chemistry of unsaturated organic molecules. Woodward and Hoffmann first proposed the concept of pericyclic reactions in 1965 [1]. These are reactions that are thought to occur via a concerted process involving a cyclic transition state, without the formation of intermediates. As a result of their concerted nature, pericyclic reactions are known to provide high stereoselectivity [2]. This chapter focuses on a particular type of pericyclic reaction, cycloaddition reactions, in which π electrons from two or more unsaturated reactant molecules form new σ bonds between reactants to create a cyclic product. Due to the aforementioned stereoselectivity and their ability to form new carbon–carbon bonds and new carbon rings, cycloadditions play a central role in organic synthesis [3, 4]. The two types of cycloaddition reactions considered in this chapter are the [2+2] cycloaddition and [4+2] cycloaddition, where the reaction is named for the number of π electrons of each reactant molecule. [4+2] cycloaddition is also known as the Diels–Alder reaction.

The Woodward–Hoffmann selection rules describe the reactivity of such cycloadditions. Symmetry analysis of the parity of the highest occupied molecular orbital (HOMO) and lowest unoccupied molecular orbital (LUMO), as depicted in Fig. 4.1, indicates that the [2+2] reaction is symmetry-forbidden due to a mismatch in parity whereas the Diels–Alder reaction is symmetry-allowed [5]. In fact, [2+2] cycloadditions in organic chemistry typically require photochemical activation to excite one of the reactants to a state in which the parity of the HOMO and LUMO orbitals that form the new bonds is matched, and the reaction is symmetry-allowed.

[1]This chapter was completed in 2009 and therefore does not capture new developments in the field occurring since 2009.

Functionalization of Semiconductor Surfaces, First Edition.
Edited by Franklin (Feng) Tao and Steven L. Bernasek.

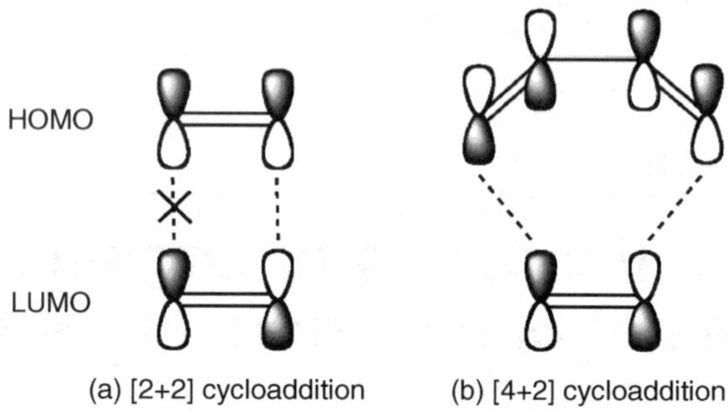

FIGURE 4.1 Schematic depicting the HOMOs and LUMOs of ethylene reacted by (a) [2+2] cycloaddition with another ethylene molecule and (b) [4+2] cycloaddition with 1,3-butadiene. The mismatched parity of the HOMO and LUMO of ethylene leads to [2+2] cycloaddition being symmetry-forbidden according to the Woodward–Hoffmann selection rules.

Diels–Alder reactions, on the other hand, occur readily at room temperature and are commonly used in organic syntheses [2–4].

Since the dimers of the 2×1 reconstructed Group IV semiconductor surfaces (C(100)-(2×1) (diamond), Si(100)-(2×1), and Ge(100)-(2×1)) can be described as having partial π bonds and, thus, alkene character, cycloaddition products may be expected to form on these surfaces. Indeed, numerous unsaturated organic molecules have been found to generate [4+2] or [2+2] cycloaddition products on these surfaces; such reactions are the subject of this chapter. Evidence exists indicating that cycloadditions on semiconductor surfaces do not necessarily proceed via a concerted process with a cyclic transition state; as will be discussed in this chapter, stepwise mechanisms involving short-lived intermediates have been proposed instead [6–9]. Thus, these reactions may not be truly pericyclic, and the term *cycloaddition*, as it is used in this chapter, is not meant to imply a truly pericyclic mechanism but is convenient nomenclature that highlights the analogies that can be made to organic chemistry. Other pericyclic-like reactions such as 1,3-dipolar cycloaddition and group transfer reactions can occur at semiconductor surfaces; however, these reactions are somewhat less common and less studied than [2+2] and [4+2] cycloadditions at semiconductor surfaces and, thus, will not be addressed in detail here. Additional discussion of pericyclic reactions at semiconductor surfaces can be found in the literature [10–20].

Although it has been the subject of fewer studies than the Si(100)-(2×1) surface, cycloaddition-like reactions are also known to occur on the Si(111)-(7×7) surface, and several examples will be discussed in this chapter. The structure of the Si(111)-(7×7) surface can be described by the widely accepted dimer–adatom–stacking-fault structure [21]. This structure does not contain partial π bonds like the 2×1 reconstructed (100) surfaces; therefore, direct comparison to an alkene cannot be made. However, each unit cell contains seven distinct types of surface atoms. Charge transfer among the dangling bonds at these surface atoms has been theoretically predicted and experimentally observed [22–26]. The resulting uneven distribution of electrons at the surface is believed to enable cycloaddition-like reactions on the

Si(111)-(7×7) surface similar to cycloaddition reactions at the buckled dimers of the Si (100)-(2×1) and Ge(100)-(2×1) surfaces.

This chapter first describes the [2+2] cycloadditions of alkenes and alkynes on Si (100)-(2×1)—historically, the first cycloadditions on a semiconductor surface studied. This topic is followed by a discussion of [4+2] reactions of dienes, which adds the possibility of multiple cycloaddition products ([4+2] and [2+2] products). Finally, more recent work on cycloadditions of unsaturated molecules containing N, O, or S heteroatoms is reviewed. In each section, selected adsorbates that highlight particular features of cycloaddition at semiconductor surfaces are discussed. Where possible and informative, comparisons are made between the reactivity on Si(100)-(2×1) and that on Ge(100)-(2×1), C(100)-(2×1), or Si(111)-(7×7).

4.2 [2+2] CYCLOADDITION OF ALKENES AND ALKYNES

Although the [2+2] cycloaddition reaction is symmetry-forbidden according to classic theory (Woodward–Hoffmann selection rules), many molecules have been found to readily form [2+2] cycloaddition products, originally labeled as di-σ or bridge-bonded products, on semiconductor surfaces [11–14, 16, 18]. This observation highlights some of the important differences between cycloaddition reactions in solution and at semiconductor surfaces.

4.2.1 Ethylene

Ethylene is the simplest organic molecule that can undergo a [2+2] cycloaddition reaction; it serves as a model for understanding many [2+2] cycloadditions at semiconductor surfaces. Studies in the late 1980s of adsorption of ethylene on Si (100)-(2×1) were among the earliest examples of cycloadditions at a semiconductor surface, although the product was initially labeled as a di-σ product [27, 28]. These and later studies of ethylene adsorption on Si(100)-(2×1) using temperature programmed desorption (TPD), scanning tunneling microscopy (STM), scanning tunneling spectroscopy (STS), high-resolution electron energy loss spectroscopy (HREELS), low-energy electron diffraction (LEED), near edge X-ray absorption fine structure (NEXAFS) spectroscopy, ultraviolet photoelectron spectroscopy (UPS), and density functional theory (DFT) calculations indicated that two new Si–C σ bonds are formed, the alkene π bond is broken and the silicon dimer bond remains intact [27–43]. The results show that the product of ethylene reacted at Si(100)-(2×1) can also be labeled as a [2+2] cycloaddition product.

The sticking probabilities of ethylene and other simple alkenes on Si(100)-(2×1) are near unity, indicating facile reaction with the surface [44]. This result is interesting since, as noted previously, the [2+2] cycloaddition reaction is classically symmetry-forbidden and typically occurs slowly in solution [3, 5]. To reconcile the different reactivity between the heterogeneous reaction at Si(100)-(2×1) and the homogeneous reaction in solution, low-symmetry stepwise mechanisms with little or no activation barrier(s) were proposed for the reaction on Si(100)-(2×1) of both ethylene and acetylene, which may also undergo [2+2] cycloaddition, as discussed later [7, 8]. These mechanisms entail the formation of three-centered π-complex (Fig. 4.2d) and/or diradical (Fig. 4.2c) states as precursors or intermediates. Using a CASSCF(6,6) wave function to model [2+2] cycloaddition of 1,3-cyclohexadiene on Si(100)-(2×1), Choi and Gordon have suggested that the π-complex intermediate is an artifact of using a single-configurational wave function [6], although additional calculations by the same group show a relatively flat potential energy surface associated with dimer buckling [45], which may facilitate initial attack of adsorbates. In agreement with the originally proposed mechanisms, more recent theoretical studies of ethylene adsorption on Si(100)-(2×1) have also found pathways involving a π-complex precursor and diradical intermediate [46–48], and Nagao et al. [48] reported the observation of a π-complex precursor by HREELS at low temperature. Low-symmetry pathways are believed to be possible on the Si(100)-(2×1) surface

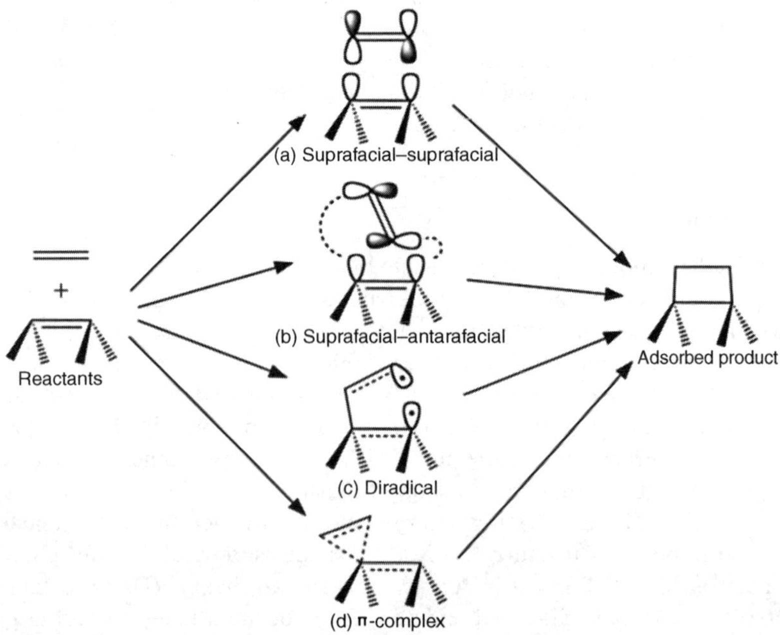

FIGURE 4.2 Schematic representation of commonly discussed reaction pathways for the [2+2] cycloaddition of ethylene on a Group IV (100)-(2×1) surface dimer: (a) concerted suprafacial–suprafacial reaction, which is symmetry-forbidden according to the Woodward–Hoffmann selection rules; (b) concerted suprafacial–antarafacial reaction, which may be sterically hindered; (c) reaction by a diradical intermediate or precursor; and (d) reaction by a π-complex intermediate or precursor. Note that analogous reaction mechanisms are discussed for [2+2] cycloaddition of other unsaturated organic molecules.

due to the dynamic tilting of dimers at room temperature, the relative weakness of the dimer π bond and the small (compared to alkenes) separation of the π and π^* frontier orbitals. Liu et al. [7] used *cis*- and *trans*-dideuteroethylene and the splitting of ν_s(C–H) and ν_{as}(C–H) modes to determine stereoselectivity upon adsorption on Si(100)-(2×1). They find that the symmetry of the original reactant is retained indicating that the symmetry-allowed suprafacial–antarafacial reaction mechanism (Fig. 4.2b) does not occur, whereas nonconcerted (i.e., nonpericyclic), low-symmetry mechanisms involving diradical or π-complex intermediates are possible. The mechanism of [2+2] cycloaddition on Si(100)-(2×1) remains a topic of interest; yet more theoretical studies have recently suggested that a concerted reaction mechanism (similar to that in Fig. 4.2a) may compete with low-symmetry mechanisms [49, 50]. The viability of a concerted mechanism has been explained in terms of the crossing of energy bands associated with the surface frontier orbitals through the Fermi level [49] or the presence of crystal wave functions with the appropriate symmetry for bonding interaction with an alkene [50], thus creating a symmetry-allowed pathway.

Another interesting feature of the [2+2] cycloaddition product of ethylene on Si (100)-(2×1) is its thermodynamic stability. The binding energy is found to be approximately 40 kcal/mol by both experiment and theory [30, 33, 35, 40]. This is somewhat surprising given the significant ring strain, as evidenced by the Si–Si–C bond angle of 78.2°—far from the tetrahedral angle—calculated by Konecny and Doren [30]. Thus, the thermodynamic stability of the [2+2] product must arise primarily from the strength of the two Si–C bonds formed. Indeed, a bond energy of 81.7 kcal/mol has been calculated for the Si–C bond of methylsilane [51]. The strength of the bonds formed with the surface and the resulting overall thermodynamic stability of the ethylene [2+2] product have attracted much interest in the use of [2+2] cycloaddition for functionalizing the Si(100)-(2×1) surface.

Unlike on Si(100)-(2×1) where ethylene is believed to adsorb in a single configuration (on top of a single Si dimer) [52, 53], ethylene is found to form at least two molecular adsorption states on Ge(100)-(2×1) [54–56]. The first experimental studies of ethylene adsorption on Ge(100)-(2×1) performed by Lal et al. [56] using multiple internal reflection Fourier transform infrared spectroscopy (MIR-FTIR) and TPD were consistent with the formation of a [2+2] cycloaddition product. However, time- and coverage-dependent behavior and the presence of two molecular desorption peaks in TPD measurements indicate the existence of at least two molecular adsorption states [56]. TPD and angle-resolved UPS experiments by Fink et al. [54] attributed the minor peak in TPD measurements to adsorption at step sites. STM images, such as that shown in Fig. 4.3, later showed the presence of two distinct bonding geometries at room temperature, which are attributed to a [2+2] product across a single dimer (on-top configuration; shown in Fig. 4.4a) and two ethylene molecules bridging between two dimers in the same row (pair end-bridge configuration; shown in Fig. 4.4c) [55]. Calculations for acetylene on Ge (100)-(2×1) show that the pair end-bridge configuration is most stable [57, 58], suggesting that the same may be the case for ethylene; however, according to theoretical calculations of ethylene on Ge(100)-(2×1), the on-top [2+2] product is

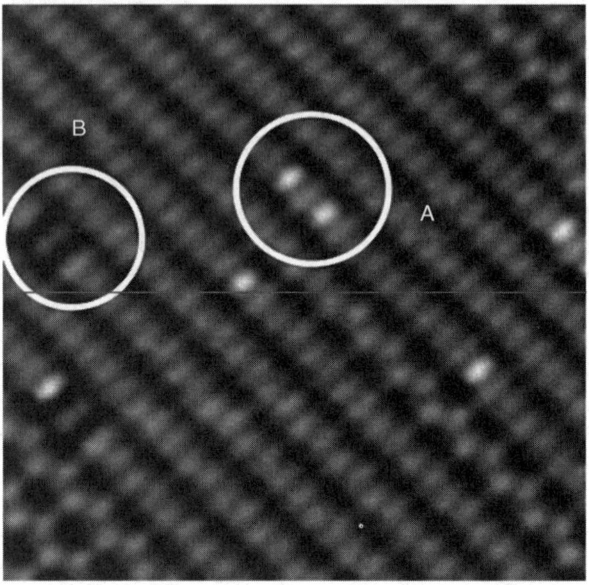

FIGURE 4.3 Filled state STM image of a Ge(100)-(2×1) surface exposed to a low dose of ethylene. Two distinct bonding configurations can be distinguished and are labeled A and B. The authors attributed feature A to on-top adsorption and feature B to pair end bridge adsorption. A 10×10 nm^2 area is imaged at -1.8 V sample bias and 100 pA tunneling current. Reprinted with permission from Ref. 55. Copyright 2004 American Chemical Society.

most stable [59]. Though there is debate over the exact adsorption states of ethylene on Ge(100)-(2×1), this system clearly demonstrates several important concepts: (1) similar but not necessarily identical reaction products are observed for cycloadditions at the Group IV semiconductor surfaces; (2) interdimer reactions can be important even for small molecules such as ethylene; and (3) multiple reaction products can form even for the simplest cycloaddition systems.

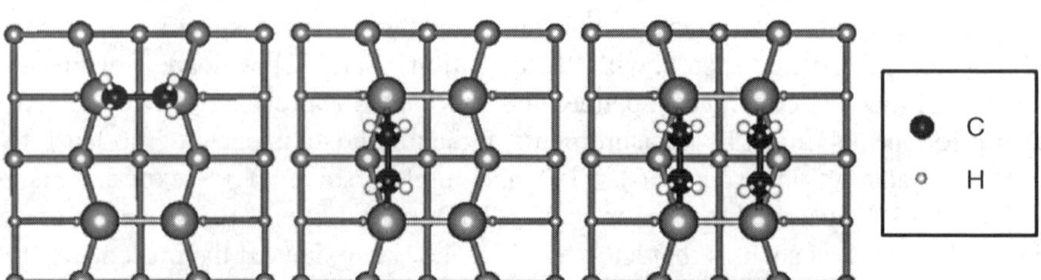

FIGURE 4.4 Top view of adsorption configurations of ethylene on a Group IV (100)-(2 × 1) surface: (a) on-top; (b) end-bridge; and (c) pair end-bridge. Surface dimer atoms are represented by the largest gray circles, and subsurface atoms are represented by increasingly small gray circles. Adapted with permission from Ref. 55. Copyright 2004 American Chemical Society.

Finally, ethylene is found to desorb molecularly from the Ge(100)-(2×1) surface, whereas decomposition occurs on Si(100)-(2×1) [14, 18, 54–56]. Similarly, reversible adsorption on Ge(100)-(2×1) versus decomposition on Si(100)-(2×1) has been observed for other systems including the [2+2] product of acetylene and the [4+2] product of 1,3-butadiene [18, 51]. This difference in stability can be attributed to the difference in the strength of bonds formed with the surface. The bond energy of the Ge–C bond in methylgermane was calculated to be 72.7 kcal/mol, which is 9 kcal/mol less than the Si–C bond energy in methylsilane (81.7 kcal/mol) [51]. Thus, the binding energy of adsorbates on Ge is expected to be significantly less than on Si, allowing for molecular desorption to occur upon heating before decomposition. This result highlights the important influence the surface can have on the stability of adsorbates. On the basis of this concept, it has been suggested that reactions at the Si surface are more often under kinetic control, whereas reactions at the Ge surface are more often under thermodynamic control since less favorable adsorption products may have such low binding energies on Ge that they readily desorb [16].

4.2.2 Acetylene

The behavior of acetylene, C_2H_2, at Si(100)-(2×1) and Ge(100)-(2×1) mimics that of ethylene to some degree, although its triple bond adds the possibility of additional tetra-σ bonded products in which an acetylene molecule undergoes two [2+2]-like cycloadditions to form four σ bonds with the surface (two per carbon atom; shown in Fig. 4.5c). Like ethylene adsorption on Ge(100)-(2×1), the bonding configuration of acetylene on Si(100)-(2×1) has been debated at length in the literature with tetra-σ, end-bridge [2+2] and on-top [2+2] products (Fig. 4.5)

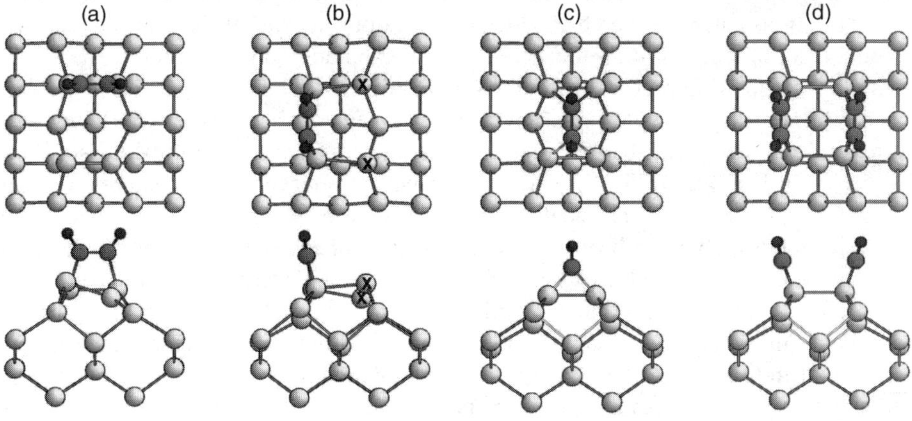

FIGURE 4.5 Top and side views of optimized structures of acetylene on Si(100)-(2×1): (a) on-top; (b) end-bridge; (c) tetra-σ; and (d) pair end-bridge. Light gray, medium gray, and dark gray circles correspond to silicon, carbon, and hydrogen, respectively. Note that tetra-σ configurations with the acetylene molecule perpendicular (as shown) and parallel (not shown) to the surface dimer have both been proposed. Figure adapted from Ref. 84. Copyright 2004 by The American Physical Society.

all suggested [8, 30, 32, 35, 37, 38, 43, 60–85]. It is generally agreed that multiple of these products form in appreciable quantity at room temperature (perhaps with different coverage dependences), thus reinforcing the assertions made earlier that cycloadditions across multiple dimers can be important and even simple molecules may form multiple products at the surface.

Comparatively few studies have been published regarding acetylene adsorption on Ge(100)-(2×1). Theoretical calculations initially suggested that the on-top [2+2] product should be observed [58, 86]. The presence of both the on-top di-σ and tetra-σ products was suggested by STM studies [87]; however, these results were later reinterpreted following additional DFT calculations to show that acetylene adsorbs on Ge(100)-(2×1) in on-top di-σ and pair end-bridge configurations, similar to ethylene [57]. Also like ethylene, acetylene binds to Ge(100)-(2×1) more weakly than to Si(100)-(2×1) according to calculations [88]. This is confirmed experimentally, as mentioned previously, by the dominance of dissociative desorption of acetylene from Si(100)-(2×1) [61] versus molecular desorption from Ge(100)-(2×1) [87], which again highlights the difference in thermodynamic stability typically observed for cycloaddition products on the Si versus Ge surface.

4.2.3 *Cis*- and *Trans*-2-Butene

Issues of stereoselectivity or stereospecificity are further explored by the reactions of *cis*- and *trans*-2-butene at Si(100)-(2×1). The structures of *cis*- and *trans*-2-butene are equivalent to the aforementioned molecules *cis*- and *trans*-dideuteroethylene, respectively, with the substitution of methyl groups in the place of deuterium. Thus, the 2-butene stereoisomers can similarly be used to shed light on the mechanism of [2+2] cycloaddition at Si(100)-(2×1) with the additional advantage that different adsorption configurations can be probed by STM. Indeed, Lopinski et al. [9, 89] showed that two binding configurations—one observed for *cis*-2-butene and the other for *trans*-2-butene—can be differentiated using STM. However, they found that 2–3% of both stereoisomers isomerized upon adsorption to the clean surface (not at defect sites) [9]. An example of a *trans*-2-butene molecule isomerized to *cis*-2-butene upon adsorption is circled in Fig. 4.6. In Fig. 4.6, the dimer rows are imaged as gray features running diagonally across the image, and the paired light-colored protrusions correspond to the methyl groups on both ends of adsorbed 2-butene molecules. Most of the adsorbed 2-butene molecules are angled slightly with respect to the underlying dimers, consistent with adsorption of *trans*-2-butene without isomerization, but the circled feature is parallel to the dimer direction, indicating isomerization to *cis*-2-butene. Since earlier work had indicated that the *trans* adsorbate is 4 kcal/mol more stable than the *cis* adsorbate [90], equal isomerization of both isomers indicates the process is under kinetic rather than thermodynamic control. Their results add to the previous studies of dideuteroethylene [7] by showing that adsorption by [2+2] cycloaddition at Si(100)-(2×1) is highly stereoselective but not truly stereospecific. They conclude that the mechanism must proceed in a stepwise fashion on a timescale of picoseconds in order to allow for isomerization by rotation about the C–C bond [9].

Interestingly, Madachik and Teplyakov [91] report that 2,3-dimethyl-2-butene does not react with the Si(100)-(2×1) surface at room temperature. No signs of

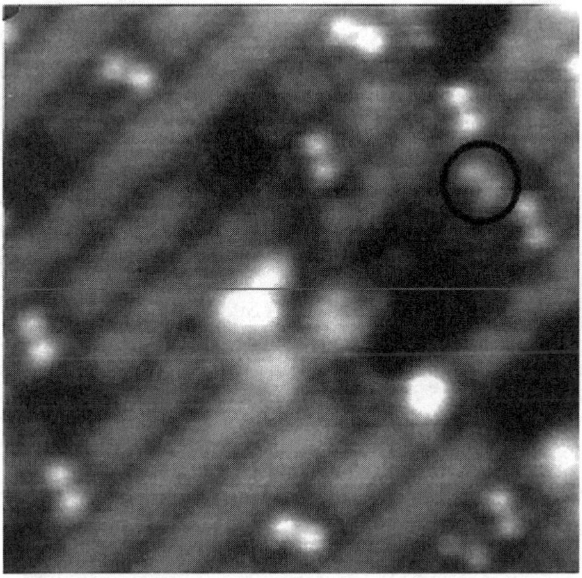

FIGURE 4.6 Empty state STM image of a Si(100)-(2×1) surface exposed to *trans*-2-butene. Most *trans*-2-butene molecules lead to paired light-colored protrusions angled with respect to the dimers, but the circled feature shows a pair of protrusions that is aligned with the underlying dimers, which corresponds to a *trans*-2-butene molecule that isomerized to *cis*-2-butene upon adsorption. A 75 × 75 Å area is imaged at a sample bias of 2 V and tunneling current of 40 pA. Reprinted with permission from Ref. 9. Copyright 2000 American Chemical Society.

chemisorption are present in the IR spectra after large doses at room temperature or heating multilayers adsorbed at 100K to room temperature. In agreement with the experimental results, the authors' DFT calculations indicate that formation of the expected [2+2] cycloaddition product is kinetically unfavorable compared to desorption from the two π-complex precursor states considered. Steric hindrance due to the additional methyl groups (as compared to 2-butene) is suggested as the reason for the large barrier to [2+2] cycloaddition. Due to its lack of reactivity with the Si(100)-(2×1) surface, it is suggested that 2,3-dimethyl-2-butene could be used as an inert carrier gas or precursor ligand.

4.2.4 Cyclopentene

Cyclopentene has served as a model system for exploring ordering of [2+2] cycloaddition products at semiconductor surfaces. Due to the nature of the [2+2] cycloaddition product, which contains two directional bonds to the surface per molecule, Hamers et al. [11] suggest that [2+2] cycloaddition provides a means to control both translational and orientational ordering of adsorbates at the interface. Indeed, the STM images in Fig. 4.7 demonstrate such ordering of cyclopentene adsorbed on Si(100)-(2×1). These images and others obtained by the same authors

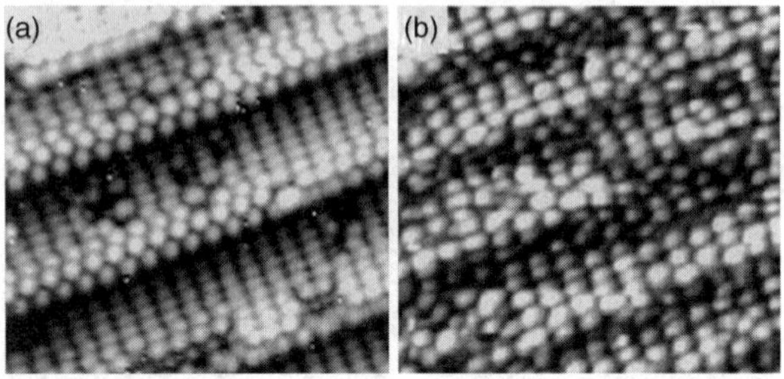

FIGURE 4.7 STM images of (a) a clean vicinal Si(100)-(2×1) surface (4° miscut toward the ⟨1$\bar{1}$0⟩ direction) showing terraces on the surface with 8–12 dimers in each row and (b) the same surface after exposure to cyclopentene. Figure adapted with permission from Ref. 95. Copyright 1998 Elsevier.

show the formation of a highly ordered monolayer with each cyclopentene molecule aligned over a Si dimer [92–95]. FTIR spectroscopy and X-ray photoelectron spectroscopy (XPS) confirm that cyclopentene adsorbs to form a [2+2] cycloaddition product [92, 93, 95–97], and anisotropy in the C–H stretching region of the FTIR spectra suggests that the ordering observed by STM is present over macroscopic (centimeter) length scales [92, 93, 95]. The formation of ordered monolayers has also been demonstrated by STM for 1,5-cyclooctadiene [98] and 1,3,5,7-cyclooctatetraene [99], which form single- and dual-[2+2] cycloaddition products on Si(100)-(2×1), respectively. Later analysis by photoemission spectroscopy (PES) and NEXAFS indicated the presence of some dual-[4+2] products in the reaction of 1,3,5,7-cyclooctatetraene with Si(100)-(2×1) [100]. Nonetheless, both 1,3-cyclooctadiene and 1,3,5,7-cyclooctatetraene provide examples of generating a functionalized semiconductor surface with ordered arrays of unreacted alkenes that may be apt for further functionalization [101].

Cyclopentene adsorption has also been studied on Ge(100)-(2×1) and C(100)-(2×1), thus providing for useful comparison among the Group IV semiconductor surfaces. The lack of high-frequency alkene C–H stretching modes in the FTIR spectra indicates the formation of a [2+2] cycloaddition product on all three surfaces [92, 102, 103]. Hovis et al. [102] found the interesting result that the sticking coefficients are approximately 1, 0.1 and 10^{-3} on Si(100)-(2×1), Ge(100)-(2×1), and C(100)-(2×1), respectively. They proposed several explanations for this trend. First, the sticking coefficients correlate negatively with the π–π^* splitting of the surfaces. Large π–π^* splitting is expected to slow the rate of the symmetry-forbidden suprafacial–suprafacial reaction mechanism. Second, the relative strength of the C dimer π bond compared to that of Si and Ge may slow the reaction rate on C(100)-(2×1). Finally, the geometry of the surface dimers may impact the reaction rate. Tilting out of plane of the Si and Ge dimers enables a low-symmetry pathway with little or no activation barrier [102, 104] equivalent to π-complex mechanisms that have been suggested for other adsorbates [7, 8, 46–48].

The clean C(100)-(2×1) surface does not exhibit the same dimer tilting, and calculations show that the presence of an alkene does not induce tilting [102]. The lack of dimer tilt results in a much higher activation barrier and is suggested as the primary explanation for the much lower sticking coefficient on C(100)-(2×1) [11]. The Si(100)-(2×1) and Ge(100)-(2×1) surfaces are both comprised of similarly tilted dimers yet the sticking coefficients of cyclopentene on these surfaces differ by about an order of magnitude. Lee et al. [103] suggest that the slightly larger dimer bond length on the Ge surface (about 0.2 Å longer than on Si) may increase the energy of the transition state between the π-complex intermediate and the product, leading to lower sticking probability. Thus, cyclopentene provides an excellent example for exploring the effects of surface properties on reactivity of adsorbates by [2+2] cycloaddition.

4.2.5 [2+2]-Like Cycloaddition on Si(111)-(7×7)

Adsorption of ethylene [105] and acetylene [106] on Si(111)-(7×7) were first investigated beginning in the 1980s. As with early studies of these molecules on Si(100)-(2×1) around the same time, the primary adsorption products were initially termed di-σ products. Later studies made the analogy to [2+2] cycloaddition and such products are now sometimes referred to as [2+2]-like cycloaddition products. [2+2]-like cycloaddition products have also been suggested for adsorption on Si(111)-(7×7) of other nonconjugated unsaturated hydrocarbons such as cyclopentene, cyclohexene, and 1,4-cyclohexadiene [107, 108], but most research has focused on ethylene and acetylene.

Early studies by Yoshinobu et al. [105, 106] used HREELS to show that both ethylene and acetylene primarily adsorb without dissociating and that they rehybridize to near sp^3 and between sp^2 and sp^3 hybridizations, respectively. They suggested that both molecules adsorb to form primarily di-σ bonded products (i.e., [2+2]-like products) on the basis of these results. A study of ethylene adsorption on Si(111)-(7×7) by STM and STS suggested that ethylene molecules bridge adjacent adatom–rest atom pairs [109]. Furthermore, it was shown that adsorption of an ethylene molecule changes the local density of states at neighboring surface atoms in addition to the surface atoms to which it bonds, but no changes to the surface reconstruction were observed. Later study by XPS agreed with the assignment of the primary product to a di-σ bonded product and was able to differentiate a minor product attributed to dissociative adsorption [110]. By comparison to ethylene adsorption on Si(100)-(2×1), the authors estimate that the maximum coverage of ethylene on Si(111)-(7×7) is several times higher than would be possible with binding only at adatom–rest atom bridge sites. Based on this result and the LEED pattern, which is reported to change from 7×7 to 7×1 with increasing ethylene exposure, it is suggested that a rearrangement of the surface atoms does occur. A more recent DFT study found a barrierless diradical reaction pathway with no intermediate or transition states for the adsorption of ethylene on Si(111)-(7×7) [111]. Reaction by such a pathway should occur readily, and the authors suggest that significant reconstruction of the surface should not be induced by this reaction.

The binding energy of ethylene on Si(111)-(7×7) is calculated to be 47.9 kcal/mol—slightly larger than values typically obtained for ethylene adsorption on Si(100)-(2×1).

Acetylene adsorption on Si(111)-(7×7) has been the focus of more studies than that of ethylene. Formation of a di-σ bonded product, or [2+2]-like cycloaddition product, upon acetylene adsorption on Si(111)-(7×7) was supported by early theoretical studies, but it was suggested that restructuring of surface atoms accompanies adsorption in order for the spacing between dangling bonds of surface Si atoms to be close enough to interact with acetylene [112, 113]. STM results again agreed with the di-σ product, but no restructuring of the surface was observed [114]. MacPherson et al. [115] attempted to resolve the discrepancy between theory and experiment by suggesting that easily resolved features of the 7×7 unit cell (e.g., corner holes and dimer walls) remain even though restructuring within the unit cell that cannot be resolved by STM occurs. Later studies using XPS, X-ray absorption spectroscopy (XAS), UPS, and HREELS were consistent with the di-σ model bridging an adatom–rest atom pair [116, 117]. Changes in the angular dependence of the HREELS spectrum with increasing exposure suggested that a different adsorption geometry may dominate at higher coverage, possibly due to restructuring of the surface [117]. Calculations indicated that an upright adsorption geometry (in which acetylene is bound to the surface through only one carbon atom) may also explain the HREELS angular dependence [118]. A pathway similar to some suggested for [2+2] cycloaddition on Si(100)-(2×1) involving two diradical intermediate states and three transitions states was calculated for adsorption of acetylene to form a di-σ bonded product on Si(111)-(7×7) [111].

4.3 [4+2] CYCLOADDITION OF DIENES

The [4+2] cycloaddition, or Diels–Alder, reaction between a conjugated diene and alkene (the alkene is referred to in this sense as the dienophile) is well known to occur readily in solution. In 1997, Konecny and Doren predicted on the basis of DFT calculations that the dimers of the Si(100)-(2×1) surface could act as a dienophile and react with a diene with minimal activation barrier [119]. Their calculations for adsorption of 1,3-cyclohexadiene on an Si_9H_{12} cluster representing a single surface dimer result in a large 54.0 kcal/mol binding energy for the [4+2] product and no significant activation barrier. Moreover, reaction of one double bond of 1,3-cyclohexadiene to yield a [2+2] product is calculated to be 15.2 kcal/mol less exothermic. The difference in stability is attributed primarily to the difference in ring strain; Si–Si–C bond angles of approximately 78° and 97° are predicted for the [2+2] and [4+2] products, respectively, yielding higher ring strain for the [2+2] product. [4+2] cycloadditions on Ge(100)-(2×1) and C(100)-(2×1) were subsequently investigated theoretically and shown to be similarly favorable [51, 120]. Besides the greater product stability and minimal activation barrier predicted for [4+2] cycloaddition, this reaction inherently leaves an alkene moiety on the surface. It was suggested that this could be the starting point for further functionalization [119]; however, the alkene moieties at the surface have proven to have low reactivity, and work to date on layer-by-layer growth of organic films has focused instead on other chemistries [121–142]. Since conjugated dienes are capable of forming both [2+2] and [4+2] cycloaddition products, they present a more complicated chemistry than simple alkenes and alkynes because competition and selectivity among multiple reaction channels is often important.

4.3.1 1,3-Butadiene and 2,3-Dimethyl-1,3-Butadiene

Following Konecny and Doren's theoretical predictions, the first experimental evidence of [4+2] reactions at the Si(100)-(2×1) surface was provided by Teplyakov et al. [143]. In this and a later [144] study by the same authors, adsorption of 1,3-butadiene and 2,3-dimethyl-1,3-butadiene were studied by MIR-FTIR, NEXAFS, and TPD. Lack of absorption in the Si–H stretching region of the IR spectra for both molecules and presence of features attributed to both single bonds and nonconjugated double bonds in the NEXAFS spectrum of 2,3-dimethyl-1,3-butadiene indicate that these molecules do not react by C–H dissociation at room temperature. Figure 4.8 shows the IR spectra of chemisorbed and physisorbed 1,3-butadiene and two of its deuterated derivatives. Vibrational modes consistent with a terminal $=CH_2$ group near $3080\,cm^{-1}$ are present in the low-temperature multilayer IR

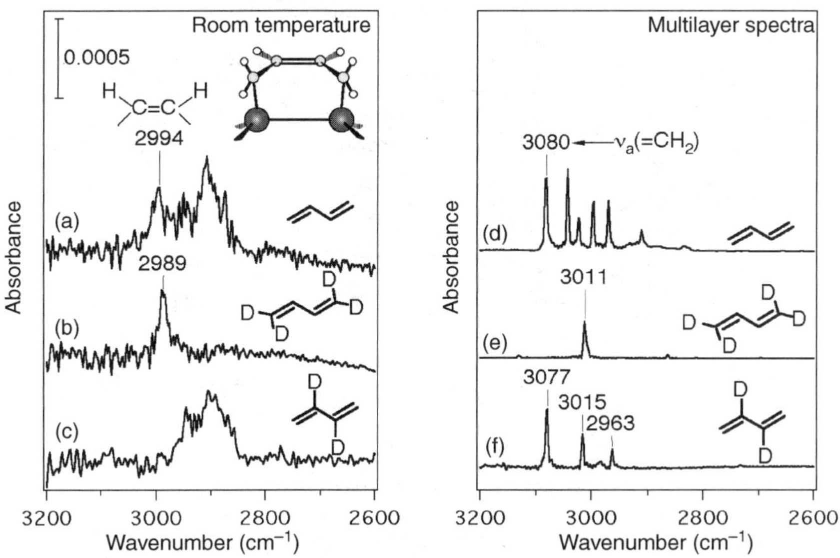

FIGURE 4.8 Infrared spectra showing the C–H stretching region of 1,3-butadiene adsorbed on Si(100)-(2×1): (a) 1 L (1 L = 10^{-6} Torr · s) of 1,3-butadiene adsorbed at 300K; (b) 1 L of 1,3-butadiene-1,1,4,4-d_4 adsorbed at 300K; (c) 1000 L of 1,3,-butadiene-2,3-d_2 adsorbed at 300K; (d) multilayers of 1,3-butadiene (1000 L) at 100K; (e) multilayers of 1,3-butadiene-1,1,4,4-d_4 (1000 L) at 100K; and (f) multilayers of 1,3-butadiene-2,3-d_2 (1000 L) at 100K. Adapted with permission from Ref. 144. Copyright 1998 American Institute of Physics.

spectra (Fig. 4.8d and f), and are clearly identified as they shift out of the plotted range upon substitution of deuterium for the terminal hydrogens (Fig. 4.8e). No absorption near 3080 cm^{-1} is discernable upon adsorption at room temperature (Fig. 4.8a–c). Olefinic and aliphatic C–H stretches, however, are clearly observed at room temperature near 2990 cm^{-1} and 2900 cm^{-1} (Fig. 4.8a), respectively, and can again be identified by comparison to the deuterated derivatives (Fig. 4.8b and c). NEXAFS spectra of 2,3-dimethyl-1,3-butadiene are shown in Fig. 4.9. There is a reduction in the number of π^* transitions from four in the physisorbed spectra (Fig. 4.9a and b) to only one in the room temperature spectrum (Fig. 4.9c). Additionally, the energy of the single π^* transition at room temperature is consistent with that of monounsaturated compounds. Both IR and NEXAFS results are consistent with the formation of a [4+2] product at the surface since this product contains a C=C group within the ring rather than terminal $=CH_2$ groups (as indicated by IR results), and the conjugation that leads to multiple π^* transitions in the NEXAFS spectra at low temperature is no longer present in the [4+2] product.

To further characterize the adsorption of 2,3-dimethyl-1,3-butadiene on Si(100)-(2×1), Teplyakov et al. [144] used NEXAFS to determine the angle between

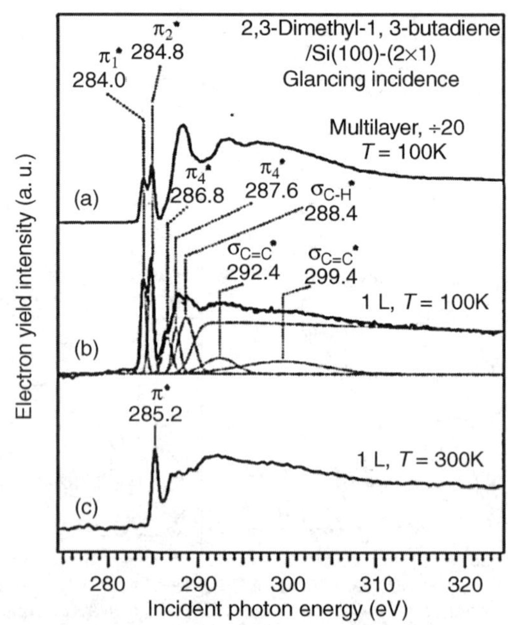

FIGURE 4.9 NEXAFS spectra of 2,3-dimethyl-1,3-butadiene adsorbed on Si(100)-(2×1): (a) multilayers of 2,3-dimethyl-1,3-butadiene physisorbed at 100K; (b) submonolayer (1 L exposure) of 2,3-dimethyl-1,3-butadiene physisorbed at 100K; and (c) submonolayer (1 L exposure) of 2,3-dimethyl-1,3-butadiene chemisorbed at room temperature. Spectra were obtained at glancing incidence of the incoming photon flux. Reprinted with permission from Ref. 144. Copyright 1998 American Institute of Physics.

the transition moment of the π^* orbital and the surface normal. Their experimentally determined value of $41 \pm 2°$ differs slightly from Konecny and Doren's theoretical prediction of 30°. The discrepancy may result from interactions with adjacent dimers not captured by the single-dimer theoretical model, surface defects present experimentally or the formation of side products with different adsorption geometries. Hovis et al. [145] investigated the latter possibility by analyzing STM images of Si(100)-(2×1) exposed to 2,3-dimethyl-1,3-butadiene. Their images clearly show the presence of two products: a primary product accounting for about 80% of adsorbates was assigned to the [4+2] product and a side product accounting for the remaining adsorbates was assigned to a [2+2] product involving reaction of only one of the C=C double bonds. However, it has also been suggested by Doren that a [4+2] product bridging neighboring dimers may be the source of the side product seen in STM images [12, 16]. The assignment by Hovis et al. [145] of the side product to a [2+2] product was supported by the presence of a small C–H stretch in the IR spectrum at 3093 cm^{-1}—in the region expected for terminal $=CH_2$. Given the significantly greater thermodynamic stability (about 15 kcal/mol for cyclohexadiene [119]) of the [4+2] product over the [2+2] product, the results of Hovis et al. suggest that the reaction of 2,3-dimethyl-1,3-butadiene is under kinetic control and that, despite being classically symmetry-forbidden, the [2+2] reaction at the Si surface competes to some extent with the [4+2] reaction for this molecule. This result implies that obtaining high selectivity between [2+2] and [4+2] products, as is typical in solution organic chemistry, may be difficult to achieve at semiconductor surfaces.

The adsorption of 1,3-butadiene has also been studied on the analogous Ge [51, 146] and diamond [120, 147] surfaces. The IR spectrum of 1,3-butadiene on Ge (100)-(2×1) closely resembles that on Si(100)-(2×1); hence, it was concluded that the [4+2] product is also the dominant species on Ge [146]. The spectral features of C(100)-(2×1) differ noticeably from those of the other two surfaces [147]. In particular, the peaks are shifted up in frequency from their positions on the other surfaces. This shift was hypothesized to result from different electron donation effects on the various surfaces or the smaller bond length of the dimers on the diamond surface leading to greater ring strain. Nonetheless, isotopic studies led to the conclusion that 1,3-butadiene predominantly forms the [4+2] product on the C(100)-(2×1) surface as well [147]. It has also been noted that ~1–2 orders of magnitude larger doses of 1,3-butadiene are required to saturate the C(100)-(2×1) surface than either the Si(100)-(2×1) or Ge(100)-(2×1) surfaces [12]. This result is in agreement with calculated activation barriers [120] and largely mimics the result noted previously for the sticking coefficient of cyclopentene on these surfaces. Of the proposed explanations for the trend in [2+2] reactivity toward cyclopentene—based on differing π–π^* splitting, dimer π bond strength, and dimer tilt—only the dimer bond strength is expected to impact the reactivity of the surface toward a diene since the [4+2] cycloaddition reaction is symmetry-allowed [12].

Also of interest is the differing thermal chemistry exhibited by 1,3-butadiene on the Si(100)-(2×1) and Ge(100)-(2×1) surfaces. Upon heating, 1,3-butadiene on Si(100)-(2×1) primarily decomposes to form surface carbon and Si–H species [144].

On Ge(100)-(2×1), however, molecular desorption is the predominant pathway upon thermal annealing [146]. Again, this is similar to results presented earlier for ethylene and acetylene on these surfaces, and the difference has been attributed to the approximately 10 kcal/mol weaker Ge–C bond than Si–C bond. Molecular desorption of 1,3-butadiene from Ge(100)-(2×1) indicates the occurrence of the reverse of the Diels–Alder reaction, that is, the retro-Diels–Alder reaction. It has been suggested that this reversibility provides a possible means for spatially controlled modification by using photo- or electron-induced reaction to selectively react another precursor with the adsorbed dienes, followed by removal of unreacted dienes by retro-Diels–Alder reaction.

4.3.2 1,3-Cyclohexadiene

1,3-Cyclohexadiene was the subject of the first theoretical prediction of [4+2] cycloaddition on Si(100)-(2×1) [119] as well as several later experimental and theoretical investigations [6, 30, 145, 148]. STM images obtained by Hovis et al. [145] show the presence of three surface products accounting for 55%, 35%, and 10% of adsorbates. Likewise, at least four peaks are discernable by infrared spectroscopy in the alkene C–H stretching region. The three products observed by STM were attributed to [4+2] cycloaddition (55%), [2+2] cycloaddition (35%) and an unidentified product (10%) based on the shape of the observed protrusions and their alignment with the underlying dimers. IR and NEXAFS spectroscopy data consistent with the presence of a mix of [4+2] and [2+2] products was later provided by Kong et al. [148]. Given the theoretical predictions of significantly greater thermodynamic stability (~15 kcal/mol) of the [4+2] product [6, 30], which would suggest several orders of magnitude higher abundance of the [4+2] product if the reaction were under thermodynamic control, the observed product distribution implies that the reaction of 1,3-cyclohexadiene is under kinetic control on the Si(100)-(2×1) surface. The same phenomenon has been noted for several other previously discussed adsorbates. Later STM images obtained by Teague and Boland [149, 150] distinguish multiple bonding configurations for both [2+2] and [4+2] products, as shown in Fig. 4.10. In addition to intradimer products for both cycloaddition reactions (Fig. 4.10a, d, and e), interdimer products bridging dimers in the same row (Fig. 4.10b) or adjacent rows (Fig. 4.10c) are identified for [4+2] cycloaddition. DFT calculations by the same group agree with the finding that the reaction of 1,3-cyclohexadiene is under kinetic control on Si(100)-(2×1) [151].

Due to the greater stability of the [4+2] products, annealing may be able to skew the product distribution in favor of the [4+2] products. Upon heating above approximately 400K, C–H stretching modes in the IR spectra are found to decrease in intensity without noticeable change in shape, concurrent with an increase in Si–H stretching modes, as shown in Fig. 4.11 [145, 148]. These IR results as well as TPD spectra suggest that 1,3-cyclohexadiene primarily desorbs or decomposes before isomerization can occur, in agreement with the large activation barrier (>40 kcal/mol) calculated for isomerization between [4+2] and [2+2] products [6].

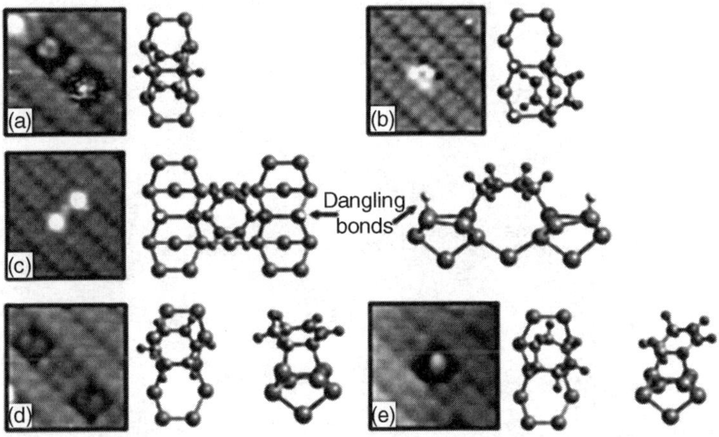

FIGURE 4.10 High-resolution empty state STM images showing multiple distinct adsorption configurations of 1,3-cyclohexadiene on Si(100)-(2×1) at 300K and models of the corresponding proposed bonding configuration. The features in the STM images are proposed to correspond to: (a) [4+2] cycloaddition on a single dimer; (b) [4+2] cycloaddition bridging two dimers in the same row; (c) [4+2] cycloaddition bridging two dimers in adjacent rows; and (d) and (e) two conformations of the [2+2] cycloaddition product. Approximately a 4×4 nm area is imaged at a tip bias of −1.3 V. Adapted with permission from Ref. 149. Copyright 2003 American Chemical Society.

4.3.3 Cyclopentadiene

An interesting study of cyclopentadiene and dicyclopentadiene adsorption on Si(100)-(2×1) was carried out by Wang et al. [152]. Dicyclopentadiene is formed by the spontaneous dimerization of cyclopentadiene by Diels–Alder reaction in solution. The spectral features in the C–H stretching region of chemisorbed cyclopentadiene at room temperature align well with calculated frequencies for the [4+2] product. The authors concluded that cyclopentadiene yields primarily the [4+2] product although the presence of a small amount of [2+2] side product could not be ruled out.

Dicyclopentadiene was also studied as it was hypothesized that the surface might catalyze the retro-Diels–Alder reaction (yielding two cyclopentadiene monomers per dicyclopentadiene dimer), as the formation of two [4+2] adducts (of cyclopentadiene) is thermodynamically much more favorable than formation of a single [2+2] adduct (of dicyclopentadiene). If this were the case, the IR spectrum after exposing the surface to dicyclopentadiene should be nearly, if not exactly, identical to that of cyclopentadiene. In fact, the spectra of chemisorbed cyclopentadiene and dicyclopentadiene chemisorbed on Si(100)-(2×1) at room temperature, shown in Fig. 4.12, differ significantly in both peak positions and intensities in the C–H stretching region, and chemisorbed dicyclopentadiene was instead assigned to a [2+2] product. The authors' calculations show a thermodynamic favorability of the surface-catalyzed retro-Diels–Alder reaction resulting in [4+2] cycloaddition of

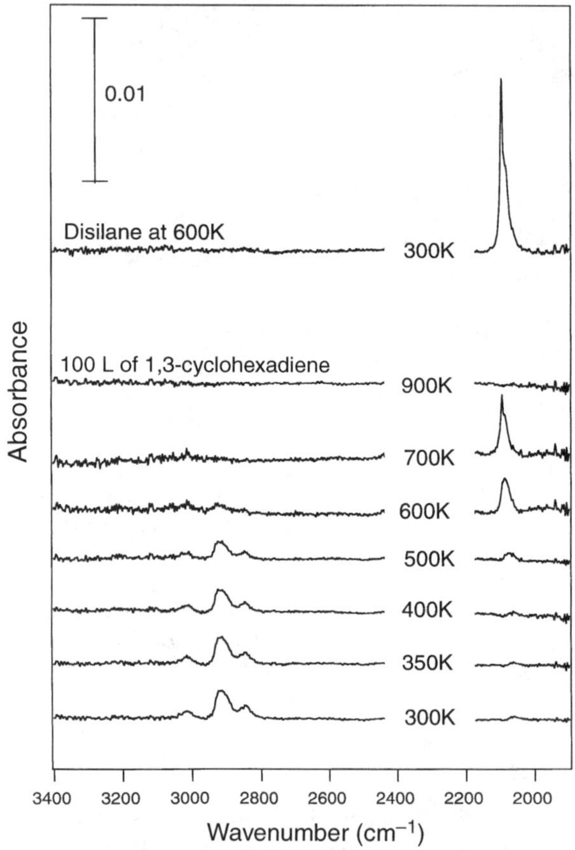

FIGURE 4.11 IR spectra for an annealing series of a Si(100)-(2×1) surface exposed to 100 L of 1,3-cyclohexadiene. The top spectrum corresponds to a well-ordered monohydride-terminated surface resulting from the exposure of Si(100)-(2×1) to disilane at 600K. Reprinted with permission from Ref. 148. Copyright 2000 American Chemical Society.

cyclopentadiene of nearly 46 kcal/mol over the [2+2] cycloaddition of intact dicyclopentadiene; thus, the experimental evidence indicates that kinetics controls the outcome of reaction. Though they do not calculate an activation barrier for surface-catalyzed retro-Diels–Alder reaction, it is noted that the energy barrier in the gas phase (33.7 kcal/mol) provides an upper bound on the energy barrier for the reaction catalyzed by the surface. Nonetheless, the energy barrier at the surface may still be much greater than that of the [2+2] reaction, which, as discussed earlier, occurs readily at the Si(100)-(2×1) surface. Moreover, geometric constraints for a transition state involving interaction of a dicyclopentadiene molecule with two surface dimers may yield a low pre-exponential factor for the surface-catalyzed reaction.

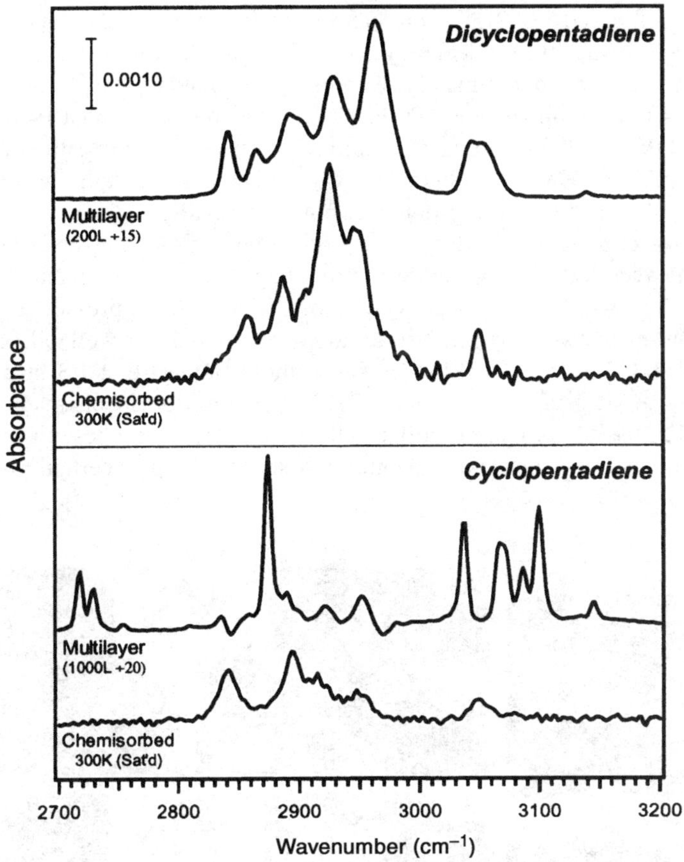

FIGURE 4.12 IR spectra of multilayers (unpolarized) and chemisorbed (p-polarized) dicyclopentadiene and cyclopentadiene on Si(100)-(2×1). The saturated chemisorbed spectra were collected after doses of 1500 and 500 L of dicyclopentadiene and cyclopentadiene, respectively. Reprinted with permission from Ref. 152. Copyright 1999 American Chemical Society.

4.3.4 [4+2]-Like Cycloaddition on Si(111)-(7×7)

Tao et al. [108] show using HREELS, XPS, and UPS that 1,3-cyclohexadiene binds to the Si(111)-(7×7) surface to form a cyclohexene-like adduct containing only a single C=C π bond. DFT calculations indicate that a [4+2]-like product is over 24 kcal/mol more stable than a [2+2]-like product [108]. Further experimental studies are necessary to differentiate between the [4+2]-like and [2+2]-like configurations of 1,3-cyclohexadiene adsorbed on Si(111)-(7×7).

Reaction pathways for 1,3-butadiene reaction with an adatom–rest atom pair on Si (111)-(7×7) were investigated using DFT calculations by Lu et al. [111]. For the more stable *s-trans* conformation of 1,3-butadiene, reaction pathways leading to both [4+2]-like and [2+2]-like products were investigated. The authors found a barrierless pathway to [4+2]-like cycloaddition passing through a diradical intermediate. A barrier of approximately 5 kcal/mol was found for the [2+2]-like reaction pathway, and the binding energy for the [2+2]-like product is about 19 kcal/mol less than that of the [4+2]-like product. A similar barrierless reaction pathway for [4+2]-like cycloaddition of *s-cis*-1,3-butadiene was calculated. The authors conclude that formation of the [4+2]-like product is both kinetically and thermodynamically favored. An experimental study of 1,3-butadiene adsorption on Si(111)-(7×7) found that reaction at adatom–rest atom pairs is the dominant pathway for adsorption, but STM images such as those in Fig. 4.13 provide evidence for 1,3-butadiene reaction at symmetric adatom–adatom pairs as well (adsorption sites labeled B and C in Fig. 4.13) [153]. The authors suggest that 1,3-butadiene may react at these sites via a concerted [4+2] cycloaddition mechanism due to the symmetry of the adatom pair (similar to the symmetry of alkenes or C(100)-(2×1) surface dimers). Further experimental work may help ascertain the bonding

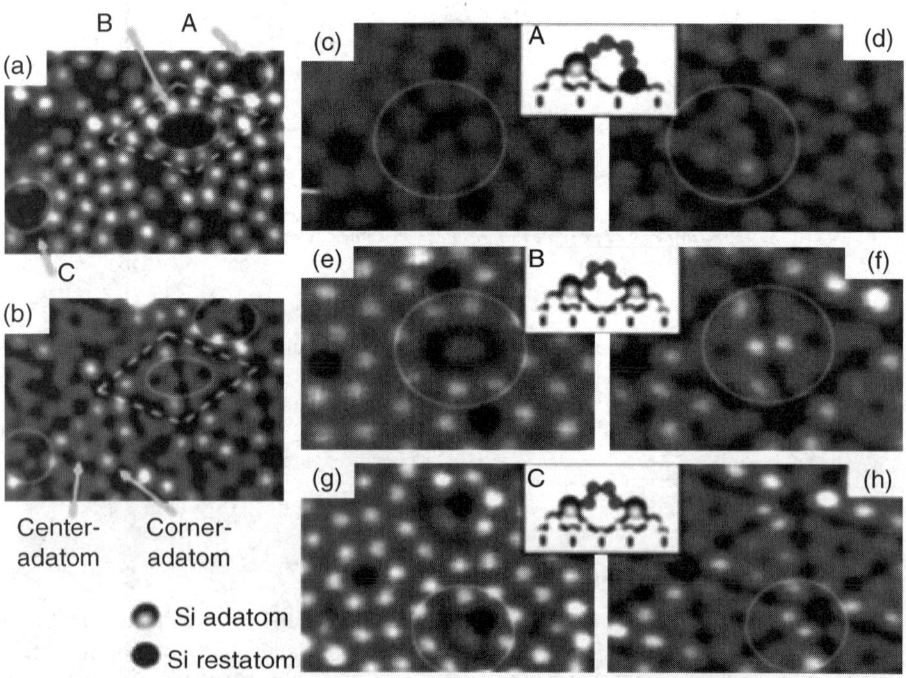

FIGURE 4.13 STM images of a Si(111)-(7×7) surface exposed to 0.1 L of 1,3-butadiene at room temperature. A sample bias of 0.5 and −1.0 V was used to collect images (a) and (b), respectively, in which the unit cell is outlined by dashed lines, and adsorbed 1,3-butadiene molecules are indicated by circles. Images (c)–(h) show empty and filled state images of three adsorption sites collected at 1.0 and −2.0 V sample bias, respectively. Schematics of the proposed adsorption configurations are also shown. Reprinted with permission from Ref. 153. Copyright 2006 American Chemical Society.

configuration at symmetric adatom pairs, and calculations are necessary to determine whether reaction by a concerted mechanism is, in fact, feasible at symmetric sites on the Si(111)-(7×7) surface.

4.4 CYCLOADDITION OF UNSATURATED ORGANIC MOLECULES CONTAINING ONE OR MORE HETEROATOM

Cycloaddition at semiconductor surfaces is not limited to organic molecules containing only carbon and hydrogen. For example, Ellison et al. [154] showed that azo-*tert*-butane primarily forms a [2+2] cycloaddition product on Si(100)-(2×1) by reacting across the N=N bond in a similar manner to the alkenes discussed earlier. The presence of one or more heteroatoms in a functional group can create uneven charge distribution, facilitating low-symmetry reaction pathways, and lone pairs on heteroatoms such as N, O, or S enable reactions to proceed through a dative-bonded precursor state. Dative bonding, which is discussed in more detail in a later chapter, is generally not possible for the simple alkenes, alkynes, and dienes discussed thus far in this chapter. In this section, reactivity of the Si(100)-(2×1) and Ge(100)-(2×1) surfaces toward heteroatom-containing organic molecules to form cycloaddition products is discussed, including mechanistic implications of dative bonding.

4.4.1 C=O-Containing Molecules

Armstrong et al. [155, 156] investigated adsorption of acetaldehyde, acetone, and biacetyl on Si(100)-(2×1) using HREELS, XPS, and TPD. The authors concluded that all three molecules form di-σ products at low temperature in which the C=O and surface dimer π bonds are broken and new Si–C and Si–O bonds are formed, as in [2+2] cycloaddition. In the case of the bifunctional molecule, biacetyl, the authors conclude that only one of the ketone moieties interacts with the surface due to geometric constraints; however, Barriocanal and Doren [157] later reinterpreted the HREELS data using calculated vibrational frequencies to conclude that the thermodynamically favored [4+2] product is instead the major product. In all three cases, decomposition products are also observed even at the lowest dosing temperatures, and annealing leads to further decomposition of the adsorbates to produce both gas-phase and surface products [155, 156]. Theoretical calculations have indicated that the formation of the [2+2] product across a C=O bond passes through a dative-bonded precursor state in which the oxygen atom donates a lone pair of electrons to the electrophilic "down" dimer atom at the surface [157, 158]. For the reaction of glyoxal, a dialdehyde, with Si(100)-(2×1), Barriocanal and Doren [157] calculated mechanisms for both [2+2] and [4+2] cycloaddition that pass through a precursor state in which the oxygen lone pair is directed toward the dangling bond of the Si surface, and the nodal plane of the nearest π orbital is aligned along the Si–O axis, as shown in Fig. 4.14a and b. In contrast to alkenes and dienes in which only the π orbitals are available to interact with the surface, these results indicate the possibility of a reaction pathway mediated by dative bonding through an oxygen lone pair. In

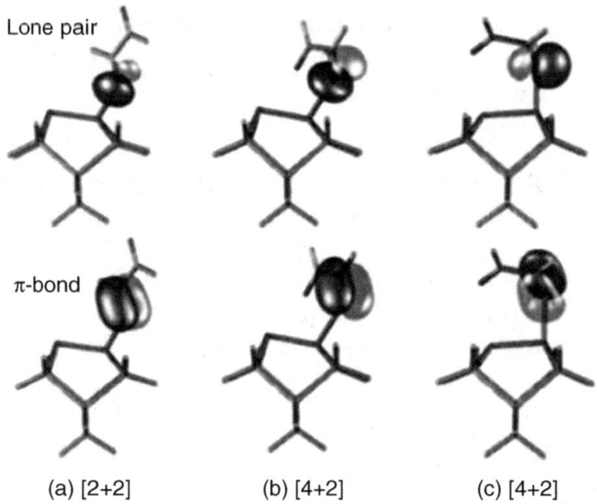

FIGURE 4.14 Calculated lone pair (top) and π (bottom) orbitals along three reaction pathways for the reaction of glyoxal with a Si_9H_{12} cluster representing a single dimer on the Si(100)-(2×1) surface: (a) [2+2] cycloaddition of the *trans* isomer of glyoxal; and (b) and (c) two pathways for [4+2] cycloaddition of the *cis* isomer of glyoxal. The alignment of the oxygen lone pair orbital toward an Si surface atom and alignment of the π orbital nodal plane along the Si–O axis indicate a dative bond-like interaction with the surface. Reprinted with permission from Ref. 157. Copyright 2001 American Chemical Society.

addition, they show optimization steps that pass through this dative-bonded state and continue without any activation barrier to both the [2+2] and [4+2] cycloaddition products. No such barrierless pathway could be found for [2+2] cycloaddition of acetone on Si(100)-(2×1) in calculations performed by Wang et al. [158]. It is hypothesized that the small (5 kcal/mol) barrier to from the dative-bonded state of acetone may arise from steric hindrance of the two methyl groups, that the conjugation of glyoxal may weaken the C=O bond allowing for a barrierless transition, or that a similar barrierless pathway exist for acetone but was not found by the calculations. Regardless of the uncertainty as to the presence or origin of an activation barrier, both studies identify the presence of dative-bonded precursor states and propose mechanisms for cycloaddition that pass through this state—a key theme in adsorption of heteroatom-containing molecules.

In their study of acetone adsorption, Wang et al. [158] shed light on important differences between the Si(100)-(2×1) surface and the Ge(100)-(2×1) surface. Based on MIR-FTIR data, they find that acetone transfers an α-hydrogen to the Ge surface to produce an enol-like adduct at room temperature. This resembles an "ene" reaction, which falls under the category of group transfer reaction—another type of pericyclic reaction. Figure 4.15 shows a reaction coordinate diagram for acetone reaction on Ge(100)-(2×1) from DFT calculations performed by Wang et al. [158]. Based on a combination of theoretical and experimental results, they reach the conclusion that reaction of acetone on the Ge surface is under

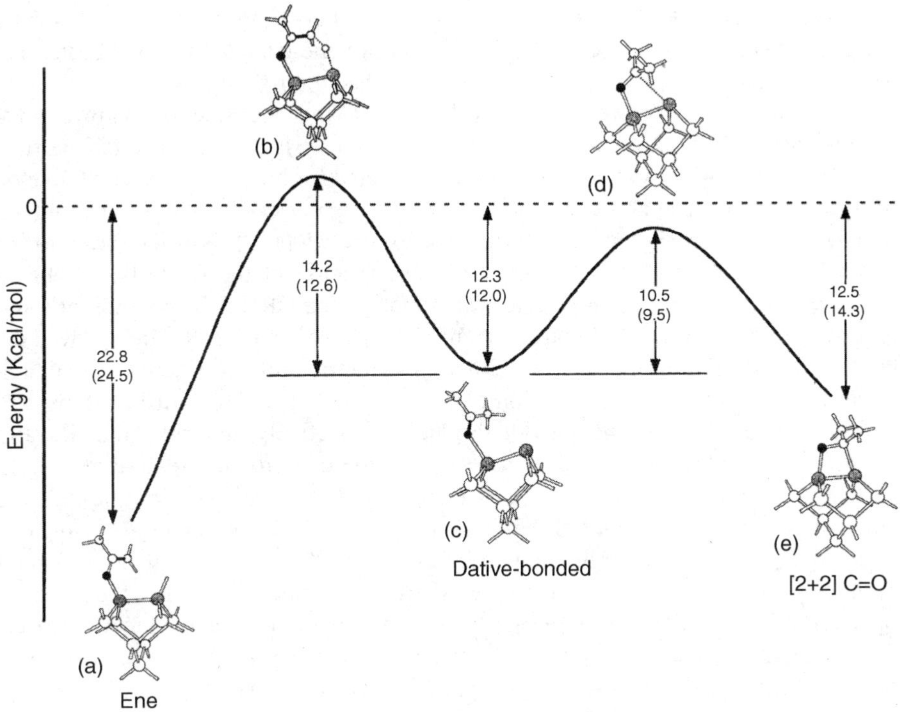

FIGURE 4.15 Calculated critical point energies on the ene and [2+2] reaction pathways of acetone on Ge(100)-(2×1). The dative-bonded precursor state shown in the center (c) can proceed through transition state (b) to the ene product (a) or through transition state (d) to the [2+2] product (e). Energies shown are in kcal/mol and were calculated using the 6-311++G(d,p) basis set (no parentheses) and 6-311+G(2df,pd) basis set (in parentheses). Reprinted with permission from Ref. 158. Copyright 2001 American Chemical Society.

thermodynamic control. Calculations show that although both the [2+2] cycloaddition and ene reaction pathways traverse the same oxygen dative-bonded precursor state (Fig. 4.15c), the transition states lie 2 kcal/mol below the vacuum level for [2+2] cycloaddition and 2 kcal/mol above the vacuum level for ene reaction at the Ge(100)-(2×1) surface (Fig. 4.15d and b, respectively). Wang et al. conclude that acetone molecules may undergo [2+2] cycloaddition at the Ge surface, as it is kinetically favored, but because the binding energy of this product is only 12.5 kcal/mol (Fig. 4.15e), reversible desorption will quickly occur at room temperature. Eventually, the thermodynamically favored product of the ene reaction (with a 22.8 kcal/mol binding energy) is formed instead. This conclusion is substantiated by the relatively large doses required to observe the ene product at room temperature, which is consistent with the presence of an activation barrier lying slightly above the vacuum level. In an attempt to skew the product distribution toward the dative-bonded precursor state or the [2+2] cycloaddition product, the authors investigated acetone adsorption at low temperature. The IR spectrum of acetone exposed to

Ge(100)-(2×1) at 115K closely resembles the multilayer spectrum of physisorbed acetone with the exception of a red-shifted C=O stretching mode, which indicates that acetone primarily adsorbs by dative bonding at this temperature.

DFT calculations for the [2+2] cycloaddition and ene reaction pathways for acetone on Si(100)-(2×1) were also carried out [158]. Similar reaction barriers (within the error of the calculations) were calculated for both pathways, and, like on the Ge surface, the ene product is thermodynamically favored by about 9 kcal/mol on Si(100)-(2×1). Experimental results of Armstrong et al. [156] show the formation of a [2+2] product and a product involving dissociation of the C=O bond, but no evidence for an ene reaction product was found in the HREELS spectrum at room temperature. However, STM images obtained by Hamai et al. [159] later showed the presence of adsorbates appearing as either a symmetric protrusion centered over a Si dimer or a protrusion and depression on a Si dimer. The authors attributed these to [2+2] cycloaddition and ene reaction products, respectively, and found that the ratio between the products can be changed to favor the ene product upon annealing to 420K. According to the calculations by Wang et al. [158], this result implies that the reaction is under kinetic control. Unlike on Ge(100)-(2×1), both the [2+2] and ene products have high enough binding energies on Si(100)-(2×1) (35.5 and 44.7 kcal/mol, respectively) to be observed on the experimental time scale. Thus, the presence of both products on the surface in significant quantity implies that the kinetic barriers determine the product distribution. Wang et al. [158] suggest that a barrierless pathway similar to that found for glyoxal [2+2] cycloaddition on Si(100)-(2×1) [157] may not have been located in their calculations for the reaction of acetone. If such a pathway exists, formation of the [2+2] product would be kinetically favored, thus explaining the larger percentage of [2+2] product at room temperature observed by Hamai et al. [159]. Moreover, the finding that annealing skews the product distribution toward ene product is suggested as evidence for thermodynamic control at high temperature [159], although the change in product distribution may also be rationalized in terms of the kinetics.

Wang et al. [158] note that the dative-bonded precursor states are of very similar energy (within 2 kcal/mol) on the Si(100)-(2×1) and Ge(100)-(2×1) surfaces. Compilation of data from a number of studies has shown that dative bonds are consistently only slightly stronger (~1–7 kcal/mol) on Si(100)-(2×1) than on Ge(100)-(2×1) [160]. Given this similarity in dative bond strength and the weaker bonds being formed in the reaction at the Ge surface, it is expected that higher activation barriers will generally be seen for reaction on the Ge surface. Most importantly, the studies discussed above provide clear experimental and theoretical evidence for one of the key differences between adsorption on Si(100)-(2×1) versus Ge(100)-(2×1): the strong bonds formed with Si favor kinetic control of product distribution, whereas the weaker bonds formed with Ge can lead to thermodynamic control for some reactions. This difference between the reactivity of Si and Ge surfaces may allow for much higher selectivity and the ability to control product distributions by thermal annealing on Ge(100)-(2×1).

To further investigate this difference in reactivity between the Si and Ge surfaces, a series of unsaturated ketones were studied [161]. Ethylvinylketone

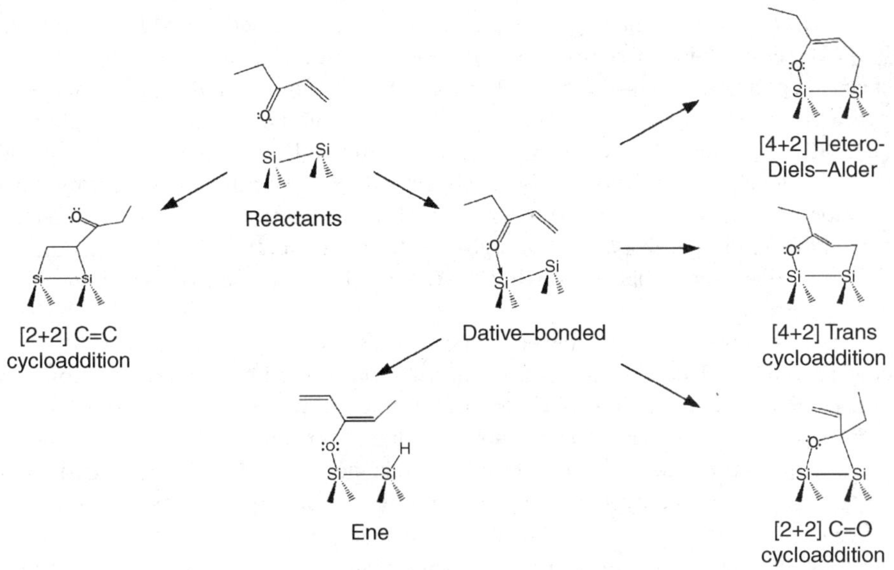

FIGURE 4.16 Schematic of possible intradimer reaction pathways for ethylvinylketone on Si(100)-(2×1). Analogous reactions pathways are also possible on Ge(100)-(2×1). Reprinted with permission from Ref. 161. Copyright 2002 American Chemical Society.

can potentially form a number of products at the surface due to its conjugated C=C and C=O bonds. Dative bonding through the oxygen lone pair, [2+2] cycloaddition across either the C=C or C=O bonds, [4+2] hetero-Diels–Alder cycloaddition (with ethylvinylketone in the *s-cis* conformation), [4+2] "*trans* cycloaddition" (with ethylvinylketone in the *s-trans* conformation) and the ene reaction are, in principle, all possible for the reaction of ethylvinylketone at Si(100)-(2×1) and Ge(100)-(2×1), as shown in Fig. 4.16. Using MIR-FTIR, it was determined that ethylvinylketone primarily forms a [4+2] cycloaddition product on Ge(100)-(2×1) at room temperature [161]. Although the two possible [4+2] products—hetero-Diels–Alder and *trans*—cannot be distinguished on the basis of IR spectra, calculations indicate that the hetero-Diels–Alder product should be kinetically favored at room temperature; *trans* cycloaddition products bridging two dimers may be present initially but may convert to the hetero-Diels–Alder product at the surface. Further study by STM may be able to definitively differentiate between these products but no such study has been reported. On Si (100)-(2×1), the IR spectrum provides evidence for the presence of a significant amount of ene product as well as [4+2] cycloaddition products. On both surfaces [2+2] cycloaddition across the C=O bond cannot be ruled out on the basis of IR spectroscopy, but the DFT calculations discussed below suggest it is unlikely to be present in significant quantity on Ge.

With the help of critical point energies calculated by DFT for the various reaction pathways, the difference in selectivity observed for reaction of ethylvinylketone with these two surfaces is once again explained in terms of kinetics and thermodynamics [161]. For all of the reaction pathways passing through a dative-bonded precursor state except hetero-Diels–Alder reaction, which was found to go through a barrierless pathway, higher barriers and smaller product binding energies are observed on Ge (100)-(2×1), similar to previously discussed systems. Hetero-Diels–Alder cycloaddition is clearly favored kinetically consistent with the experimental observation that it is the primary product on both surfaces. However, whereas the larger activation barriers for other pathways lead to high selectivity on Ge(100)-(2×1), the comparatively small activation barriers on Si(100)-(2×1) allow other pathways to compete, as seen experimentally.

The adsorption of 2-cyclohexen-1-one was also investigated as its ring structure constrains the molecule to the *s-trans* conformation, thus eliminating the possibility of hetero-Diels–Alder reaction [161]. IR spectra show similar results as ethylvinylketone for adsorption at room temperature: [4+2] cycloaddition on Ge(100)-(2×1) (although presumably by *trans* cycloaddition rather than hetero-Diels–Alder cycloaddition) versus primarily [4+2] cycloaddition with significant quantity of ene side product on Si (100)-(2×1). Using the calculated dative-bonded and transition state energies for ethylvinylketone, which are expected to be very similar to those for 2-cyclohexen-1-one, the *trans* cycloaddition pathway is kinetically favored by a similar amount on both surfaces when comparing to the second most favored pathway (ene reaction on Si and C=O [2+2] cycloaddition on Ge). In Fig. 4.17, this is apparent by comparing the transition state energies of *trans* cycloaddition (solid lines) to the next lowest transition state energy for each surface. In both cases there is a similar (~5–6 kcal/mol) difference in transition state energies between the two most kinetically favorable pathways. Thus, kinetic arguments alone cannot be used to rationalize the difference in selectivity observed experimentally for 2-cyclohexen-1-one on Si(100)-(2×1) and Ge (100)-(2×1). However, if one also considers thermodynamics, Fig. 4.17 shows that the C=O [2+2] cycloaddition product of 2-cyclohexen-1-one has a low binding energy on Ge (11.9 kcal/mol) and, thus, is not expected to be stable at room temperature. Similar to the case of acetone adsorption on Ge(100)-(2×1), the C=O [2+2] product may form temporarily but is expected to desorb on the time scale of the experiment at room temperature. The next most kinetically favored pathway on Ge(100)-(2×1) is the ene reaction, which leads to a more thermodynamically stable product (23.1 kcal/mol binding energy) that should be observable at room temperature. However, Fig. 4.17 shows that the transition state for ene reaction (the dashed line in Fig. 4.17) is significantly higher on Ge(100)-(2×1) than Si(100)-(2×1), leading to selective formation of *trans* cycloaddition product on Ge and a combination of *trans* cycloaddition and ene products on Si. Thus, Wang et al. conclude that both kinetic and thermodynamic factors play a role in determining product distribution on Ge(100)-(2×1), whereas kinetics are the dominant factor on Si(100)-(2×1) owing to the greater thermodynamic stability of all products on Si.

Adsorption of ethylvinylketone on Si(111)-(7×7) was later studied by Tang et al. [162]. HREELS and XPS evidence suggest that the conjugated C=C and C=O bonds of ethylvinylketone are both involved in the adsorption process with high selectivity, and a new peak in the HREELS spectrum at 1660 cm^{-1} is assigned to an

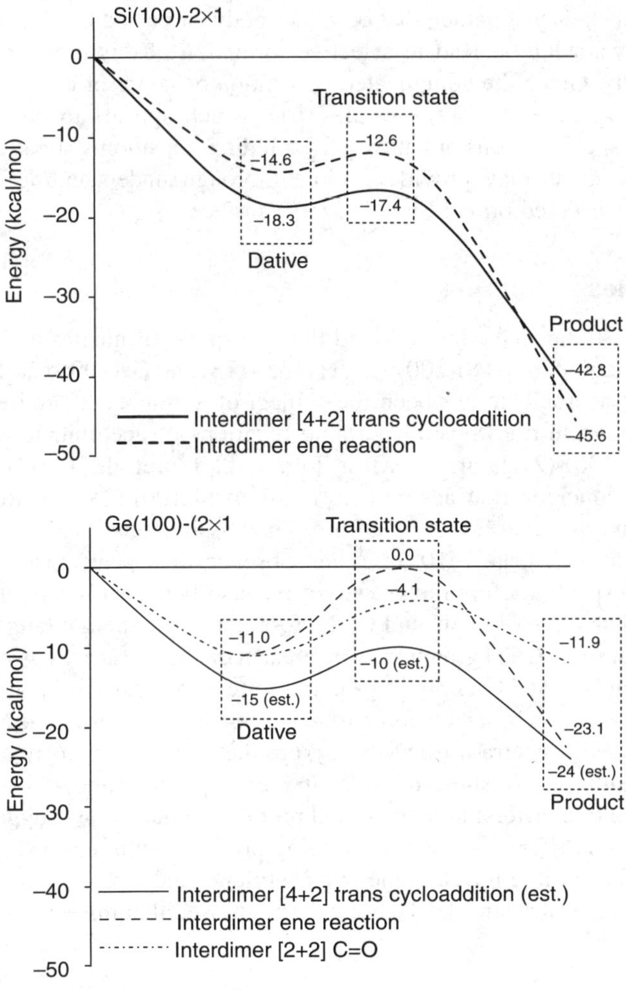

FIGURE 4.17 Calculated critical point energies on the potential energy surface for relevant reactions of 2-cyclohexen-1-one on Si(100)-(2×1) (top) and Ge(100)-(2×1) (bottom) [161]. Note that energies for *trans* cycloaddition on Ge(100)-(2×1) are estimated based on intradimer calculations, and energies on all pathways for dative-bonded and transition states are for ethylvinylketone, which are expected to be similar to the values for 2-cyclohexen-1-one.

internal C=C stretching mode. DFT calculations show that the [4+2]-like product has higher binding energy than the ene or [2+2]-like products, although binding energies for all products are high enough that they could be observed on the experimental time scale. By analogy to similar reactions on Si(100)-(2×1), the authors suggest that ethylvinylketone likely forms a dative-bonded precursor state by donation of a lone pair of electrons to an electrophilic Si adatom, and a barrierless or low activation pathway from the dative-bonded state to [4+2]-like cycloaddition may exist. Thus, they conclude that both thermodynamic and kinetic favorability of [4+2]-like cycloaddition lead to selective formation of this product, as observed experimentally. Given the complicated adsorption behavior of ethylvinylketone on Si(100)-(2×1) studied by Wang et al. [161], which appears to be dominated by kinetic factors, calculations of pathways (including transition states) for the various adsorption products may provide a more thorough understanding of why high selectivity is observed on the Si(111)-(7×7) surface.

4.4.2 Nitriles

A number of studies have investigated the adsorption of nitriles and other C≡N-containing compounds on Si(100)-(2×1) [163–173] and Ge(100)-(2×1) [171, 174]. In particular, acetonitrile has been the subject of a number of studies as it is the simplest organic nitrile. Several possible reactions of acetonitrile with Si(100)-(2×1) or Ge(100)-(2×1) are shown in Fig. 4.18. Tao et al. [173] used XPS and HREELS to conclude that acetonitrile binds to Si(100)-(2×1) to form a [2+2] cycloaddition product at low temperature. Their assignment is substantiated by the presence of a peak near $1610\,cm^{-1}$ and absence of a peak near $2220\,cm^{-1}$ in the HREELS spectra before the growth of physisorbed multilayers, indicating the presence and absence of C=N and C≡N, respectively, although later studies have suggested that some C≡N groups may be present on the surface, as discussed below. Large shifts in C 1s and N 1s binding energies are also consistent with the proposed binding configuration based on comparison to results for acetonitrile adsorption on metal substrates. The tetra-σ (dual-[2+2]) product was also ruled out on the basis of HREELS results. TPD showed molecular desorption peaks at 400 and 467K attributable to two different chemisorbed products. The authors suggest that these may result from different on-top or bridging products within a row, similar to the configurations possible for ethylene and acetylene, and a product bridging neighboring rows; however, later DFT calculations showed that this explanation cannot

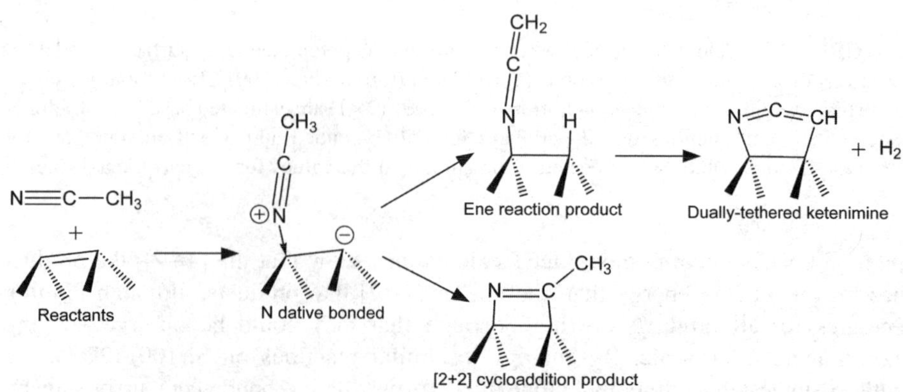

FIGURE 4.18 Schematic of possible intradimer reaction pathways for acetonitrile on Si(100)-(2×1) or Ge(100)-(2×1).

account for the experimental results, as the binding energy of a product bridging rows is too low to account for either of the observed TPD peaks [175]. DFT calculations performed by Lu et al. [170] demonstrated that [2+2] cycloaddition of acetonitrile on top of a single Si dimer may occur and showed that the reaction can proceed through a dative-bonded precursor state. A study by Bournel et al. [168] using PES and NEXAFS provided evidence for the existence of multiple products on the surface at room temperature, some containing a C≡N bond tilted with respect to the substrate.

While the presence of the [2+2] product is generally agreed upon, the configurations of additional products formed upon adsorption of acetonitrile on Si(100)-(2×1) at room temperature has been the topic of some debate. A stable N dative-bonded state was predicted by Miotto et al. [176], while Mui et al. [171] suggest based on DFT calculations that α-hydrogen elimination (ene reaction) would also occur at room temperature leading to a cumulated C=C=N structure (ketenimine) bound to the surface through the nitrogen (structure labeled "Ene reaction product" in Fig. 4.18). Polarization-dependent NEXAFS studies [166] again showed the presence of multiple products but did not support the theoretical prediction of Mui et al. [171]. To reconcile the discrepancy, Schwartz and Hamers [167] more recently proposed that a C=C=N species bound to the surface at both ends can form by loss of two hydrogens (structure labeled "Dually-tethered ketenimine" in Fig. 4.18). Formation of this species is consistent with the NEXAFS polarization-dependence, and further evidence for its presence is provided by polarized-FTIR experiments showing a C=C=N mode at about 1950 cm^{-1} polarized parallel to the surface dimers and in the plane of the surface. Additional DFT calculations by Carniato et al. [177] agree with the Si−C=C=N−Si structure proposed by Schwartz and Hamers (although Carniato et al. note that the structure must be twisted from its unstrained geometry due to the directionality of the bonds with the surface) and suggest the presence of additional products containing C=N and C≡N bonds. Their calculations show that modes associated with both of these groups (especially the C≡N group) have significantly lower cross sections than the C=C=N mode, which may explain why the C≡N stretching mode was not observed in the HREELS spectra by Tao et al. [173] while Bournel et al. [168] found evidence for the presence of such a C≡N bond by NEXAFS. Although the most recent works by Schwartz and Hamers [167] and Carniato et al. [177] reconcile apparent discrepancies among IR, NEXAFS, and HREELS results, neither specifically address whether the proposed Si−C=C=N−Si structure is consistent with previous TPD results.

In contrast to the results on Si(100)-(2×1), Filler et al. [174] showed experimentally that acetonitrile does not react with the Ge(100)-(2×1) surface at room temperature. These results agree with DFT calculations, which predict that the dative-bonded precursor, ene reaction product and [2+2] cycloaddition product will all desorb readily at room temperature [171]. It has been suggested that the nitrile functional group could serve as a protecting group for functionalization of Ge(100)-(2×1). Upon adsorption through a second functional group, the nitrile may remain unreacted on the surface. Hydrogenation of the nitrile to an amine could then be used to convert it to a more reactive moiety for further functionalization. Filler et al. [174]

also investigated adsorption on Ge(100)-(2×1) of the bifunctional nitriles 2-propenenitrile, 3-butenenitrile, and 4-pentenenitrile. Not surprisingly, the second functionality enables these molecules to adsorb on the Ge surface at room temperature. IR spectra show that the conjugated molecule 2-propenenitrile forms primarily the hetero-Diels–Alder [4+2] product and a small amount of [2+2] cycloaddition product reacted across the C=C bond. [4+2] cycloaddition of 2-propenenitrile creates a ring at the surface that includes a C=C=N ketenimine group, similar to that seen for the ene reaction product of acetonitrile. This again leads to absorption near 1950 cm^{-1} in the IR spectrum. Surprisingly, this mode is also observed when the nonconjugated molecules 3-butenenitrile and 4-pentenenitrile are exposed to the Ge(100)-(2×1) surface at room temperature, but Filler et al. tentatively attribute this to isomerization of these molecules to their conjugated forms before dosing them on the surface. Like acetonitrile, none of the bifunctional nitriles reacted with the Ge(100)-(2×1) surface directly through the C≡N bond at room temperature.

HREELS, XPS, and UPS results suggest that acetonitrile binds to Si(111)-(7×7) by [2+2]-like cycloaddition [178]. STM images were also used to show that acetonitrile binds at adatom–rest atom pairs, and to compare the reactivity of various sites in the unit cell. However, a later study by Bournel et al. [179] using synchrotron-based XPS and NEXAFS found evidence for two bonding configurations including one with a free C≡N group. By comparison of experimental results to DFT calculations, the authors assign one product to [2+2]-like cycloaddition bridging an adatom-rest atom pair in agreement with the previous study. The free C≡N group observed by NEXAFS is attributed to a second product involving bonding to the surface through the terminal carbon via α-hydrogen dissociation. Although this is not equivalent to the ene reaction proposed for acetonitrile adsorption on Si(100)-(2×1), and no evidence for dissociation of a second hydrogen (to form a C=C=N structure) has been reported, acetonitrile adsorption on Si(111)-(7×7) bears some similarity to that on Si(100)-(2×1): on both surfaces [2+2] cycloaddition appears to be the dominant pathway and secondary products are formed by dissociation of relatively acidic α-hydrogens.

4.4.3 Isocyanates and Isothiocyanates

The cumulative double bond of isocyanate and isothiocyanate functional groups provides for interesting chemistry at semiconductor surfaces, as they are able to undergo a number of reactions. In principal, two different dative-bonded states, two different [2+2] cycloaddition products and a 1,3-dipolar cycloaddition product [180] are possible for both functional groups, as shown in Fig. 4.19. Ellison and Hamers [181] reported the selective formation of a [2+2] product across the C=N bond for phenyl isothiocyanate exposed to Si(100)-(2×1) at room temperature based on XPS, IR spectroscopy, and STM. No evidence for interaction through the phenyl ring was observed. Phenyl isocyanate, on the other hand, was reported to decompose upon bonding to Si(100)-(2×1), although detailed studies of this system have not been published [11]. It is suggested that the difference between isothiocyanate and isocyanate reactivity may result from oxygen's propensity to insert into the

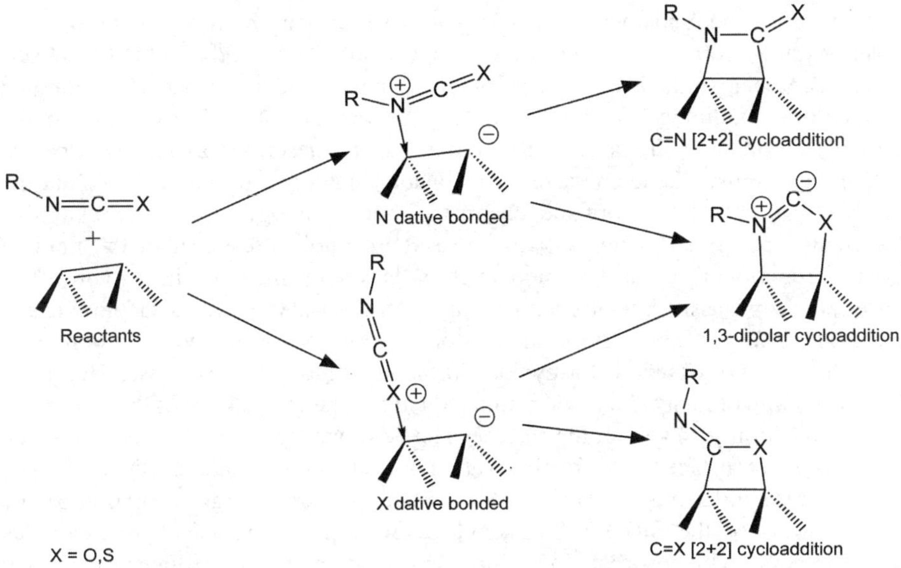

FIGURE 4.19 Schematic of possible intradimer reaction pathways for an isocyanate (X = O) or isothiocyanate (X = S) functional group on Si(100)-(2×1) or Ge(100)-(2×1).

Si–Si bonds due to its smaller size than sulfur and the larger difference in electronegativity between Si and O than between Si and S.

MIR-FTIR spectra for 1,4-phenylene diisocyanate exposed to Ge(100)-(2×1) show that it adsorbs primarily by [2+2] cycloaddition across the C=N bond and at least some of the isocyanate functional groups remain unreacted [132]. No evidence for significant decomposition on this surface was reported. As this molecule was being investigated for use as a molecular layer deposition precursor, adsorption of 1,4-phenylene diisocyanate was followed by exposure to ethylenediamine. IR spectra after this exposure indicated some desorption of 1,4-phenylene diisocyanate from the surface in addition to the intended reaction between the isocyanate and amine, highlighting the relatively weak bonding between the surface and isocyanate. Unlike many adsorbates on Ge(100)-(2×1) that exhibit fairly high selectivity, phenyl isocyanate and phenyl isothiocyanate are found to form multiple adsorption products on the Ge surface and exhibit interesting coverage or time dependences [182]. Further studies are currently underway to fully understand the adsorption behavior of these molecules on Ge(100)-(2×1), but this example demonstrates that although thermodynamic factors may play a role in determining adsorption products on Ge, high selectivity does not always result.

4.5 SUMMARY

Cycloaddition reactions, including [2+2] and [4+2] Diels–Alder reactions, comprise an important means for functionalization of semiconductor surfaces and understanding the fundamental chemistry at these surfaces. Many examples of these reactions at Si(100)-(2×1), Ge(100)-(2×1), C(100)-(2×1), and Si(111)-(7×7) surfaces have been studied for alkenes, alkynes, dienes and a number of unsaturated heteroatom-containing functional groups. The fundamental reactive unit on the Group IV (100)-(2×1) surfaces, the dimer, can be described as having a partial π bond and, thus, alkene character. This enables analogy between reactions at the surface and in solution organic chemistry. Such analogy allows the extensive knowledge of organic chemistry to be used in guiding research in the field of organic functionalization of semiconductors. However, there are important differences between classic organic chemistry and chemistry at the surface. Geometric and electronic effects of the surface enable, for example, low-symmetry reaction pathways that are not observed for cycloadditions in solution or gas phase. The facile [2+2] reaction of many alkenes at semiconductor surfaces exemplifies the possibility of such reaction pathways at the surface. The heterogeneity of surface atoms on the Si(111)-(7×7) surface has led to observation of cycloaddition-like reactions believed to occur by similar mechanisms on this surface as well. Charge separation on the tilted dimers of the Si(100)-(2×1) and Ge(100)-(2×1) surfaces also facilitates dative bonding with adsorbate lone pairs, thus creating an intermediate or precursor state that is important for cycloaddition reactions of many heteroatom-containing adsorbates.

Comparing adsorption of various molecules on the three Group IV (100)-(2×1) surfaces leads to several conclusions. Adsorption on Si(100)-(2×1), the subject of the majority of studies, very often produces multiple products, the distribution of which is largely controlled by kinetic factors. These products are typically strongly bound to the surface and, upon heating, decompose before desorbing. The dimers of C(100)-(2×1) are the most direct analogue of an alkene. Although cycloadditions also occur on this surface, lower sticking coefficients are typically observed, especially in the case of [2+2] cycloaddition. This is believed to result from the symmetry of the dimers on the diamond surface and highlights the influence of the surface geometry on cycloaddition reactivity. Adsorbates tend to form weaker bonds with the Ge(100)-(2×1) surface resulting in lower binding energies and higher activation barriers. The lower binding energies have been found in many cases to favor thermodynamic control over kinetic control on the Ge surface, and higher selectivity is often achieved.

Much recent work has focused on adsorption of heteroatom-containing compounds, since the ability to dative bond to the Si(100)-(2×1) and Ge(100)-(2×1) surfaces facilitates cycloaddition. Adsorption behavior of multifunctional compounds is also of growing interest, because multiple functionalities may allow subsequent reaction; however, use of multifunctional compounds is accompanied by the challenge of achieving controlled, selective attachment to the surface. While many useful attributes have been identified in separate systems, such as higher selectivity on Ge(100)-(2×1) for some adsorbates and well-ordered adsorption of cyclopentene on Si(100)-(2×1), future work may focus on bringing together such features to functionalize semiconductor surfaces in a highly controlled fashion. As many unsaturated organic compounds readily undergo cycloaddition reactions at these surfaces, understanding the nature of cycloadditions is necessary to achieve such control.

ACKNOWLEDGMENT

This work was supported by the National Science Foundation (Grants CHE 0615087 and CHE 0910717).

REFERENCES

1. Woodward, R. B.; Hoffmann, R. *J. Am. Chem. Soc.* **1965**, *87*, 395.
2. Fleming, I. *Pericyclic Reactions*. Oxford University Press, Oxford, 1999.
3. Carruthers, W. *Cycloaddition Reactions in Organic Synthesis*. Pergamon Press, New York, 1990.
4. Fallis, A. G. *Acc. Chem. Res.* **1999**, *32*, 464.
5. Woodward, R. B.; Hoffmann, R. *The Conservation of Orbital Symmetry*. Academic Press, New York, 1970.
6. Choi, C. H.; Gordon, M. S. *J. Am. Chem. Soc.* **1999**, *121*, 11311.
7. Liu, H.; Hamers, R. J. *J. Am. Chem. Soc.* **1997**, *119*, 7593.
8. Liu, Q.; Hoffmann, R. *J. Am. Chem. Soc.* **1995**, *117*, 4082.
9. Lopinski, G. P.; Moffatt, D. J.; Wayner, D. D. M.; Wolkow, R. A. *J. Am. Chem. Soc.* **2000**, *122*, 3548.
10. Buriak, J. M. *Chem. Commun.* **1999**, 1051.
11. Hamers, R. J.; Coulter, S. K.; Ellison, M. D.; Hovis, J. S.; Padowitz, D. F.; Schwartz, M. P.; Greenlief, C. M.; Russell, J. N. *Acc. Chem. Res.* **2000**, *33*, 617.
12. Bent, S. F. *J. Phys. Chem. B* **2002**, *106*, 2830.
13. Bent, S. F. *Surf. Sci.* **2002**, *500*, 879.
14. Buriak, J. M. *Chem. Rev.* **2002**, *102*, 1271.
15. Lu, X.; Lin, M. C. *Int. Rev. Phys. Chem.* **2002**, *21*, 137.
16. Filler, M. A.; Bent, S. F. *Prog. Surf. Sci.* **2003**, *73*, 1.
17. Tao, F.; Xu, G. Q. *Acc. Chem. Res.* **2004**, *37*, 882.
18. Loscutoff, P. W.; Bent, S. F. *Annu. Rev. Phys. Chem.* **2006**, *57*, 467.
19. Ma, Z.; Zaera, F. *Surf. Sci. Rep.* **2006**, *61*, 229.
20. Leftwich, T. R.; Teplyakov, A. V. *Surf. Sci. Rep.* **2008**, *63*, 1.
21. Takayanagi, K.; Tanishiro, Y.; Takahashi, M.; Takahashi, S. *J. Vac. Sci. Technol. A* **1985**, *3*, 1502.
22. Hamers, R. J.; Tromp, R. M.; Demuth, J. E. *Phys. Rev. Lett.* **1986**, *56*, 1972.
23. Northrup, J. E. *Phys. Rev. Lett.* **1986**, *57*, 154.
24. Hamers, R. J.; Tromp, R. M.; Demuth, J. E. *Surf. Sci.* **1987**, *181*, 346.
25. Wolkow, R.; Avouris, P. *Phys. Rev. Lett.* **1988**, *60*, 1049.

26. Stich, I.; Payne, M. C.; King-Smith, R. D.; Lin, J. S.; Clarke, L. J. *Phys. Rev. Lett.* **1992**, *68*, 1351.
27. Cheng, C. C.; Choyke, W. J.; Yates, J. T., Jr. *Surf. Sci.* **1990**, *231*, 289.
28. Yoshinobu, J.; Tsuda, H.; Onchi, M.; Nishijima, M. *J. Chem. Phys.* **1987**, *87*, 7332.
29. Birkenheuer, U.; Gutdeutsch, U.; Rosch, N.; Fink, A.; Gokhale, S.; Menzel, D.; Trischberger, P.; Widdra, W. *J. Chem. Phys.* **1998**, *108*, 9868.
30. Konecny, R.; Doren, D. J. *Surf. Sci.* **1998**, *417*, 169.
31. Casaletto, M. P.; Zanoni, R.; Carbone, M.; Piancastelli, M. N.; Aballe, L.; Weiss, K.; Horn, K. *Phys. Rev. B* **2000**, *62*, 17128.
32. Chua, L. H.; Jackman, R. B.; Foord, J. S. *Surf. Sci.* **1994**, *315*, 69.
33. Clemen, L.; Wallace, R. M.; Taylor, P. A.; Dresser, M. J.; Choyke, W. J.; Weinberg, W. H.; Yates, J. T. *Surf. Sci.* **1992**, *268*, 205.
34. Craig, B. I. *Surf. Sci.* **1995**, *329*, 293.
35. Fisher, A. J.; Blochl, P. E.; Briggs, G. A. D. *Surf. Sci.* **1997**, *374*, 298.
36. Huang, C.; Widdra, W.; Weinberg, W. H. *Surf. Sci.* **1994**, *315*, L953.
37. Matsui, F.; Yeoim, H. W.; Imanishi, A.; Isawa, K.; Matsuda, I.; Ohta, T. *Surf. Sci.* **1998**, *401*, L413.
38. Matsui, F.; Yeom, H. W.; Matsuda, I.; Ohta, T. *Phys. Rev. B* **2000**, *62*, 5036.
39. Mayne, A. J.; Avery, A. R.; Knall, J.; Jones, T. S.; Briggs, G. A. D.; Weinberg, W. H. *Surf. Sci.* **1993**, *284*, 247.
40. Pan, W.; Zhu, T.; Yang, W. *J. Chem. Phys.* **1997**, *107*, 3981.
41. Widdra, W.; Huang, C.; Briggs, G. A. D.; Weinberg, W. H. *J. Electron Spectrosc. Relat. Phenom.* **1993**, *64-65*, 129.
42. Widdra, W.; Huang, C.; Weinberg, W. H. *Surf. Sci.* **1995**, *329*, 295.
43. Xu, S. H.; Keeffe, M.; Yang, Y.; Chen, C.; Yu, M.; Lapeyre, G. J.; Rotenberg, E.; Denlinger, J.; Yates, J. T. *Phys. Rev. Lett.* **2000**, *84*, 939.
44. Wolkow, R. A. *Annu. Rev. Phys. Chem.* **2003**, *50*, 413.
45. Shoemaker, J.; Burggraf, L. W.; Gordon, M. S. *J. Chem. Phys.* **2000**, *112*, 2994.
46. Lu, X. *J. Am. Chem. Soc.* **2003**, *125*, 6384.
47. Lu, X.; Zhu, M.; Wang, X. *J. Phys. Chem. B* **2004**, *108*, 7359.
48. Nagao, M.; Umeyama, H.; Mukai, K.; Yamashita, Y.; Yoshinobu, J.; Akagi, K.; Tsuneyuki, S. *J. Am. Chem. Soc.* **2004**, *126*, 9922.
49. Fan, X. L.; Zhang, Y. F.; Lau, W. M.; Liu, Z. F. *Phys. Rev. B* **2005**, *72*, 165305.
50. Ryan, P. M.; Teague, L. C.; Boland, J. J. *J. Am. Chem. Soc.* **2009**, *131*, 6768.
51. Mui, C.; Bent, S. F.; Musgrave, C. B. *J. Phys. Chem. A* **2000**, *104*, 2457.
52. Hennies, F.; Fohlisch, A.; Wurth, W.; Witkowski, N.; Nagasono, M.; Novella Piancastelli, M. *Surf. Sci.* **2003**, *529*, 144.
53. Seo, J. H.; Park, J. Y.; Whang, C. N.; Kim, S. S.; Choi, D. S.; Chae, K. H. *Surf. Sci.* **2005**, *582*, L129.
54. Fink, A.; Huber, R.; Widdra, W. *J. Chem. Phys.* **2001**, *115*, 2768.
55. Kim, A.; Choi, D. S.; Lee, J. Y.; Kim, S. *J. Phys. Chem. B* **2004**, *108*, 3256.
56. Lal, P.; Teplyakov, A. V.; Noah, Y.; Kong, M. J.; Wang, G. T.; Bent, S. F. *J. Chem. Phys.* **1999**, *110*, 10545.

57. Cho, J.-H.; Kleinman, L. *J. Chem. Phys.* **2003**, *119*, 2820.
58. Miotto, R.; Ferraz, A. C. *Surf. Sci.* **2002**, *513*, 422.
59. Miotto, R.; Ferraz, A. C.; Srivastava, G. P. *Surf. Sci.* **2002**, *507-510*, 12.
60. Nishijima, M.; Yoshinobu, J.; Tsuda, H.; Onchi, M. *Surf. Sci.* **1987**, *192*, 383.
61. Taylor, P. A.; Wallace, R. M.; Cheng, C. C.; Weinberg, W. H.; Dresser, M. J.; Choyke, W. J.; Yates, J. T. *J. Am. Chem. Soc.* **1992**, *114*, 6754.
62. Carmer, C. S.; Weiner, B.; Frenklach, M. *J. Chem. Phys.* **1993**, *99*, 1356.
63. Cheng, C. C.; Taylor, P. A.; Wallace, R. M.; Gutleben, H.; Clemen, L.; Colaianni, M. L.; Chen, P. J.; Weinberg, W. H.; Choyke, W. J.; YatesJr., J. T. *Thin Solid Films* **1993**, *225*, 196.
64. Huang, C.; Widdra, W.; Wang, X. S.; Weinberg, W. H. *J. Vac. Sci. Technol. A* **1993**, *11*, 2250.
65. Imamura, Y.; Morikawa, Y.; Yamasaki, T.; Nakatsuji, H. *Surf. Sci.* **1995**, *341*, L1091.
66. Yan, C.; Zhaohui, L. *Appl. Phys. Lett.* **1995**, *67*, 2936.
67. Widdra, W.; Huang, C.; Yi, S. I.; Weinberg, W. H. *J. Chem. Phys.* **1996**, *105*, 5605.
68. Dyson, A. J.; Smith, P. V. *Surf. Sci.* **1997**, *375*, 45.
69. Li, L.; Tindall, C.; Takaoka, O.; Hasegawa, Y.; Sakurai, T. *Phys. Rev. B* **1997**, *56*, 4648.
70. Meng, B.; Maroudas, D.; Weinberg, W. H. *Chem. Phys. Lett.* **1997**, *278*, 97.
71. Sorescu, D. C.; Jordan, K. D. *J. Phys. Chem. B* **2000**, *104*, 8259.
72. Terborg, R.; Baumgärtel, P.; Lindsay, R.; Schaff, O.; Giessel, T.; Hoeft, J. T.; Polcik, M.; Toomes, R. L.; Kulkarni, S.; Bradshaw, A. M.; Woodruff, D. P. *Phys. Rev. B* **2000**, *61*, 16697.
73. Cho, J.-H.; Kleinman, L.; Chan, C. T.; Kim, K. S. *Phys. Rev. B* **2001**, *63*, 073306.
74. Hofer, W. A.; Fisher, A. J.; Wolkow, R. A. *Surf. Sci.* **2001**, *475*, 83.
75. Kim, W.; Kim, H.; Lee, G.; Hong, Y.-K.; Lee, K.; Hwang, C.; Kim, D.-H.; Koo, J.-Y. *Phys. Rev. B* **2001**, *64*, 193313.
76. Mezhenny, S.; Lyubinetsky, I.; Choyke, W. J.; Wolkow, R. A.; Yates, J. T. *Chem. Phys. Lett.* **2001**, *344*, 7.
77. Morikawa, Y. *Phys. Rev. B* **2001**, *63*, 033405.
78. Silvestrelli, P. L.; Toigo, F.; Ancilotto, F. *J. Chem. Phys.* **2001**, *114*, 8539.
79. Kim, W.; Kim, H.; Lee, G.; Chung, J.; You, S.-Y.; Hong, Y.-K.; Koo, J.-Y. *Surf. Sci.* **2002**, *514*, 376.
80. Terborg, R.; Polcik, M.; Hoeft, J. T.; Kittel, M.; Sayago, D. I.; Toomes, R. L.; Woodruff, D. P. *Phys. Rev. B* **2002**, *66*, 085333.
81. Wang, F.; Sorescu, D. C.; Jordan, K. D. *J. Phys. Chem. B* **2002**, *106*, 1316.
82. Di Felice, R.; Pignedoli, C. A.; Bertoni, C. M.; Catellani, A.; Silvestrelli, P. L.; Sbraccia, C.; Ancilotto, F.; Palummo, M.; Pulci, O. *Surf. Sci.* **2003**, *532–535*, 982.
83. Silvestrelli, P. L.; Pulci, O.; Palummo, M.; Del Sole, R.; Ancilotto, F. *Phys. Rev. B* **2003**, *68*, 235306.
84. Cho, J.-H.; Kleinman, L. *Phys. Rev. B* **2004**, *69*, 075303.
85. Mineva, T.; Nathaniel, R.; Kostov, K. L.; Widdra, W. *J. Chem. Phys.* **2006**, *125*, 194712.
86. Toscano, M.; Russo, N. *J. Mol. Catal.* **1989**, *55*, 101.
87. Kim, A.; Maeng, J. Y.; Lee, J. Y.; Kim, S. *J. Chem. Phys.* **2002**, *117*, 10215.

88. Cho, J.-H.; Kim, K. S.; Morikawa, Y. *J. Chem. Phys.* **2006**, *124*, 024716.
89. Lopinski, G. P.; Moffatt, D. J.; Wayner, D. D. M.; Wolkow, R. A. *Nature* **1998**, *392*, 909.
90. Kiskinova, M.; Yates, J. T. *Surf. Sci.* **1995**, *325*, 1.
91. Madachik, M. R.; Teplyakov, A. V. *J. Vac. Sci. Technol. A* **2008**, *26*, 1241.
92. Hamers, R. J.; Hovis, J. S.; Lee, S.; Liu, H.; Shan, J. *J. Phys. Chem. B* **1997**, *101*, 1489.
93. Hovis, J. S.; Lee, S.; Liu, H.; Hamers, R. J. *J. Vac. Sci. Technol. B* **1997**, *15*, 1153.
94. Hovis, J. S.; Liu, H.; Hamers, R. J. *J. Appl. Phys. A* **1998**, *66*, S553.
95. Hovis, J. S.; Liu, H.; Hamers, R. J. *Surf. Sci.* **1998**, *402–404*, 1.
96. Liu, H.; Hamers, R. J. *Surf. Sci.* **1998**, *416*, 354.
97. Yamashita, Y.; Nagao, M.; Machida, S.; Hamaguchi, K.; Yasui, F.; Mukai, K.; Yoshinobu, J. *J. Electron Spectrosc. Relat. Phenom.* **2001**, *114–116*, 389.
98. Hovis, J. S.; Hamers, R. J. *J. Phys. Chem. B* **1997**, *101*, 9581.
99. Hovis, J. S.; Hamers, R. J. *J. Phys. Chem. B* **1998**, *102*, 687.
100. Rochet, F.; Bournel, F.; Gallet, J. J.; Dufour, G.; Lozzi, L.; Sirotti, F. *J. Phys. Chem. B* **2002**, *106*, 4967.
101. Yates, J. T., Jr. *Science* **1998**, *279*, 335.
102. Hovis, J. S.; Coulter, S. K.; Hamers, R. J.; D'Evelyn, M. P.; Russell, J. N.; Butler, J. E. *J. Am. Chem. Soc.* **2000**, *122*, 732.
103. Lee, S. W.; Hovis, J. S.; Coulter, S. K.; Hamers, R. J.; Greenlief, C. M. *Surf. Sci.* **2000**, *462*, 6.
104. Cho, J.-H.; Kleinman, L. *Phys. Rev. B* **2003**, *67*, 115314.
105. Yoshinobu, J.; Tsuda, H.; Onchi, M.; Nishijima, M. *Solid State Commun.* **1986**, *60*, 801.
106. Yoshinobu, J.; Tsuda, H.; Onchi, M.; Nishijima, M. *Chem. Phys. Lett.* **1986**, *130*, 170.
107. MacPherson, C. D.; Hu, D. Q.; Leung, K. T. *Surf. Sci.* **1992**, *276*, 156.
108. Tao, F.; Wang, Z. H.; Xu, G. Q. *Surf. Sci.* **2003**, *530*, 203.
109. Piancastelli, M. N.; Motta, N.; Sgarlata, A.; Balzarotti, A.; De Crescenzi, M. *Phys. Rev. B* **1993**, *48*, 17892.
110. Rochet, F.; Jolly, F.; Bournel, F.; Dufour, G.; Sirotti, F.; Cantin, J.-L. *Phys. Rev. B* **1998**, *58*, 11029.
111. Lu, X.; Wang, X.; Yuan, Q.; Zhang, Q. *J. Am. Chem. Soc.* **2003**, *125*, 7923.
112. Chu, S.-Y.; Anderson, A. B. *Surf. Sci.* **1988**, *194*, 55.
113. Weiner, B.; Carmer, C. S.; Frenklach, M. *Phys. Rev. B* **1991**, *43*, 1678.
114. Yoshinobu, J.; Fukushi, D.; Uda, M.; Nomura, E.; Aono, M. *Phys. Rev. B* **1992**, *46*, 9520.
115. MacPherson, C. D.; Hu, D. Q.; Doan, M.; Leung, K. T. *Surf. Sci.* **1994**, *310*, 231.
116. Rochet, F.; Dufour, G.; Prieto, P.; Sirotti, F.; Stedile, F. C. *Phys. Rev. B* **1998**, *57*, 6738.
117. De Renzi, V.; Biagi, R.; del Pennino, U. *Phys. Rev. B* **2001**, *64*, 155305.
118. Sbraccia, C.; Pignedoli, C. A.; Catellani, A.; Di Felice, R.; Silvestrelli, P. L.; Toigo, F.; Ancilotto, F.; Bertoni, C. M. *Surf. Sci.* **2004**, *557*, 80.
119. Konecny, R.; Doren, D. J. *J. Am. Chem. Soc.* **1997**, *119*, 11098.
120. Fitzgerald, D. R.; Doren, D. J. *J. Am. Chem. Soc.* **2000**, *122*, 12334.
121. Yoshimura, T.; Tatsuura, S.; Sotoyama, W. *Appl. Phys. Lett.* **1991**, *59*, 482.
122. Yoshimura, T.; Tatsuura, S.; Sotoyama, W.; Matsuura, A.; Hayano, T. *Appl. Phys. Lett.* **1992**, *60*, 268.

123. Kubono, A.; Yuasa, N.; Shao, H.-L.; Umemoto, S.; Okui, N. *Thin Solid Films* **1996**, *289*, 107.
124. Bitzer, T.; Richardson, N. V. *Appl. Phys. Lett.* **1997**, *71*, 662.
125. Shao, H.-I.; Umemoto, S.; Kikutani, T.; Okui, N. *Polymer* **1997**, *38*, 459.
126. Bitzer, T.; Richardson, N. V. *Appl. Surf. Sci.* **1999**, *144–145*, 339.
127. Haq, S.; Richardson, N. V. *J. Phys. Chem. B* **1999**, *103*, 5256.
128. Usui, H. *Thin Solid Films* **2000**, *365*, 22.
129. Nagai, A.; Shao, H.; Umemoto, S.; Kikutani, T.; Okui, N. *High Perform. Polym.* **2001**, *13*, S169.
130. Miyamae, T.; Tsukagoshi, K.; Matsuoka, O.; Yamamoto, S.; Nozoye, H. *Jpn. J. Appl. Phys.* **2002**, *41*, 746.
131. Lee, J. S.; Lee, Y.-J.; Tae, E. L.; Park, Y. S.; Yoon, K. B. *Science* **2003**, *301*, 818.
132. Kim, A.; Filler, M. A.; Kim, S.; Bent, S. F. *J. Am. Chem. Soc.* **2005**, *127*, 6123.
133. Du, Y.; George, S. M. *J. Phys. Chem. C* **2007**, *111*, 8509.
134. Putkonen, M.; Harjuoja, J.; Sajavaara, T.; Niinisto, L. *J. Mater. Chem.* **2007**, *17*, 664.
135. Adamczyk, N. M.; Dameron, A. A.; George, S. M. *Langmuir* **2008** *24*, 2081.
136. Dameron, A. A.; Seghete, D.; Burton, B. B.; Davidson, S. D.; Cavanagh, A. S.; Bertrand, J. A.; George, S. M. *Chem. Mater.* **2008**, *20*, 3315.
137. George, S. M.; Yoon, B.; Dameron, A. A. *Acc. Chem. Res.* **2009**, *42*, 498.
138. Lee, B. H.; Im, K. K.; Lee, K. H.; Im, S.; Sung, M. M. *Thin Solid Films* **2009**, *517*, 4056.
139. Loscutoff, P. W.; Zhou, H.; Clendenning, S. B.; Bent, S. F. *ACS Nano* 2009.
140. Peng, Q.; Gong, B.; VanGundy, R. M.; Parsons, G. N. *Chem. Mater.* **2009**, *21*, 820.
141. Yoon, B.; O'Patchen, J. L.; Seghete, D.; Cavanagh, A. S.; George, S. M. *Chem. Vap. Deposition* **2009**, *15*, 112.
142. Yoshimura, T.; Kudo, Y. *Appl. Phys. Expr.* **2009**, *2*, 015502.
143. Teplyakov, A. V.; Kong, M. J.; Bent, S. F. *J. Am. Chem. Soc.* **1997**, *119*, 11100.
144. Teplyakov, A. V.; Kong, M. J.; Bent, S. F. *J. Chem. Phys.* **1998**, *108*, 4599.
145. Hovis, J. S.; Liu, H.; Hamers, R. J. *J. Phys. Chem. B* **1998**, *102*, 6873.
146. Teplyakov, A. V.; Lal, P.; Noah, Y. A.; Bent, S. F. *J. Am. Chem. Soc.* **1998**, *120*, 7377.
147. Wang, G. T.; Bent, S. F.; Russell, J. N.; Butler, J. E.; D'Evelyn, M. P. *J. Am. Chem. Soc.* **2000**, *122*, 744.
148. Kong, M. J.; Teplyakov, A. V.; Jagmohan, J.; Lyubovitsky, J. G.; Mui, C.; Bent, S. F. *J. Phys. Chem. B* **2000**, *104*, 3000.
149. Teague, L. C.; Boland, J. J. *J. Phys. Chem. B* **2003**, *107*, 3820.
150. Teague, L. C.; Boland, J. J. *Thin Solid Films* **2004**, *464-465*, 1.
151. Teague, L. C.; Chen, D.; Boland, J. J. *J. Phys. Chem. B* **2004**, *108*, 7827.
152. Wang, G. T.; Mui, C.; Musgrave, C. B.; Bent, S. F. *J. Phys. Chem. B* **1999**, *103*, 6803.
153. Baik, J.; Kim, M.; Park, C.-Y.; Kim, Y.; Ahn, J. R.; An, K.-S. *J. Am. Chem. Soc.* 2006, *128*, 8370.
154. Ellison, M. D.; Hovis, J. S.; Liu, H.; Hamers, R. J. *J. Phys. Chem. B* **1998**, *102*, 8510.
155. Armstrong, J. L.; Pylant, E. D.; White, J. M. *J. Vac. Sci. Technol. A* **1998**, *16*, 123.
156. Armstrong, J. L.; White, J. M.; Langell, M. *J. Vac. Sci. Technol. A* **1997**, *15*, 1146.
157. Barriocanal, J. A.; Doren, D. J. *J. Am. Chem. Soc.* **2001**, *123*, 7340.

158. Wang, G. T.; Mui, C.; Musgrave, C. B.; Bent, S. F. *J. Phys. Chem. B* **2001**, *105*, 12559.
159. Hamai, C.; Takagi, A.; Taniguchi, M.; Matsumoto, T.; Kawai, T. *Angew. Chem. Int. Ed.* **2004**, *43*, 1349.
160. Kachian, J. S.; Wong, K. T.; Bent, S. F. *Acc. Chem. Res.* **2010**, *43*, 346.
161. Wang, G. T.; Mui, C.; Musgrave, C. B.; Bent, S. F. *J. Am. Chem. Soc.* **2002**, *124*, 8990.
162. Tang, H. H.; Cai, Y. H.; Ning, Y. S.; Lai, Y. H.; Xu, G. Q. *Surf. Sci.* **2007**, *601*, 3293.
163. Bacalzo-Gladden, F.; Lu, X.; Lin, M. C. *J. Phys. Chem. B* **2001**, *105*, 4368.
164. Schwartz, M. P.; Hamers, R. J. *Surf. Sci.* **2002**, *515*, 75.
165. Choi, C. H.; Gordon, M. S. *J. Am. Chem. Soc.* **2002**, *124*, 6162.
166. Rangan, S.; Bournel, F.; Gallet, J. J.; Kubsky, S.; Le Guen, K.; Dufour, G.; Rochet, F.; Sirotti, F.; Carniato, S.; Ilakovac, V. *Phys. Rev. B* **2005**, *71*, 165319.
167. Schwartz, M. P.; Hamers, R. J. *Surf. Sci.* **2007**, *601*, 945.
168. Bournel, F.; Gallet, J. J.; Kubsky, S.; Dufour, G.; Rochet, F.; Simeoni, M.; Sirotti, F. *Surf. Sci.* **2002**, *513*, 37.
169. Bu, Y.; Ma, L.; Lin, M. C. *J. Phys. Chem.* **1993**, *97*, 7081.
170. Lu, X.; Xu, X.; Wu, J.; Wang, N.; Zhang, Q. *New J. Chem.* **2002**, *26*, 160.
171. Mui, C.; Filler, M. A.; Bent, S. F.; Musgrave, C. B. *J. Phys. Chem. B* **2003**, *107*, 12256.
172. Tao, F.; Sim, W. S.; Xu, G. Q.; Qiao, M. H. *J. Am. Chem. Soc.* **2001**, *123*, 9397.
173. Tao, F.; Wang, Z. H.; Qiao, M. H.; Liu, Q.; Sim, W. S.; Xu, G. Q. *J. Chem. Phys.* **2001**, *115*, 8563.
174. Filler, M. A.; Mui, C.; Musgrave, C. B.; Bent, S. F. *J. Am. Chem. Soc.* **2003**, *125*, 4928.
175. Cho, J.-H.; Kleinman, L. *J. Chem. Phys.* **2003**, *119*, 6744.
176. Miotto, R.; Oliveira, M. C.; Pinto, M. M.; de León-Pérez, F.; Ferraz, A. C. *Phys. Rev. B* **2004**, *69*, 235331.
177. Carniato, S.; Rochet, F.; Gallet, J. J.; Bournel, F.; Dufour, G.; Mathieu, C.; Rangan, S. *Surf. Sci.* **2007**, *601*, 5515.
178. Tao, F.; Chen, X. F.; Wang, Z. H.; Xu, G. Q. *J. Phys. Chem. B* **2002**, *106*, 3890.
179. Bournel, F.; Carniato, S.; Dufour, G.; Gallet, J. J.; Ilakovac, V.; Rangan, S.; Rochet, F.; Sirotti, F. *Phys. Rev. B* **2006**, *73*, 125345.
180. Barriocanal, J. A.; Doren, D. J. *J. Vac. Sci. Technol. A* **2000**, *18*, 1959.
181. Ellison, M. D.; Hamers, R. J. *J. Phys. Chem. B* **1999**, *103*, 6243.
182. Loscutoff, P. L.; Wong, K. T.; Bent, S. F. *J. Phys. Chem. C* **2010** *114*, 14193.

CHAPTER 5

Chemical Binding of Five-Membered and Six-Membered Aromatic Molecules

FRANKLIN (FENG) TAO AND STEVEN L. BERNASEK

5.1 INTRODUCTION

Understanding the mechanism of reaction of five- and six-membered aromatic organics with semiconductor surfaces at the atomic level provides the foundation for further chemical attachment of multilayer organic materials to these semiconductor surfaces. These chemical reactions and related biofunctionalization of semiconductor surfaces is important for the development of sensing technology. This chapter focuses on the reaction mechanism of these representative categories of organic molecules. It is organized by starting from simple five-membered ring molecules, benzene, six-membered aromatic molecules with one heteroatom, and ending at six-membered aromatic molecules with two heteroatoms. For each category of these molecules, their reaction mechanisms on different semiconductor surfaces are reviewed and compared in order to understand the electronic and structural factors that affect the chemical functionalization.

5.2 FIVE-MEMBERED AROMATIC MOLECULES CONTAINING ONE HETEROATOM

Thiophene, furan, and pyrrole are three representative five-membered ring aromatic molecules with one heteroatom, as shown in Fig. 5.1. Their chemical binding on Si(111)-(7×7), Si(100), and Ge(100) have been extensively studied. Each of the three molecules exhibits different reaction mechanisms in the modification and functionalization of semiconductor surfaces due to the participation of their heteroatoms.

Functionalization of Semiconductor Surfaces, First Edition.
Edited by Franklin (Feng) Tao and Steven L. Bernasek.

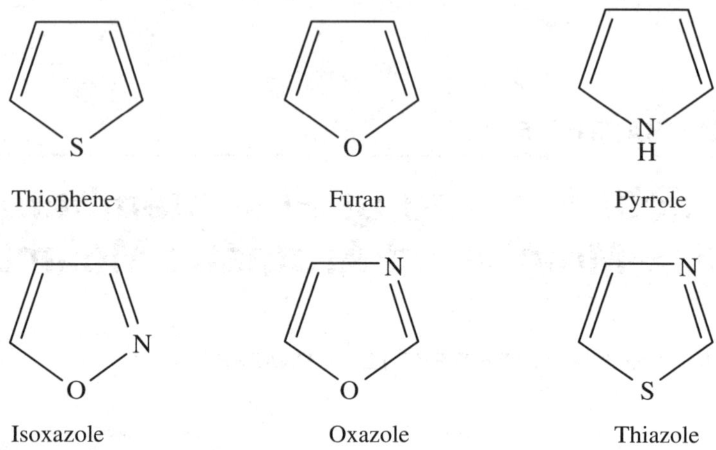

FIGURE 5.1 Molecular structure of thiophene, furan, pyrrole, isoxazole, oxazole, and thiazole.

5.2.1 Thiophene, Furan, and Pyrrole on Si(111)-(7×7)

Compared to the homogeneous distribution of electronic density on benzene, the electron density on the molecular ring of thiophene is unevenly distributed due to its heteroatom, sulfur. The electron density of the HOMO is mostly concentrated at the α-position, suggesting a highly nucleophilic nature for the C^1 and C^4 atoms. Thus, they are expected to interact with electrophilic dangling bonds on adatoms of Si (111)-(7×7). Both experimental techniques and theoretical approaches reveal that thiophene can be chemically bound to one adatom–rest atom pair via two Si–C sigma bonds at the C^1 and C^4 atoms through a [4+2]-like addition [1–3]. STM studies [1, 4] show the higher reactivity of the center-adatom in contrast to the corner-adatom due to its geometric arrangement in a unit cell and a smaller steric strain induced by thiophene bonded at the center adatom site. In addition, the faulted half-unit cell exhibits higher reactivity than the unfaulted half, due to a relatively higher electrophilicity of the adatom sites on the faulted subunit.

In contrast to thiophene, furan (with an oxygen heteroatom) exhibits a significantly different reaction mechanism on interaction with Si(111)-(7×7) (Fig. 5.2). Both high-resolution electron energy loss spectroscopy (HREELS) and TDS experiments revealed two chemisorbed states (β1 and β2) at both low temperature (110 K) and room temperature [5, 6]. β1 is assigned to a [4+2]-like adduct linked to an adatom–rest atom pair through two Si–C sigma bonds (Fig. 5.2a). β2 is a di-sigma-linked dimerized furan formed through a di-radical mechanism (Fig. 5.2b). Both adstates are formed from the same intermediate, a mono-sigma complex. Due to the different adsorption energy of the two products, the product distribution exhibits strong temperature dependence. Distinctly different from thiophene and furan, pyrrole is chemically bonded on Si(111)-(7×7) through a dissociation of the N–H bond [7]. It forms silicon-based pyrroyl and Si–H species. Compared to the

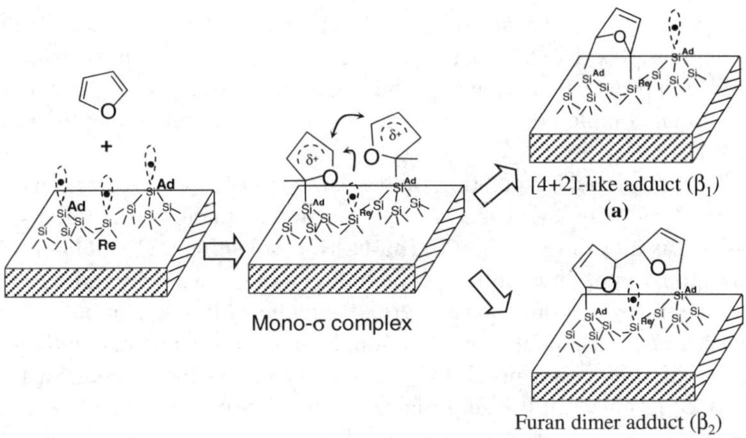

FIGURE 5.2 Scheme showing the mechanism for the formation of [4+2]-like adduct (β1) (a) and dimerized product (β2) (b) of furan on Si(111)-(7×7).

dissociation at the N–H bond, the [4+2]-like addition is expected to be thermodynamically and kinetically unfavorable for pyrrole.

Theoretical simulations show that there is a common reaction pathway for thiophene, benzene, and 1,3-butadiene on Si(111)-(7×7) [2]. Isolated reactants initially form a mono-sigma bonded radical-like intermediate via a transition state. Passing another transition state, the mono-sigma intermediate converts into the final [4+2]-like adduct. The stability of the intermediate depends on the height of the barrier between the intermediate and final [4+2]-like adduct. This barrier increases from *cis*-1,3-butadiene, benzene to thiophene because the increased separation between C^4 and the rest atom due to the decreased C^1–C^4 distance in their mono-σ intermediates, results in a higher energy transition state. Following this dependence, furan may have a more stable intermediate due to its smaller C^1–C^4 distance (2.20 Å) in contrast to benzene and thiophene.

Based on the theoretical simulation of conjugated dienes including 1,3-butadiene, benzene, and thiophene [2], all of them initially adsorb onto electrophilic adatoms by the electrostatic interaction with a nucleophilic C^{α} atom to form a mono-sigma-bonded intermediate. The unstable mono-sigma-bonded intermediate has two possible reaction channels: binding to an adjacent rest atom (Fig. 5.2a) or coupling with an adjacent mono-sigma-bonded complex (Fig. 5.2b). The mono-sigma-bonded complex of furan is expected to be more stable than that of benzene or thiophene, due to its smaller C^1 and C^4 separation, lower aromaticity, or lower resonance energy. This suggests the possible concurrent operation of these two reaction channels for furan on Si(111)-(7×7). At a temperature of 110 K, the mono-σ-bonded furan intermediate is thermally stabilized and a high population of this intermediate can be built up on the surface. Two adjacent radical-like mono-sigma-bonded species can readily couple to each other to form a new C–C bond, giving rise to the β2 state (Fig. 5.2b), the dimerized furan on Si(111)-(7×7). Figure 5.2 is the scheme of the two reaction channels for furan on Si(111)-(7×7). In fact, the formation of the more stable dimer complex (β2 state) for furan on Si(111)-(7×7) was further supported by a PM3 semiempirical calculation [5]. The calculation results show that the resulting dimerized furan complex (β2) is about 23 kcal/mol more stable than the [4+2]-like adduct (β1).

There is no theoretical simulation for the pathways of N–H dissociation of pyrrole on Si(111)-(7×7). On the basis of theoretical calculations of the dissociation pathways of this molecule on Si(100) [8], N–H dissociation on Si(111)-(7×7) could occur through two possible pathways, including a direct pathway through an initial binding at the nitrogen atom and an alternative pathway through an initial binding at the C^{α} atom and a subsequent isomerization. Notably, this chemical binding through N–H dissociation does not break the aromaticity of pyrrole in contrast to loss of aromaticity of products in the addition reactions of benzene, thiophene, and furan. The preserved aromatic π conjugation allows a weak π–π electronic coupling between two adjacent pyrroyl groups bonded on the surface. This weak electronic coupling makes the next molecule preferentially bind to an unreacted site adjacent to the bonded pyrroyl ring, evidenced in the observation of molecular wire-like binding behavior at low exposure (Fig. 5.3) [7].

5.2.2 Thiophene, Furan, and Pyrrole on Si(100) and Ge(100)

Both thiophene and furan molecularly chemisorb on Si(100) through formation of Si–C sigma bonds. Both experimental studies and theoretical calculations show that the chemisorbed thiophene molecules have similar multiple binding configurations as benzene on Si(100) [9–12]. Compared to the [4+2]-like addition on Si(111)-(7×7) [1, 4], thiophene forms multiple products, including 2,3-dihydrothiophene-like species and 2,5-dihydrothiophene-like species through di-sigma binding, as well as twist bridge-like and tight bridge-like species with tetra-sigma binding [9–11]. For furan on Si(100) [10, 12], no dimerized complex was observed, in

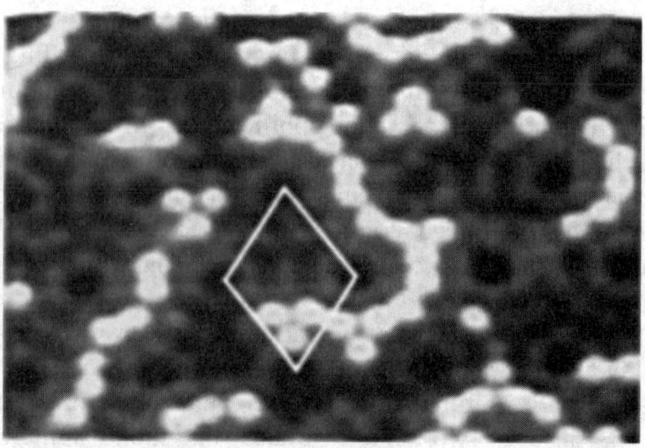

FIGURE 5.3 STM image of Si(111)-(7 × 7) with the chemisorbed pyrrole showing the formation of molecular chain-like structure on this surface.

contrast to the coexistence of [4+2]-like adduct and dimerized complex on Si(111)-(7×7) [5]. Furan is chemisorbed on Si(100) through a [4+2]-like addition pathway [10, 12]. This is probably due to the absence of a reactive site on Si(100), similar to two neighboring adatom–adjacent rest atom sites with a reasonable separation on Si(111)-(7×7), which allows the electronic coupling of two adjacent mono-sigma complexes. The chemical binding of thiophene and furan on Si(100) was investigated by hybrid density functional (B3LYP) calculation in combination with a cluster model approach [12]. The calculations show that the [4+2]-like addition is barrierless and favorable over the [2+2]-like addition, since the barriers of a [2+2]-like addition for thiophene and furan are 2.6 and 1.2 kcal/mol, respectively.

Compared to thiophene and furan, pyrrole exhibits a major N–H dissociation pathway as on Si(111)-(7×7) [7] and a minor C–H dissociation pathway [13] on Si(100) [7, 8, 13, 14]. A direct dissociation pathway (Fig. 5.4a) and an alternative dissociation channel (Fig. 5.4b) through N–H cleavage were theoretically simulated [8, 15, 16]. DFT calculations suggest that this reaction takes place via a barrierless addition of the pyrrole molecule at the C^{α} position to a Si=Si dimer, followed by N–H dissociation and isomerization to form a pyrroyl group bonded to Si(100) through a Si–N covalent linkage [8, 15, 16]. Basically, this reaction can be divided into three steps [8] as shown in Fig. 5.4b. The first step is the molecular adsorption of gaseous pyrrole onto the silicon surface through one of its C^{α} atoms to form a stable adsorbate (iii in Fig. 5.4b) (−6.6 kcal/mol referred to the reactants) without a barrier. In fact, this step is similar to the electrophilic attack of thiophene and furan at a C^{α}

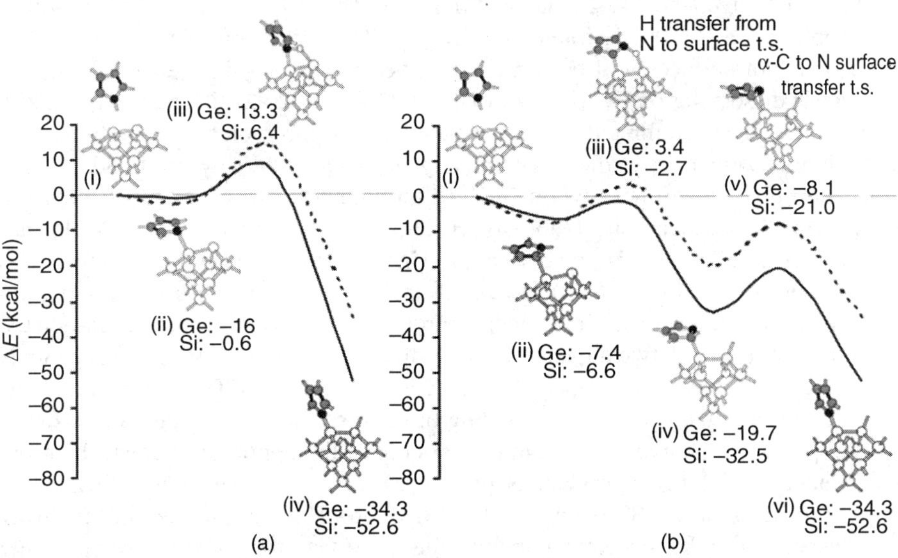

FIGURE 5.4 Critical points on the potential energy surfaces of the N–H dissociation reaction of pyrrole on Si(100) and Ge(100) through direct pathway by an initial binding at nitrogen atom (a) and alternative pathway by an initial binding at C^{α} (b). Adapted from Ref. 8.

atom to form a mono-sigma intermediate for [4+2]-like addition on Si(111)-(7×7). The second step is the dissociation of the intermediate formed in the first step to give $C_4H_4N^{\bullet}_{ads}+H_{ads}$ through a five-membered ring transition state (iv in Fig. 5.4b) with a barrier of 3.9 kcal/mol. The last step is an isomerization of the dissociated intermediate to form final product C_4H_4N–Si (vi in Fig. 5.4b) (−52.6 kcal/mol) through a transition state with a barrier of 11.5 kcal/mol. Notably, this reaction has no overall barrier for the dissociation of pyrrole to form C_4H_4N–Si and H–Si species. Thus, the alternative dissociation channel is more kinetically favorable than a direct dissociative adsorption of pyrrole (Fig. 5.4a) by the interaction of the lone pair electrons of the nitrogen atom with the unoccupied orbital of the Si=Si dimer, as the direct pathway has an overall energy barrier of 6.4 kcal/mol. In addition, compared to the dissociation channel through the formation of an intermediate binding at C^{α}, the [2+2]- and [4+2]-like additions of pyrrole on Si(100) are not favorable, either kinetically or thermodynamically.

Interestingly, thiophene chemisorbs onto Ge(100) through the formation of a Ge· · ·S dative bond at a coverage lower than 0.25 monolayer (ML) and through the [4+2]-like addition at higher coverage at room temperature [11, 17, 18]. In contrast to the [4+2]-like addition of thiophene on Si(100), the [4+2]-like addition reaction on Ge(100) has a higher activation barrier for the transition states [18], as the major contribution to the barrier, the energy difference between the buckled and symmetric dimers of Ge(100) (6.9 kcal/mol), is larger than that for Si(100) (3.2 kcal/mol) [19]. Alternatively, dative bonding is barrierless, suggesting that Ge· · ·S dative bonding is kinetically favorable [18]. Thus, dative-bonded thiophene is formed at low coverage. At high coverage, the [4+2]-like addition similar to the formation of 2,5-dihydrothiophene on Si(100) is found on Ge(100). The formation of dative bonds on semiconductor surfaces will be further reviewed in Chapter 8.

Pyrrole dissociates on Ge(100) without [4+2]-like or [2+2]-like additions [8, 20]. STM studies revealed three dissociation pathways for this molecule [20]. In the first dissociation pathway, pyrrole is chemically bound to two dangling bonds of two adjacent dimers in a dimer row on Ge(100) through both Ge–N and Ge–C covalent bonds via the dissociation at both N–H and C^{α}–H of one molecule. Molecular aromaticity is still retained upon dissociation in this pathway. Therefore, it appears as a bright protrusion in the STM image (A in Fig. 5.5b). In the second and third pathways, pyrrole forms a tilted species through N–H dissociation; the dissociated pyrrolyl interacts with two adjacent Ge=Ge dimer rows through both a strong Ge–N bond and a relatively weak Ge–C^{β} interaction. The only difference between the second and third pathways is the bonding of the dissociated hydrogen atom. In the second pathway, the hydrogen atom diffuses out of the region of interest; therefore the bonded pyrrolyl-like product is present as a complete hexagonal flower-like bright protrusion in the STM image (B in Fig. 5.5b). However, in the third pathway the dissociated hydrogen atom bonds to the other Ge atom of the Ge=Ge dimer interacting with the dissociated pyrrolyl, making the product observed as a flower-like image with a dark site contributed from the bonded hydrogen atom (C in Fig. 5.5b). Notably, all three pathways involve at least two Ge=Ge dimers for each molecule as the arrangement of two adjacent dimers makes them accessible to one

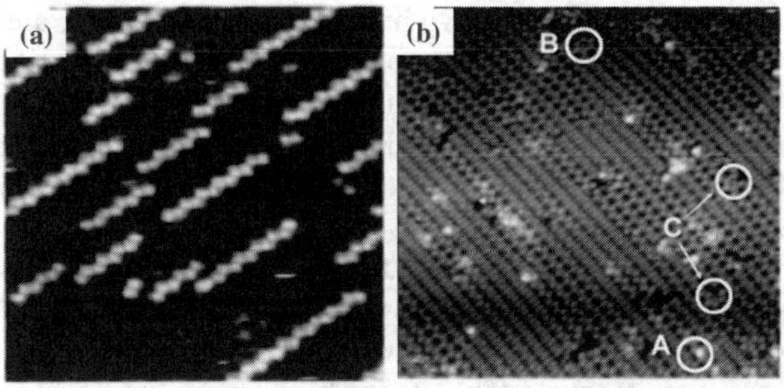

FIGURE 5.5 (a) Occupied state STM image of Ge(100) with the chemisorbed thiophene. (b) Occupied state STM image of Ge(100) with the chemisorbed pyrrole.

molecule simultaneously. The aromaticity of the products from the second and third pathways is weakened to some extent because the molecular C^{β} atom interacts with the Ge atom of the adjacent dimer row. DFT calculations show the first pathway forms the most stable product [21].

Pyrrole does not react with Ge(100) through [4+2]- or [2+2]-like addition. Similar to the higher barrier for the transition state of [4+2]-like addition of thiophene on Ge(100) compared to that on Si(100), it is expected that the energy barrier for [4+2]-like addition of pyrrole on Ge(100) is higher than that on Si(100). DFT calculation [21] shows that [4+2]-like addition is not favorable thermodynamically and kinetically, in contrast to N–H dissociation via initial attachment to the C^{α} atom. Recent theoretical studies [8] show that N–H dissociation on Ge(100) forms a product bonded to Ge(100) through a Ge–N bond with an adsorption energy of -34.3 kcal/mol, through a direct dissociation pathway with an energy barrier of 14.9 kcal/mol (Fig. 5.4a) or through an alternative barrierless dissociation channel (Fig. 5.4b).

5.3 FIVE-MEMBERED AROMATIC MOLECULES CONTAINING TWO DIFFERENT HETEROATOMS

Isoxazole, oxazole, and thiazole are three representative five-membered ring aromatic molecules containing two different heteroatoms (Fig. 5.1). They can be considered as aromatic molecules formed by replacing two carbon atoms of benzene with an oxygen or sulfur atom, and one carbon atom of benzene with a nitrogen atom. The oxygen/sulfur atom and the nitrogen atom contribute two and one electrons, respectively, for the formation of an aromatic π-conjugation of $4n+2$ electrons in the ring. In the three molecules, each nitrogen atom has sp^2 hybridization and contributes one electron of the unhybridized 2p orbital into the formation of the six-electron π-conjugation. The lone pair localized in one sp^2 hybridized orbital of the nitrogen atom can be donated to form a dative bond. Thus, the electronic structure of the nitrogen atom in each of the 3 five-membered ring aromatic molecules is different from that of nitrogen in pyrrole. Compared to the two heteroatoms separated by carbon in oxazole and thiazole, the two heteroatoms in isoxazole are adjacent. The different geometric arrangement of heteroatoms in oxazole and isoxazole results in a different distribution of electron density on the aromatic ring. In addition, the different heteroatoms in oxazole and thiazole also induce a slightly different distribution of electron density. Chemical binding of the three molecules on Si (111)-(7×7) was studied using HREELS.

Isoxazole has two heteroatoms with different electronic structures. Similar to pyridine, the lone pair of the nitrogen atom in isoxazole does not participate in the formation of the aromatic π-conjugation, though the lone pair is slightly withdrawn by its neighboring oxygen atom. Therefore, the nitrogen atom of isoxazole has an electron density higher than the oxygen atom and all the carbon atoms of this molecule. It can act as an electron donor to form a dative bond with an electron-deficient adatom of Si(111)-(7×7) [22] (Fig. 5.6a). Oxazole is an isomer of isoxazole. The difference between the two isomers is the geometry of the nitrogen atom on the five-membered ring. For isoxazole, the nitrogen atom is adjacent to the oxygen atom. However, it is separated by one carbon atom from the oxygen atom in oxazole. The electron-withdrawing effect of oxygen for the nitrogen atom in oxazole is weaker than that for the case of isoxazole. Therefore, the nitrogen atom of oxazole has an electron density similar to that of pyridine, implying capability for a significant donation of electron density. Similar to oxazole, thiazole also exhibits the capability of donating electron density to form a Si· · ·N dative bond. These molecules exhibit different pathways on Si(111)-(7×7) as shown in Fig. 5.6c.

The lone pair in an unhybridized p orbital of the oxygen atom of oxazole, which contributes to the formation of aromatic π-conjugation, in fact plays a different role

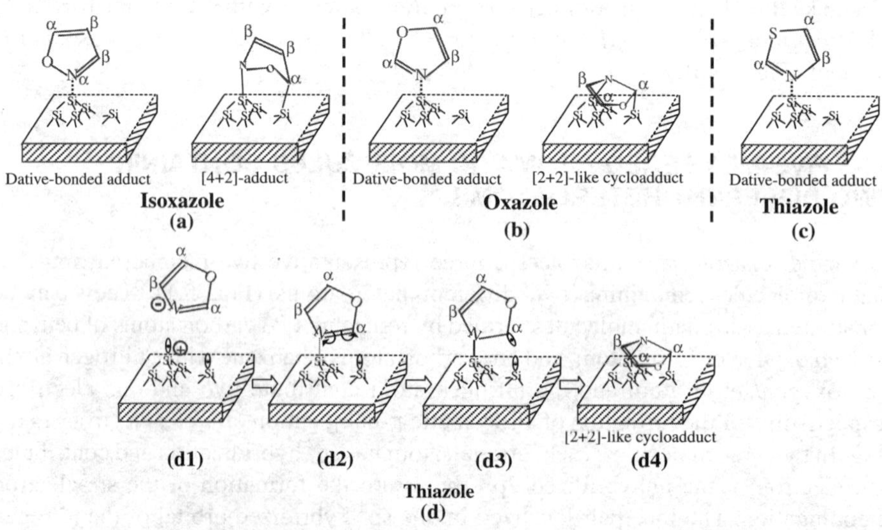

FIGURE 5.6 Binding modes of isoxazole (a), oxazole (b), and thiazole (c) on Si(111)-(7 × 7) at low temperature. (d) Schematic showing pathway for the formation of [2 + 2]-like adduct of oxazole on Si(111)-(7 × 7) at low temperature.

in the formation of the aromatic π-conjugation than the two electrons in the two p orbitals of two carbon atoms of a pyridine molecule. To some extent, oxazole may be considered as a hybrid of furan and pyridine. In fact, oxazole can be chemically attached to Si(111)-(7×7) through both the covalent addition channel similar to thiophene [4] and furan [5], as well as the dative-bond addition similar to pyridine [23]. Oxazole carries out the addition reaction at low temperature through a [2+2]-like addition at N and C^{α} atoms, instead of a [4+2]-like addition. This is determined by its characteristic electronic structure. It is understandable if we consider this addition reaction as a step-wise reaction mechanism as schematically shown in Fig. 5.6d. The attack of a nitrogen atom on an adatom forms an intermediate with a lower energy (Fig. 5.6d2) than other possible intermediates because the nitrogen atom and silicon adatom are electron-rich and electron-deficient, respectively, and geometrically the adatom is at a favorable outward position in contrast to the inward rest atom. In addition, the interaction of the radical of an electron-rich rest atom with a C^{α} atom with lower electron density due to the electron-withdrawing effect of its adjacent oxygen atom (Fig. 5.6d3), will facilitate the subsequent formation of a Si–C sigma bond. Therefore, [2+2]-like addition is a kinetically favorable pathway for oxazole at low temperature. However, a [4+2]-like addition at two C^{α} atoms is not kinetically favorable for oxazole due to a high barrier for an initial binding at the C^{α} atom, in contrast to an initial binding at the N atom in the [2+2]-like addition. For pyridine, [2+2]-like addition is not kinetically favorable over [4+2]-like addition, as both transition states are formed through an initial binding at the N atom. Thus, pyridine is chemisorbed on Si(111)-(7×7) through a thermodynamically favorable [4+2]-like addition mechanism at C^1 and C^4 in addition to the formation of a Si· · ·N dative bond.

For isoxazole, an isomer of oxazole, its electron-rich nitrogen atom can form an intermediate with the electron-deficient adatom of Si(111)-(7×7). Its C^{α} atom has lower electron density than the C^{β} atom due to the strong electron-withdrawing effect of the oxygen atom on its adjacent C^{α} atom. It can interact with the radical of the rest atom of this surface, forming a Si–C sigma bond [22]. Thus, the low electron density and favorable geometric arrangement of the C^{α} atom of isoxazole makes [4+2]-like addition thermodynamically and kinetically favorable (Fig. 5.6b).

The chemisorption of thiazole on Si(111)-(7×7) (Fig. 5.6c) is different from the simultaneous dative-bonded addition and [4+2]-like or [2+2]-like addition for isoxazole and oxazole at low temperature (Fig. 5.6a and b). This difference can be understood in the contrast of their electronic structures. For both isoxazole and oxazole, the C^{α} atom has a low electron density due to the electron-withdrawing effect of the electronegative oxygen atom. Thus, for isoxazole the binding between the electron-rich N atom and the electron-deficient C^{α} atom and the adatom–rest atom pair of Si(111)-(7×7) (Fig. 5.6b) is thermodynamically and kinetically favorable; for oxazole the attachment of the electron-rich N atom and the electron-deficient C^{α} to the adatom–rest atom pair is kinetically favorable at low temperature. Compared to isoxazole and oxazole, the C^{α} atom of thiazole has a relatively high electron density due to the absence of a strong electron-withdrawing effect from the sulfur atom with a relatively lower electronegativity than the oxygen atom.

Thus, referring to the [2+2]-like addition of oxazole at low temperature (Fig. 5.6b), for thiazole a chemical binding between the nitrogen atom and the C^{α} atom of thiazole and the adatom–rest atom pair to form a [2+2]-like adduct is neither kinetically nor thermodynamically favorable. Alternatively, dative-bond addition between the electron-rich nitrogen atom of thiazole and the electron-deficient adatom site is kinetically favorable at low temperature (Fig. 5.6c). Therefore, the dative-bonded thiazole is the major product at low temperature.

5.4 BENZENE

5.4.1 Different Binding Configurations on (100) Face of Silicon and Germanium

Benzene is the prototype molecule for aromatic systems. Using both vibrational EELS and TDS, Taguchi et al. explored the chemical binding of benzene on Si(100) at 300 K early on [24]. Two chemisorption states with desorption peaks at ~432–460 and ~500K were identified [24, 25]. They are attributed to benzene adsorbed on defect-free and near-defect regions, respectively. Several binding configurations were proposed for the molecularly chemisorbed benzene on Si(100) [24–29]. Two di-sigma binding modes were initially proposed for benzene adsorbed on the defect-free region [24]. They are schematically shown in Fig. 5.7a and b. One is formed via two

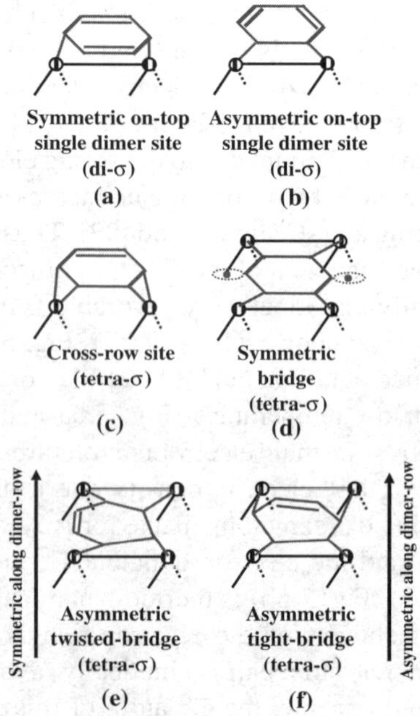

FIGURE 5.7 Six possible binding configurations of benzene molecules chemisorbed on Si(100).

Si–C covalent linkages at C^1 and C^4 atoms, producing a 1,4-cyclohexadiene-like product (Fig. 5.7a) and the other involving the bonding at C^5 and C^6, giving rise to a 1,3-cyclohexadiene-like product (Fig. 5.7b). Besides the two di-sigma binding modes, four tetra-sigma binding modes involving two pairs of Si=Si dimers in each case giving three bridge-like configurations as shown in Fig. 5.7d, e, and f were proposed [26–29]. Figure 5.7c is a tetra-sigma binding mode involving two dangling bonds from two adjacent dimer rows. A symmetric bridge binding configuration was proposed as shown in Fig. 5.7d [27, 28]. Figure 5.7e and f are asymmetric twisted-bridge and tight-bridge binding modes on two adjacent dimers in a dimer row [28].

STM and FTIR experimental techniques [28, 30] revealed the coexistence of symmetric on-top single dimer binding (Fig. 5.7a), asymmetric twisted-bridge binding (Fig. 5.7e), and asymmetric tight-bridge binding (Fig. 5.7f) on Si(100). Notably, the on-top di-sigma bound molecules can be converted into asymmetric bridge bound molecules by thermal promotion, suggesting that asymmetric tetra-sigma bridge binding is thermodynamically more favorable than the on-top di-sigma binding [28, 29].

Compared to the di-sigma binding of benzene on Si(100), benzene weakly binds to Ge(100). TDS revealed two desorption peaks for benzene on Ge(100) at 234 and 252K [25, 31], corresponding to molecular adsorption at terrace and step sites, respectively. However, the adsorbed benzene on Si(100) desorbs at ~432–460 and ~500 K [24, 25], corresponding to the chemisorbed benzene at terrace and step sites, respectively. DFT calculations [31] rationalized the experimentally observed difference in the adsorption of benzene on Ge(100) and Si(100). The calculated adsorption energies for benzene on Ge(100) and Si(100) are 1.4 and 20.0 kcal/mol, respectively. This significant difference in adsorption energy could partially result from the larger Ge–C bond length compared to the Si–C bond length in the adsorbed benzene. In fact, a similar difference in molecular adsorption energy on the two semiconductor surfaces has been seen for other organic molecules [8].

5.4.2 Di-Sigma Binding on Si(111)-(7×7)

On Si(111)-(7×7), early investigation using vibrational EELS with low resolution indicated that benzene chemisorbs on this surface at room temperature through a π-interaction [32]. Recent studies clearly show that benzene chemically binds to Si (111)-(7×7) with two Si–C sigma bonds formed through a [4+2]-like pericyclic addition mechanism [33, 34]. DFT calculations show that the [4+2]-like addition is both thermodynamically and kinetically preferred [2, 6, 35]. The binding site, an adatom and its adjacent rest atom, where the C^1 and C^4 atoms terminate on an adjacent adatom–rest atom pair, was confirmed in recent STM studies [34, 36]. Compared to the multiple binding configurations of benzene on Si(100), the di-sigma bonded 1,4-cyclohexadiene-like product is the only one seen on Si(111)-(7×7). This difference results from the accessible multiple reactive sites on Si(100). The difference among these binding sites on Si(100) that have different geometric and even electronic structures is mainly their spatial arrangement. Alternatively, in general Si(111)-(7×7) has only one reactive site, consisting of an adatom and its adjacent rest atom, which forms a binding configuration.

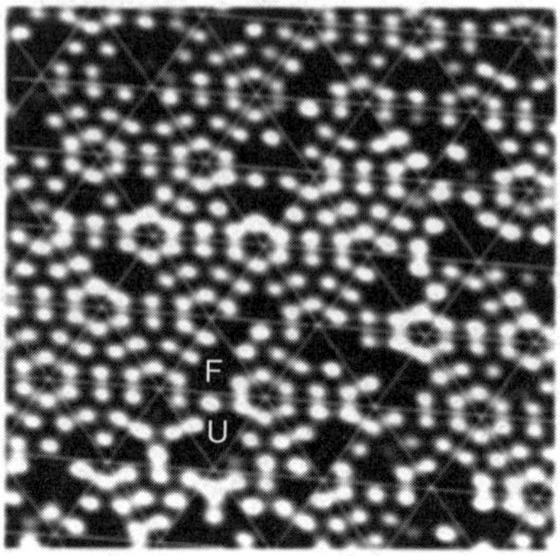

FIGURE 5.8 STM image of Si(111)-(7 × 7) with the chemisorbed benzene molecules. F and U represent the faulted and unfaulted unit cells, respectively. Adapted from Ref. 45.

Figure 5.8 is an STM image of benzene on Si(111)-(7×7). Clearly, each half unit cell contains at most three reacted adatoms appearing as dark features, which do not align in a straight-line. This observation provides evidence for a nondissociated, di-sigma binding of benzene on an adatom–rest atom pair on Si(111)-(7×7) [34, 36]. Each unit cell of the Si(111)-(7×7) has six corner-adatoms and six center-adatoms. On average, each center-adatom–rest atom pair has more opportunity to react with benzene in contrast to a corner-adatom–rest atom pair. This is because each center-adatom faces two adjacent rest atoms in contrast to only one for the corner-adatom due to the geometric arrangement of adatoms on Si(111)-(7×7). A similar difference in binding density was also observed for the binding of thiophene on Si(111)-(7×7) by STM [1].

5.5 SIX-MEMBERED HETEROATOM AROMATIC MOLECULES

Pyridine is a tertiary amine with an aromatic ring. Compared to pyrrole, the nitrogen atom of pyridine has significantly different electronic density and structure though the nitrogen atoms of both molecules are sp^2 hybridized. In pyridine, the sp^2 hybridization of the nitrogen atom is different from that of pyrrole. For pyrrole, the three sp^2-hybridized orbitals are equivalent, forming two N–C bonds and one N–H bond. In pyridine, the nitrogen atom is inequivalently sp^2 hybridized (Fig. 5.9f). The remaining unhybridized p orbital of the nitrogen atom has only one electron that combines with the other five p orbitals of the carbon atoms to form an aromatic sextet. Two sp^2 orbitals with one electron in each orbital form two sigma bonds with two C^{α} atoms. Notably, one sp^2 orbital has a lone pair that does not participate in the aromatic π-conjugation system. Thus, compared to the nitrogen atom of pyrrole, the nitrogen atom of pyridine is electron rich. More importantly, the lone pair on the nitrogen atom of pyridine is localized on this atom (Fig. 5.9b). Thus, the lone pair could be donated to the electron-deficient adatom of the Si(111)-(7×7) or the buckled down atoms of Si=Si or Ge=Ge dimers to form a dative bond.

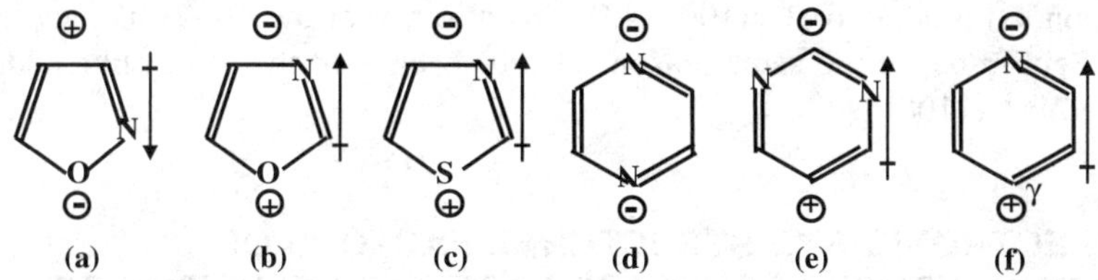

FIGURE 5.9 Polarity of isoxazole (a), oxazole (b), thiazole (c), pyrazine (d), pyrimidine (e), and pyridine (f).

Pyridine forms both the dative-bonded adduct and the [4+2]-like adduct on Si(111)-(7×7) [23] and Si(100) [37, 38] at low temperature. At room temperature, the dative-bonded pyridine desorbs or partially converts into the [4+2]-like adduct [23, 37]. There is no dative-bonded adduct of pyridine formed on the two silicon single crystal surfaces at room temperature. Compared to Si(100), pyridine forms a nearly complete dative-bonded monolayer on Ge(100) at room temperature [38, 39]. The difference in reaction mechanisms of pyridine on silicon and germanium surfaces reflects the trend of increased electron affinity along the periodic table from Si(100) to Ge(100).

The reasons for the absence of a [4+2]-like addition for pyridine on Ge(100) could be both thermodynamic and kinetic. From the viewpoint of thermodynamics, the relatively weak Ge–N sigma bond in contrast to Si–N sigma bond results in a lower adsorption energy for [4+2]-like addition on Ge(100) than on Si(100). The weak Ge–N bond stems from the large strain in the [4+2]-like adduct due to the larger Ge–Ge bond length of the Ge=Ge dimer. In addition, the larger energy difference between the buckled Ge=Ge dimer and the symmetric Ge=Ge dimer [19] results in a higher energy barrier for the transition state of the [4+2]-like addition pathway on Ge(100) than on Si(100); therefore from the point of kinetics pyridine prefers the dative-bonding pathway on Ge(100), which is barrierless.

5.6 SIX-MEMBERED AROMATIC MOLECULES CONTAINING TWO HETEROATOMS

Pyrazine (Fig. 5.9d) is a six-membered aromatic molecule containing two nitrogen atoms at two opposite ends of the aromatic ring. Recent studies show that pyrazine can be chemically bound to Si(100) and Si(111)-(7×7) via a [4+2]-like addition through binding the two *para*-nitrogen atoms to a Si=Si dimer on Si(100) [40] or a pair of adjacent adatom–rest atoms on Si(111)-(7×7) [41]. This indicates that the carbon atoms of the ring are not directly involved in any chemical binding with the two surfaces.

Pyrimidine (Fig. 5.9e) is an isomer of pyrazine. Compared to pyrazine, the two nitrogen atoms are not at opposite positions on the aromatic ring. It has two nitrogen atoms at α and γ positions, which have electronic structure similar to that of pyridine. Interestingly, pyrimidine forms a [4+2]-like adduct on Si(111)-(7×7) [42], but a dative-bonded product on Ge(100) [43]. The absence of the [4+2]-like addition on Ge(100) could be for the same reasons as the absence of the [4+2]-like adduct for pyridine on Ge(100).

5.7 ELECTRONIC AND STRUCTURAL FACTORS OF THE SEMICONDUCTOR SURFACES FOR THE SELECTION OF REACTION CHANNELS OF FIVE-MEMBERED AND SIX-MEMBERED AROMATIC RINGS

Thiophene forms a dative bond on Ge(100) but not on Si(100) because [4+2]-like addition on Ge(100) is unfavorable due to the higher energy barrier mainly contributed by the larger energy difference between the buckled and symmetric Ge=Ge dimers [44]. On the other hand, the strong electron affinity in terms of a larger amount of transferred electron density in the formation of the buckled Ge=Ge dimer than that in the Si=Si dimer, makes the buckled-down Ge atom more electron-deficient and therefore forms a relatively more stable dative-bonded product.

Interestingly, pyrrole carries out N–H dissociation on both Si(100) and Ge(100) surfaces as the N–H bond of the molecule can dissociate through an alternative pathway in which the electron-deficient substrate atom is initially attached to the electron-rich C^{α} atom of pyrrole rather than the nitrogen atom. An additional five-membered ring transition state is involved for transferring the initial binding at the C^{α} to the N atom [8, 15]. This transition state has a lower energy than the first one. In the first transition state, the adsorption of pyrrole at the C^{α} atom is facilitated in this aromatic molecule due to its capability for delocalizing the π-electron on the aromatic ring. Thus, the first transition state has a lower barrier. The overall process of the alternative dissociation pathway is barrierless. However, for pyrrolidine and 3-pyrroline, a transition state formed via initial attachment at the carbon atom could be highly unstable as it results in a pentavalent carbon due to the absence of electronic delocalization in the nonaromatic rings. Thus, the two nonaromatic molecules cannot process N–H dissociation through the alternative pathway in terms of an initial binding at the carbon atom adjacent to the nitrogen atom. For a potential direct dissociation pathway for pyrrolidine and 3-pyrroline on Ge(100), the barriers are 10.2 and 8.2 kcal/mol higher than the same barriers on Si(100), suggesting that the N–H dissociation on Ge(100) is suppressed [8]. Thus, the two molecules adopt a kinetically favorable pathway on Ge(100), formation of dative bonds. However, they follow a thermodynamically favorable pathway on Si(100), N–H dissociation, though the formation of dative bonds on Si(100) for these molecules is also barrierless.

Furan dimerizes on Si(111)-(7×7), but not on Si(100) and Ge(100) as a reasonable arrangement of reactive sites accessible for a subsequent coupling of two mono-sigma bonded intermediates is absent on the (100) surfaces. Alternatively, it bonds to Si(100) through a [4+2]-like addition reaction.

Pyridine, pyrazine, and pyrimidine all have nitrogen atoms in the aromatic ring, which are electron-rich with a formal lone pair. In the case of these molecules, the electronic structure of the semiconductor surface controls the reaction channel for adsorption. In each case, donation of the lone pair electrons to the electron-deficient Si(111)-(7×7) adatom, or the buckled down Si or Ge atoms of the dimer rows on the (100) surfaces controls the reaction pathway. Differences in reactivity for these molecules on the elemental semiconductor surfaces follow the differences in electron affinity for the elemental surfaces.

REFERENCES

1. Cao, Y.; Yong, K. S.; Wang, Z. H. et al. *J. Chem. Phys.* **2001**, *115* (7), 3287.
2. Lu, X.; Wang, X. L.; Yuan, Q. H. et al. *J. Am. Chem. Soc.* **2003**, *125* (26), 7923.
3. Lu, X.; Lin., M. C. *Int. Rev. Phys. Chem.* **2002**, *21* (1), 137.
4. Cao, Y.; Yong, K. S.; Wang, Z. Q. et al. *J. Am. Chem. Soc.* **2000**, *122* (8), 1812.
5. Cao, Y.; Wang, Z. H.; Deng, J. F. et al. *Angew. Chem. Int. Ed.* **2000**, *39* (15), 2740.
6. Wang, Z. H.; Cao, Y.; Xu., G. Q. *Chem. Phys. Lett.* **2001**, *338* (1), 7.
7. Yuan, Z. L.; Chen, X. F.; Wang, Z. H. et al. *J. Chem. Phys.* **2003**, *119* (19), 10389.
8. Wang, G. T.; Mui, C.; Tannaci, J. F. et al. *J. Phys. Chem. B* **2003**, *107* (21), 4982.
9. (a) Isobe, N.; Shibayama, T.; Mori, Y. et al. *Chem. Phys. Lett.* **2007**, *443* (4–6), 347. (b) Qiao, M. H.; Cao, Y.; Tao, F. et al. *J. Phys. Chem. B* **2000**, *104* (47), 11211. (c) Jeong, H. D.; S Lee, Y.; Kim., S. *J. Chem. Phys.* **1996**, *105* (12), 5200.
10. (a) Qiao, M. H.; Tao, F.; Cao, Y. et al. *J. Chem. Phys.* **2001**, *114* (6), 2766. (b) Lee, H. K.; Kim, K. J.; Kang, T. H. et al. *Surf. Sci.* **2008**, *602* (4), 914.
11. Rousseau, G. B. D.; Dhanak, V.; Kadodwala., M. *Surf. Sci.* **2001**, *494* (3), 251.
12. Lu, X.; Xu, X.; Wang, N. Q. et al. *J. Phys. Chem. B* **2001**, *105* (41), 10069.
13. Cao, X. P.; Coulter, S. K.; Ellison, M. D. et al. *J. Phys. Chem. B* **2001**, *105* (18), 3759.
14. (a) Qiao, M. H.; Cao, Y.; Deng, J. F. et al. *Chem. Phys. Lett.* **2000**, *325* (5–6), 508. (b) Kim, K.; Han, J.; Kang, T. H. et al. *J. Electron Spectrosc. Rel. Phenom.* **2005**, *144*, 429. (c) Qiao., M. H.; Tao., F.; Cao, Y. et al. *Surf. Sci.* **2003**, *544* (2–3), 285. (d) Barteau., M. A. *Chem. Rev.* **1996**, *96*, 1413.
15. Luo, H. B.; Lin., M. C. *Chem. Phys. Lett.* **2001**, *343* (3–4), 219.
16. Seino, K.; Schmidt, W. G.; Furthmuller, J. et al. *Surf. Sci.* **2003**, *532*, 988.
17. (a) Lee, H.; Jeon, S. M.; Kim, H. D. et al. *J. Phys.: Condens. Matter* **2008**, *20* (13), 135006. (b) Jeon, S. M.; Jung, S. J.; Lim, D. K. et al. *J. Am. Chem. Soc.* **2006**, *128* (19), 6296.
18. Jeon, S. M.; Jung, S. J.; Kim, H. D. et al. *J. Phys. Chem. B* **2006**, *110* (43), 21728.
19. Kruger, P.; Pollmann., J. *Phys. Rev. Lett.* **1995**, *74* (7), 1155.
20. (a) Kim, D. H.; Choi, D. S.; Kim, A. et al. *J. Phys. Chem. B* **2006**, *110* (15), 7938. (b) Kim, D. H.; Choi, D. S.; Kim, A. et al. *Jpn. J. Appl. Phys. Part 1* **2006**, *45* (3B), 2148.
21. Kim, D. H.; Choi, D. S.; Hong, S. et al. *J. Phys. Chem. C* **2008**, *112* (19), 7412.
22. Tao, F.; Bernasek., S. L. *J. Am. Chem. Soc.* **2007**, *129* (15), 4815.
23. Tao, F.; Lai, Y. H.; Xu., G. Q. *Langmuir* **2004**, *20* (2), 366.
24. Taguchi, Y.; Fujisawa, M.; Takaoka, T. et al. *J. Chem. Phys.* **1991**, *95* (9), 6870.
25. Fink, A.; Menzel, D.; Widdra., W. *J. Phys. Chem. B* **2001**, *105* (18), 3828.

26. (a) Craig., B. I. *Surf. Sci.* **1993**, *280* (3), L279. (b) Wolkow, R. A.; Lopinski, G. P.; Moffatt., D. J. *Surf. Sci.* **1998**, *416* (3), L1107.
27. Jeong, H. D.; Ryu, S.; Lee, Y. S. et al. *Surf. Sci.* **1995**, *344* (3), L1226.
28. Lopinski, G. P.; Fortier, T. M.; Moffatt, D. J. et al. *J. Vac. Sci. Technol.* **1998**, *16* (3), 1037.
29. Lopinski, G. P.; Moffatt, D. J.; Wolkow., R. A. *Chem. Phys. Lett.* **1998**, *282* (3–4), 305.
30. Hofer, W. A.; Fisher, A. J.; Lopinski, G. P. et al. *Surf. Sci.* **2001**, *482*, 1181.
31. Cho, J. H.; Kim, K. S.; Morikawa., Y. *J. Chem. Phys.* **2006**, *124* (2), 024716.
32. Taguchi, Y.; Fujisawa, M.; Nishijima., M. *Chem. Phys. Lett.* **1991**, *178* (4), 363.
33. (a) Cao, Y.; Wei, X. M.; Chin, W. S. et al. *J. Phys. Chem. B* **1999**, *103* (27), 5698. (b) Tomimoto, H.; Sekitani, T.; Sumii, R. et al. *Surf. Sci.* **2004**, *566*, 664.
34. (a) Kawasaki, T.; Sakai, D.; Kishimoto, H. et al. *Surf. Inter. Anal.* **2001**, *31* (2), 126. (b) Yong, K. S.; Yang, S. W.; Zhang, Y. P. et al. *Langmuir* **2008**, *24* (7), 3289. (c) Horn, S. A.; Patitsas., S. N. *Surf. Sci.* **2008**, *602* (2), 630.
35. (a) Petsalakis, I. D.; Polanyi, J. C.; Theodorakopoulos., G. *Israel J. Chem.* **2005**, *45* (1–2), 111. (b) Li, Y. C.; Wang, W. N.; Cao, Y. et al. *Acta Chim. Sinica* **2002**, *60* (4), 653.
36. Tomimoto, H.; Takehara, T.; Fukawa, K. et al. *Surf. Sci.* **2003**, *526* (3), 341.
37. (a) Tao, F.; Qiao, M. H.; Wang, Z. H. et al. *J. Phys. Chem. B* **2003**, *107* (26), 6384. (b) Miwa, J. A.; Eves, B. J.; Rosei, F. et al. *J. Phys. Chem. B* **2005**, *109* (43), 20055.
38. Kim, H. J.; Cho., J. H. *J. Chem. Phys.* **2004**, *120* (17), 8222.
39. (a) Hong, S.; Cho, Y. E.; Maeng, J. Y. et al. *J. Phys. Chem. B* **2004**, *108* (39), 15229. (b) Cho, Y. E.; Maeng, J. Y.; Kim, S. et al. *J. Am. Chem. Soc.* **2003**, *125* (25), 7514.
40. (a) Huang, H. G.; Huang, J. Y.; Ning, Y. S. et al. *J. Chem. Phys.* **2004**, *121* (10), 4820. (b) Lu, X.; Xu, X.; Wu, J. M. et al. *New J. Chem.* **2002**, *26* (1), 160. (c) Shimomura, M.; Ichikawa, D.; Fukuda, Y et al. *Phys. Rev. B* **2005**, *72* (3),
41. Huang, H. G.; Wang, Z. H.; Xu., G. Q. *J. Phys. Chem. B* **2004**, *108* (33), 12560.
42. Huang, H. G.; Huang, J. Y.; Wang, Z. H. et al. *Surf. Sci.* **2007**, *601* (5), 1184.
43. Lee, J. Y.; Jung, S. J.; Hong, S. et al. *J. Phys. Chem. B* **2005**, *109* (1), 348.
44. (a) Wolkow., R. A. *Phys. Rev. Lett.* **1992**, *68* (17), 2636.(b) Weakliem, P. C.; Carter., E. A. *J. Chem. Phys.* **1992**, *96* (4), 3240. (c) Kubby, J. A.; Griffith, J. E.; Becker, R. S. et al. *Phys. Rev. B* **1987**, *36* (11), 6079.

CHAPTER 6

Influence of Functional Groups in Substituted Aromatic Molecules on the Selection of Reaction Channel in Semiconductor Surface Functionalization

ANDREW V. TEPLYAKOV

6.1 INTRODUCTION

6.1.1 Scope of this Chapter

Aromaticity has long been a fascinating topic in the entire field of chemistry [1]. Since the discovery of benzene that in a few years will celebrate its bicentennial anniversary, scientists and engineers have put this concept to a good use. As the measurements and applications have evolved though, the definition of aromaticity has also undergone dramatic changes. The original formulations of Kekulé (1865) and Erlenmeyer (1866) have become much more general and are based mostly on experimental and computational observations of the exceptional stability of aromatic compounds. The unusually high degree of symmetry of their molecular and often crystal structures and, as a consequence, often a highly symmetric wave function describing their quantum mechanical behavior lead to unusual magnetic and spectroscopic characteristics of aromatic compounds. It should be pointed out that the initially accepted description of aromaticity as a characteristic of compounds whose molecules are built upon flat ring-like structures with $4n+2$ π electrons has long been abandoned as the aromaticity in some species has been demonstrated to involve σ orbitals, or only $4n$ electrons instead of $4n+2$, or even be observed in transition metal sandwich structures and in three-dimensional clusters.

Functionalization of Semiconductor Surfaces, First Edition.
Edited by Franklin (Feng) Tao and Steven L. Bernasek.

Compared to such a long and storied history of aromaticity, chemistry of semiconductor surface modification is still in its infancy, despite a myriad of practical applications. This chapter will take a full advantage of the use of aromatic compounds for surface modification but we will mostly stick to the nearly two centuries old definitions of aromaticity to avoid the additional complexity. Most of what we will examine are the aromatic compounds that really are based on stable planar ring-like structures with reasonably well-characterized electronic properties. That said, we will also consider examples of heterocyclic compounds that are not aromatic within the classical definition but that may lead to the formation of aromatic structures in a reaction with the surface or that may help us understand the differences between reactivity of "aromatic" and "nonaromatic" compounds. One should hope that the amazing newly developed aromatic structures with extraordinary geometric, electronic, and magnetic properties will also at some point soon be used in realistic applications that are based on semiconductor substrates.

There are several main reasons why aromatic compounds are becoming such popular surface modifiers. With recent development of such fields as microelectronics, communication, and sensing, the interface between organic layers and semiconductor materials has truly become the focus of molecular-level research. The miniaturization of electronic components all the way down to subnanometer scale drives both fundamental research and engineering solutions to build such interfaces at a truly molecular level. Since silicon still remains a cornerstone of modern technology, it will be the main focus of this chapter. It is still one of the cheapest, most abundant, easily prepared semiconductor materials with electronic properties easily tuned by well-developed doping methods. The problem with this material is that it is very reactive, regardless of whether the starting point is a clean surface or it is modified with some functionality. This is where aromatic compounds can become especially valuable. On one hand, the stable aromatic groups possess sufficient reactivity to chemically interact with a very reactive clean silicon surface. On the other hand, they can act as relatively inert spacers if any other functional group is available on an aromatic compound. If the starting surface is functionalized, for example hydrogen terminated, the aromatic spacers prove to be exceptionally reliable inert geometric linkers between the desired functional groups: one needed to react with the surface and the other available for further modification.

With the intriguing possibilities of using molecules as potential components of molecular electronics, where conduction and switching is performed by means of a single molecule or small molecular agglomerates, the use of aromatic compounds on semiconductor surfaces received an additional boost, once the possibility of lateral conduction along the π-stacked aromatic rings was considered.

Although structurally diverse semiconductor surfaces are currently available, the very idea of creating a defect-free interface relies heavily on the ideally ordered starting surface. The easily accessible single crystalline elemental semiconductor surfaces make them an even more attractive playing field. Both fundamental studies and practical applications make a full use of single crystalline semiconductors and the surface chemistry of single crystalline silicon is by far the most studied. The next section will introduce the most commonly studied silicon surfaces and they will indeed be the focus of this chapter. Occasionally, where appropriate, the surfaces of other elemental semiconductors, particularly germanium, will also be considered.

6.1.2 Structure of Most Common Elemental Semiconductor Surfaces: Comparison of Silicon with Germanium and Carbon

One of the main reasons why silicon single crystals became the focus of fundamental scientific research in 1980s and 1990s was that they could provide an idealized starting point with exact positions of the surface atoms being known. The two major reviews, one by Waltenburg and Yates [2] and the other by Yoshinobu et al. [3], have indicated that silicon surface chemistry has reached the potential of a separate and highly regarded scientific field. Detailed reviews in nearly every aspect of this field started appearing regularly in the scientific literature and this chapter will not attempt to summarize all the enormous amount of work in this field. What we will try to achieve is to place the most recent work involving aromatic compounds, mostly on the traditionally best understood silicon single crystal (100) and (111) surfaces, in context of general rules and guidelines to design an appropriate pathway for semiconductor surface functionalization. The clean silicon surfaces of the (100) and (111) orientations can exhibit several different surface reconstructions, and the specifics of surface chemistry of aromatic compounds can definitely depend on the specific reconstruction of the same surface. Chapter 3 of this book focuses on the structures and origins of reactivity of elemental semiconductor surfaces. We will only highlight a few key features of the two most commonly studied reconstructions of clean silicon surfaces, Si(100)-(2×1) and Si(111)-(7×7) that will be used to illustrate the basic rules governing interactions of multifunctional aromatic molecules with silicon.

The structure of the clean Si(100)-(2×1) surface has been a subject of more than a decade-long debate, which is mostly resolved now. The tetracoordinate atoms of the bulk silicon are lacking saturation on the freshly cleaved (100) surface. This causes them to pair up and form Si–Si dimers. These dimers form long-range rows, with the orientation of these rows switching to perpendicular at the single atomic height step edges. Despite this reconstruction, each atom of the silicon surface still has formally one unpaired electron, making this surface very reactive. Interestingly enough, each of these silicon surface dimers is slightly buckled, with electrons redistributed between the two participating atoms. This buckling is in a dynamic equilibrium at room temperature but can be "frozen" at cryogenic conditions. The low barrier for a dimer inversion and the fact that the overall potential for the process is rather flat, as discussed by Jung et al. [4], indeed caused substantial debate, since the specific temperature of this "freeze" was difficult to determine; but for the purposes of this chapter it is of less importance, since most surface modification chemistry is performed at room temperature or above. Figure 6.1 shows the STM studies of the formation of the silicon surface dimers and their buckling at cryogenic temperatures.

As was noted above, this reconstructed surface still has one unpaired electron per surface silicon atom, which can definitely cause a set of very interesting relations between its structure and reactivity [6]. In fact, the redistribution of the

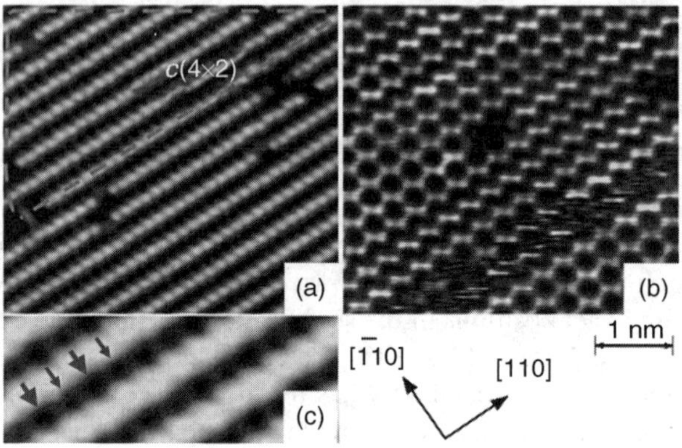

FIGURE 6.1 STM images taken below 10K with a sample bias voltage of −2.2 V (a) and + 1.6 V (b). Tunneling current is set at 38 and 55 pA, respectively. The filled-state image (a) apparently shows symmetric dimers, while the empty-state image (b) shows buckled dimers. In an area surrounded by a dashed green line in image (a) (zoomed in (c)) a faint trace of the *c*(4×2) structure, alternative array of dark spots between the dimer rows, is observed, as pointed with arrows, indicating an existence of the *c*(4×2) periodicity and the buckled dimers there. Figure reprinted with permission from Ref. 5. Copyright 2009 The American Physical Society.

electronic density between the two silicon atoms of the surface dimer draws parallels with the chemistry of biradical species, zwitterionic compounds, and even the carbon–carbon double bond in alkenes [7]. The arguments in favor of one of these descriptions over another sometimes hinge on the specifics of the proposed surface reaction mechanisms, which are still often debated. However, most of the time, the final species, the results of the proposed surface transformations, are very stable at room temperature and can be undoubtedly characterized by common spectroscopic and microscopic techniques.

In case of elemental group IV semiconductors, there is also an additional axis for comparison of the electronic properties for the (100) surfaces. Ge(100) and C(100) can be prepared and characterized as well. They also both exhibit a stable 2×1 reconstruction. However, in case of Ge(100)-(2×1), the surface germanium dimers are permanently buckled even at room temperature, while for the diamond surface, these dimers are symmetric even at cryogenic temperatures. These subtle differences add to the fundamentals of the comparison of all three semiconductors, as their dimers of the (100)-(2×1) surfaces illustrated in Fig. 6.2 can contribute to the differences in reactivity with respect to similar test reactions by altering the mechanisms of these transformations.

Unlike the Si(100)-(2×1) surface, Si(111) has a more complex reconstruction, which has no known direct analogues for other group IV semiconductors. The well-studied 7×7 reconstruction of the Si(111) surface is shown schematically in Fig. 6.3.

The unit cell of this complex structure has 12 adatoms with available dangling bonds, which are the obvious origin of chemical reactivity, 6 rest atoms, and 1 corner atom. Very useful information about the electronic properties of this surface can be obtained with the help of scanning tunneling microscopy with typical images shown

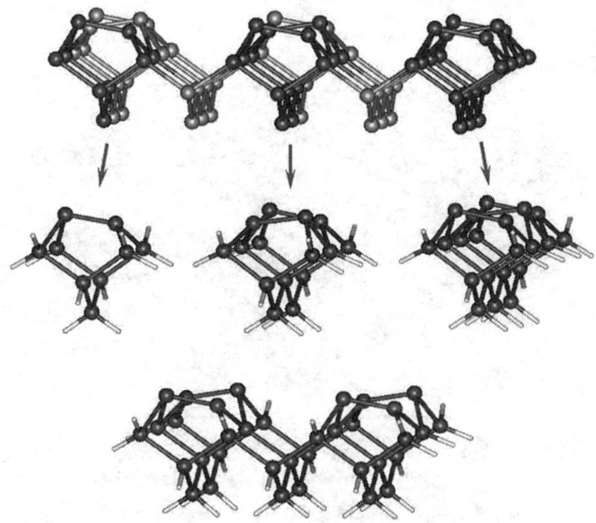

FIGURE 6.2 A representative view of a Si(100)-(2×1) surface reconstruction and representative clusters used in density functional theory (DFT) calculations to model the silicon dimers of this surface. On Ge(100)-(2×1) these dimers are permanently buckled and on C (100)-(2×1) they are symmetric at room temperature.

in Fig. 6.4. Interestingly, despite a variety of nonequivalent surface reactive sites and unusual electronic properties, the chemistry exhibited by the Si(111)-(7×7) surface is largely determined by the reactivity of an adatom–rest atom pair, which nicely mimics the structure and reactivity of a silicon dimer of the Si(100)-(2×1) surface. In the case of Si(111)-(2×1), of course, the two atoms involved remain in fixed positions determined by the overall reconstruction. There are several different adatom–rest atom pairs on the Si(111)-(7×7) surface, which can result in very large clusters for its modeling; however, most often simplified cluster models can be used, as suggested in Fig. 6.5 [8].

As some of the studies summarized in this chapter will involve hydrogen-covered Si(100) and Si(111) surfaces, it should be noted that the simplest form of hydrogenation yields the structures shown in Fig. 6.6, where hydrogen atoms are simply added to the two dangling bonds of the silicon dimer of the Si(100)-(2×1) surface or to every dangling bond of the unreconstructed Si(111).

Since the procedure to produce a well-defined flat Si(111) surface terminated with hydrogen that is stable in ambient for several hours was suggested by Chabal et al. [10], it has undergone several improvements [11, 12] and currently serves as the main way to prepare this substrate. There are also a number of alternative methods to prepare both H–Si(100) and H–Si(111) in ambient and in vacuum, as summarized in an excellent review by Wayner and Wolkow [13], a report by Baluch et al. [14], and a recent report by Mayne et al. [15].

The H-terminated silicon surfaces are stable in air for up to several hours but longer exposure leads to the formation of oxide layers. In vacuum these surfaces are

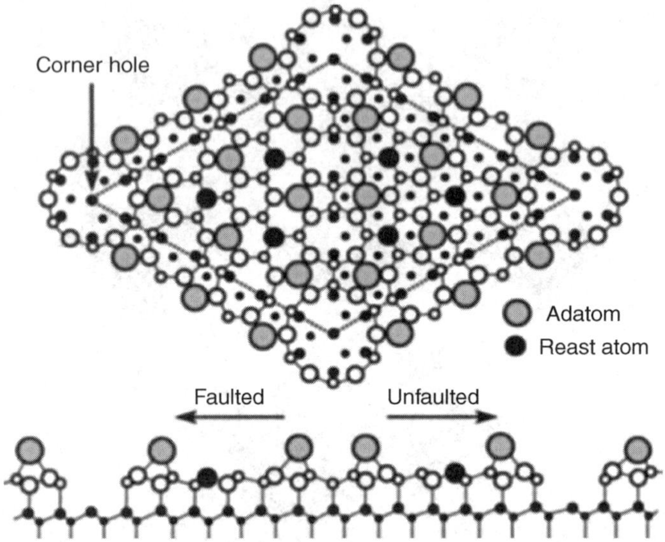

FIGURE 6.3 A representative view of the Si(111)-(7×7) surface with corner, adatoms and rest atoms marked on the reconstructed view, faulted and unfaulted halves of the unit cell shown as a cross section.

very stable for extended periods of time. It should be pointed out that unlike ultrahigh vacuum preparation, the traditional etching-based methods for preparing hydrogen-terminated Si(100) surface produce a mixture of surface reactive sites with various chemical reactivity; however, it is indeed possible to prepare the nearly ideal monohydride surface termination shown in Fig. 6.6 (left) in ultrahigh vacuum [16].

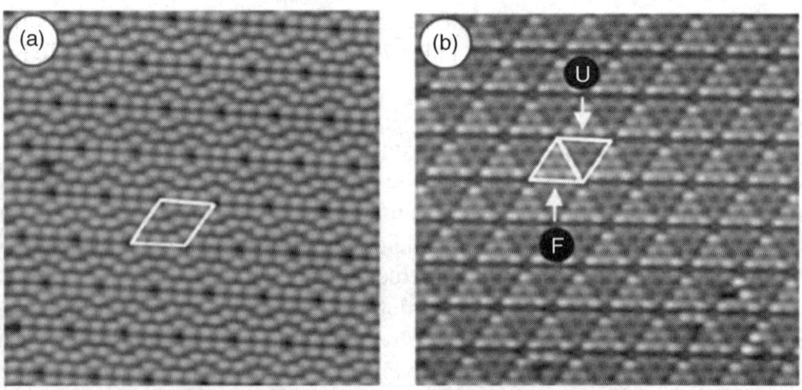

FIGURE 6.4 STM overviews comparing the clean 7×7 and the hydrogen-exposed surfaces both filled- and empty-state images. (a) Empty-state STM image of the Si(111)-(7×7) surface recorded with a tip voltage of −2 V. (b) Filled-state STM image of the Si(111)-(7×7) surface recorded with a tip voltage of +1.4 V. The size of images is 180 Å×180 Å. The tunneling current was 0.3 nA. Figure reprinted with permission from Ref. 9. Copyright 2009 The American Physical Society.

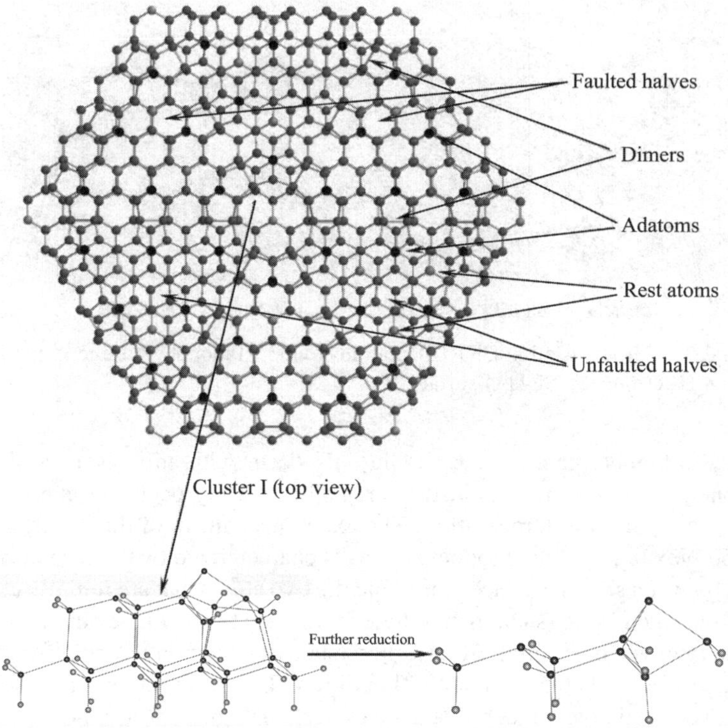

FIGURE 6.5 A representative view of a Si(111)-(7×7) surface and representative clusters used in DFT calculations to model this surface. Cluster models II and III are enhanced for clarity. Figure reprinted with permission from Ref. 8. Copyright 2009 American Chemical Society.

6.1.3 Brief Overview of the Types of Chemical Reactions Relevant for Aromatic Surface Modification of Clean Semiconductor Surfaces

The intrinsic multifunctionality of most aromatic compounds makes the world of reactions of aromatic compounds on semiconductor surfaces very complex. In fact, they often undergo a combination of several reactions simultaneously. While below we will consider multiple specific examples in detail, it is important to place the possible reactions considered in a context of several major classes of surface transformations. In order to understand the multifunctional reactivity on semiconductor surfaces, we should remember three classes of major reactions that are known to occur on the clean semiconductor surfaces: (a) formation of a dative bond (e.g., for substituted amines; this class of surface reaction may also lead to nonlocal surface effects following the initial adsorption); (b) dissociative addition (e.g., for alcohols); and (c) cycloaddition. The most common and probably the best understood type of cycloaddition is the [2+2] addition, where two electrons from the silicon atoms forming the surface dimer interact with the two electrons of the adsorbate, such as ethylene. This would be a

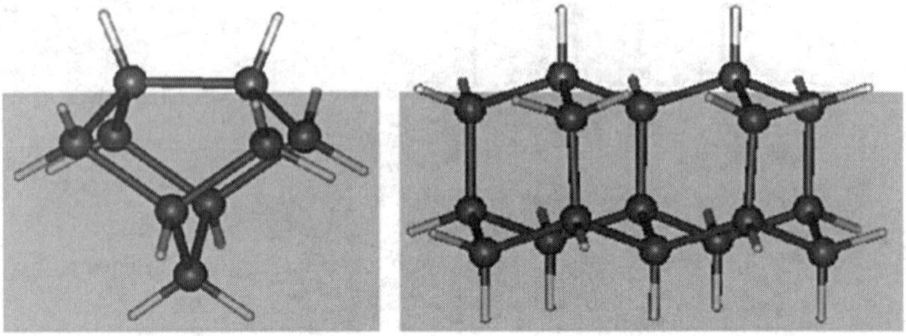

H-terminated Si(100) H-terminated Si(111)

FIGURE 6.6 Clusters used in DFT calculations to model monohydride-terminated Si(100) surface and H-terminated Si(111) surface.

symmetry forbidden reaction that should only occur with an assistance of external excitation, such as UV. However, these reactions readily occur on many semiconductor surfaces at room temperature. The exact mechanism of this reaction may be rather complex but the final product is always characterized by the formation of two σ-bonds between silicon surface atoms and the two atoms that are initially connected by an unsaturated bond, such as in ethylene, and if at least one of the atoms of the adsorbate involved is different from carbon (such as in carbonyls), this process could be characterized as heterocyclo [2+2] addition. The second type of cycloaddition was discovered relatively recently and its closest analogue is a common Diels–Alder reaction, such as the one between ethylene and 1,3-butadiene. If this process involves two electrons of the Si(100)-(2×1) surface dimer (instead of ethylene) and four electrons of the diene entity, then the observed process can be best described as a [4+2] cycloaddition and is observed for many dienes on Si(100)-(2×1), Ge(100)-(2×1), and C(100)-(2×1), as well as for the adatom–rest atom combination of the Si(111)-(7×7) surface. Finally, dipolar cycloadditions (or 1,3-cycloaddition processes) were observed for multiple groups that can be represented by a resonance structure with a 1,3-dipole, such as $-NO_2$ or $-N_3$. There are also special types of reactions that can follow some of the more common types, such as group-transfer that can sometimes follow dative bonding. Interestingly, the structure of aromatic compounds is such that *all of these processes* could be potentially observed for them to occur simultaneously. One can imagine a datively bonded benzene molecule that can dissociate its C–H bond, react using one −C=C− part of the aromatic ring, participate in a Diels–Alder [4+2] process with its −C=C−C=C− diene-like entity, or even to undergo some type of 1,3-cycloaddition process, especially if one of the carbon atoms of the ring is replaced with a heteroatom. That is why it is important to understand the ground rules that govern relative reactivity of the aromatic compounds with respect to the elemental group IV semiconductor surfaces.

Two recent reviews have addressed multifunctionality in reactions on semiconductors. Filler and Bent [17] compared the reactivity of Si(100)-(2×1) and Ge(100)-(2×1) pointing out that on average the more stable products and more closely spaced activation barriers are typically observed for multifunctional compounds on

Si(100)-(2×1) as compared to the same reagents on Ge(100)-(2×1), which can explain higher selectivity in some processes on germanium. However, the work of Leftwich and Teplyakov [7] also examined the cases on Si(100)-(2×1) where kinetics (the barrier height) rules in favor of the much less stable product, meaning that the temperature of the experiment can dictate the product distribution. The detailed discussion below will address both approaches.

If the semiconductor surface is fully or partially hydrogenated, there could be potentially other types of reactions observed for aromatic compounds. The brief summary of the examples below will include the mechanism of radical chain reaction of alkenes with a H-terminated Si(111) surface, where the reaction starts at a defect, such as a dangling bond and the cyclocondensation reaction that does not require the defect presence but is rather thermally driven [18]. These two types of reactions will be considered for specific cases only as discussed below.

6.2 MULTIFUNCTIONAL AROMATIC REACTIONS ON CLEAN SILICON SURFACES

6.2.1 Homoaromatic Compounds Without Additional Functional Groups

The next chapter will focus specifically on the reactions of polycyclic aromatic hydrocarbons on semiconductors. However, in order to understand and predict the reactions of multifunctional aromatic compounds, the reactivity of the simplest aromatic substituents will be considered here. It is most important to realize that within the framework of this chapter, aromatic groups are not just the relatively inert "linkers" but fully functional chemical groups. We start with a simple generic energy diagram shown in Fig. 6.7. It is in fact rather characteristic of many multichannel chemical processes on clean silicon surfaces. Reactant R may form one or more types of surface precursors, SP1 and SP2, which then can be converted into products of different stability, P1 and P2 via transition states TS1 and TS2. The energy gain in the formation of surface-bound precursors is generally in the order of 10–40 kJ/mol; however there are examples of reactions without known precursor states and also with extremely strongly bound precursors, well outside this approximate range. Since the precursor states have been reported both for the more or less obvious functionalities, such as ketones and amines, and for the compounds that were less expected to form these states, such as ethylene, the formation of such a bound state should be considered as a prerequisite for most transformations on clean silicon surfaces. It will also be described later that even for the reactions shown to proceed straight downhill in computational studies, experimental barriers were found on the way to the formation of such precursor states.

A brief examination of this generic reaction diagram suggests the possibility of kinetic control, where the chemical reaction proceeds preferentially via the pathway with the lower kinetic barrier, or thermodynamic control, where all the states of the system can be sampled and the most thermodynamically stable product is formed. On

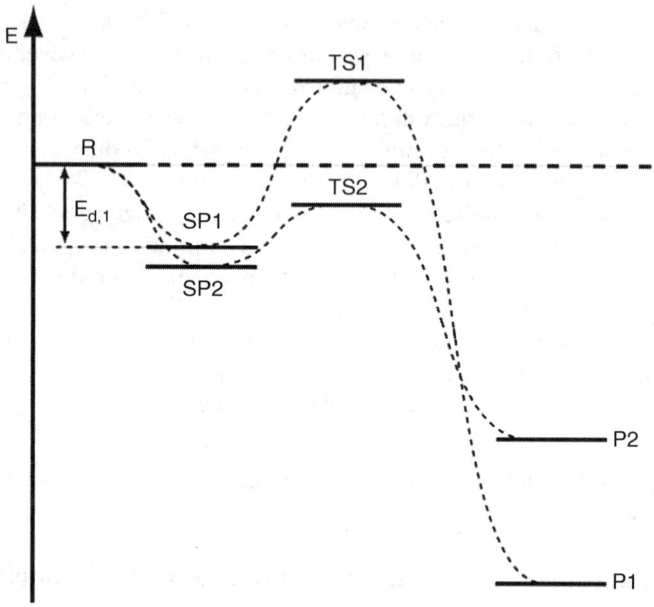

FIGURE 6.7 Comparison of the potential energy profiles for competing chemical reactions on silicon. Most of the time the difference in energy between TS1 and TS2 is substantially smaller than the difference in energy between P1 and P2. The thermodynamic stability of the products is normally hundreds of kJ/mol, while the differences between TS1 and TS2 are substantially smaller, so the scheme is not to scale.

silicon, these two control regimes often give very different results and since the products for all the reactions described above are rather stable, their formation is essentially irreversible.

Of course, the case presented in Fig. 6.7 is very simplistic. Often, a combination of two different functionalities on the same molecule can produce very stable surface species or the reaction of one may trigger the reaction of the other. Reactions with metal-containing aromatic compounds may proceed by an entirely different scheme. However, for the purposes of this chapter, it is important to illustrate the fact that even the simplest aromatic substituent alters the energy landscape of the surface reaction and should be considered a functional group. Thus, the reactivity of the simplest aromatic homonuclear compounds will be considered first, with the most emphasis on the recent developments.

The reactivity of a phenyl group should generally follow the trends previously discovered for this type of chemistry and delivering this group on silicon may serve as a starting point for further modification [19, 20]. Thus, we will briefly review the reactions of benzene on silicon and germanium. Despite its seeming simplicity, benzene is one of those compounds that could potentially undergo all the classes of reactions listed above. Its chemical transformations on clean silicon surfaces have eluded scientists for well over a decade. A detailed review by Wolkow highlights the most important results of the benzene surface transformations of a clean Si(100)-(2×1) surface [21]. After the chemisorption of benzene was demonstrated by Taguchi et al. [22], the work from several different research groups involved microscopy, spectroscopy, and DFT studies to discern the initial formation of the metastable [4+2], or Diels–Alder, cycloaddition product between benzene and a single silicon dimer of the Si(100)-(2×1) surface. This state slowly converts to the asymmetric tight state bridging two neighboring silicon dimers via the activation energy barrier of 0.94 eV and with a rate constant of $(1.5 \pm 0.05) \times 10^3\ s^{-1}$ [23–30]. Some recent studies suggest that at high coverage the initial [4+2] adduct is stabilized and in fact forms the most stable structure [31–36]. There are two major conclusions that can be drawn from this research: (1) benzene (and by extension phenyl group) are indeed reactive entities and should be treated as such when considering reactions on clean silicon; and (2) if phenyl group in a substituted compound is found to react with a clean silicon surface, multiple reaction pathways and kinetic studies should be undertaken. These observations can be generalized for polyaromatic compounds. Naphthalene [37], pentacene [38–46], tetracene [47, 48], perylene [48, 49], and coronene derivatives [50–52] were all found to chemisorb on a clean Si(100)-(2×1) surface in a rather complex fashion that sometimes can even involve silicon dimers from different dimer rows. Pentacene was suggested to cleave a C–H bond leading to the irreversible chemisorption [38, 53–55] and similar reactions have been explored computationally for benzene [56] but no experimental evidence for the C–H dissociation in benzene has been reported so far. Interestingly, on Ge(100), only weakly bound benzene was reported to desorb at 234 and 252K, the desorption regimes corresponding to binding to terraces and steps [57]. The very weak Ge–C bond, compared to the strong Si–C bond driving the addition processes on silicon, is the most likely explanation for such different reactivities, as was examined in a computational study [58].

Although the amount of work on a clean Si(111)-(7×7) surface is much less extensive compared to the studies of the Si(100)-(2×1) surface, there are certainly parallels that can be drawn between these two surfaces. There are several recent reviews highlighting the chemical reactivity of Si(111)-(7×7) and comparing it to Si(100)-(2×1) [7, 8, 59, 60]. Benzene reaction could serve as a segue into this comparison. It is important to realize that the dynamic dipole within the dimer of the Si(100)-(2×1) has very similar zwitterionic properties if compared to the stationary charge distribution between the adatom and the rest atom of the Si(111)-(7×7) surface. Since the distances between these atoms differ on these two substrates and also since the dipole nature is not identical, one should be careful, however, transferring the chemistry discovered on Si(100)-(2×1) directly to Si(111)-(7×7). For example, benzene attachment to the adatom–rest atom pair of the Si(111)-(7×7) does result in a [4+2] Diels–Alder-type product; however this reaction was suggested to proceed via a very different, diradical mechanism, on Si(111)-(7×7) and to form the majority stable species [61]. A theoretical study by Petsalakis and Polanyi followed this process and also compared the calculations with the radiation-induced attachment of benzene on this surface [62]. Similarly, STM and DFT studies of naphthalene suggested that this compound forms nearly exclusively the [4+2] adduct involving the opposite carbon atoms of the same ring of naphthalene but this highly selective chemistry exhibits complex coverage dependence and preference for selected surface adatom–rest atom pairs [63]. A recent study by Guan et al. explored the adsorption of a large polyaromatic tetracene on Si(111)-(7×7) by ultraviolet photoelectron spectroscopy (UPS) and DFT and proposed a plausible model for the most stable adsorption configuration [64] near the top of a center adatom with the longer molecular axis along the (110) direction. Previous tetracene studies [65, 66] seem to suggest that the tetracene acts as an acceptor of electrons with respect to the Si(111)-(7×7) surface and that it does participate in covalent bonding with the several combination of adatom–rest atom pairs on this substrate. Because of the larger size, pentacene actually loses all its aromatic character when reacted with Si(111)-(7×7) leading to the formation of multiple Si–C bonds [53, 67].

Despite the rich chemistry of benzene and polyaromatic compounds, biphenyl will perhaps show more similar reactivity to *substituted* benzene. Biphenyl was suggested to adsorb across two adjacent dimers within the same row [68, 69]. Several recent studies have addressed different possible adsorption configurations [70–72]; however, an interesting aspect of the biphenyl behavior is the observation that at 5K, biphenyl could be trapped in a bistable configuration pivoting about a rotational axis, and that the STM could be used to switch the pivot reproducibly [73]. Interesting electronic properties [74, 75] coupled with the most recent STM-induced chemistry studies [76] summarized in Fig. 6.8 keep fueling the scientific interest toward biphenyl. In fact, the STM manipulation shown in Fig. 6.8 could serve as a stepping stone to developing molecular switches for practical applications because of such a rich set of electronic properties of biphenyl on Si(100)-(2×1). Even more intriguing behavior was recorded for *para*-hexaphenyl molecules, where six phenyl rings are coupled to one another sequentially in *para*-positions, forming a line of aromatic structures on Si(100)-(2×1) [77]. Two different states are observed by STM at room temperature and a very strong angular dependence is found between the axis of this molecule and the direction of propagation of the dimer lines of the Si(100)-(2×1) surface.

6.2.2 Functionalized Aromatics

6.2.2.1 Dissociative Addition With all the reactivity exhibited by the benzene and phenyl groups described above, it may seem impossible to deliver an unreacted phenyl substituent to the silicon substrate. There are, however, at least two general approaches that help address this problem. One is based on decreasing the reactivity of the silicon surface, for example, by saturating the available surface silicon dangling bonds with hydrogen. This approach will be addressed later. The other is based on selecting appropriate chemical reactivity of a polyfunctional aromatic molecule. This can be illustrated by following the previously discussed energy diagram in Fig. 6.7. Since the formation of any covalent bonds with the silicon surface is very often irreversible, the preferential selective attachment schemes should either be based on the thermodynamically controlled regime where the target

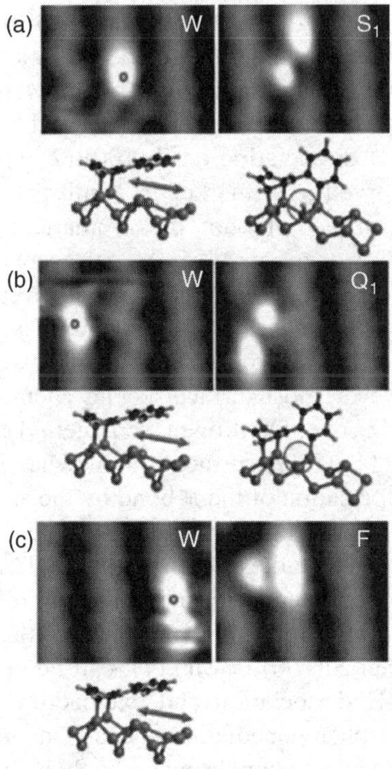

FIGURE 6.8 Changes of adsorption configuration by STM manipulation with positive surface voltage pulses. All $2.2 \times 1.8\ \mathrm{nm}^2$ STM topographies are recorded at $V_S = -2$ V and $I =$ 240 pA. Left-hand and right-hand side STM topographies are recorded before and after STM manipulation, respectively. The dots indicate the location of the STM tip during the surface voltage pulse at $V_S = +4$ V. (a) Change from the weakly chemisorbed (W) to the strongly chemisorbed bistable (S_1) configuration. (b) Change from the weakly chemisorbed (W) to the strongly chemisorbed quadristable (Q_1) configuration. (c) Change from the weakly chemisorbed (W) to the strongly chemisorbed fixed (F) configuration. The adsorption configuration of the biphenyl molecule is shown below each STM topography. For the S_1 configuration, the hydrogen atom dissociated from the fixed phenyl ring is circled. In the Q_1 configuration, the hydrogen atom has been desorbed. Figure reprinted with permission from Ref. 76. Copyright 2009 American Chemical Society.

surface species is the most stable, or on a kinetically controlled approach to reach the less stable species over the lower energy barrier. Fortunately, both approaches can work on silicon. The attachment chemistries described below involve halogens (X), oxygen, nitrogen, or sulfur and often lead to the formation of two bonds at the expense of one, meaning that the processes leading to the formation of Si–O, Si–N, or Si–X bonds and accompanied by the formation of Si–H surface sites should be thermodynamically preferred compared to the reaction of the phenyl group. In addition, most of these processes do have lower reaction barriers compared to the phenyl reactions on silicon. There are plenty of exceptions from these rules, as will be outlined below.

A substantial number of experimental and theoretical studies do suggest that the reactivity of phenyl group in singly substituted aromatic molecules is lower than the reactivity of almost any functional group. In fact even in toluene and xylenes, where phenyl group competes in reactivity with very stable methyl groups, there has been reported methyl C–H dissociation on Si(100)-(2×1) [78], where isotopically substituted compounds served as a marker to confirm that the source of surface hydrogen is the methyl group and not the aromatic ring. The first principles calculation studies by Costanzo et al. [79, 80] were in complete agreement with the experiment. In a spectroscopic and DFT investigation of *para*-xylene (1,4-dimethylbenzene), it was suggested that some *para*-xylene molecules desorb intact at 400 and 470K after chemisorption pathways that were proposed to be similar to those for benzene; however, most surface species chemisorb via the formation of the Si–C bond following the loss of hydrogen from methyl groups [81]. The behavior of toluene on a clean Si(111)-(7×7) surface is somewhat different, as a photoemission study suggested the formation of the π-bond by the aromatic electrons [82] and STM studies followed the adsorption of both benzene and toluene at cryogenic temperature [83, 84]; however, studies of further thermal transformations of toluene on this substrate are also consistent with methyl group dissociation [85].

Among the functional groups relevant to this chapter, the types of reactions with a clean silicon surface, either Si(100) or Si(111), can be generally divided into two major categories: Ph–Y–H dissociation and cycloaddition. Reactions of phenol, benzenethiol, aniline, 1,4-phenylenediamine, and benzoic (and 4-aminobenzoic) acid, can cleanly deliver an intact phenyl ring to the Si(100)-(2×1) surface by Ph–Y–H dissociation forming –O–[86], –S– [87], –NH– [88–91], and –O–CO– [90, 92, 93] linkages, respectively, and the corresponding Si–H species. A detailed review by Mui et al. analyzes the parallels between silicon and germanium chemistry with respect to these reactions [94]. For the case of aniline, multiple bonding configurations following dissociation of one and both N–H bonds have been considered and there are several possibilities for the exact long-range structures formed [88–91, 95]; however, both these studies and the investigation of aniline adsorption on the Si(5 5 12) surface [96] suggest that the nitrogen atom is connected to silicon and that the aromatic ring structure is retained. If needed, the formation of an extra surface hydrogen atom evidenced in the dissociative addition processes can be avoided by employing the reaction of, for example, disulfide compounds, as was shown for diphenyl disulfide (C_6H_5S–SC_6H_5) [87]), where two C_6H_5S– surface groups were cleanly formed. Most of these reactions are relatively easy to follow spectroscopically, as the phenyl ring, in addition to having its own set of spectroscopic signatures (such as vibrational characteristics) helps to identify the chemical state of the elements in the adjacent functional groups. For example, upon dissociation of phenol on a clean Si(100)-(2×1), this reaction can be confirmed by following the photoemission spectra, as illustrated in Fig. 6.9 for the C 1s core-level spectrum of the submonolayer coverage of phenol [86]. Here, the carbon atom bound to oxygen can be easily distinguished from the lower binding energy component of carbon atoms in the phenyl ring [86]. The quantitative analysis confirms this assessment. A detailed computational study by Carbone et al. [97, 98] explored

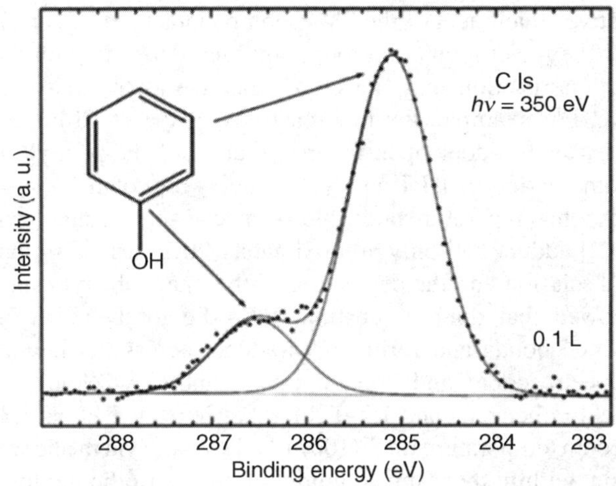

FIGURE 6.9 Line-shape analysis of the C 1s core-level spectrum after exposure to 0.1 L phenol, recorded at a photon energy of 350 eV. Figure reprinted with permission from Ref. 86. Copyright 2009 Elsevier B.V.

the energy landscape for the phenol adsorption on Si(100)-(2×1) and also found the dissociation channel to lead to the most thermodynamically stable structure. The propensity of these dissociated molecules to form molecular lines on Si(100)-(2×1) has also been explored computationally [98]. Similar chemical behavior was observed for phenol on a clean Si(111)-(7×7) surface. It was shown to dissociate at temperatures as low as 50K [99]. Near-edge X-ray absorption fine structure (NEXAFS) studies of phenol, 4-fluorophenol, and 2,3,4-trifluorophenol on Si(111)-(7×7) confirm that this reaction is rather general [100]. Benzenethiol and 1,4-benzenethiol also dissociate one or both –SH functional groups, respectively, on Si(111)-(7×7) [101].

In some cases, the specific silicon surface reaction, and as a result the chemical nature of the surface-bound group of the aromatic compound can be altered by using a metal promoter. The Richardson group reported an investigation of the interaction of benzoic acid with a Si(100)-(2×1) surface covered with 0.5 monolayer of Na atoms, adsorbed on cave sites between silicon dimers [92, 102]. Promoting interaction of benzoic acid with sodium caused the formation of a bidentate ligand instead of the Si–O–(C=O)–C linker observed for the reaction on an unpromoted surface. The observation of the Si–H stretching mode at 2090 cm^{-1} and the benzoate functionality characterized by ν(OCO) = 1425 cm^{-1} in HREELS studies confirmed this assessment and was supplemented by other analytical techniques. In this specific example, the selectivity of the bifunctional 4-aminobenzoic acid in a reaction with the Na-promoted Si(100)-(2×1) surface does not change [93], as the carboxylic group reacts predominantly with silicon regardless of the presence of metal promoter; however, the formation of the benzoate bidentate linker leads to the formation of a much more stable adduct, which is stable up to 300°C [92].

The dissociative attachment of the Ph–X compounds to the clean silicon surfaces can also deliver the intact aromatic group; however, depending on the reactivity of the C–X bond, this reaction may not be as selective as the dissociative reactions described above. For example, the low reactivity of the C–Cl bond on a Si(100)-(2×1) surface leaves the door open for the reaction of the phenyl group with this surface. The combination of DFT and STM studies by Polanyi's group [103, 104] suggests that at room temperature both chlorobenzene and 1,2-dichlorobenzene yield 45–50% of [4+2] adduct and only approximately 20% surface products consistent with C–Cl bond scission and the deposition of the intact phenyl ring. However, the same study showed that doubly substituted 1,4-dichlorobenzene selectively dissociates both C–Cl bonds and forms a 1,4-adduct across the neighboring silicon dimer rows. Spectroscopic and computational study by Zhou and Leung also suggested that chlorobenzene and 1,2-dichlorobenzene, as well as 1,3-dichlorobenzene adsorb at room temperature on Si(100)-(2×1) mostly via the benzene ring [105]. The comparison within the same column of the periodic table followed 1,2-difluoro-, 1,2-dichloro-, and 1,2-dibromobenzenes [106]. The adsorption of 1,2-difluorobenzene was found to be completely reversible, as it desorbed intact in a molecular state at elevated temperatures, 25% of 1,2-dichlorobenzene was found to undergo dechlorination and 1,2-dibromobenzene was shown to undergo irreversible double debromination forming 1,2-phenylene species on Si(100)-(2×1) [106].

It seems that in a reaction of the phenyl ring with the adatom–rest atom pair of the clean Si(111)-(7×7) surface, the formation of the [4+2] surface adducts should be even more preferential as compared to the Si(100)-(2×1). Consistent with this assessment, no direct dissociation of the C–Cl bond in chlorobenzene was observed on Si(111)-(7×7) [107–110]. The physisorbed chlorobenzene and toluene were proposed as reversible molecular switches [111]. In fact, only bromo derivatives were shown to dissociate on this surface at room temperature, overpowering the reactivity of the phenyl ring [112] and more recently a detailed computational study of 1,2-dibromobenzene on Si(111)-(7×7) helped explain the experimental results [113–115]. Interestingly, the reactivity of the 1,2- and 1,4-dibromoxylenes was found to be lower than that for the corresponding substituted benzenes [116].

6.2.2.2 Cycloaddition The cycloaddition processes on silicon can be generally grouped into 1,2-, 1,3-, and 1,4-cycloaddition classes of reactions. The 1,2- (or [2+2]) and 1,4- (or [4+2]) processes were mentioned above for the reactions of the phenyl ring. However, similar attachment chemistry can be used to deliver an intact phenyl group to the silicon surface.

The reactions of ethylene and acetylene described earlier in this book proceed by [2+2] pathways. Despite continuous studies of these seemingly simple hydrocarbons on clean silicon surfaces, there is still some debate over the exact mechanism of the attachment process. However, for the purposes of this chapter, the result of the attachment reaction, the Si–C–C–Si cycle, is fully described and agreed upon in the surface science community. Substituting one of the hydrogen atoms in ethylene or acetylene by a phenyl group should provide a convenient vehicle to deliver the aromatic functionality onto a silicon surface by this 1,2-cycloaddition chemistry.

In fact, both styrene (Ph–CH=CH_2) and phenylacetylene (Ph–CCH) react with the Si(100)-(2×1) surface by [2+2] cycloaddition. Considering all the possibilities for styrene, one could suggest that interaction of the vinyl group with the phenyl functionality could potentially lead to complex reactions including [4+2] addition; however, all the current studies suggest that a simple [2+2] process across a single silicon dimer dominates the reactions of this compound with a clean Si(100)-(2×1) surface [117–120]. As was done previously many times, reactions on the silicon surface can be paralleled with the test reactions with disilenes, and the reaction of styrene with Si(100)-(2×1) is a perfect example. As shown schematically in Fig. 6.10, the structure of the proposed attachment product for styrene on Si (100)-(2×1) is the same as for a variety of disilenes [121–123], suggesting the possible parallels in electronic structure of a Si–Si surface dimer of this substrate and a Si=Si double bond. The same addition reaction of styrene seems to take place on the permanently buckled dimer of the Ge(100)-(2×1) surface [124] or on Ge-covered Si(100) [125].

Since the need to deliver intact phenyl groups to a clean silicon surface was in part dictated by the possible electronic and switchable behavior of the aromatic compounds, it is important to highlight that the switching between two rotational states in a chemisorbed styrene molecule on a clean Si(100)-(2×1) surface could be followed experimentally by STM, however, this motion could be suppressed by chemical modification and by using 5-vinyl-2,3-dihydrobenzofuran instead of styrene, as summarized by schematics and STM images in Figs. 6.11 and 6.12 [126].

Despite such a high selectivity on Si(100)-(2×1), the studies of the isotope-substituted styrene on Si(111)-(7×7) [127] by HREELS clearly show rehybridization of the phenyl ring carbon atoms from sp^2 to sp^3, which would support the Diels–Alder-like chemistry involving the phenyl ring. Since the reactivity of the Si(111)-(7×7) seems to favor the [4+2] additions whenever possible, it is likely that the potential for conjugation between the vinyl group and the phenyl mentioned above for the reaction with Si(100)-(2×1) is actually a reality in a reaction with Si(111)-(2×1).

The geometrical restrictions of a C≡C bond make surface chemistry of phenylacetylene on silicon more complex compared to that of styrene. A simple [2+2] addition similar to that of styrene was proposed by Tao et al. [128] to dominate the

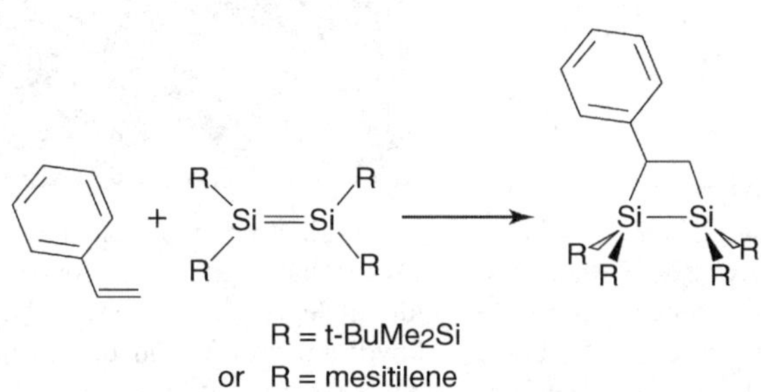

FIGURE 6.10 Reaction of styrene with tetramesityldisilene and tetradisilylene. Based on a schematic from Ref. 117.

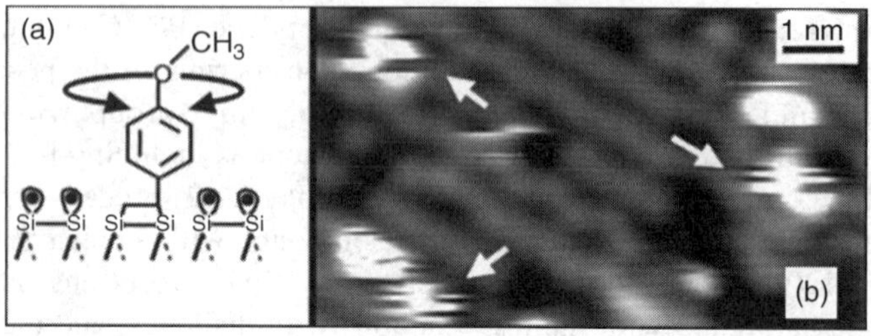

FIGURE 6.11 (a) Schematic of [2+2] cycloaddition of a 4-methoxystyrene molecule on a dimer on the clean Si(100) surface. The methoxy group is expected to have the rotational degree of freedom indicated in the figure; (b) UHV STM image of 4-methoxystyrene molecules on the clean Si(100) surface following a dose of 0.7 L. Several of the molecules (indicated by white arrows) exhibit multiple switching events. Imaging conditions are sample bias of −2.0 V and tunneling current of 0.1 nA. Figure reprinted with permission from Ref. 126. Copyright 2009 American Vacuum Society.

reaction of phenylacetylene with Si(100)-(2×1), while possible dissociation of a weak ≡C−H bond was not observed. STM and DFT investigation reported by Kim et al. [129] used phenylacetylene and 1-phenyl-1-propyne (a molecule with an extra methyl marker group to help distinguish structural differences in STM images) to observe two different surface structures. One of these structures was proposed to be the same as the one reported by Tao et al. [128], the result of [2+2] addition. The other was proposed to involve both functionalities, with phenyl group bound to one silicon dimer row and acetylene entity interacting with the other silicon dimer row, as shown in Fig. 6.13. Table 6.1 summarizes the predicted stabilities of these

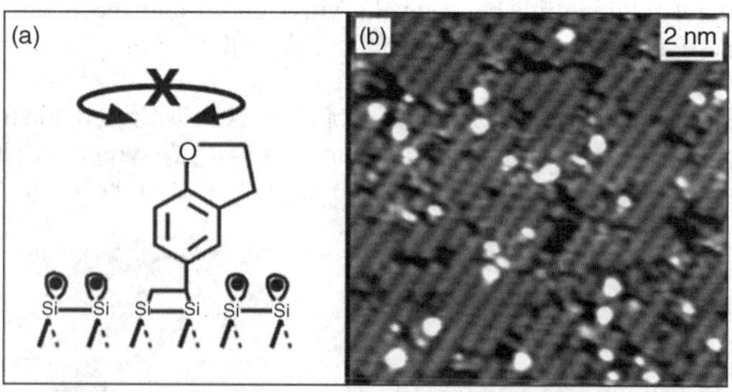

FIGURE 6.12 (a) Schematic of a 5-vinyl-2,3-dihydrobenzofuran bound to Si(100). This molecule is analogous to 4-methoxystyrene except that its methoxy group is covalently linked to the aromatic ring in order to inhibit its rotational degree of freedom; (b) UHV STM image of a submonolayer coverage of 5-vinyl-2,3-dihydrobenzofuran on the clean Si(100) surface following a dose of 0.4 L. No switching events are observed in this case. Imaging conditions are sample bias of −2.0 V and tunneling current of 0.1 nA. Figure reprinted with permission from Ref. 126. Copyright 2009 American Vacuum Society.

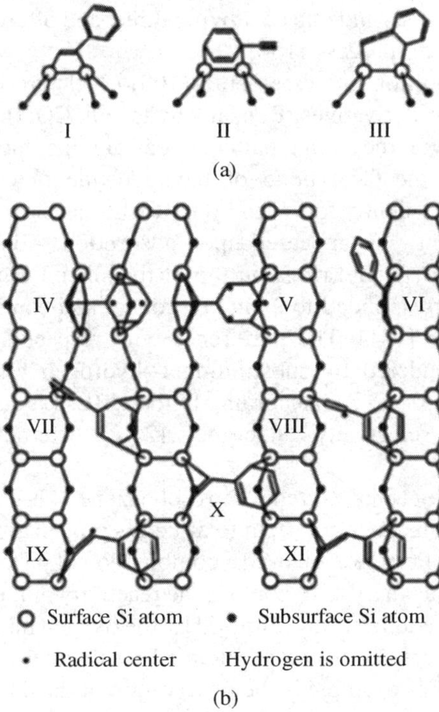

FIGURE 6.13 Schematic presentation of the proposed binding structures of phenylacetylene on Si(100)-(2×1). (a) Side view for structures I–III. (b) Top view for structures IV–XI. The structure of phenylacetylene is drawn only for the carbon skeleton without showing hydrogen atoms. Figure reprinted with permission from Ref. 129. Copyright 2009 American Chemical Society.

structures, with the [2+2] structure I indeed being the most stable; however, several other bridging structures were also observed experimentally. A more recent STM study combined with Monte Carlo simulation also confirmed the presence of two different adsorption modes for phenylacetylene on Si(100)-(2×1), one occupying

TABLE 6.1 Adsorption Energy of the Proposed Binding Structures in Fig. 6.13 (in eV/molecule) for Phenylacetylene on Si(100) Calculated from DFT (slab-GGA) for a Coverage of 1/8 ML [129]

Binding Structure	Adsorption Energy	Binding Structure	Adsorption Energy
I	2.44–2.67	VII	Converted to VIII
II	0.81	VIII	1.59
III	0.66	IX	Converted to X
IV	Unstable	X	0.45
V	Unstable	XI	1.09
VI	2.11–2.28		

two surface dangling bonds and one occupying three, and offered an accurate kinetic model for the adsorption process [130].

The [2+2] cycloaddition processes on Si(100)-(2×1) are not restricted to only ethylene and acetylene derivatives. Benzaldehyde (Ph–COH), a heteroanalogue of styrene, also reacts via the same pathway despite the fact that it could also potentially dissociate the C–H bond of the aldehyde functionality [131]. This reaction was very conveniently followed by HREELS, as the very intense stretching mode $\nu_{C=O} = 1713\,cm^{-1}$ disappeared upon adsorption. Additionally, bright protrusions assigned to the phenyl rings on top of the silicon dimers in low-coverage STM investigation corresponded to a single product and confirmed the absence of the inter-row reactions [131]. The preference with respect to the [2+2] addition could not even be hindered by substituting α-hydrogen for a methyl group in acetophenone (C_6H_5–CO–CH_3). According to the HREELS and DFT investigation, this compound still selectively undergoes [2+2] heterocycloaddition via its C=O bond [132].

Another heteroanalogue of styrene, nitrosobenzene (Ph–NO), has both carbon atoms substituted but this does not seem to affect its propensity to [2+2] cycloaddition on Si(100)-(2×1). Perrine et al. used a combination of infrared spectroscopy and XPS with computational analysis to confirm the reactivity of the –N=O group [133]. The initial [2+2] cycloaddition was followed by the oxygen migration subsurface, as expected, and this process could be accelerated by thermal annealing. One of the interesting aspects of this example is the observation of the datively bonded Ph–NO at cryogenic temperatures, as suggested by the infrared studies summarized in Fig. 6.14 and substantiated by computational investigation.

A nitrogen-substituted heteroanalogue of phenylacetylene, benzonitrile (C_6H_5–C≡N), has also been investigated on Si(100)-(2×1) and it was originally suggested to react exclusively by [2+2] addition via the nitrile functionality [134]. The easily distinguishable $\nu_{C\equiv N}$ absorption band at $2247\,cm^{-1}$ disappeared upon chemisorption and another intense $\nu_{C=N}$ absorption band at $1629\,cm^{-1}$ appeared when reaction was carried out at 110K [134] and followed by HREELS. A combination of XPS analysis and DFT investigation agreed with these results. It should be pointed out that in the case of benzonitrile, the reversibility of the adsorption process was confirmed by thermal desorption as the molecular desorption was recorded at ~490K [134]. The [2+2] addition of benzonitrile found even more support in a computational study by Qu and Han [135] but careful investigation combining computational studies, XPS, and NEXAFS by Rangan et al. [136] suggested that in addition to the [2+2] addition product, another adduct bonded via the phenyl ring is observed at room temperature.

On Si(111)-(7×7) the unsaturated phenylacetylene [127] and benzonitrile [137] keep their phenyl groups intact. The HREELS studies observe the formation of styrene-like surface adducts. The high selectivity on this surface is likely aided both by high reactivities of both C≡C and C≡N in electrophilic addition reactions and by the fact that the geometry of the triple bond forces the phenyl ring to point away from the surface upon reaction with the adatom–rest atom pair, making it virtually impossible for any subsequent surface transformations to occur.

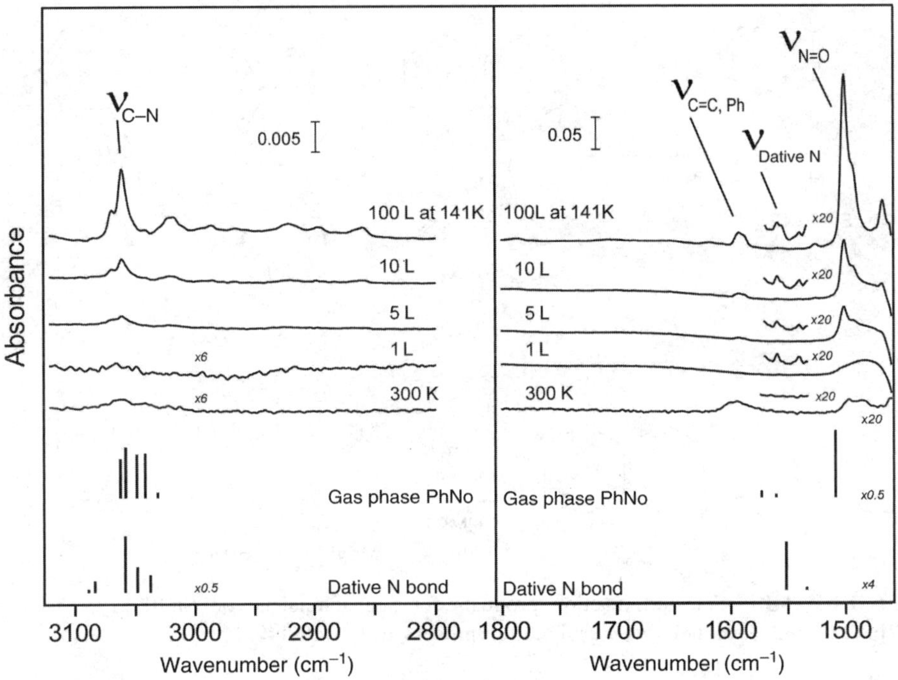

FIGURE 6.14 MIR–FTIR coverage profile of nitrosobenzene reacted on Si(100)-(2×1) at cryogenic and room temperatures. Left panel: C–H stretch region. Right panel: C=C and N=O stretch region. The exposure of the nitrosobenzene on Si(100)-(2×1) at 300K results in saturation of surface sites, based on IR signal. Below the experimental spectra are shown the computationally predicted infrared absorption bands corresponding to structures of a single molecule of nitrosobenzene in the gas phase and to the datively bonded nitrosobenzene on a Si (100)-(2×1) surface. The computationally predicted frequencies are scaled by 0.9564. Figure reprinted with permission from Ref. 133. Copyright 2009 American Chemical Society.

Finally, larger aromatic systems can also be delivered onto a silicon surface via [2+2] cycloaddition reaction. The investigation reported by Schwarts et al. focused on a possibility to deliver a stackable derivative of naphthalene, acenaphthalene, to react selectively with the silicon dimers of the Si(100)-(2×1) surface, as summarized in Fig. 6.15 [138]. A detailed set of STM and infrared spectroscopic studies of selectively deuterated acenaphthalene drew on a comparison with silylation chemistry to produce the acenaphthalene [2+2] product over a dimer row with selectivity over 90%, thus creating a sharp interface between silicon and assembled polycyclic aromatic molecules. This work was also complemented by computational studies by Nunzi et al. that confirmed the stability and preferential formation of the [2+2] cycloaddition product [139].

Within the general class of cycloaddition reactions on silicon, reactions of 1,3-cycloaddition have received a very substantial attention over the last decade. The reaction initially considered for nitromethane and nitroethane was proposed to lead

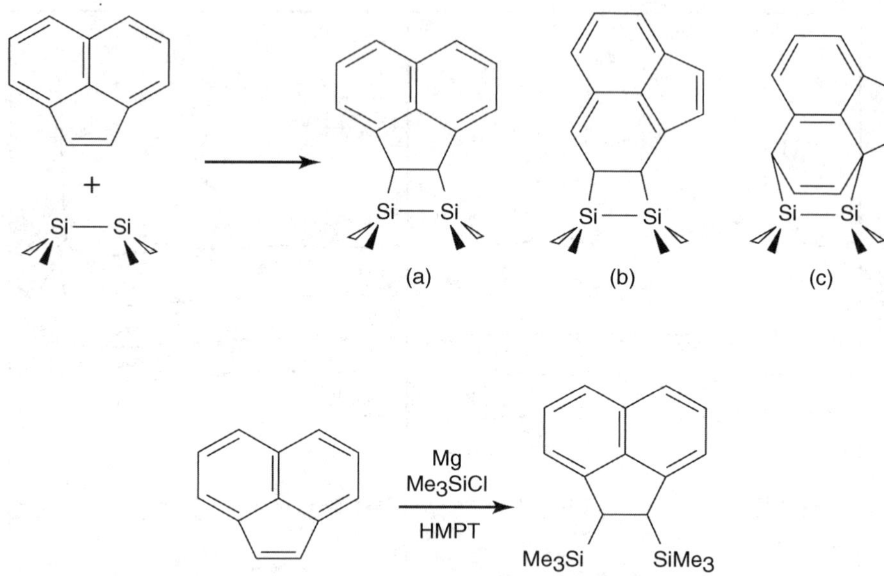

FIGURE 6.15 Selected reaction products for acenaphthalene on Si(100)-(2×1) from Ref. 138 and a parallel with the silylation process from Ref. 140.

to the formation of the Si–O–C–O–Si with subsequent oxygen migration into the silicon back-bond [141–144]. A detailed set of experimental and computational studies suggested that nitrobenzene ($PhNO_2$) would also react with a clean Si(100)-(2×1) surface via 1,3-dipolar cycloaddition [145, 146], as summarized in Fig. 6.16. Similar to other nitro compounds, nitrobenzene was shown to react with its $–NO_2$ group exclusively, as no remaining nitrofunctionality was observed by MIR–FTIR following nitrobenzene reaction at room temperature and the binding between two different silicon dimers was prohibitively expensive as confirmed in the computational studies. Interestingly, a barrier was observed experimentally for this cycloaddition reaction despite the fact that the entire transformation was shown theoretically to proceed downhill.

As a sidenote, the initial five-membered ring adduct was suggested computationally to have a potential to work as a molecular switch. The geometry of the nitrogen center could be switched between two positions, as shown in Fig. 6.17. The phenyl ring could change its position from being nearly parallel to the surface (IB) to almost exactly perpendicular (IA or IC) [146]. This potentially useful process could benefit from low-temperature STM studies. The cryogenic experiments would be required since the barriers found for this "flip" are rather low; however, experimentally found barrier for the attachment (40.8 kJ/mol) [145] may make these studies difficult.

In many molecules amenable to 1,3-dipolar cycloaddition, there also exists a possibility for 1,2-cycloaddition reactions for the same functional group. For example, phenyl isothiocyanate (Ph–N=C=S) could potentially react with Si (100)-(2×1) via [2+2] cycloaddition either to the N=C bond or to the C=S bond,

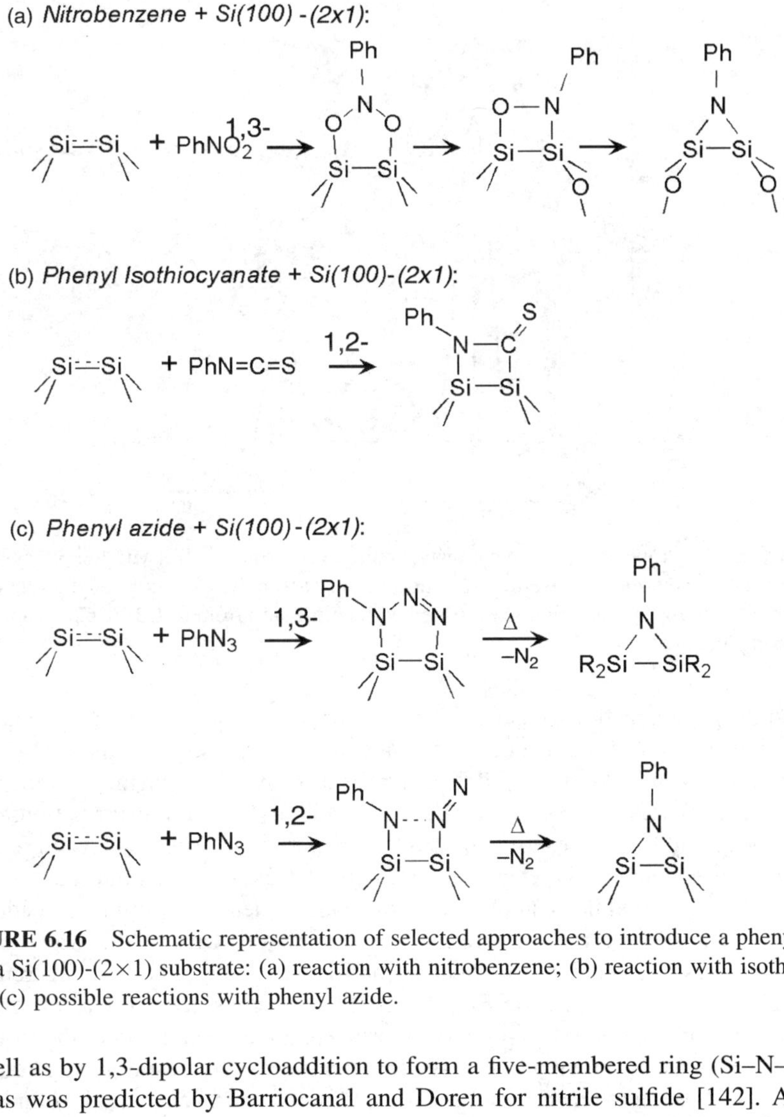

FIGURE 6.16 Schematic representation of selected approaches to introduce a phenyl ring onto a Si(100)-(2×1) substrate: (a) reaction with nitrobenzene; (b) reaction with isothiocyanate; (c) possible reactions with phenyl azide.

as well as by 1,3-dipolar cycloaddition to form a five-membered ring (Si–N–C–S–Si), as was predicted by Barriocanal and Doren for nitrile sulfide [142]. A very thorough investigation reported by Ellison and Hamers [147] used FTIR, XPS, STM, and DFT to prove that the major surface reaction is the 1,2-cycloaddition via the N=C group, as shown in Fig. 6.16b above. This reaction would also be expected based on the previous studies of solution silylation chemistry [148, 149]. The reaction of the isoelectronic functional group was tested in a study on Ge(100), where Kim et al. used ethylenediamine coupling with 1,4-phenylene diisocyanate ($O{=}C{=}N{-}C_6H_4{-}N{=}C{=}O$) to initiate a layer-by-layer growth [150]. This investigation employed infrared spectroscopy to prove that the reaction occurs via 1,2-cycloaddition to form Si–N and Si–C bonds, very similarly to the reaction of phenyl isothiocyanate on Si(100)-(2×1).

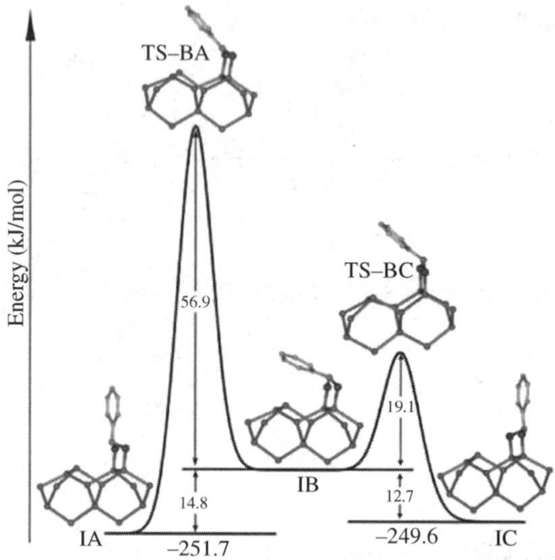

FIGURE 6.17 Proposed reaction pathway leading from structure IB (with the phenyl group nearly parallel to the surface) to either structure IA or structure IC with the phenyl group nearly perpendicular to the surface. Figure reprinted with permission from Ref. 146. Copyright 2009 American Chemical Society.

Despite great reaction selectivity and exceptional stability of the products of interaction of both nitrobenzene [145, 146] and phenyl isothiocyanate [147], the nature of the chemical reactions delivering the aromatic functionality to the silicon surface in both cases implies that there is a potential for further transformations of the initial adducts in both cases. Specifically, the presence of sulfur or oxygen near the silicon surface can alter the electronic properties of the produced interface tremendously. Thus, it would be desirable to have the high selectivity of this cycloaddition chemistry without the risk of interface contamination. One of the ways around this problem was offered by Barriocanal and Doren [142]. The general trends in 1,3-cycloaddition chemistry on Si(100)-(2×1) were considered for azido groups, $–N_3$, which could undergo this process to form a five-membered ring of the structure Si–N–N=N–Si, as shown in Fig. 6.16c for phenylazide. Interestingly, regardless of the exact surface reaction, 1,2- or 1,3-cycloaddition, if nitrogen could be eliminated in a thermal reaction, both pathways would lead to the formation of identical stable three-membered ring structures Si–N–Si. Despite the fact that phenylazide is explosive under certain experimental conditions, a successful study of this compound and its comparison with another azide homologue, benzylazide ($Ph–CH_2–N_3$), which is more stable, suggest that both 1,2- and 1,3- cycloadditions likely occur for these compounds [151, 152]. Leftwich and Teplyakov compared the chemistry of phenylazide and benzylazide using a combination of thermal desorption, MIR–FTIR, and DFT and found that 1,3-cycloaddition is favored in both cases but 1,2-addition was observed spectroscopically for benzylazide and not for phenylazide. The infrared spectroscopy experiments summarized in Fig. 6.18 clearly follow the

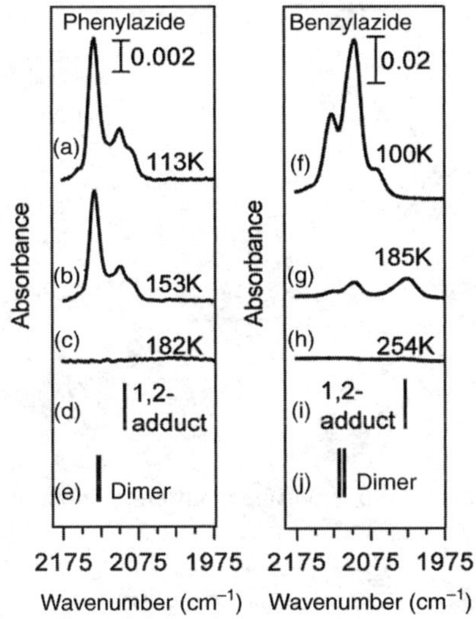

FIGURE 6.18 Infrared studies of phenylazide, left, and benzylazide, right, on Si(100)-(2×1): (a) 5L submonolayer coverage of phenylazide condensed on a Si(100)-(2×1) surface at 113K; (b) same as (a) but briefly annealed to 153K; (c) same as (a) but briefly annealed to 182K; (d) predicted infrared spectrum for a 1,2-adduct of phenylazide on a Si(100)-(2×1) surface; (e) predicted infrared spectrum for a dimer of phenylazide; (f) 10 L submonolayer coverage of benzylazide condensed on a Si(100)-(2×1) surface at 100K; (g) same as (f) but briefly annealed to 185K; (h) same as (f) but briefly annealed to 254K; (i) predicted infrared spectrum for a 1,2-adduct of benzylazide on a Si(100)-(2×1) surface; (j) predicted infrared spectrum for a dimer of benzylazide. The frequencies of all predicted spectra are scaled by 0.95. The spectra corresponding to the adsorbed benzylazide and phenylazide are reported with a deuterium-substituted silicon cluster to avoid interference from Si–H vibrations. Figure reprinted with permission from Ref. 152. Copyright 2009 American Chemical Society.

computationally predicted frequencies for 1,2-cycloaddition processes; however, computational investigation suggested that this process is much less likely in case of phenylazide. Also consistent with the computational predictions, thermal release of nitrogen from benzylazide, presumably from the 1,3-adduct, was found to occur at higher temperature compared to the same process for phenylazide [152]. A recent computational study of reaction of nitrones, including *N*-phenyl nitrone (Ph–N(OH)–CH–CH_2) explored the propensity of the 1,3-cycloaddition on the C(100)-(2×1) surface and found that it is indeed the preferred cycloaddition pathway because of very substantial kinetic barriers for the cycloaddition processes [153].

Although involving multiple chemical functionalities in a formation of a single cyclic product seems to possibly lead to multiple surface binding configurations, sometimes a clever design can produce very well characterized and expected product with exceptional selectivity. The Hamers group used 9,10-phenanthrenequinone on Si(100)-(2×1) [154, 155] in hopes of selectively inducing a [4+2] hetero-Diels–Alder reaction leading to the formation of two Si–O–C linkages, and retaining an

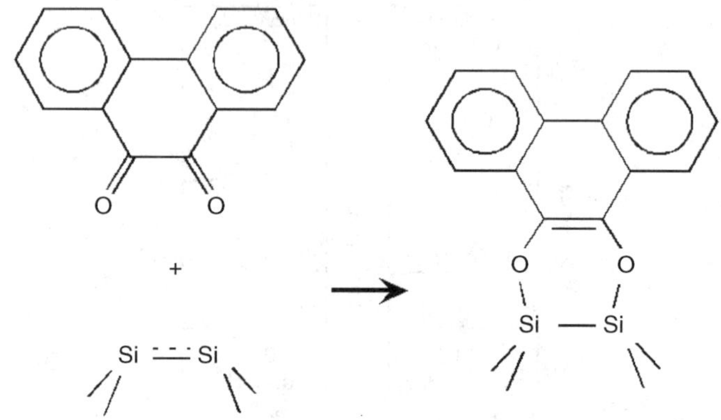

FIGURE 6.19 Reaction of 9,10-phenanthrenequinone with Si(100)-(2×1).

intact aromatic system. This very selective reaction forms a cyclic ring as shown in Fig. 6.19. Stretching vibrations within the 3000–3200 cm^{-1} spectral range observed with FTIR and detailed investigations of O 1s and C 1s core levels confirmed the presence of an intact aromatic ring and the Si–O–C bridges [154]. The STM studies of this system summarized in Fig. 6.20 show clear nodal structure corresponding to a single molecule of 9,10-phenanthrenequinone chemisorbed on Si (100)-(2×1) [154].

6.2.3 Heteroaromatics: Aromaticity as a Driving Force in Surface Processes

In most cases, the reactions of heteroaromatic compounds on clean semiconductor surfaces are rather complex and understandably so, as multiple reaction pathways are generally available. This is why there is a very limited effort to use these reactions for *attaching* target groups to semiconductors. However, aromaticity is a very important factor to consider while designing surface modification schemes because very often transformations that require breaking up an aromatic system are kinetically slow and at the same time if a stable aromatic system is created as a result of surface transformation, this could be used as a driving force in increasing selectivity of surface processes. Chapter 5 in this book details the binding of five- and six-membered aromatic molecules on semiconductors and this section will only briefly consider the influence of aromaticity on the reactions that either start with an aromatic system or whose resulting product is an aromatic system. Thus, the effect of aromaticity on the selectivity of the surface processes will be considered from the opposite points of view. Most of the compounds discussed in this section are shown in Fig. 6.21.

A recent review by Tao et al. [156] has considered electronic and structural factors determining the reactions of various heteroaromatics on silicon and germanium. One of the most studied classes of heteroaromatic compounds is derivatives of pyrrole. The rich multiple functionalities of pyrrole imply the variety of possible surface reactions on Si(100)-(2×1). However, as illustrated in Fig. 6.22, the dominant

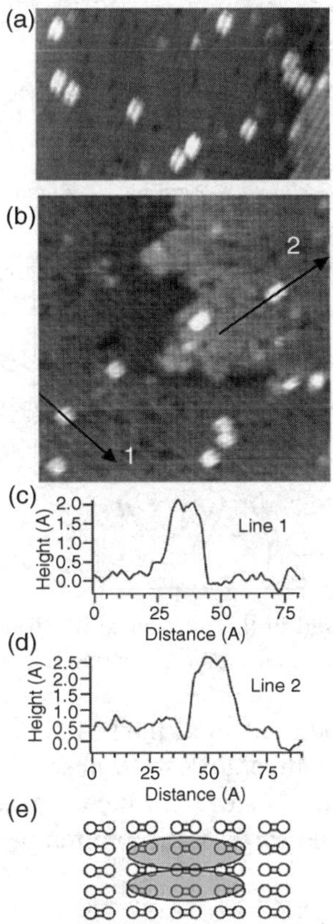

FIGURE 6.20 (a) STM image of Si(001) surface after it was exposed to 0.03 L 9,10-phenanthrenequinone. Molecular features are observed on one surface domain. (b) STM image obtained from another Si(001) surface exposed to 0.03 L 9,10-phenanthrenequinone. Molecular features are observed on two adjacent surface domains. (c) Height profile along line labeled "1" in (b). (d) Height profile along line labeled "2" in (b). (e) Schematic illustration showing the bonding location of the observed features with respect to the underlying surface reconstruction, based upon high-resolution images. Figure reprinted with permission from Ref. 154. Copyright 2009 Elsevier B.V.

pathway for the chemical reaction of this molecule (a) with the silicon dimers of the Si(100)-(2×1) surface is the one that retains the aromatic stabilization, leading to the N–H dissociation and the formation of product (b) [89, 157–160]. Further reaction following N–H dissociation involves C–H dissociation shown as (c), as was confirmed by the studies of deuterium-substituted pyrrole [89]. Thus, the retention of aromaticity dictates the selectivity of the reaction. At the same time, the reaction of a nonaromatic compound, 3-pyrroline (d) with the same surface is driven toward attaining aromatic character by the surface product. Consistent with the computational studies [161], a mixture of possible adducts is formed initially in this case. Most notably, N–H dissociation (e) competes with [2+2] cycloaddition of the C=C

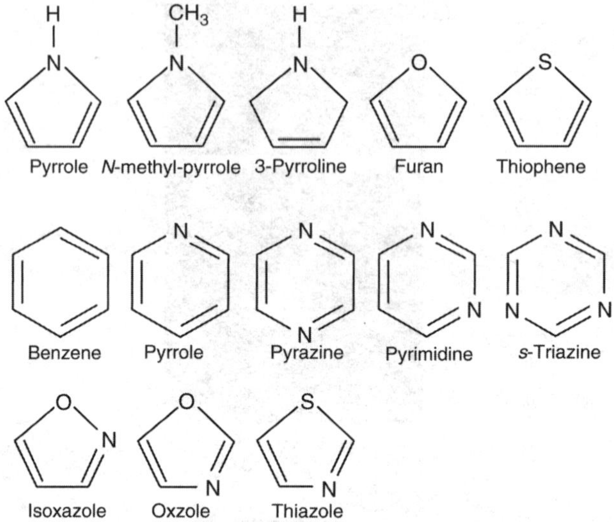

FIGURE 6.21 Compounds used in this section to illustrate the effect of aromaticity as a driving force for reactions of heteroaromatic compounds with silicon.

bond (g) [89]. However, if the product of the [2+2] pathway was suggested to lose extra hydrogen even at room temperature to form species (h), surface annealing past 550K leads to the formation of species (f), which is exactly the same as the result of pyrrole reaction with the same surface at room temperature [89].

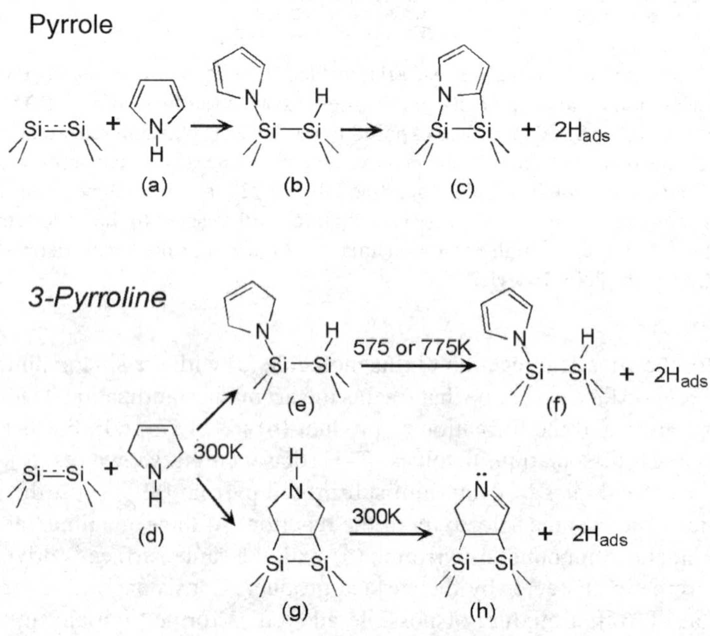

FIGURE 6.22 Reactions of pyrrole and 3-pyrroline with Si(100)-(2×1).

Similarly to the reaction on Si(100)-(2×1), pyrrole also dissociates on Si(111)-(2×1), as was confirmed by following the disappearance of the N–H stretching at 3384 cm^{-1} upon adsorption at room temperature in a set of vibrational spectroscopy studies [162]. This process occurred concurrently with the appearance of the vibrational signatures corresponding to Si–N and Si–H. However, if N–H functionality was replaced by C–CH_3 in *N*-methylpyrrole, [2+2] and [4+2] cycloaddition products dominated the surface reaction both on Si(100)-(2×1) and on Si(111)-(7×7) [163], as the C–N bond has a very high dissociation barrier compared to that for N–H.

The parallel between pyrrole and *N*-methylpyrrole suggests that the suppression of the N–H dissociation channel should lead to the cycloaddition processes. Thus, in the absence of the possibility for the dissociative addition available for pyrrole and pyrroline, other heterocycles, such as furan and thiophene, predominantly react with the Si(100)-(2×1) via [4+2] cycloaddition as shown in Fig. 6.23.

A set of detailed studies by Qiao et al. used HREELS to detect the conversion of sp^2 carbon into sp^3 for both furan [164] and thiophene [165] on Si(100)-(2×1) and also observed peaks at 284.7 and 285.5 eV in the C 1s XPS spectra corresponding to α- and β-carbon atoms, respectively. The singular features at 532.3 eV for O 1s in furan [164] and at 227.4 eV for S 2s in thiophene [165] observed following the chemisorption of these compounds at room temperature agree with the structure proposed for the products of [4+2] cycloaddition reactions for both compounds. Thermal annealing causes further transformation of the adsorbed species including oxygen (or sulfur) migration to the neighboring silicon surface dimer [166–169]. Lee et al. confirmed the [4+2] addition of furan on Si(100)-(2×1) by NEXAFS

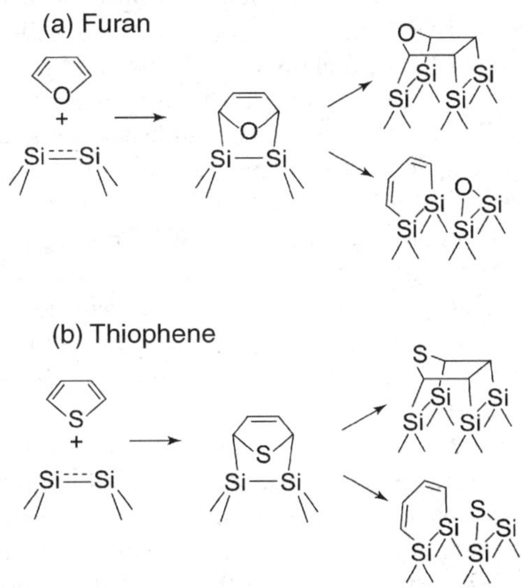

FIGURE 6.23 General reaction pathways of (a) furan and (b) thiophene with Si(100)-(2×1).

investigation [170]. Recent STM studies combined with DFT analysis confirmed the presence of the di-σ-bonded structure for thiophene that further converts to more tightly bonded structures even at room temperature [171]. It should be pointed out that on Si(111)-(7×7), where there is a higher propensity for the [4+2] cycloadditions, the aromaticity of furan is also broken. However, in addition to the expected cycloaddition reactions, thermal desorption and HREELS studies also showed that at high coverages, there is a formation of *mono*-σ-bonded species resting on two adjacent adatoms, that can dimerize via α-carbon coupling [172]. This type of chemical transformation is impossible on Si(100)-(2×1).

It can be mentioned here that the reaction of maleic anhydride could potentially produce an aromatic surface adduct structurally similar to furan; however, a set of detailed studies by the Richardson group employed HREELS, NEXAFS, photoluminescence, STM, and computational investigations to show that this seemingly complex compound reacts exclusively by its C=C function [173–176]. [2+2] cycloaddition pathways dominate at low coverage and involve either one surface silicon dimer or two dimers from the neighboring rows, while the high coverage provides an opportunity for C–H dissociation. Both regimes retain the –CO–O–CO– entity. Nor does cyclopentadiene form a potential aromatic adduct on Si(100)-(2×1), as it initially undergoes almost exclusively a [4+2] cycloaddition [177] similarly to furan and thiophene.

Although the formation of a datively bonded surface intermediate between heteroaromatic compounds and the silicon surface is often considered to be the starting point for many computational investigations, it is sometimes difficult to confirm this state experimentally. However, in case of nitrogen-containing aromatics, it is indeed the case. As was mentioned above, the existence of the N-datively bonded surface intermediate was shown experimentally for nitrozobenzene with the help of infrared spectroscopy and DFT calculations [133]. For pyridine, extensive studies both on Si(100)-(2×1) [178–180] and on Si(111)-(7×7) [181] confirm the existence of the intermediate datively bonded through the nitrogen atom, along with the [4+2] cycloaddition product. The datively bonded adduct undergoes cycloaddition or desorbs if the surface temperature is increased. The [4+2] cycloaddition products were shown theoretically to further convert to more stable configurations involving neighboring silicon dimers on Si(100)-(2×1) [180]. This suggestion was confirmed experimentally by surface differential reflectance spectroscopy studies [182, 183] and by a detailed temperature dependent molecularly resolved STM investigation by Miwa et al. [179]. Further heating above room temperature causes pyridine to undergo several complex reactions including dissociation, fragmentation, and dissociative desorption [184].

The analysis by Lu and Lin [169] compared the behavior of pyridine, pyrazine, and *s*-triazine on Si(100)-(2×1) and placed the chemistry observed in a context of general understanding of chemical transformations of different heterocycles on this surface. Selected surface transformations available for these compounds are shown in Fig. 6.24. The DFT investigation suggests that all three molecules could form datively bonded intermediates via Si–N linkage [185]. In the case of pyridine, as suggested above, [4+2] cycloaddition follows the formation of the datively bonded

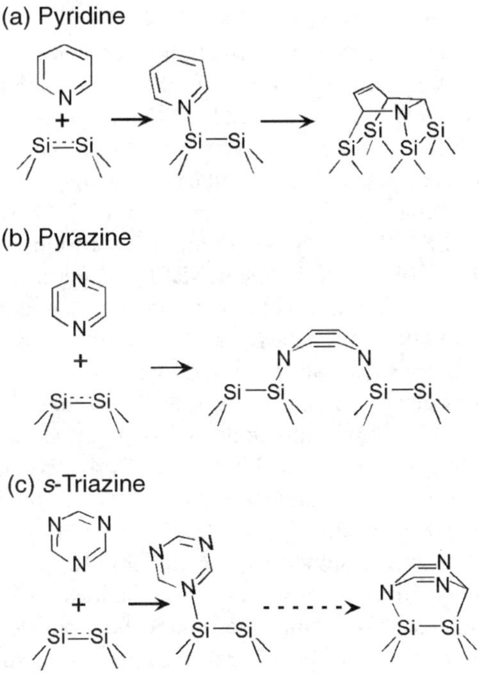

FIGURE 6.24 Reactions of (a) pyridine, (b) pyrazine, and (c) *s*-triazine with Si(100)-(2×1).

intermediate. Because of the symmetry of the molecule, this cycloaddition process can either involve only the formation of Si–C bonds or nitrogen atom can participate in the [4+2] scheme, which seems to be the dominant process, as would be expected considering that the reaction starts with the datively bonded intermediate. Compared to pyridine, pyrazine has two nitrogen atoms within a six-membered aromatic structure and its [4+2] cycloaddition that follows the formation of the datively bonded adduct could form either a product involving a single silicon dimer or neighboring silicon dimers (as shown in Fig. 6.24) of the Si(100)-(2×1) surface [185–187] or involve the adatom–rest atom pair of the Si(111)-(7×7) surface [188]. Theoretical studies initially suggested that the [4+2] adduct that forms two Si–C σ bonds would be favorable [169, 185]; however, experimental investigations involving HREELS and XPS confirmed binding via *para*-nitrogens [189]. If instead of *para*-heteroatoms, two nitrogen atoms are located at *meta*-positions, as in pyrimidine, [4+2] cycloaddition on Si(111)-(7×7) involves a carbon atom and a nitrogen atom in a *para*-position with respect to it, both reacting with the adatom–rest atom pair [190]. Out of the two possibilities for this [4+2] cycloaddition processes, N bonded to the adatom, and N bonded to the rest atom, the product with the $N–Si_{adatom}$ bond is more stable by more than 7.5 eV and is indeed the most stable of all the products considered. It is likely that the kinetics also favors this cycloaddition pathway since its starting point is the original surface adduct datively bonded via nitrogen to the adatom of the Si(111)-(7×7) surface. Neither pyrimidine nor *s*-triazine were confirmed to form the datively bonded adducts as intermediates on the way to further transformations; however, for these compounds the dative bonding is weaker than that for pyridine on Si(111)-(7×7) by 4.0 and 21.6 kcal/mol, respectively [190]. Finally, only one possible [4+2] bonding configuration is available for *s*-triazine because of the symmetry of this molecule. In a set of studies by Bu et al. [191], the behavior of *s*-triazine on Si(100)-(2×1) was investigated by a combination of HREELS, UPS, XPS, and thermal desorption. Since the asymmetric ring-breathing mode at 145 meV in the HREELS spectra above 200K was not observed, it was suggested that at this temperature *s*-pyrazine molecules lie flat on this surface. A weakly bound species was found to desorb at 235K from the monolayer, while decomposition was observed by 550K [191]. A more recent study of substituted triazines also combined several analytical techniques and suggested that a saturation coverage of one molecule per two surface silicon dimers of the Si(100)-(2×1) sample can be reached [192]. A derivative of *s*-triazine, 2,4,6-triphenyl-1,3,5-triazine was also studied on Si(111)-(7×7) by STM and suggested to exhibit nondissociative adsorption [193]. Despite the fact that the specific nature of the bound configuration was disputed by Lu and Lin [169], this microscopy study seems to be consistent with general behavior of triazines on silicon.

Thus, in six-membered N-containing aromatic heterocycles the aromaticity is broken in a thermodynamic regime, where the most stable surface structures are formed; however, kinetic trapping at low temperatures can be used to deliver a datively bonded aromatic ring to the silicon surface. The comparison of the computational investigations suggests that this kinetic trapping could work best for pyridine [185]. The appearance of the vertically oriented aromatic rings in an STM study of pyridine on a high-index silicon surface, Si(5 5 12)-(2×1) [194, 195], confirms the existence of the datively bonded species in this reaction and also suggests that geometric restriction may play a role in stabilizing aromatic systems delivered to the semiconductor surfaces. Finally, it should be noted that Ge(100)-(2×1) substrate that has similar structure and often follows the same reactions with the hydrocarbons as Si(100)-(2×1), reacts with pyridine very differently. In this case the datively bonded state is formed almost exclusively leading to a stable $c(4 \times 2)$ pattern observed in STM [196] and consistent with the instability of the two-dimer bridge structures on Ge(100)-(2×1) predicted by DFT [180]. Thus, in the case of germanium, the aromaticity of the pyridine ring is retained at higher temperature compared to that on silicon. In general, datively bound surface adducts are stable on the Ge(100) surface and often determine the outcome of surface reactions, especially for the thermodynamic regime. These reactions are described in much more detail in Chapters 5 and 8 of this book.

Recent computational exploration of the potential use of pyridine on the selectively hydrogen-covered Si(100) surface shows that this approach also stabilizes the datively bonded configuration and pyridine molecules may form continuous lines of aromatic functionalities on this substrate [197]. More detailed description of the organized delivery of aromatic systems onto the hydrogen-precovered surface is presented below.

A recent review by Tao et al. [156] has analyzed the reaction of one more class of aromatic heterocyclic compounds on silicon. If a five-membered aromatic ring contains two *different* heteroatoms, the effect of the electronic structure on surface adsorption and reactions can be explored systematically. The complete analysis of the electronic effects in the three molecules considered in this review [156] and in the original paper by Tao and Bernasek [59] is based on specific distribution of the molecular electronic density, polarity, geometry, electronegativity of the heteroatoms, and their contribution to the driving of the resulting surface adduct toward the stable "formal" ($4n+2$) π-electron aromatic systems. The reactions of thiazole, oxazole, and isoxazole were followed by HREELS as a function of temperature and compared to those known for other 5- and 6-membered aromatic systems [59]. The most relevant to the material covered in this chapter is the fact that although isoxazole and oxazole form a mixture of surface products even at 100K, thiazole initially forms a stable datively bonded adduct, which later converts into a [4+2]-type product, opening the possibility of kinetic trapping of an aromatic system similarly to that for pyridine.

6.2.4 Chemistry of Aromatic Compounds on Partially Hydrogen-Covered Silicon Surfaces

As was briefly mentioned in the previous section, surface hydrogen may play a stabilizing role in delivering aromatic groups to semiconductor surfaces. The chemistry of fully hydrogenated surfaces is linked directly to this statement as the reactions on fully hydrogen-covered silicon surfaces most of the time start at the defect sites, which can be simply unoccupied silicon atoms. While the chemistry of hydrogen-covered surfaces has been reviewed extensively before, this section will focus on the information that can be obtained from investigating the behavior of aromatic compounds on hydrogen-covered surfaces where selected hydrogen atoms are removed, creating very specific surface sites with well-defined (and often limited) reactivity.

Since the understanding of the reactivity of the specific silicon surface reactive sites "unblocked" from precovered hydrogen literally has to be performed on the atomic and molecular level, this section will rely on the single crystalline silicon surfaces carefully prepared and studied in vacuum by scanning tunneling microscopy. This technique allows both for the atomic resolution and for the removal of a selected hydrogen atom from the H-covered silicon. This idea, proposed in the late 1990s [198, 199], was developed by multiple research groups to become an entirely novel approach to understanding chemical reactions by observing and manipulating single molecules and atoms [198, 200–214]. The reactive features produced on silicon surfaces in an unbelievably controlled fashion provided the opportunities to template the reactive sites on the scale ranging from a single silicon dangling bond to lines of empty dimers on Si(100) to objects of nearly macroscopic sizes [15, 215], and to ask fundamental questions never posed before.

This atomic template approach is illustrated in Fig. 6.25, where the STM tip was used to selectively remove 1, 2, 3, or 4 surface hydrogen atoms to create the reactive sites of a truly atomic pattern [216].

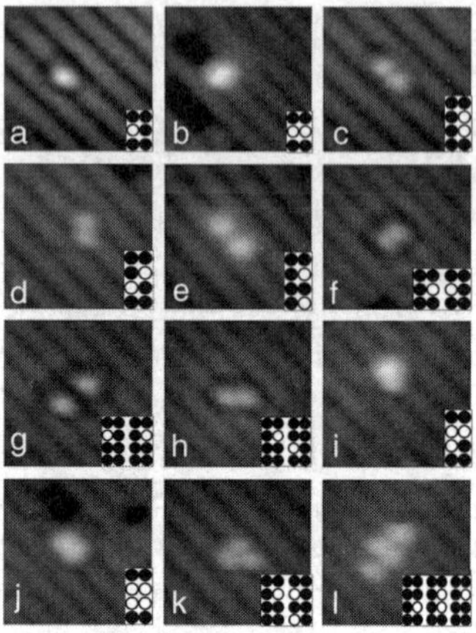

FIGURE 6.25 Shows the detailed configurations of typical elements of the array of dangling bonds fabricated on a H-terminated Si(100) with a current–voltage pulse of +3.5 V/4.0 A, 10 ms. The corresponding sketches of each configuration are plotted inside of each figure; a filled circle denotes a hydrogen-terminated Si atom, an unfilled circle denotes a dangling bond. Figure reprinted with permission from Ref. 216. Copyright 2009 Elsevier B.V.

The atomic templating based on creating well-defined surface reactive sites by the removal of hydrogen atoms blocking specific dangling bonds on silicon has been explored to understand binding and electronic structures of many different molecules, including norbornadiene (NBE), copper phthalocyanine (CuPc), and C_{60} fullerene [217]. For example, in a molecule as complex as CuPc, molecular rotation and charge transfer from the semiconductor substrate have been investigated. Here, we will attempt to summarize some of the rules for manipulating chemical reactions involving aromatic compounds.

As was mentioned above, biphenyl is a very interesting compound with unusual adsorption configuration available on a clean Si(100)-(2×1) surface, with a structure pivoting around one of the phenyl groups at 5K [68, 69, 73]. This is a perfect structure to investigate on a hydrogen-terminated Si(100) surface, if selected hydrogen atoms can be removed with the help of the STM tip. To determine the size limitations of the adsorption site, the experiments were performed on a template with 1, 2, 3, 4 (two dimers), and more adjacent silicon atoms stripped of surface hydrogen. The room temperature adsorption required at least four surface atoms on two adjacent silicon dimers to be cleared and available for binding [218]. This template, described by the authors as a "molecular mold," directed the biphenyl binding into a fixed configuration, on a clean Si(100)-(2×1) surface that was achieved with the help of the STM tip. The effects of the hydrogen-templated artificial surface site on the binding of biphenyl are illustrated in Fig. 6.26, where a chronological sequence of events illustrates binding of this molecule onto a template of three silicon surface dimers stripped of hydrogen protection.

The design of molecular wires relies on the templating and assembling of the appropriate compounds in a highly ordered fashion. Sometimes this can be achieved

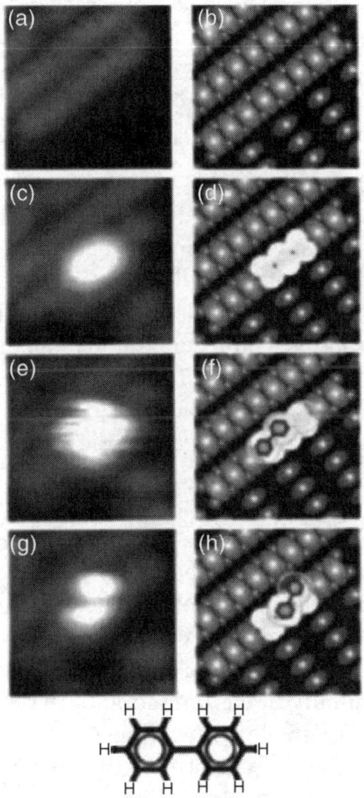

FIGURE 6.26 Chronological sequence of four filled-state STM topographies (each 50 Å×50 Å, −1.5 V, and 0.5 nA) of the same area of the hydrogenated Si(100)-(2×1) surface. The left-hand column shows (a) the clean hydrogenated surface, (c) the surface after fabricating the molecular mold, (e) a moving biphenyl molecule adsorbed in the mold, and (g) the biphenyl molecule is fixed in the mold. These four stages are shown schematically in the right-hand column (b), (d), (f), and (h). Note in (f), only one of the bistable states of the molecule is shown. The structure of the molecule is shown at the bottom. Figure reprinted with permission from Ref. 218. Copyright 2009 American Institute of Physics.

by chemical methods adapted to self-assembly, as described later in this chapter; however, the alternative approach is to design the wires based on the large molecular structures and then "staple" them to the semiconductor surface using appropriate chemistry or an STM tip. Similarly to the concept of using *para*-hexaphenyl molecules described above [77], other molecules can also be used. The main problem is that the more complex molecules tend to also have a very complex set of chemical interactions with clean semiconductor surfaces. For example, 1,4″-paratriphenyl-dimethylacetone, whose structure is presented in Fig. 6.27 adsorbs on a clean Si(100)-(2×1) in a rather well-defined direction. The adsorption of this molecule on a clean Si(100)-(2×1) surface requires five sequential silicon

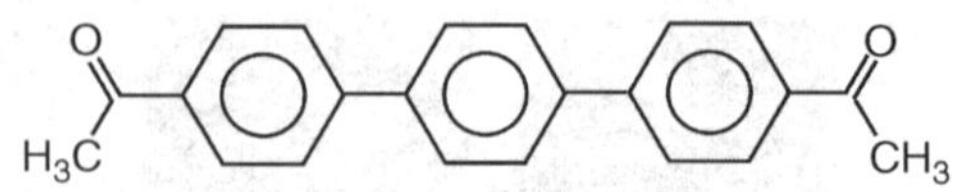

FIGURE 6.27 Structure of 1,4″-paratriphenyl-dimethylacetone.

dimers [210, 218]. However, the specific thermal or STM tip-induced chemistry may be very complex, as was suggested in a recent theoretical investigation [219]. Nevertheless, a combination of chemical design, microscopy studies, computational investigation, and possibly selective blocking of the potential reactive sites with hydrogen by creating appropriate "molecular molds" should ultimately lead to well-controlled electronic behavior. The applications may not just be limited to the molecular wires but could also extend to other fields, such as photonics, as a photochromic behavior of a complex *cis*-1,2-dicyano-1,2-bis(2,4,5-trimethyl-3-thienyl)ethene (CMTE) molecule was also demonstrated on Si(100) with the help of STM [219a].

The examples above showed that some of the ensemble effect limitations can be studied and understood directly with the help of STM, especially when it is obvious that more than one surface dangling bond is needed for the bonding to occur. However, some reactions could potentially need only one available surface dangling bond. For example, as was discussed above for pyridine, there is only one surface dangling bond involved initially in the formation of the Si–N linkage and later this structure converts into more stable adducts on the Si(100)-(2×1) surface. Thus, it could be expected that the individual dangling bonds available as defects on H-covered Si(100) should readily bind pyridine as well as prevent its conversion by cycloaddition pathways, leaving an intact aromatic ring available for investigations or further modification of this substrate. Miwa et al. posed exactly this question in their STM studies [179]. However, as shown in Fig. 6.28, pyridine did not bind to readily available isolated dangling bonds on H-covered Si(100). Instead, it selectively formed the Si–N linkages with the surface sites where both hydrogen atoms of the same silicon surface dimer were removed. The authors explained this phenomena by the electron-accepting role of the "down" atom of the silicon surface dimer, while an individual surface dangling bond on a silicon surface could not possibly act in this manner [179]. On the other hand, the adsorption of an iron-based porphyrin heme(b) on the H-covered Si(111) studied computationally suggested only weak adsorption on the ideal surface; however, the removal of a single hydrogen atom to create a dangling bond increased the binding energy to 0.543 eV and a pronounced interaction between the dangling bond of the silicon substrate and the orbitals of the iron atom in the porphyrin frame was noted [220].

Starting with an individual dangling bond on otherwise ideally H-terminated silicon, the entire surface could be modified. The complete analysis of the STM manipulation of the molecular systems on silicon is described in detail in Chapter 10 and wet chemical methods to create organic layers on semiconductors are highlighted in Chapters 11 and 12, so the next few paragraphs will only focus on the specifics of the delivery and assembly of aromatic structures by these approaches.

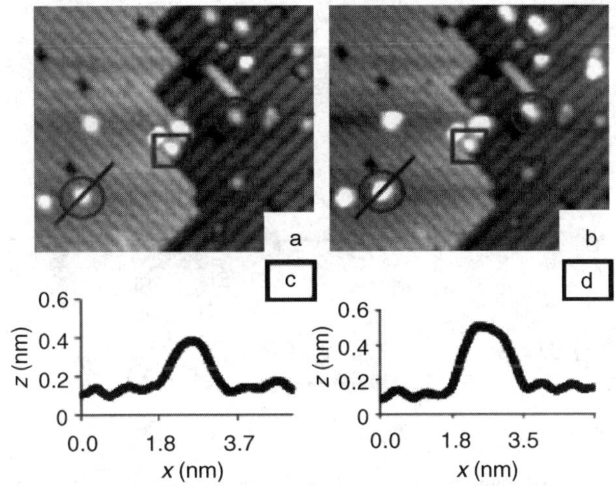

FIGURE 6.28 (a) STM image (17 nm×17 nm, −2.1 V, 26 pA) of a hydrogen-terminated Si(100) surface with a small number of reactive sites. Examples of clean dimers (circles) and single dangling bonds (squares) are indicated. (b) Shows the same area after dosing the surface with pyridine. The molecules only adsorb to clean dimers and appear as asymmetric protrusions across a single dimer. Cross sections through the clean dimer before (c) and after (d) the reaction are also shown. Figure reprinted with permission from Ref. 179. Copyright 2009 American Chemical Society.

On a well-defined H-terminated Si(111) surface that can be prepared by wet methods and on H-terminated Si(100) that can be prepared by controlled-temperature reaction of hydrogen molecules dissociated on a hot tungsten filament with a clean Si(100)-(2×1) surface, the predominant surface site is a monohydride Si–H. The modification approach suggested in the early 1990s by the Chidsey group used the chain reaction illustrated in Fig. 6.29 [221, 222] to build self-assembled monolayers by chain reaction involving an alkene function.

This process can be initiated at a dangling bond that is either naturally available on the surface as a defect or induced by radical precursors or UV exposure [223, 224]. The reaction scheme is similar to what is known for organosilane chemistry [225]. What is most relevant to the material covered in this chapter is whether aromatic functionality could be introduced onto a well-defined semiconductor by such an approach. It was indeed the fact, as a stable, cheap, and easy to manipulate and monitor styrene molecule reacted with H-terminated Si(100) or Si(111) as was followed by STM. The studies by Cicero et al. [226] presented in Fig. 6.30 follow the formation of exactly one styrene molecule high islands on a H-terminated Si(111) and confirmed that the growth of each of these islands started at an isolated silicon dangling bond.

The functionality of the aromatic compounds used to grow these islands can be altered proving the robustness of this surface functionalization approach. For example, a monolayer of 4-bromostyrene was also grown and characterized on Si(111) [227, 228].

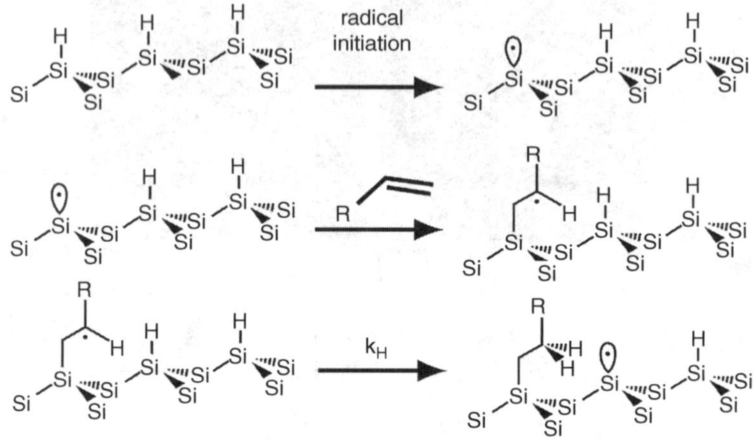

FIGURE 6.29 The mechanism of radical chain reaction of alkenes with a H-terminated Si(111) surface. Figure based on Fig. 7 from Ref. [13].

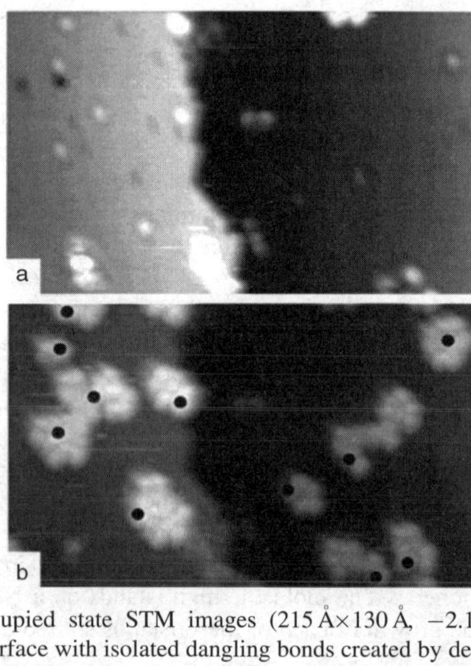

FIGURE 6.30 Occupied state STM images (215 Å×130 Å, −2.1 V, 44 pA) of an H-terminated Si(111) surface with isolated dangling bonds created by desorption activated with the STM tip (a) before dosing with styrene and (b) after exposure to 12 Langmuirs of styrene. The black dots in (b) mark the positions of the initial dangling bonds, showing that these sites serve to nucleate the growth of styrene islands. The gray scale ranges in the two images are different; the maximum height range in (a) is ~5 Å and ~10 Å in (b). Figure reprinted with permission from Ref. 226. Copyright 2009 American Chemical Society.

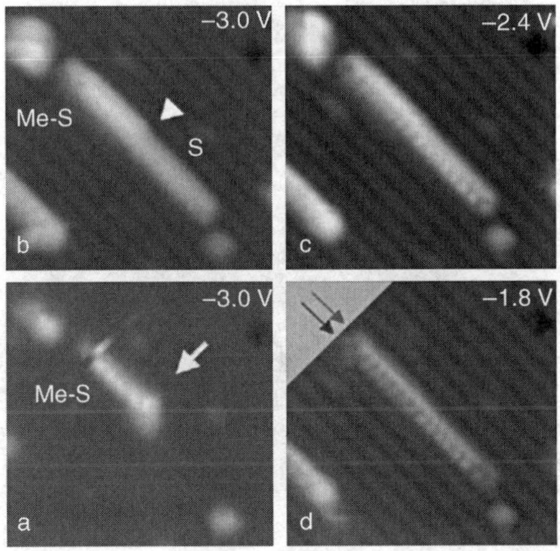

FIGURE 6.31 Constant current filled-state STM imaging of a methylstyrene/styrene heterostructure on the H:Si(100) surface. (a) Growth of the first line segment of methylstyrene (8 L exposure). The chemically reactive silicon radical that initiated the line growth (not shown) is now located at the end of the CH_3-styrene segment (indicated by white arrow). This silicon radical will serve as the initiation site for the growth of the second line segment. (b) Following a 78 L exposure of styrene, the methylstyrene segment in panel (a) has been extended to form a molecular heterostructure. The white wedge indicates the location of the heterointerface. At $V_s = -3.0$ V, the methylstyrene line segment images higher than the styrene segment. (c) $V_s = -2.4$ V, the methylstyrene and styrene line segments image with similar height. (d) $V_s = -1.8$ V, the methylstyrene and styrene line segments reveal different molecular contrast. Images were acquired using a constant tunnel current of 60 pA. Image areas: ~$9 \times 9\,nm^2$ [229]. Figure reprinted with permission from Ref. 229. Copyright 2009 The American Physical Society.

A brief examination of the chemistry of this growth process suggests that as the styrene molecule starts the reaction at an individual dangling bond, it can only create the required surface-bound Ph–CH_2–CH_2– group by plucking a hydrogen atom from a neighboring silicon site. However, on Si(111) there are six directions of propagation for such a process so the shapes of the islands created lack specific directionality. On Si(100) covered with monohydride, the directionality of the silicon surface dimer rows implies the possibility of growing molecular lines along the direction of these rows, which is exactly the case. The STM image of a molecular line and a junction produced when the dosing compound is switched from styrene to methylstyrene is presented in Fig. 6.31 and the molecular model is given in Fig. 6.32 [229]. The fact that the compound delivered onto the silicon surface as a molecular line is aromatic triggered a very substantial interest in this approach as the overlap of aromatic π-systems gives this molecular line a possible attribute of a conducting molecular wire [226, 229–242]. Different methods of initiation of this reaction, and various measurements relevant to the potential applications of these molecular lines as electron

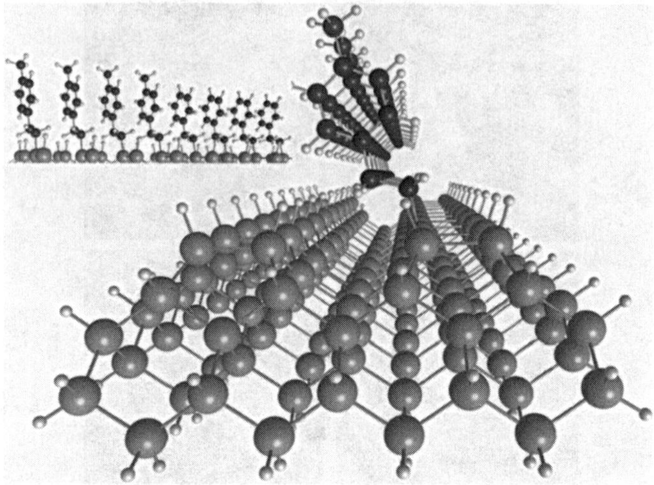

FIGURE 6.32 Model structure of a heterojunction between methylstyrene and styrene chains (each of four molecules) on a hydrogen-passivated silicon cluster representing a dimerized H-terminated Si(001) surface. Carbon atoms are black, silicon is gray, hydrogen is white. Inset: Another view of the molecules and nearby substrate atoms. Figure reprinted with permission from Ref. 229. Copyright 2009 The American Physical Society.

conductors on different substrates have been reported [126, 227, 238, 243–256]. One of the possible initiators, a stable radical species TEMPO, could actually serve both as an initiator of this reaction by removing a single hydrogen atom from H-precovered silicon, and as an inhibitor by passivating available dangling bonds [237, 257].

In the context of potential future applications it should be pointed out that the specific electronic structure of the wires produced may be complex and the aromatic rings, especially at the end of the lines could be fluxional [258]. In addition, the silicon substrate plays a very important role in the electron transport phenomena using molecular ethylbenzene wires [259]. One of the plausible pathways to decouple the properties of the silicon substrate from the electronic conductance through the π-stacked aromatics was offered in a theoretical study by Rochefort and Beausoleil [260], who considered π-stacking in a molecular line of 1-hexyl-naphthalene molecules on Si(100)-(2$\times$1). Here, the stack of aromatic molecules is separated from the silicon surface by a hexyl linker; however, experimental study of this specific system may be difficult. Considering all these factors in a combination with designing more functional molecular lines should ultimately lead to the design of a functional device.

The mechanism of the surface chain reaction proposed for styrene proved to be quite robust. In addition to adding a functionalized junction with methylstyrene [229] and building the self-aligned structures from bromostyrene [227], dimethylstyrene was also shown to react with the dangling bond on the Si(100)-(2$\times$1)-H surface similarly to styrene [233]. However, some structurally similar molecules, including

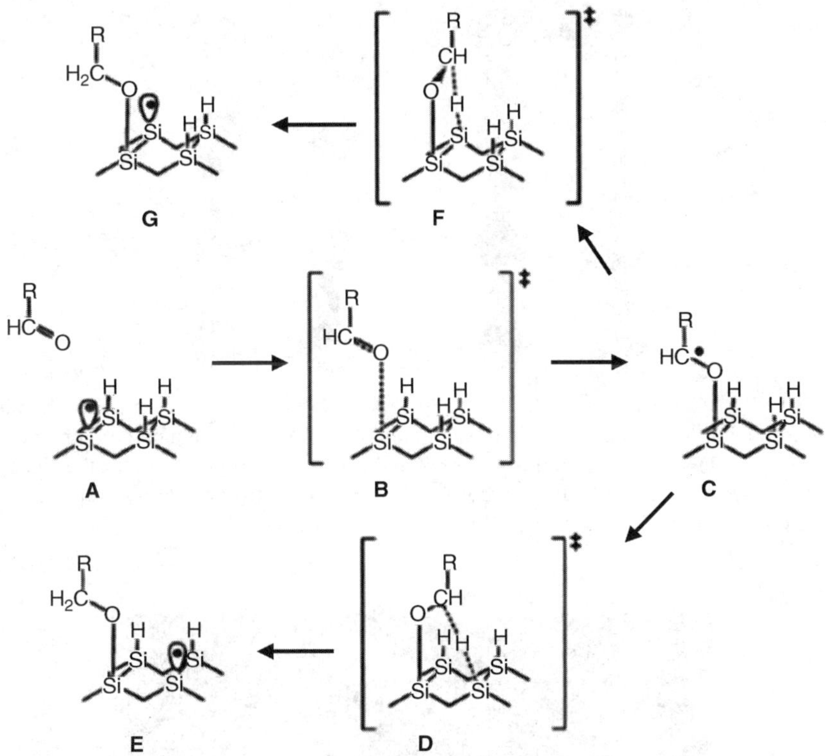

FIGURE 6.33 Reaction mechanisms for the self-directed growth of aldehyde lines on H-Si (100)-(2 × 1). For benzaldehyde, R = C_6H_5, and for acetaldehyde, R = CH_3. The calculated C· · ·H distances in the transition-state structures are 1.65–1.70 Å. Figure reprinted with permission from Ref. 252. Copyright 2009 American Chemical Society.

the aromatic 4-vinylpyridine did not show any observable line growth [233]. On the other hand, functionalized aromatics such as benzonitrile [248] and benzaldehyde [252] follow the same overall chemistry as styrene on an H-terminated Si(100)-(2×1) surface. In the case of benzaldehyde [252], the chemical functionality actually provides two alternative possibilities for the reaction leading to the formation of the surface species of a general formula Ph–O–CH_2–. As illustrated in Fig. 6.33 for an aldehyde of a general formula R–C=O, the attachment to an individual dangling bond A on an ideally H-terminated Si(100)-(2×1) surface leads to the formation of intermediate C, which in turn can abstract a hydrogen atom from the same dimer (F-G) or from the neighboring dimer (E-D), potentially causing two different mechanisms for the growth of the molecular chain, as depicted in Fig. 6.34 [252]. The STM studies of benzaldehyde presented in Fig. 6.35 show the existence of both single and double lines, including the transformation of a single line into a double line, confirming the possibility of these mechanisms [252].

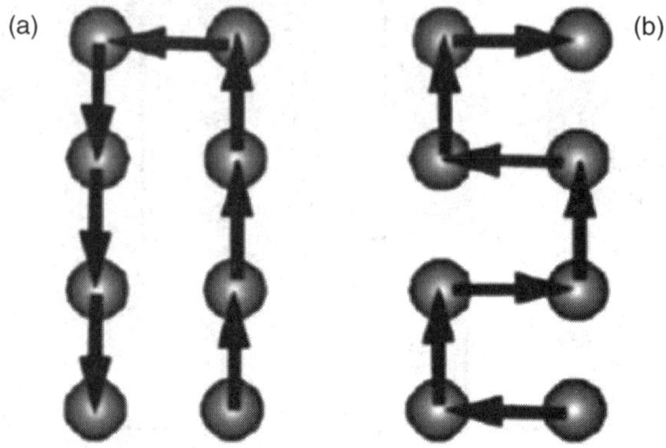

FIGURE 6.34 Cartoon representing (a) antiparallel and (b) square wave growth mechanisms for the formation of double lines. Horizontal pairs of circles represent a silicon dimer on the H-Si(100)-(2 × 1) surface. The tail of the arrow indicates the adsorption site for the aldehyde, while the head of the arrow indicates the site of hydrogen abstraction. Figure reprinted with permission from Ref. 252. Copyright 2009 American Chemical Society.

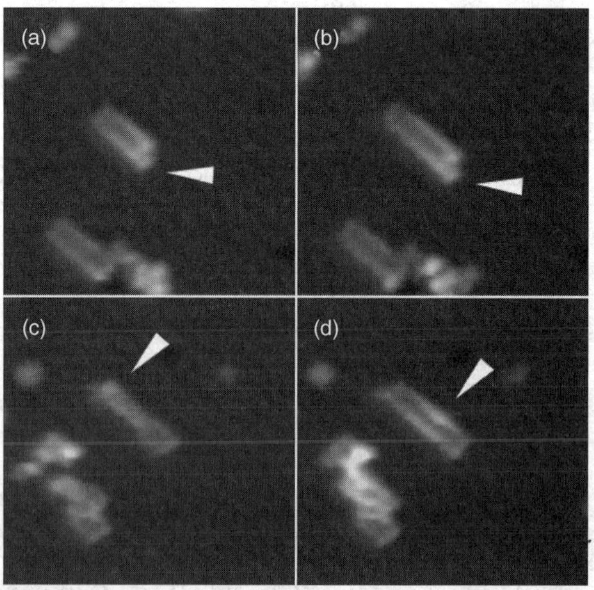

FIGURE 6.35 STM images demonstrating the mechanisms for aldehyde line growth. (a) Benzaldehyde double line that has a dangling bond at its end extends its growth (b) via the square wave mechanism upon additional exposure to benzaldehyde. (c) Nanostructure composed of double and single line portions with a dangling bond at the end of the single line. Further exposure to benzaldehyde leads to the formation of a complete double line through the antiparallel growth mechanism (d). Wedges indicate the location of the dangling bonds. STM images: 12 nm × 12 nm, −3.5 V, 0.1 nA. Figure reprinted with permission from Ref. 252. Copyright 2009 American Chemical Society.

The continuing STM studies proved that the chain reaction of H-terminated Si(100) with styrene can be reversed at elevated temperatures [261] and also explored the possibility of cross-linking neighboring silicon dimer rows with chemical functionalities [233, 262, 263]; however, the linking with aromatic compounds to attempt to bridge the neighboring rows of silicon dimers electronically has not been attempted thus far. To conclude this section, it can be summarized that despite a very rich chemistry for the aromatic molecules on semiconductor surfaces, decreasing the ensemble of surface sites available for the interaction of the aromatics with clean semiconductor surfaces helps to retain aromaticity in surface functionalization processes, providing the means to use the aromatic groups both as stable linkers and as building blocks for molecular electronic structures and future devices.

6.2.5 Delivery of Aromatic Groups onto a Fully Hydrogen Covered Silicon Surface

6.2.5.1 Hydrosilylation Chapters 11 and 12 will describe all the subtle points of using wet chemistry methods for hydrosilylation of H-terminated silicon and germanium surfaces and for chemical modification of the Cl-terminated surfaces that can be prepared starting with H-terminated substrates. Because of stability, wide variety of produced functionalities, cheapness and attractive electronic properties, aromatic linkers are often used in these reactions. Other chapters in this book will delineate the mechanistic details of these general processes. Here it will be sufficient to say that in all those reactions aryl groups are delivered to the surface intact and do not participate in surface bonding. They are mostly used as stable linkers. The potential to use these linkers for applications other than just the *para*-position functionality does exist. However, it should be pointed out that most of the time the monolayers created by these methods, despite making the silicon surface very resistive to oxidation in ambient conditions, do not form dense monolayers. Coverages of only about 20–30% are found to be reached at full saturation, meaning that there are remaining Si–H or Si–Cl species available for further reactions. Nevertheless, recent advances in alkylation procedures can safely deliver phenyl- [264] and benzyl- [265] groups by chlorination/alkylation protocol, *para*-functionalized aryls in diazonium reactions starting with tetrafluoroborates [266], and sterically hindered aromatics also by diazonium chemistry [267]. A photoinduced reaction of Cl-terminated layers on Si(111) could be performed with benzonitrile [268]. However, in addition to the fact that not every reactive surface site can be modified with a desired group, it should also be pointed out that the need to use solvents in the wet chemistry preparation schemes often plays a very substantial role in determining surface modification reaction. For example, silicon hydrosilylation of both H-terminated Si(111) and Si(100) with the use of Spier catalyst, $H_2[PtCl_6]$, leads exclusively to the alcohol solvent grafting instead of the intended attachment of the complex "indene ligand" (3-[2′(1*H*-inden-3″-yl)ethyl]-5-(4‴-vinylphenyl)-1*H*-indene) containing aromatic moieties. Only UV light initiation and only on the H-terminated Si(111) surface produced the target result [269].

6.2.5.2 Cyclocondensation Despite being commonly used in homogenous chemistry, cyclocondensation processes in the chemistry of H-terminated silicon surfaces, where the surface provides the hydrogen for the elimination process, have only been discovered very recently. As shown in Fig. 6.36, dehydrative cyclocondensation can deliver a nitroso-fragment onto an H-terminated silicon surface. A detailed set of studies compared the reaction of the nitrozobenzene with H-terminated Si(100) and Si(111) and also with the reaction of the nitrozobenzene on a clean Si(100)-(2×1) in vacuum (used as a spectroscopic marker) to confirm that this reaction occurs with a retention of the aromatic ring.

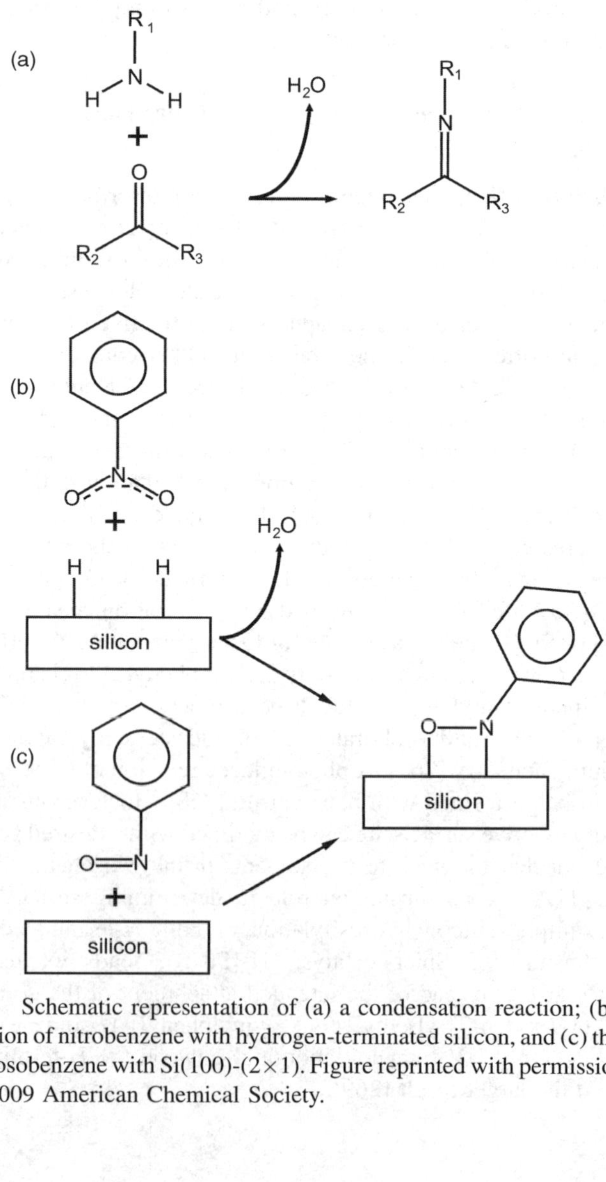

FIGURE 6.36 Schematic representation of (a) a condensation reaction; (b) the cyclocondensation reaction of nitrobenzene with hydrogen-terminated silicon, and (c) the cycloaddition reaction of nitrosobenzene with Si(100)-(2×1). Figure reprinted with permission from Ref. 18. Copyright © 2009 American Chemical Society.

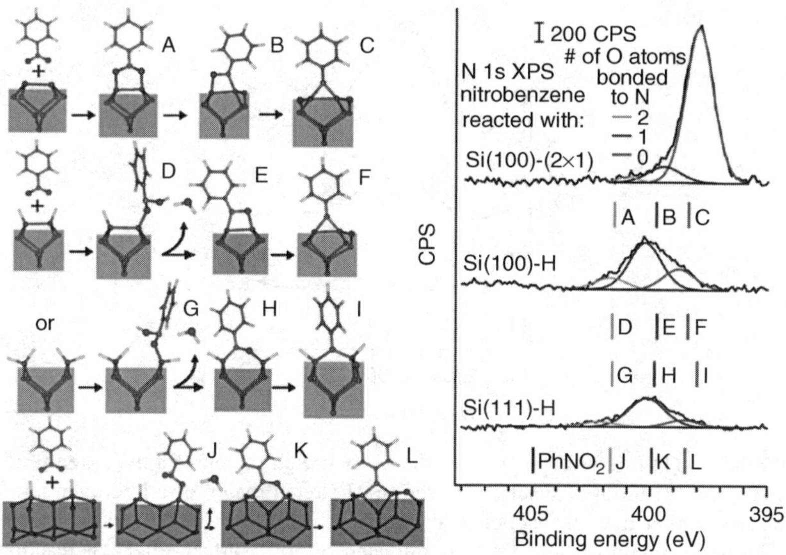

FIGURE 6.37 Comparison of the 1,3-cycloaddition reaction of nitrobenzene with a Si(100)-(2×1) surface with the condensation reactions of nitrobenzene with an ideal monohydride-terminated Si(100) surface, a dihydride-terminated Si(100) surface, and an ideal monohydride-terminated Si(111) surface, descending order on the left. The *ex situ* high-resolution N 1s XPS spectra, with the predicted N 1s binding energy of the corresponding products, after these reactions occur, and nitrobenzene itself, are presented on the right of the respective reaction. Computations were preformed at the B3LYP/6-311+G(d,p) level of theory. Organic fragments are represented as follows: black, oxygen; dark gray, nitrogen; light gray, carbon; white, hydrogen. The nitrobenzene reaction with the Si(100)-(2×1) surface was performed in ultrahigh vacuum as described previously and the XPS spectrum was collected *ex situ*. The condensation reaction was performed under nitrogen *ex situ*. On the schematic, hydrogen atoms terminating the silicon clusters are omitted for clarity. Figure reprinted with permission from Ref. 18. Copyright © 2009 American Chemical Society.

A set of infrared spectroscopy experiments proved the formation of the aromatic adduct on a surface following the reaction and were also used to follow the reduction of the intensity of the Si–H stretching vibration around 2100 cm^{-1} [18]. In addition, as shown in Fig. 6.37, XPS spectra for a variety of possible surface reactions were analyzed with the help of computational methods [18, 270]. Despite a variety of possible reactions with different reactive sites on Si(100) and Si(111) surfaces terminated with hydrogen, these studies suggest that the primary product of the dehydrative cyclocondensation process is nitrosoadduct (E, H, or K in Fig. 6.37) [18]. Further heating of the silicon substrate past the surface modification temperature may drive the oxygen of the nitroso-group subsurface [133] to form a phenylnitrene adduct (C, F, I, L in Fig. 6.37).

Thermal control is one of the most attractive aspects of this reaction. It is slow at room temperature but can be completed within hours at 200°C. A computational investigation was used to follow the possible reaction pathways for the dehydrative

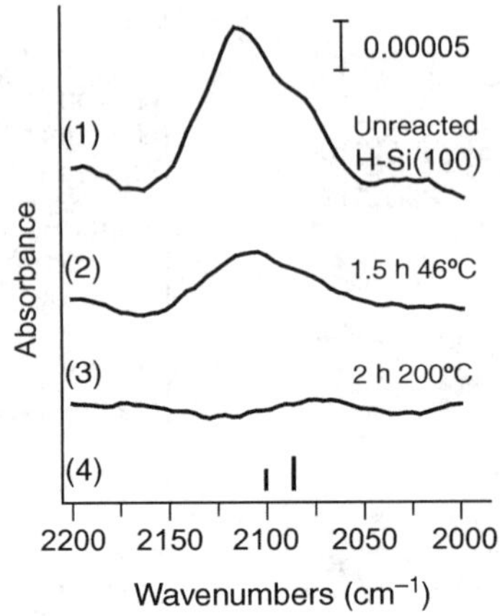

FIGURE 6.38 (1) Infrared spectrum of the Si–H stretch region of a hydrogen-terminated Si(100) surface. (2) Infrared spectrum of the Si–H stretch region of a hydrogen-terminated Si(100) surface after a 1.5 h reaction with nitrobenzene at 46°C. (3) Infrared spectrum of the Si–H stretch region of a hydrogen-terminated Si(100) surface after 2 h reaction with nitrobenzene at 200°C. (4) Predicted infrared spectrum of the Si–H stretch region of structure H. Figure reprinted with permission from Ref. 18. Copyright © 2009 American Chemical Society.

cyclocondensation scheme, and discovered a substantial barrier for the first step of surface modification. The opportunity to use thermal control for surface coverage dependence is illustrated in Fig. 6.38, where the decrease of the intensity of the infrared absorption band corresponding to the Si–H stretch shows that partial coverage can be obtained by varying reaction time and temperature [18].

The use of this dehydrative cyclocondensation technique is amenable for silicon surface functionalization in a controlled manner but the current model also delivers an additional surface oxygen atom. To combat this process, which can be detrimental to the interface formed during the reaction, we have also attempted to use nitrosobenzene as a starting point; however, for the moment, alternative reaction pathways seem to preclude us from creating an oxygen-free interface [271].

6.2.6 Delivery of Aromatic Compounds onto Protected Silicon Substrates

In some cases, covalent binding of aromatic molecules delivered to the semiconductor surface may be undesired. In fact, a chemically inert layer can be employed to prevent this chemisorption. For example, hydrogen passivation may provide this buffer zone with limited chemical interaction, as was suggested for the iron-based porphyrin described above [220]. Of course, in that case the presence of a surface dangling bond at a defect site was found to increase the binding energy substantially. In fact, even an ideally H-terminated silicon surface may bind aromatic compounds very strongly via dispersion interactions, as was found for pentacene [272]. The covalent bonding might also occur by an unintended catalysis, as was proposed in a computational study of phenylacetylene reaction with hydrogenated silicon surfaces in the presence of Lewis acids [273].

In order to avoid these interactions, the silicon surface could be premodified before introducing the aromatic functionality. For example, Wakatsuchi et al. used the reaction of 1,4-cyclohexadiene with a clean Si(100)-(2×1) surface to produce a monolayer that served as a buffer zone to prevent the chemisorption of the subsequently used pentacene, which was allowed to self-assemble in the target surface structures as a result of this modification [274]. Metal passivation of the silicon surface has been demonstrated to promote the diffusion of perylene tetracarboxylic dianhydride (PTCDA) and tetraaminobenzene (TAB) on the Ag/Si (111)-($\sqrt{3}\times\sqrt{3}$) R30° that form stable hydrogen-bonded structures [275] suggesting that the choice of metal may not only promote the chemical interaction between aromatic molecules and silicon, as was the case for Na-promoted adsorption of benzoic acid [92, 93, 102], but also block the chemisorption entirely.

Alternatively, the strong and well-defined bonding on a premodified silicon surface may be desired as the building of molecular structures extends into the 3rd dimension. Some of the building blocks used in this modification approach are based on aromatic systems. For example, Bitzer and Richardson [276, 277] suggested to use the initial surface reaction of 1,4-phenylene diamine followed by imide coupling to build a 3D network in a controllable layer-by-layer method [276, 278]. This approach, bearing close resemblance to the atomic layer deposition chemistry, can be termed molecular layer deposition and is currently being developed for multiple types of reactions. The coupling of amino-terminated monolayers on Si(111) with benzaldehyde was recently used by Eves et al. [279]. Azide reactions with acetylene-terminated monolayers on Si(100) [280] and the attachment of branched substituents to the organic layer on an ester-terminated Si(111) surface [281] have also been performed. More elaborate growth strategies involve covalently linked multiporphyrin-imide architectures on Si(100), as reported by Jiao et al. [282]. One could hope that inclusion of aromatic functionalities into the 3D building approaches will introduce something more than just relatively inert linker into these novel structures, but also bring the electronic structure of these functionalities into play for designing materials with unusual electronic, photochemical, and photochromic properties.

6.3 SUMMARY

This chapter focused on the role of aromaticity as a factor in determining surface reactions on group IV semiconductor surfaces. The retention of aromaticity and the drive toward the formation of an aromatic adduct as a result of chemical reactions leading to surface functionalization most of the time govern the chemical processes in these systems. Thus, the chemistry that has been previously discovered for other organic compounds on silicon can often be applied (although with some degree of caution) to compounds containing aromatic functionality. The chemistry of silicon, germanium, and diamond surfaces can be utilized for these purposes, but while some reactions can be directly replicated on all these substrates, many chemical transformations will depend drastically on the specific surface and specific element. The aromatic groups attached to the semiconductor surface may further serve as the molecular electronic components, structures, and devices, which makes the in-depth study of this field ever so important.

ACKNOWLEDGMENTS

AVT acknowledges financial support from the National Science Foundation (CHE-313803, CHE-0415979, and CHE-1057374) and from the Donors of the American Chemical Society Petroleum Research Fund for making this work possible. I greatly appreciate the help of the following researchers who graciously provided the permission to use their figures in this review: Maria Pia Casaletto, Gérald Dujardin, Göran Hansson, Yukio Hasegawa, Robert J. Hamers, Mark Hersam, Heon Kang, Gregory P. Lopinski, Andrew J. Mayne, Robert A. Wolkow, Qin Xu.

REFERENCES

1. Schleyer, P. v. R. Aromaticity (Editorial), *Chem. Rev.* **2001**, *101*, 1115.
2. Waltenburg, H. N.; Yates, J. T. Jr., *Chem. Rev.* **1995**, *95*, 1589.
3. Yoshinobu, J.; Tanaka, S.; Nishijima, M. *Jpn. J. Appl. Phys.* **1993**, *32*, 1171.
4. Jung, Y.; Shao, Y.; Gordon, M. S.; Doren, D. J.; Head-Gordon, M. *J. Chem. Phys.* **2003**, *119*, 10917.
5. Ono, M.; Kamoshida, A.; Matsuura, N.; Ishikawa, E.; Eguchi, T.; Hasegawa, Y. *Phys. Rev. B* **2003**, *67*, 201306.
6. Yoshinobu, J. *Prog. Surf. Sci.* **2004**, *77*, 37.
7. Leftwich, T. R.; Teplyakov, A. V. *Surf. Sci. Rep.* **2008**, *63*, 1.
8. Tao, F.; Xu, G. Q. *Acc. Chem. Res.* **2004**, *37*, 882.
9. Razado, I. C.; Zhang, H. M.; Uhrberg, R. I. G.; Hansson, G. V. *Phys. Rev. B* **2005**, *71*, 235411.
10. Higashi, G. S.; Chabal, Y. J.; Trucks, G. W.; Raghavachari, K. *Appl. Phys. Lett.* **1990**, *56*, 656.
11. Wade, C. P.; Chidsey, C. E. D. *Appl. Phys. Lett.* **1998**, *71*, 1679.
12. Allongue, P.; de Villeneuve, C. H.; Morin, S.; Boukerroub, R.; Wayner, D. D. M. *Electrochim. Acta* **2000**, *45*, 4591.
13. Wayner, D. D. M.; Wolkow, R. A. *J. Chem. Soc., Perkin Trans.* **2002**, *2*, 23.
14. Baluch, A. S.; Guisinger, N. P.; Basu, R.; Foley, E. T.; Hersam, M. C. *J. Vac. Sci. Technol. A* **2004**, *22*, L1.
15. Mayne, A. J.; Riedel, D.; Comtet, G.; Dujardin, G. *Prog. Surf. Sci.* **2006**, *81*, 1.
16. Faggin, M. F.; Green, S. K.; Clark, I. T.; Queeney, K. T.; Hines, M. A. *J. Am. Chem. Soc.* **2006**, *128*, 11455.

17. Filler, M. A.; Bent, S. F. *Prog. Surf. Sci.* **2003**, *73*, 1.
18. Leftwich, T. R.; Madachik, M. R.; Teplyakov, A. V. *J. Am. Chem. Soc.* **2008**, *130*, 16216.
19. Morrison, R. T.; Boyd, R. N. *Organic Chemistry.* Boston, Allyn and Bacon, 1959.
20. Wade, L. G., Jr., *Organic Chemistry.* Prentice-Hall, Inc., NJ, 1995.
21. Wolkow, R. A. *Annu. Rev. Phys. Chem.* **1999**, *50*, 413.
22. Taguchi, Y.; Fujisawa, M.; Takaoka, T.; Okada, T.; Nishijima, M. *J. Chem. Phys.* **1991**, *95*, 6870.
23. Borovsky, B.; Krueger, M.; Ganz, E. *Phys. Rev. B* **1998**, *57*, R4269.
24. Jeong, H. D.; Ryu, S.; Lee, Y. S.; Kim, S. *Surf. Sci.* **1995**, *344*, L1226.
25. Lopinski, G. P.; Fortier, T. M.; Moffatt, D. J.; Wolkow, R. A. *J. Vac. Sci. Technol. A* **1998**, *16*, 1037.
26. Lopinski, G. P.; Moffatt, D. J.; Wolkow, R. A. *Chem. Phys. Lett.* **1998**, *282*, 305.
27. Wolkow, R. A.; Lopinski, G. P.; Moffatt, D. J. *Surf. Sci.* **1998**, *416*, L1107.
28. Gokhale, S.; Trischberger, P.; Menzel, D.; Widdra, W.; Dröge, H.; Steinrück, H.-P.; Birkenheuer, U.; Gutdeutsch, U.; Rösch, N. *J. Chem. Phys.* **1998**, *108*, 5554.
29. Birkenheuer, U.; Gutdeutsch, U.; Rosch, N. *Surf. Sci.* **1998**, *409*, 213.
30. Kong, M. J.; Teplyakov, A. V.; Lyubovitsky, J. G.; Bent, S. F. *Surf. Sci.* **1998**, *411*, 286.
31. Silvestrelli, P. L.; Ancilotto, F.; Toigo, F. *Phys. Rev. B* **2000**, *62*, 1596.
32. Witkowski, N.; Hennies, F.; Pietzsch, A.; Mattsson, S.; Föhlisch, A.; Wurth, W. *Phys. Rev. B* **2003**, *68*, 115408.
33. Shimomura, M.; Munakata, M.; Honma, K.; Widstrand, S. M.; Johansson, L.; Abukawa, T.; Kono, S. *Surf. Rev. Lett.* **2003**, *10*, 499.
34. Witkowski, N.; Pluchery, O.; Borensztein, Y. *Phys. Rev. B* **2005**, *72*, 075354.
35. Jung, Y.; Gordon, M. S. *J. Am. Chem. Soc.* **2005**, *127*, 3131.
36. Besley, N. A.; Blundy, A. J. *J. Phys. Chem. B* **2006**, *110*, 1701.
37. Okamura, K.; Ishii, H.; Kimura,Y.; Niwano, M. *Surf. Sci.* **2005**, *576*, 45.
38. Kasaya, M.; Tabata, H.; Kawai, T. *Surf. Sci.* **1998**, *400*, 367.
39. Kasaya, M.; Tabata, H.; Kawai, T. *Surf. Sci.* **1998**, *406*, 302.
40. Hughes, G.; Roche, J.; Carty, D.; Cafolla, T.; Smith, K. E. *J. Vac. Sci. Technol. B* **2002**, *20*, 1620.
41. Weidkamp, K. P.; Hacker, C. A.; Schwartz, M. P.; Cao, X.; Tromp, R. M.; Hamers, R. J. *J. Phys. Chem. B* **2003**, *107*, 11142.
42. Tejima, M.; Kita, K.; Kyuno, K.; Toriumi, A. *Appl. Phys. Lett.* **2004**, *85*, 3746.
43. Okamura, K.; Hosoi, Y.; Kimura, Y.; Ishii, H.; Niwano, M. *Appl. Surf. Sci.* **2004**, *237*, 439.
44. Choudhary, D.; Clancy, P.; Bowler, D. R. *Surf. Sci.* **2005**, *578*, 20.
45. Suzuki, T.; Sorescu, D. C.; Yates, J. T., Jr., *Surf. Sci.* **2006**, *600*, 5092.
46. Lee, H.-K.; Han, J.-H.; Kim, K.-J.; Kang, T.-H.; Kim, B. *Surf. Sci.* **2007**, *601*, 1456.
47. Rada, T.; Chen, Q.; Richardson, N. V. *J. Phys.: Condens. Matter* **2003**, *15*, S2749.
48. Rada, T.; Chen, Q.; Richardson, N. *Phys. Status Solidi B* **2004**, *241*, 2353.
49. Preuss, M.; Miotto, R.; Bechstedt, F.; Rada, T.; Richardson, N. V.; Schmidt, W. G. *Phys. Rev. B* **2006**, *74*, 115402/1.

50. Suzuki, T.; Sorescu, D. C.; Jordan, K. D.; Yates, J. T., Jr., *J. Chem. Phys.* **2006**, *124*, 224708/1.
51. Suzuki, T.; Sorescu, D. C.; Jordan, K. D.; Levy, J.; Yates, J. T., Jr., *J. Chem. Phys.* **2006**, *124*, 054701/1.
52. Suzuki, T.; Levy, J.; Yates, J. T., Jr., *Nano Lett.* **2006**, *6*, 138.
53. Hughes, G.; Carty, D.; Cafolla, A. A. *Surf. Sci.* **2005**, *582*, 90.
54. Weidkamp, K. P.; Tromp, R. M.; Hamers, R. J. *J. Phys. Chem. C* **2007**, *111*, 16489.
55. Ruiz, R.; Choudhary, D.; Nickel, B.; Toccoli, T.; Chang, K.-C.; Mayer, A. C.; Clancy, P.; Blakely, J. M.; Headrick, R. L.; Iannotta, S.; Malliaras, G. G. *Chem. Mater.* **2004**, *16*, 4497.
56. Nunzi, F.; Sgamellotti, A.; Re, N. *J. Phys. Chem. C* **2007**, *111*, 1392.
57. Fink, A.; Menzel, D.; Widdra, W. *J. Phys. Chem. B* **2001**, *105*, 3828.
58. Cho, J. H.; Kim, K. S.; Morikawa, Y. *J. Chem. Phys.* **2006**, *24*, 024716.
59. Tao, F.; Bernasek, S. L. *Am. Chem. Soc.* **2007**, *129*, 4815.
60. Ma, Z.; Zaera, F. *Surf. Sci. Rep.* **2006**, *61*, 229.
61. Lu, X.; Wang, X. L.; Yuan, Q. H.; Zhang, Q. *J. Am. Chem. Soc.* **2003**, *125*, 7923.
62. Petsalakis, I. D.; Polanyi, J. C.; Theodorakopoulos, G. *Surf. Sci.* **2003**, *544*, 162.
63. Yong, K. S.; Zhang, Y. P.; Yang, S.-W.; Xu, G. Q. *Surf. Sci.* **2008**, *602*, 1921.
64. Guan, D. D.; Mao, H. Y.; Chen, M. L.; Dou, W. D.; Song, F.; Zhang, H. J.; Li, H. Y.; He, P.; Chen, H. Z.; Bao, S. N. *J. Chem. Phys.* **2009**, *130*, 174712.
65. Schedel, T.; Frank, K.-H.; Karlsson, U.; Koch, E. E. *Vacuum* **1990**, *41*, 652.
66. Yong, K. S.; Zhang, Y. P.; Yang, S.-W.; Wu, P.; Xu, G. Q. *J. Phys. Chem. A* **2007**, *111*, 12266.
67. Yong, K. S.; Zhang, Y. P.; Yang, S.-W.; Wu, P.; Xu, G. Q. *J. Phys. Chem. C* **2007**, *111*, 4285.
68. Mayne, A. J.; Lastapis, M.; Baffou, G.; Soukiassian, L.; Comtet, G.; Hellner, L.; Dujardin, G. *Phys. Rev. B* **2004**, *69*, 045409.
69. Cranney, M.; Mayne, A. J.; Comtet, G.; Dujardin, G. *Surf. Sci.* **2005**, *593*, 139.
70. Mamatkulov, M.; Stauffer, L.; Minot, C.; Sonnet, P. *Phys. Rev. B* **2006**, *73*, 035321.
71. Dubois, M.; Delerue, C.; Rubio, A. *Phys. Rev. B* **2007**, *75*, 041302(R).
72. Martin, M.; Lastapis, M.; Riedel, D.; Dujardin, G.; Mamatkulov, M.; Stauffer, L.; Sonnet, P. *Phys. Rev. Lett.* **2006**, *97*, 216103.
73. Lastapis, M.; Martin, M.; Riedel, D.; Comtet, G.; Hellner, L.; Dujardin, G. *Science* **2005** *308*, **1000**.
74. Cranney, M.; Comtet, G.; Dujardin, G.; Kim, J. W.; Kampen, T. U.; Horn, K. *Phys. Rev. B* **2007**, *76*, 075324.
75. Lima, C. P.; Miwa, R. H.; Schmidt, T. M. *Surf. Sci.* **2008**, *602*, 2634.
76. Riedel, D.; Bocquet, M.-L.; Lesnard, H.; Lastapis, M.; Lorente, N.; Sonnet, P.; Dujardin, G. *J. Am. Chem. Soc.* **2009**, *131*, 7344.
77. Cranney, M.; Chalopin, Y.; Mayne, A. J.; Dujardin, G. *Appl. Phys. A* **2009**, *94*, 767.
78. Coulter, S. K.; Hovis, J. S.; Ellison, M. D.; Hamers, R. J. *J. Vac. Sci. Technol. A* **2000**, *18*, 1965.
79. Costanzo, F.; Sbraccia, C.; Silvestrelli, P. L.; Ancilotto, F. *J. Phys. Chem. B* **2003**, *107*, 10209.

80. Costanzo, F.; Sbraccia, C.; Silvestrelli, P. L.; Ancilotto, F. *Surf. Sci.* **2004**, *566*, 971.
81. Li, Q.; He, Z. H.; Zhou, X. J.; Yang, X.; Leung, K. T. *Surf. Sci.* **2004**, *560*, 191.
82. Carbone, M.; Piancastelli, M. N.; Casaletto, M. P.; Zanoni, R.; Comtet, G.; Dujardin, G.; Hellner, L. *Surf. Sci.* **2002**, *498*, 186.
83. Tomimoto, H.; Takehara, T.; Fukawa, K.; Sumii, R.; Sekitani, T.; Tanaka, K. *Surf. Sci.* **2003**, *526*, 341.
84. Tomimoto, H.; Sekitani, T.; Sumii, R.; Sako, E. O.; Wada, S.-I.; Tanaka, K. *Surf. Sci.* **2004**, *566*, 664.
85. Macpherson, C. D.; Leung, K. T. *Surf. Sci.* **1995**, *326*, 141.
86. Casaletto, M. P.; Carbone, M.; Piancastelli, M. N.; Horn, K.; Weiss, K.; Zanoni, R. *Surf. Sci.* **2005**, *582*, 42.
87. Coulter, S. K.; Schwartz, M. P.; Hamers, R. J. *J. Phys. Chem. B* **2001**, *105*, 3079.
88. Kugler, T.; Thibaut, U.; Abraham, M.; Folkers, G.; Göpel, W. *Surf. Sci.* **1992**, *260*, 64.
89. Cao, X.; Coulter, S. K.; Ellison, M. D.; Liu, H.; Liu, J.; Hamers, R. J. *J. Phys. Chem. B* **2001**, *105*, 3759.
90. Bitzer, T.; Alkunshalie, T.; Richardson, N. V. *Surf. Sci.* **1996**, *368*, 202.
91. Rummel, R. M.; Ziegler, C. *Surf. Sci.* **1998**, *418*, 303.
92. Bitzer, T.; Richardson, N. V. *Surf. Sci.* **1999**, *427-428*, 369.
93. Lopez, A.; Bitzer, T.; Heller, T.; Richardson, N. V. *Surf. Sci.* **2001**, *480*, 65.
94. Mui, C.; Han, J. H.; Wang, G. T.; Musgrave, C. B.; Bent, S. F. *J. Am. Chem. Soc.* **2002**, *124*, 4027.
95. Kugler, T.; Ziegler, C.; Göpel, W. *Mater. Sci. Eng. B* **1996**, *37*, 112.
96. Jang, S. H.; Jeong, S.; Hahn, J. R. *J. Chem. Phys.* **2009**, *130*, 234703.
97. Carbone, M.; Meloni, S.; Caminiti, R. *Phys. Rev. B* **2007**, *76*, 085332.
98. Carbone, M.; Cazzato, P.; Caminiti, R. *Surf. Sci.* **2009**, *603*, 611.
99. Carbone, M.; Piancastelli, M. N.; Zanoni, R.; Comtet, G.; Dujardin, G.; Hellner, L. *Surf. Sci.* **1999**, *419*, 114.
100. Zheng, F.; McChesney, J. L.; Liu, X.; Himpsel, F. J. *Phys. Rev. B* **2006**, *73*, 205315.
101. Baik, J.; Park, J.; Kim, M.; Ahn, J. R.; Park, C.-Y.; An, K.-S.; Hwang, C.-C.; Hwang, H.-N.; Kim, B. *J. Korean Phys. Soc.* **2007**, *50*, 690.
102. Bitzer, T.; Richardson, N. V.; Reiss, S.; Wühn, M.; Wöll, C. *Surf. Sci.* **2000**, *458*, 173.
103. Naumkin, F. Y.; Polanyi, J. C.; Rogers, D. *Surf. Sci.* **2003**, *547*, 335.
104. Naumkin, F. Y.; Polanyi, J. C.; Rogers, D.; Hofer, W.; Fisher, A. *Surf. Sci.* **2003**, *547*, 324.
105. Zhou, X. J.; Leung, K. T. *J. Phys. Chem. B* **2006**, *110*, 9601.
106. Zhou, X. J.; Leung, K. T. *Surf. Sci.* **2006**, *600*, 3285.
107. Jiang, G. P.; Polanyi, J. C.; Rogers, D. *Surf. Sci.* **2003**, *544*, 147.
108. Matta, C. F.; Polanyi, J. C. *Phil. Trans. Royal Soc. A* **2004**, *362*, 1185.
109. Li, Z. H.; Li, Y. C.; Wang,W. N.; Cao, Y.; Fan, K. N. *J. Phys. Chem. B* **2006**, *108*, 14049.
110. Sloan, P. A.; Palmer, R. E. *J. Phys.: Condens. Matter* **2006**, *18*, S1873.
111. Lu, X.; Polanyi, J. C.; Yang, J. *Nano Lett.* **2006**, *6*, 809.
112. Dobrin, S.; Harikumar, K. R.; Polanyi, J. C. *Surf. Sci.* **2004**, *561*, 11.

113. Harikumar, K. R.; Petsalakis, I. D.; Polanyi, J. C.; Theodorakopoulos, G. *Surf. Sci.* **2004**, *572*, 162.
114. Dobrin, S. *J. Phys. Chem. B* **2005**, *109*, 22976.
115. Petsalakis, I. D.; Polanyi, J. C.; Theodorakopoulos, G. *Israel J. Chem.* **2005**, *45*, 111.
116. Dobrin, S.; Harikumar, K. R.; Matta, C. F.; Polanyi, J. C. *Surf. Sci.* **2005**, *580*, 39.
117. Schwartz, M. P.; Ellison, M. D.; Coulter, S. K.; Hovis, J. S.; Hamers, R. J. *J. Am. Chem. Soc.* **2000**, *122*, 8529.
118. Hamers, R. J.; Hovis, J. S.; Coulter, S. K.; Ellison, M. D.; Padowitz, D. F. *Jpn. J. Appl. Phys.* **2000**, *39*, 4366.
119. Calzolari, A.; Ruini, A.; Molinari, E.; Caldas, M. J. *Phys. Rev. B* **2006**, *73*, 125420.
120. Wang, Y.; Ma, J.; Inagaki, S.; Pei, Y. *J. Phys. Chem. B* **2005**, *109*, 5199.
121. Dixon, C. E.; Liu, H. W.; VanderKant, C. M.; Baines, K. M. *Organometallics* **1996**, *15*, 5701.
122. Dixon, C. E.; Baines, K. M. *Phosphorus,Sulfur Silicon Relat. Elem.* **1997**, *125*, 123.
123. Iwamoto, T.; Sakurai, H.; Kira, M. *Bull. Chem. Soc. Jpn.* **1998**, *71*, 2741.
124. Hwang, Y. J.; Kim, A.; Hwang, E. Y.; Kim, S. *J. Am. Chem. Soc.* **2005**, *127*, 5016.
125. Zhang, Y. P.; Yang, L.; Lai, Y. H.; Xu, G. Q.; Wang, X. S. *Appl. Phys. Lett.* **2004**, *84*, 401.
126. Basu, R.; Tovar, J. D.; Hersam, M. C. *J. Vac. Sci. Technol. B* **2005**, *23*, 1785.
127. Tao, F.; Wang, Z. H.; Lai, Y. H.; Xu, G. Q. *J. Am. Chem. Soc.* **2003**, *125*, 6687.
128. Tao, F.; Qiao, M. H.; Li, Z. H.; Yang, L.; Dai, Y. J.; Huang, H. G.; Xu, G. Q. *Phys. Rev. B* **2003**, *67*, 115334.
129. Kim, K.-Y.; Song, B.-K.; Jeong, S.; Kang, H. *J. Phys. Chem. B* **2003**, *107*, 11987.
130. Pluchery, O.; Coustel, R.; Witkowski, N.; Borensztein, Y. *J. Phys. Chem. B* **2006**, *110*, 22635.
131. Huang, H. G.; Zhang, Y. P.; Cai, Y. H.; Huang, J. Y.; Yong, K. S.; Xu, G. Q. *J. Chem. Phys.* **2005**, *123*, *104702-1-* 104702–6.
132. Huang, H. G.; Huang, J. Y.; Cai, Y. H.; Xu, G. Q. *Chem. Phys. Lett.* **2005**, *414*, 143.
133. Perrine, K. A.; Leftwich, T. R.; Weiland, C. R.; Madachik, M.; Opila, R. L.; Teplyakov, A. V. *J. Phys. Chem. C* **2009**, *113*, 6643.
134. Tao, F.; Wang, Z. H.; Xu, G. Q. *J. Phys. Chem. B* **2002**, *106*, 3557.
135. Qu, Y.-Q.; Han, K.-L. *J. Phys. Chem. B* **2004**, *108*, 8305.
136. Rangan, S.; Bournel, F.; Gallet, J. J.; Kubsky, S.; Le Guen, K.; Dufour, G.; Rochet, F.; Sirotti, F.; Carniato, S.; Ilakovac, V. *Phys. Rev. B* **2005**, *71*, 165319.
137. Tao, F.; Wang, Z. H.; Chen, X. F.; Xu, G. Q. *Phys. Rev. B* **2002**, *65*, 115311.
138. Schwartz, M. P.; Halter, R. J.; McMahon, R. J.; Hamers, R. J. *J. Phys. Chem. B* **2003**, *107*, 224.
139. Nunzi, F.; Sgamellotti, A.; Re, N. *J. Phys. Chem. B* **2004**, *108*, 10881.
140. Laguerre, M.; Felix, G.; Dunogues, J.; Calas, R. *J. Org. Chem.* **1979**, *44*, 4275.
141. Barriocanal, J. A.; Doren, D. J. *J. Phys. Chem. B* **2000**, *104*, 12269.
142. Barriocanal, J. A.; Doren, D. J. *J. Vac. Sci. Technol. A* **2000**, *18*, 1959.
143. Bocharov, S.; Mathauser, A. T.; Teplyakov, A. V. *J. Phys. Chem. B* **2003**, *107*, 7776.
144. Eng, J., Jr., ; Hubner, I. A.; Barriocanal, J.; Opila, R. L.; Doren, D. J. *J. Appl. Phys.* **2004**, *95*, 1963.

145. Bocharov, S.; Teplyakov, A. V. *Surf. Sci.* **2004**, *573*, 403.

146. Méndez De Leo, L. P.; Teplyakov, A. V. *J. Phys. Chem. B* **2006**, *110*, 6899.

147. Ellison, M. D.; Hamers, R. J. *J. Phys. Chem. B* **1999**, *103*, 6243.

148. Schaumann, E.; Bauch, H.-G.; Sieveking, S.; Adiwidjaja, G. *Chem. Ber.* **1982**, *115*, 3340.

149. Reddy, C. K.; Reddy, P. S. N.; Ratnam, C. V. *Ind. J. Chem.* **1987**, *26B*, 882.

150. Kim, A.; Filler, M. A.; Kim, S.; Bent, S. F. *J. Am. Chem. Soc.* **2005**, *127*, 6123.

151. Bocharov, S.; Dmitrenko, O.; Méndez De Leo, L. P.; Teplyakov, A. V. *J. Am. Chem. Soc.* **2006**, *128*, 9300.

152. Leftwich, T.; Teplyakov, A. V. *J. Phys. Chem. C* **2008**, *112*, 4297.

153. Wang, Z.; Wang, H.; Liu, Y.; Suo, Y. *J. Mol. Struct.: THEOCHEM* **2008**, *850*, 72.

154. Fang, L.; Liu, J.; Coulter, S.; Cao, X.; Schwartz, M. P.; Hacker, C.; Hamers, R. J. *Surf. Sci.* **2002**, *514*, 362.

155. Hacker, C. A.; Hamers, R. J. *J. Phys. Chem. B* **2003**, *107*, 7689.

156. Tao, F.; Bernasek, S. L.; Xu, G.-Q. *Chem. Rev.* **2009**, *109*, 3991.

157. Qiao, M. H.; Cao, Y.; Deng, J. F.; Xu, G. Q. *Chem. Phys. Lett.* **2000**, *325*, 508.

158. Wang, G. T.; Mui, C.; Tannaci, J. F.; Filler, M. A.; Musgrave, C. B.; Bent, S. F. *J. Phys. Chem. B* **2003**, *107*, 4982.

159. Qiao, M. H.; Tao, F.; Cao, Y.; Xu, G. Q. *Surf. Sci.* **2003**, *544*, 285.

160. Kim, K.-J.; Han, J.; Kang, T.-H.; Ihm, K.; Jeon, C.; Moon, S.; Hwang, C.-C.; Hwang, H.-N.; Kim, B. *J. Elect. Spectrosc. Rel. Phenom.* **2005**, *144-147*, 429.

161. Romero, A. H.; Sbraccia, C.; Silvestrelli, P. L. *J. Chem. Phys.* **2004**, *120*, 9745.

162. Yuan, Z. L.; Chen, X. F.; Wang, Z. H.; Yong, K. S.; Cao, Y.; Xu, G. Q. *J. Chem. Phys.* **2003**, *119*, 10389.

163. Tao, F.; Yuan, Z. L.; Chen, X. F.; Qiao, M. H.; Wang, Z. H.; Dai, Y. J.; Huang, H. G.; Cao, Y.; Xu, G. Q. *Phys. Rev. B* **2003**, *67*, 245406.

164. Qiao, M. H.; Tao, F.; Cao, Y.; Li, Z. H.; Dai, W. L.; Deng, J. F.; Xu, G. Q. *J. Chem. Phys.* **2001**, *114*, 2766.

165. Qiao, M. H.; Cao, Y.; Tao, F.; Liu, Q.; Deng, J. F.; Xu, G. Q. *J. Phys. Chem. B* **2000**, *104*, 11211.

166. Lu, X.; Xu, X.; Wang, N. Q.; Zhang, Q.; Lin, M. C. *J. Phys. Chem. B* **2001**, *105*, 10069.

167. Lu, X.; Lin, M. C.; Xu, X.; Wang, N. Q.; Zhang, Q. N. *Phys. Chem. Commun.* **2001**, 1.

168. Lu, X.; Lin, M. C.; Xu, X.; Wang, N. Q.; Zhang, Q. N. *Sci. Chin. Ser. B* **2001**, *44*, 473.

169. Lu, X.; Lin, M. C. *Int. Rev. Phys. Chem.* **2002**, *21*, 137.

170. Lee, H.-K.; Kim, K.-J.; Kang, T.-H.; Chung, J. W.; Kim, B. *Surf. Sci.* **2008**, *602*, 914.

171. Isobe, N.; Shibayama, T.; Mori, Y.; Shobatake, K.; Sawabe, K. *Chem. Phys. Lett.* **2007**, *443*, 347.

172. Cao, Y.; Wang, Z. H.; Deng, J. F.; Xu, G. Q. *Angew. Chem. Int. Ed.* **2000**, *39*, 2740.

173. Bitzer, T.; Dittrich, T.; Rada, T.; Richardson, N. V. *Chem. Phys. Lett.* **2000**, *331*, 433.

174. Lopez, A.; Bitzer, T.; Heller, T.; Richardson, N. V. *Surf. Sci.* **2001**, *477*, 219.

175. Bitzer, T.; Rada, T.; Richardson, N. V. *J. Phys. Chem. B* **2001**, *105*, 4535.

176. Hofer, W. A.; Fisher, A. J.; Bitzer, T.; Rada, T.; Richardson, N. V. *Chem. Phys. Lett.* **2002**, *355*, 347.

177. Wang, G. T.; Mui, C.; Musgrave, C. B.; Bent, S. F. *J. Phys. Chem. B* **1999**, *103*, 6803.
178. Tao, F.; Qiao, M. H.; Wang, Z. H.; Xu, G. Q. *J. Phys. Chem. B* **2003**, *107*, 6384.
179. Miwa, J. A.; Eves, B. J.; Rosei, F.; Lopinski, G. P. *Chem. B* **2005**, *109*, 20055.
180. Kim, H.-J.; Cho, J.-H. *J. Chem. Phys.* **2004**, *120*, 8222.
181. Tao, F.; Lai, Y. H.; Xu, G. Q. *Langmuir* **2004**, *20*, 366.
182. Pluchery, O.; Witkowski, N.; Borensztein, Y. *Phys. Status Solidi B* **2005**, *242*, 2696.
183. Borensztein, Y.; Pluchery, O.; Witkowski, N. *Phys. Rev. Lett.* **2005**, *95*, 117402.
184. Li, Q.; Leung, K. T. *Surf. Sci.* **2003**, *541*, 113.
185. Lu, X.; Xu, X.; Wu, J.; Wang, N.; Zhang, Q. *New J. Chem.* **2002**, *26*, 160.
186. Shimomura, M.; Ichikawa, D.; Fukuda, Y.; Abukawa, T.; Aoyama, T.; Kono, S. *Phys. Rev. B* **2005**, *72*, 033303.
187. Shimomura, M.; Kono, S. *Hyomen Kagaku* **2006**, *27*, 298.
188. Huang, H. G.; Wang, Z. H.; Xu, G. Q. *J. Phys. Chem. B* **2004**, *108*, 12560.
189. Huang, H. G.; Huang, J. Y.; Ning, Y. S.; Xu, G. Q. *J. Chem. Phys.* **2004**, *121*, 4820.
190. Huang, H. G.; Huang, J. Y.; Wang, Z. H.; Ning, Y. S.; Tao, F.; Zhang, Y. P.; Cai, Y. H.; Tang, H. H.; Xu, G. Q. *Surf. Sci.* **2007**, *601*, 1184.
191. Bu, Y.; Lin, M. C. *J. Phys. Chem.* **1994**, *98*, 7871.
192. Casey, S. M.; LeMond, L. M. *Silicon Surface Chemistry of Substituted Triazines.* Logan, UT, United States, June 6-9, 2004.
193. Günster, J.; Dieckhoff, S.; Souda, R. *Thin Solid Films* **1998**, *325*, 24.
194. Hahn, J. R.; Jeong, H.; Jeong, S.; Jang, S. H. *Jpn. J. Appl. Phys.* **2006**, *45*, 2175.
195. Jeong, H.; Jeong, S.; Jang, S. H.; Seo, J. M.; Hahn, J. R. *J. Phys. Chem. B* **2006**, *110*, 15912.
196. Cho, Y. E.; Maeng, J. Y.; Kim, S. *J. Am. Chem. Soc.* **2003**, *125*, 7514.
197. Choi, J.-H.; Cho, J.-H. *J. Am. Chem. Soc.* **2006**, *128*, 3890.
198. Avouris, P.; Walkup, R. E.; Rossi, A. R.; Akpati, H. C.; Nordlander, P.; Shen, T. C.; Abeln, G. C.; Lyding, J. W. *Surf. Sci.* **1996**, *363*, 368.
199. Abeln, G. C.; Hersam, M. C.; Thompson, D. S.; Hwang, S.-T.; Choi, H.; Moore, J. S.; Lyding, J. W. *J. Vac. Sci. Technol. B* **1998**, *16*, 3874.
200. Avouris, P.; Walkup, R. E.; Rossi, A. R.; Shen, T. C.; Abeln, G. C.; Tucker, J. R.; Lyding, J. W. *Chem. Phys. Lett.* **1996**, *257*, 148.
201. Foley, E. T.; Kam, A. F.; Lyding, J. W.; Avouris, P. *Phys. Rev. Lett.* **1998**, *80*, 1336.
202. Guyot-Sionnest, P.; Dumas, P.; Chabal, Y. J.; Higashi, G. S. *Phys. Rev. Lett.* **1990**, *64*, 2156.
203. Guyotsionnest, P.; Lin, P. H.; Hiller, E. M. *J. Chem. Phys.* **1995**, *102*, 4269.
204. Hashizume, T.; Heike, S.; Lutwyche, M. I.; Watanabe, S.; Nakajima, K.; Nishi, T.; Wada, Y. *Jpn. J. Appl. Phys. Part 2-Lett.* **1996**, *35*, L1085.
205. Hitosugi, T.; Hashizume, T.; Heike, S.; Kajiyama, H.; Wada, Y.; Watanabe, S.; Hasegawa, T.; Kitazawa, K. *Appl. Surf. Sci.* **1998**, *132*, 340.
206. Lyding, J. W.; Shen, T. C.; Hubacek, J. S.; Tucker, J. R.; Abeln, G. C. *Appl. Phys. Lett.* **1994**, *64*, 2010.
207. Sakurai, M.; Thirstrup, C.; Nakayama, T.; Aono, M. *Appl. Surf. Sci.* **1997**, *121*, 107.
208. Salam, G. P.; Persson, M.; Palmer, R. E. *Phys. Rev. B* **1994**, *49*, 10655.

209. Shen, T. C.; Wang, C.; Abeln, G. C.; Tucker, J. R.; Lyding, J. W.; Avouris, P.; Walkup, R. E. *Science* **1995**, *268*, 1590.

210. Soukiassian, L.; Mayne, A. J.; Comtet, G.; Hellner, L.; Dujardin, G.; Gourdon, A. *J. Chem. Phys.* **2005**, *122*, 134704.

211. Stokbro, K.; Hu, B. Y. K.; Thirstrup, C.; Xie, X. C. *Phys. Rev. B* **1998**, *58*, 8038.

212. Stokbro, K.; Thirstrup, C.; Sakurai, M.; Quaade, U.; Hu, B. Y. K.; Perez-Murano, F.; Grey, F. *Phys. Rev. Lett.* **1998**, *80*, 2618.

213. Thirstrup, C.; Sakurai, M.; Nakayama, T.; Stokbro, K. *Surf. Sci.* **1999**, *424*, L329.

214. Hersam, M. C.; Guisinger, N. P.; Lee, J.; Cheng, K.; Lyding, J. W. *Appl. Phys. Lett.* **2002**, *80*, 201.

215. Mayne, A. J.; Dujardin, G.; Comtet, G.; Riedel, D. *Chem. Rev.* **2006**, *106*, 4355.

216. Tong, X.; Wolkow, R. A. *Surf. Sci.* **2006**, *600*, L199.

217. Hersam, M. C.; Guisinger, N. P.; Lyding, J. W. *Nanotechnology* **2000**, *11*, 70.

218. Mayne, A. J.; Soukiassian, L.; Commaux, N.; Comtet, G.; Dujardin, G. *Appl. Phys. Lett.* **2004**, *85*, 5379.

219. Mamatkulov, M.; Stauffer, L.; Sonnet, P.; Mayne, A. J.; Comtet, G.; Dujardin, G. *J. Chem. Phys.* **2008**, *128*, 244710.

219a. Bellec, A.; Cranney, M.; Chalopin, Y.; Mayne, A. J., Comtet, G.; Dujardin, G. *J. Phys. Chem. C* **2007**, *111*, 14818.

220. Sena, A. M. P.; Brázdová, V.; Bowler, D. R. *Phys. Rev. B* **2009**, *79*, 245404.

221. Linford, M. R.; Chidsey, C. E. D. *J. Am. Chem. Soc.* **1993**, *115*, 12631.

222. Linford, M. R.; Fenter, P.; Eisenberger, P. M.; Chidsey, C. E. D. *J. Am. Chem. Soc.* **1995**, *117*, 3145.

223. Terry, J.; Mo, R.; Wigren, C.; Cao, R.; Mount, G.; Pianetta, P.; Linford, M. R.; Chidsey, C. E. D. *Phys. Res., Sect. B* **1997**, *133*, 94.

224. Terry, J.; Linford, M. R.; Wigren, C.; Cao, R.-Y.; Pianetta, P.; Chidsey, C. E. D. *Appl. Phys. Lett.* **1997**, *71*, 1056.

225. Chatgilialoglu, C. *Acc. Chem. Res.* **1992**, *25*, 188.

226. Cicero, R. L.; Chidsey, C. E. D.; Lopinski, G. P.; Wayner, D. D. M.; Wolkow, R. A. *Langmuir*, **2002**, *18*, 305.

227. Basu, R.; Kinser, C. R.; Tovar, J. D.; Hersam, M. C. *Chem. Phys.* **2006**, *326*, 144.

228. Basu, R.; Lin, J.-C.; Kim, C.-Y.; Schmitz, M. J.; Yoder, N. L.; Kellar, J. A.; Bedzyk, M. J.; Hersam, M. C. *Langmuir* **2007** *23*, 1905.

229. Kirczenow, G.; Piva, P. G.; Wolkow, R. A. *Phys. Rev. B* **2005**, *72*, 245306.

230. Basu, R.; Guisinger, N. P.; Greene, M. E.; Hersam, M. C. *Appl. Phys. Lett.* **2004**, *85*, 2619.

231. Cho, J. H.; Oh, D. H.; Kleinman, L. *Phys. Rev. B* **2002**, 65.

232. Guisinger, N. P.; Basu, R.; Baluch, A. S.; Hersam, M. C. *Ann. NY Acad. Sci* **2003**, *1006*, 227.

233. Hossain, M. Z.; Kato, H. S.; Kawai, M. *J. Phys. Chem. B* **2005**, *109*, 23129.

234. Kang, J. K.; Musgrave, C. B. *J. Chem. Phys.* **2002**, *116*, 9907.

235. Lopinski, G. P.; Wayner, D. D. M.; Wolkow, R. A. *Nature* **2000**, *406*, 48.

236. Pei, Y.; Ma, J. *Langmuir* **2006**, *22*, 3040.

237. Pitters, J. L.; Piva, P. G.; Tong, X.; Wolkow, R. A. *Nano Lett.* **2003**, *3*, 1431.
238. Pitters, J. L.; Wolkow, R. A. *J. Am. Chem. Soc.* **2005**, *127*, 48.
239. Takeuchi, N.; Kanai, Y.; Selloni, A. *J. Am. Chem. Soc.* **2004**, *126*, 15890.
240. Takeuchi, N.; Selloni, A. *J. Phys. Chem. B* **2005**, *109*, 11967.
241. Tong, X.; DiLabio, G. A.; Wolkow, R. A. *Nano Lett.* **2004**, *4*, 979.
242. Wang, Y.; Hwang, G. S. *Appl. Phys. Lett.* **2005**, 86.
243. Alavi, S.; Rousseau, R.; Patitsas, S. N.; Lopinski, G. P.; Wolkow, R. A.; Seideman, T. *Phys. Rev. Lett.* **2000**, *85*, 5372.
244. de Smet, L.; Zuilhof, H.; Sudholter, E. J. R.; Lie, L. H.; Houlton, A.; Horrocks, B. R. *J. Phys. Chem. B* **2005**, *109*, 12020.
245. Guisinger, N. P.; Greene, M. E.; Basu, R.; Baluch, A. S.; Hersam, M. C. *Nano Lett.* **2004**, *4*, 55.
246. Hallback, A. S.; Poelsema, B.; Zandvliet, H. J. W. *Appl. Surf. Sci.* **2007**, *253*, 4066.
247. Hofer, W. A.; Fisher, A. J.; Lopinski, G. P.; Wolkow, R. A. *Chem. Phys. Lett.* **2002**, *365*, 129.
248. Lee, J. Y.; Cho, J. H. *J. Chem. Phys.* **2004**, *121*, 8010.
249. Li, Q.; Leung, K. T. *J. Phys. Chem. B* **2005**, *109*, 1420.
250. Patitsas, S. N.; Lopinski, G. P.; Hul'ko, O.; Moffatt, D. J.; Wolkow, R. A. *Surf. Sci.* **2000**, *457*, L425.
251. Pitters, J. L.; Wolkow, R. A. *Nano Lett.* **2006**, *6*, 390.
252. Pitters, J. L.; Dogel, I.; DiLabio, G. A.; Wolkow, R. A. *J. Phys. Chem. B* **2006**, *110*, 2159.
253. Sun, Q. Y.; de Smet, L.; van Lagen, B.; Giesbers, M.; Thune, P. C.; van Engelenburg, J.; de Wolf, F. A.; Zuilhof, H.; Sudholter, E. J. R. *J. Am. Chem. Soc.* **2005**, *127*, 2514.
254. DiLabio, G. A.; Piva, P. G.; Kruse, P.; Wolkow, R. A. *J. Am. Chem. Soc.* **2004**, *126*, 16048.
255. Piva, P. G.; DiLabio, G. A.; Pitters, J. L.; Zikovsky, J.; Rezeq, M. D.; Dogel, S.; Hofer, W. A.; Wolkow, R. A. *Nature*, **2005**, *435*, 658.
256. Ratner, M. *Nature*, **2005**, *435*, 575.
257. Greene, M. E.; Guisinger, N. P.; Basu, R.; Baluch, A. S.; Hersam, M. C. *Surf. Sci.* **2004**, *559*, 16.
258. Yang, L.; Doren, D. J. *J. Phys. Chem. C* **2008**, *112*, 781.
259. Smeu, M.; Wolkow, R. A.; Guo, H. *J. Am. Chem. Soc.* **2009**, *131*, 11019.
260. Rochefort, A.; Beausoleil, A. *Chem. Phys. Lett.* **2004**, *400*, 347.
261. Hossain, M. Z.; Kato, H. S.; Kawai, M. *J. Am. Chem. Soc.* **2007**, *129*, 3328.
262. Tong, X.; DiLabio, G. A.; Clarkin, O. J.; Wolkow, R. A. *Nano Lett.* **2004**, *4*, 357.
263. Dogel, S. A.; DiLabio, G. A.; Zikovsky, J.; Pitters, J. L.; Wolkow, R. A. *J. Phys. Chem. C*, **2007**, *111*, 11965.
264. Nemanick, E. J.; Hurley, P. T.; Webb, L. J.; Knapp, D. W.; Michalak, D. J.; Brunschwig, B. S.; Lewis, N. S. *J. Phys. Chem. B* **2006**, *110*, 14770.
265. Nemanick, E. J.; Hurley, P. T.; Brunschwig, B. S.; Lewis, N. S. *J. Phys. Chem. B* **2006**, *110*, 14800.
266. Pandey, D.; Zemlyanov, D. Y.; Bevan, K.; Reifenberger, R. G.; Dirk, S. M.; Howell, S. W.; Wheeler, D. R. *Langmuir* **2007**, *23*, 4700.

267. Combellas, C.; Kanoufi, F.; Pinson, J.; Podvorica, F. I. *J. Am. Chem. Soc.* **2008**, *130*, 8576.

268. Cai, Y. H.; Shao, Y. X.; Xu, G. Q. *J. Am. Chem. Soc.* **2007**, *129*, 8404.

269. Langner, A.; Panarello, A.; Rivillon, S.; Vassylyev, O.; Khinast, J. G.; Chabal, Y. J. *J. Am. Chem. Soc.* **2005**, *127*, 12798.

270. Leftwich, T. R.; Teplyakov, A. V. *J. Electron. Spectrosc. Relat. Phenom.* **2009**, *175*, 31.

271. Leftwich, T. R.; Teplyakov, A. V. Unpublished data, **2009**.

272. DiLabio, G. A.; Johnson, E. R.; Pitters, J. *J. Phys. Chem. C* **2009**, *113*, 9969.

273. Zipoli, F.; Bernasconi, M. *J. Phys. Chem. B* **2006**, *110*, 23403.

274. Wakatsuchi, M.; Kato, H. S.; Yamada, T.; Kawai, M. *Jpn. J. Appl. Phys.* **2005**, *44*, 514.

275. Ma, J.; Rogers, B. L.; Humphry, M. J.; Ring, D. J.; Goretzki, G.; Champness, N. R.; Beton, P. H. *J. Phys. Chem. B* **2006**, *110*, 12207.

276. Bitzer, T.; Richardson, N. V. *Appl. Phys. Lett.* **1997**, *71*, 1890.

277. Bitzer, T.; Richardson, N. V. *Appl. Surf. Sci.* **1999**, *144-145*, 339.

278. Bitzer, T.; Richardson, N. V. *Appl. Phys. Lett.* **1997**, *71*, 662.

279. Eves, B. J.; Fan, C.; Lopinski, G. P. *Small* **2006**, *2*, 1379.

280. Ciampi, S.; Böcking, T.; Kilian, K. A.; James, M.; Harper, J. B.; Gooding, J. J. *Langmuir* **2007**, *23*, 9320.

281. Bin, X.; Mischki, T. K.; Fan, C.; Lopinski, G. P.; Wayner, D. D. M. *J. Phys. Chem. C* **2007**, *111*, 13547.

282. Jiao, J.; Anariba, F.; Tiznado, H.; Schmidt, I.; Lindsey, J. S.; Zaera, F.; Bocian, D. F. *J. Am. Chem. Soc.* **2006**, *128*, 6965.

CHAPTER 7

Covalent Binding of Polycyclic Aromatic Hydrocarbon Systems

KIAN SOON YONG AND GUO-QIN XU

7.1 INTRODUCTION

Polycyclic aromatic hydrocarbons (PAHs) are a class of compounds that are made up of fused aromatic rings and comprise of only carbon and hydrogen. Some examples of PAHs are shown in Fig. 7.1, which consist of the acene members, perylene, coronene, dibenzo[*a*, *j*]coronene, and acenaphthylene. The acenes are made up of linearly fused benzene rings, of which naphthalene, anthracene, tetracene, and pentacene are the most commonly studied molecules. Due to the sharing of π-orbitals between adjacent rings in PAHs, each ring segment possesses a different degree of aromaticity, according to Clar's sextet concept [1]. For instance, only one ring within naphthalene and anthracene can be assigned three double bonds without duplication and be considered as an aromatic sextet. The sextets are represented by the circles in Fig. 7.2, which are mobile within the molecules via the movements of two π-electrons as symbolized by the arrows. The highly conjugated π-electron system within the PAHs also makes them highly sought after as the active semiconducting materials in organic electronics [2]. Specifically, pentacene has aroused great attention due to its remarkable field-effect mobility that is comparable to that of amorphous silicon. Examples of organic electronic devices that incorporate pentacene are field-effect transistors [3, 4] and solar cells [5]. Besides pentacene, tetracene is another PAH that has been successfully employed for fabrication of light-emitting devices [6], photodetectors [7], and solar cells [8].

Numerous studies have been done on the adsorption of PAHs on the silicon surfaces. This is due to the interest of coupling organic functionalities with the silicon surfaces to create new technological opportunities for fabrication of silicon-based organic–inorganic hybrid devices [9]. Such devices exploit the functional flexibility of organic molecules and the well-established silicon process technology to offer

Functionalization of Semiconductor Surfaces, First Edition.
Edited by Franklin (Feng) Tao and Steven L. Bernasek.

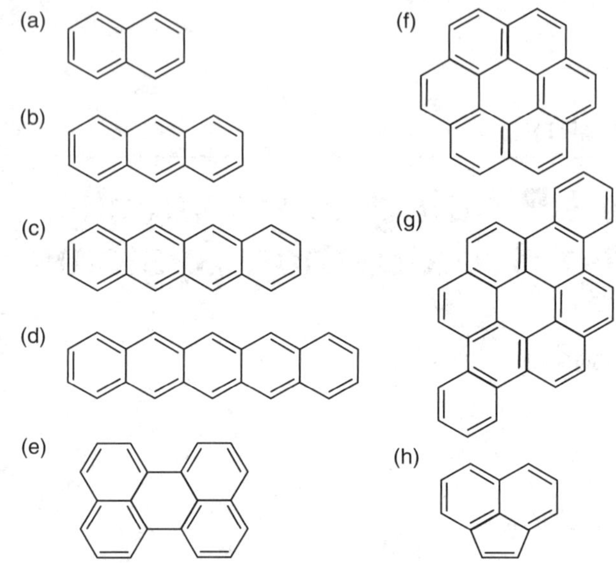

FIGURE 7.1 Some examples of polycyclic aromatic hydrocarbons: (a) naphthalene, (b) anthracene, (c) tetracene, (d) pentacene, (e) perylene, (f) coronene, (g) dibenzo[*a*, *j*] coronene, and (h) acenaphthylene.

new and enhanced properties integrated with the silicon substrate. Since PAHs exhibit potentials for applications in organic electronic devices, achieving an atomistic understanding of their binding configurations on the silicon surfaces is of industrial as well as scientific interest. Specifically, it is well known that the charge transport and optical properties of a film are influenced by the interactions between its constituent molecules and hence its morphology [10]. Hence, in order to design the optimum conditions for growing PAH films on silicon, a thorough knowledge of the interfacial chemistry between the PAHs and the silicon surface is critical

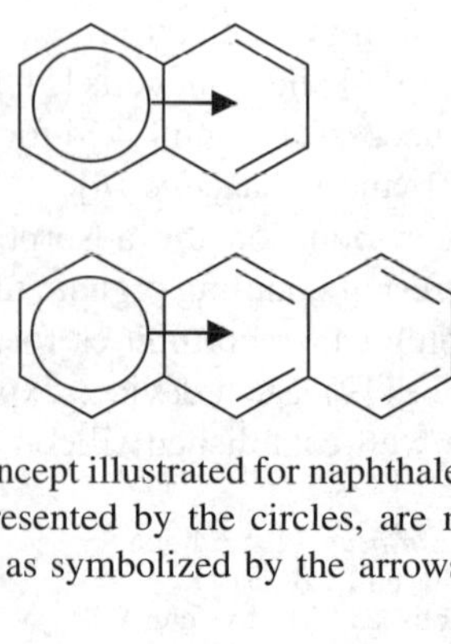

FIGURE 7.2 Clar's sextet concept illustrated for naphthalene and anthracene. The aromatic sextets of six π-electrons, represented by the circles, are mobile within the molecules via movements of two π-electrons as symbolized by the arrows.

because the first molecular layer on the substrate is going to affect the growth of subsequent PAH molecules.

The results obtained from the studies of various PAHs on the silicon substrates provide insight to the reaction mechanisms and binding configurations of these molecules on the Si(100)-(2×1) and Si(111)-(7×7) surfaces, which will be reviewed in this chapter. For practically all cases, the adsorbates form strong C–Si covalent bonds with the substrates that involve the transformations of the reacted carbon atoms from sp^2 to sp^3 hybridization. Due to the rupture of the π-conjugation in the reacted ring, the overall aromaticity of the adsorbate differs from that of the free molecule. Various binding configurations are possible in many of the systems studied and their relative binding energies are influenced by the resultant aromaticity of the molecules. Generally, adsorbates with structures that are able to sustain π-conjugation or an aromatic sextet of six π-electrons, as governed by Clar's concept, are more stable than those with bent or twisted structures with perturbed π-conjugation. Furthermore, the possible binding configurations are greatly influenced by the dimensions of the adsorbates relative to the positions of the silicon dangling bonds. For the Si(100)-(2×1) surface with rows and columns of Si dimer rows, the molecules mostly adopt orientations that are parallel or perpendicular to the dimer rows. As for the Si(111)-(7×7) surface, the adatom–rest atom pair plays an important role in the di-σ reaction with two opposite carbon atoms from the same benzene ring within the molecule. Due to the strong interactions between PAHs and the silicon surfaces, the adsorbates typically adopt an almost flat-lying orientation on the substrates. Section 7.2 focuses on the adsorption of various PAHs on Si(100)-(2×1) while the interactions of the acenes on Si(111)-(7×7) will be reviewed in Section 7.3. For easy reference, PAHs with labeled carbon atoms and rings will be shown in some of the sections and descriptions of numbered carbon atoms in the text will be made with reference to the diagrams.

7.2 PAHs ON Si(100)-(2×1)

7.2.1 Naphthalene and Anthracene on Si(100)-(2×1)

Naphthalene and anthracene are made up of, respectively, two and three linearly fused benzene rings and hence constitute the basic structures of the larger acene members. The adsorptions of naphthalene and anthracene on Si(100)-(2×1) were investigated by Okamura et al. based on infrared reflection–absorption spectroscopy (IRAS) and density functional theory (DFT) calculations [11, 12]. Similar to benzene, the adsorption of naphthalene on the surface exhibits strong coverage dependence. At low exposure, naphthalene binds to two adjacent Si dimers through four C–Si σ-bonds involving C^1, C^2, C^5, and C^6 of the adsorbate (Fig. 7.3a). This configuration, referred to as "symmetric bridge" (SB), is energetically more favorable as compared to other configurations such as "rotated symmetric bridge" (RSB) and "tight bridge" (TiB), though the aromaticity within naphthalene is completely lost in the SB

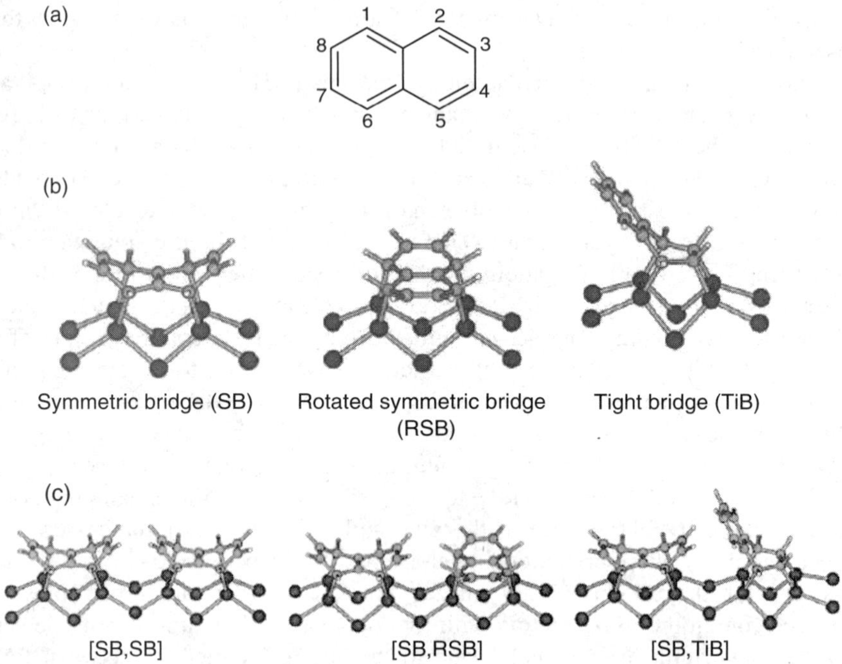

FIGURE 7.3 (a) Structure of naphthalene. Configurations of (b) one and (c) two naphthalene molecules chemisorbed on Si(100)-(2×1) (from Ref. 11).

binding mode (Fig. 7.3b). However, both IRAS and DFT calculations indicate the presence of multiple configurations of naphthalene, comprising of SB, RSB, and TiB, on Si(100)-(2×1) at high coverages. DFT calculations performed for two adsorbates on silicon models (Fig. 7.3c) show that combination of [SB,SB] is energetically less stable than [SB,RSB] and [SB,TiB]. This is attributed to the larger steric repulsion between the hydrogen atoms of the two adjacent adsorbates along a dimer row for the [SB,SB] model.

On the other hand, the adsorption of anthracene (Fig. 7.4a) on Si(100)-(2×1) exhibits less coverage dependence. This is suggested by the similar IRAS spectra peaks at different coverages of anthracene on Si(100)-(2×1), as shown in Fig. 7.4b [12]. Furthermore, the presence of peaks within the 2800–3000 cm^{-1} range of the spectra indicates the conversion of some of the anthracene carbon atoms from sp^2 to sp^3 hybridization due to their reactions with the substrate. Comparisons of these spectra with those obtained from frequency calculations (presented in the lower half of Fig. 7.4b) suggest that anthracene exists in two configurations along the dimer row, referred to as structures 1 and 2. In structure 1, anthracene forms C–Si bonds with three adjacent pairs of Si dimers via C^2, C^4, C^5, C^7, C^9, and C^{10}. In structure 2, two adjacent pairs of Si dimers interact with C^1, C^3, C^6, and C^8 of anthracene.

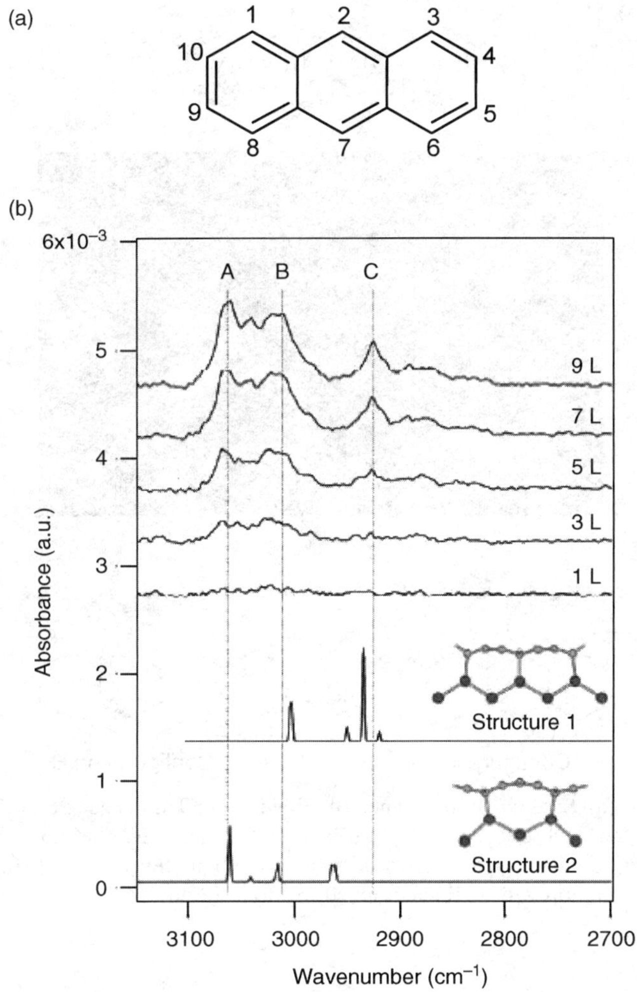

FIGURE 7.4 (a) Structure of anthracene. (b) IRAS spectra for anthracene adsorbed on Si (100)-(2×1) as a function of exposure expressed in units of langmuir. The calculated spectra for two possible adsorption configurations (structures 1 and 2) are shown in lower half of the figure (from Ref. 12).

7.2.2 Tetracene on Si(100)-(2×1)

The adsorption of tetracene (Fig. 7.5a) on Si(100)-(2×1) had been studied using scanning tunneling microscopy (STM) [13, 14], angle-resolved ultraviolet photoelectron spectroscopy (ARUPS) [15], and DFT calculations [15]. Similar to the cases for naphthalene and anthracene, the ARUPS results suggest that the tetracene adsorbates lie parallel to the substrate surface. Indeed, the filled state STM image of tetracene on Si(100)-(2×1), as shown in Fig. 7.5b, exhibits feature sizes that are

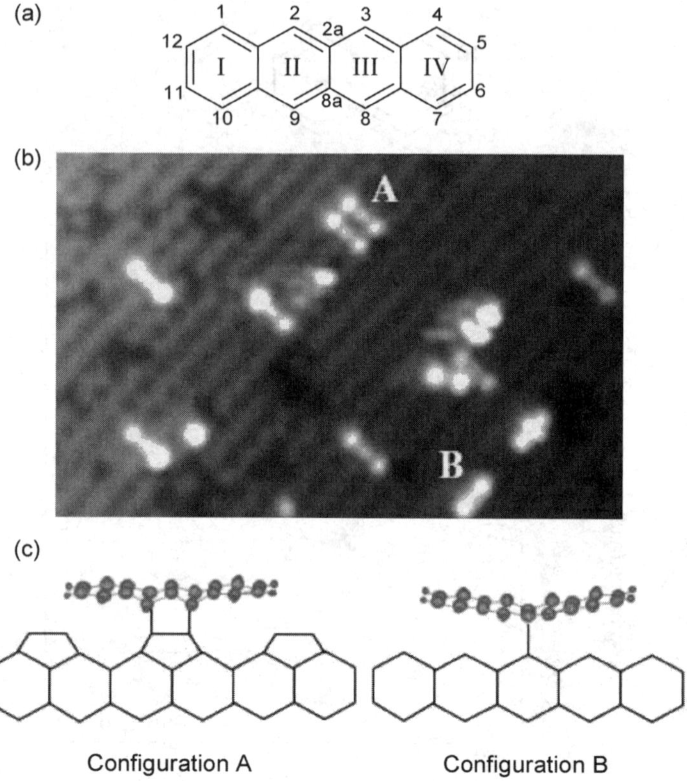

FIGURE 7.5 (a) Structure of tetracene. (b) Filled state STM image of Si(100)-(2×1) with adsorbed tetracene. Observed features are labeled "A" and "B" that corresponds to two types of binding modes. (c) Structures of binding configurations A and B of tetracene on Si(100)-(2 × 1) as proposed by Rada et al. (from Refs. 13 and 14).

consistent with the long axis of a tetracene molecule. Furthermore, surface features that correspond to two types of chemisorbed tetracene (referred to as configurations A and B) are apparent. Configuration A lies perpendicular to the Si dimer rows and comprises of three bright spots with the middle dimmer spot centered over a dimer row. On the other hand, configuration B appears as two bright spots that are aligned parallel to the dimer rows. Based on these STM images, Rada et al. proposed that tetracene in configuration A forms four C–Si bonds with two adjacent Si dimer pairs via the *meso* carbon atoms of rings II and III (C^2, C^3, C^8, and C^9), as shown in Fig. 7.5c. In configuration B, the tetracene adsorbate adopts a symmetrical structure along the dimer row as the central two carbon atoms (C^{2a} and C^{8a}) bind to a pair of Si dimer (Fig. 7.5c). However, it is noted that such a binding configuration involving the central fused carbon atoms has been

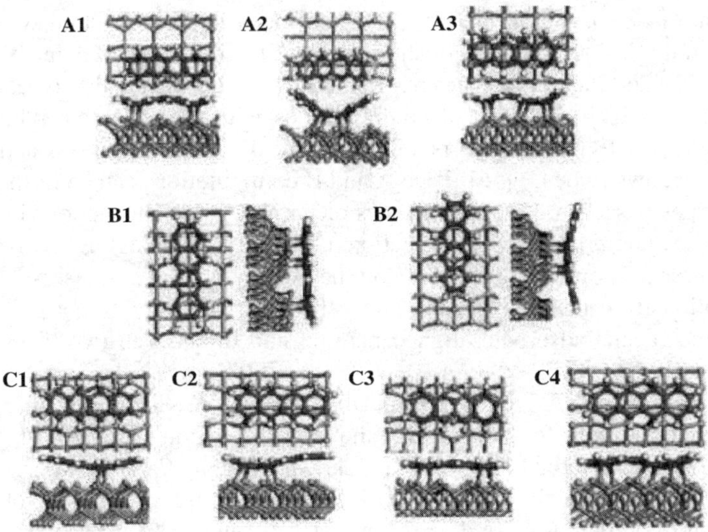

FIGURE 7.6 Different binding configurations of tetracene on Si(100)-(2×1) as calculated by DFT techniques (from Ref. 15).

shown to be thermodynamically unfavorable on Si(111)-(7×7) through theoretical calculation (see Section 7.3.2).

The binding energies of various configurations of tetracene on Si(100)-(2×1), based on DFT calculations, have been compared by Mao et al. [15]. As shown in Fig. 7.6, configurations A1–A3 represent tetracene adsorbates on the dimer row, B1 and B2 denote configurations with the long axes of tetracene aligned perpendicular to the dimer rows, and C1–C4 represent configurations for tetracene adsorbed between two dimer rows. Configuration B1, with C^1, C^2, C^4, C^7, C^9, and C^{10} of tetracene forming six C–Si bonds to the Si dimers, was found to be the most stable structure. Configuration B2, which corresponds to configuration A as proposed by Rada et al. and possesses only four C–Si bonds, is 22.4 kcal/mol less stable than B1. However, the adsorbate in configuration B2 has two aromatic sextets at the outer rings I and IV, whereas tetracene in configuration B1 has only one such sextet in the inner ring II (or ring III). On the other hand, configuration B from Rada et al. was not calculated by Mao and possible assignments of calculated configurations A1–A3 and C1–C4 to STM features observed by Rada et al. require further studies such as STM image simulation.

7.2.3 Pentacene on Si(100)-(2×1)

Due to the technological importance of the Si(100) surface as well as the potential pentacene exhibits as a material with relatively high field-effect mobility, the adsorption of pentacene on Si(100)-(2×1) has been intensively studied using both experimental [16–21] and theoretical techniques [19, 22–24]. Based on STM results, Kasaya et al. identified three types of orientations for the chemisorbed pentacene on the substrate, which comprise of configurations with the adsorbate lying parallel to the Si dimer rows (type A), perpendicular to the dimer rows (type B), and diagonal to the dimer rows (type C) [16]. Type A and B configurations can be further divided into subtypes based on their appearances or number of bright sections in the STM images. They comprise of A1 (two sections), A2 (three sections), B1 (two sections), and B2 (three sections). Hughes et al., on the other hand, did not observe the type C configuration from their STM data, but were able to detect another configuration with the pentacene adsorbate aligned parallel and in-between two Si dimer rows (configuration IB) [17].

Combining the results from STM and DFT calculations, Suzuki et al. proposed the binding configurations for A1, B1, B2, and IB, as shown in Fig. 7.7 [20]. The STM images of these four binding modes are shown in Fig. 7.7a–d(i), where the white markers indicate the positions of the Si dimers. A1, B1, B2, and IB were observed to be the major configurations and exist, respectively, with 20%, 15%, 45%, and 15% probability. Their simulated STM images, based on the proposed structures, agree well with the experimental images. Configuration A1, as shown in Fig. 7.7a(ii) and (iii), has eight carbon atoms from pentacene (lighter shade) bonded to the silicon atoms from the same dimer row. The nonplanar structure is symmetric with respect to the center of the adsorbate and possesses a binding energy of 104.5 kcal/mol. Configuration B1, which has pentacene oriented perpendicular to the Si dimer rows, possesses a higher binding energy of 127.8 kcal/mol and is depicted in Fig. 7.7b(ii) and (iii). This binding mode has the same number of C–Si bonds and positions of bonded pentacene carbon atoms (lighter shade) as A1 but has less degree of bending for the adsorbate. Figure 7.7c(ii) and (iii) shows the optimized structure for configuration B2, another subtype of configuration B. In contrast to B1, however, the structure of pentacene in B2 is asymmetric and possesses only six C–Si bonds. The binding energy of this configuration was calculated as 100.7 kcal/mol. The IB configuration, with the adsorbate aligned between two Si dimer rows, is shown in Fig. 7.7d(ii) and (iii). This configuration has the same structure as A1 and B1 except for the orientations with respect to the underlying dimer rows. The binding energy of configuration IB is 67.1 kcal/mol, which is less than both A1 and B1. The lower energy is attributed to the rearrangement of four pairs of buckled Si dimers from the down to the up positions due to the chemisorbed pentacene in the IB configuration. Besides the four major binding modes, Suzuki et al. also observed configurations A2, C, and a new A3, which together made up for the remaining 5% of the adsorbed pentacene.

The configurations that correspond to type A1 and B1, as proposed by Suzuki et al., have also been found to be the most stable structures for configurations A and B, respectively, via theoretical calculations performed by Choudhary et al. [23] and Nunzi et al. [24]. Both research groups, however, relate the stable tetradimer type A configuration to the A2 binding mode, due to the three bright sections for A2 in the STM images. Interestingly, the experimental STM images

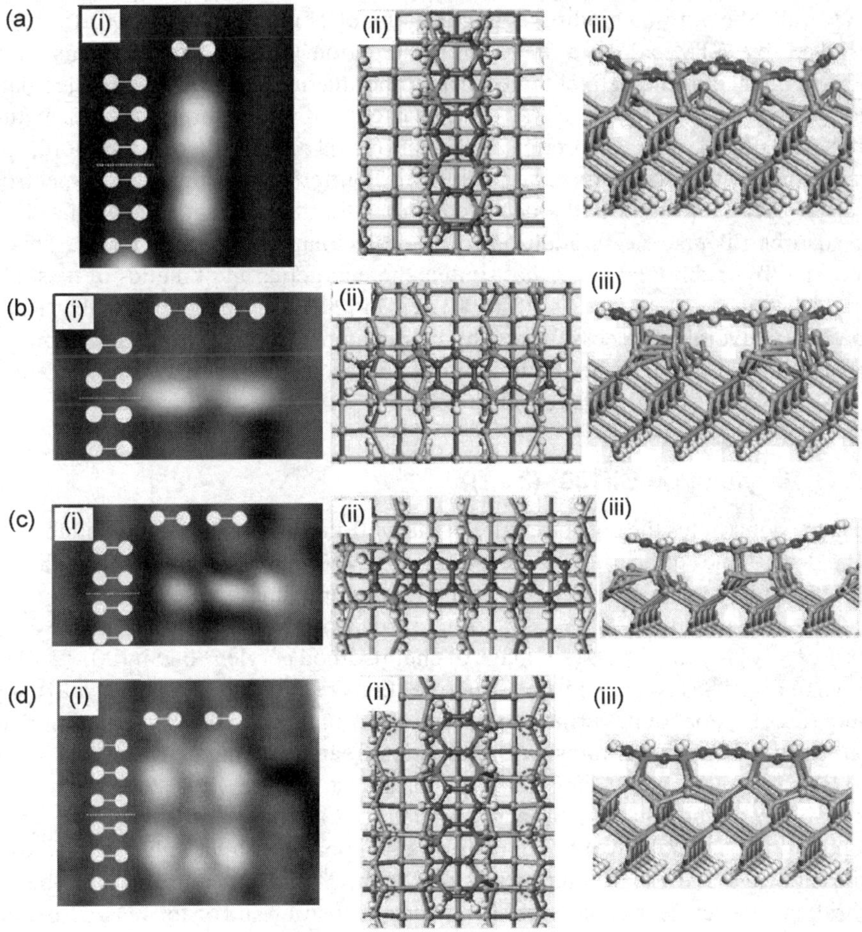

FIGURE 7.7 (i) Experimental STM images and (ii) top and (iii) side views of pentacene on Si(100)-(2×1) in binding configurations (a) A1, (b) B1, (c) B2, and (d) IB (from Ref. 20).

for A1 agree with the simulated image calculated by Suzuki et al., who further argued that the central node for the A1 STM feature originates from electronic, rather than topographic factors. As for configuration B1, Choudhary et al., Nunzi et al., and Suzuki et al. agree upon it as being the most stable configuration for the chemisorbed pentacene on Si(100)-(2×1). The high stability could be attributed to the eight C–Si bonds present as well as the snug fit of pentacene within the adsorption site that helps to minimize any bending of the molecule (Fig. 7.7b(ii) and (iii)). Furthermore, the planar and unreacted central benzene ring allows good overlapping of adjacent π-orbitals that ensures the aromaticity of the ring and hence extra stability for the whole structure.

Overall, the various binding configurations of pentacene on Si(100)-(2×1) as observed by STM adopt a flat-lying orientation on the surface. This is in correspondence to the data obtained from other techniques, such as valence band photoemission studies [17], IRAS [19], near-edge X-ray absorption fine structure (NEXAFS) [21], and theoretical calculations [19, 22–24]. Furthermore, the appearance of the Si–H vibration peak from Fourier transform infrared spectroscopy (FTIR) data suggests the occurrence of a limited amount of dissociation of the adsorbed pentacene in addition to its molecular chemisorption [18]. This is supported by theoretical work that studies the energetics and kinetics of dissociative adsorption of pentacene on Si(100)-(2×1) [24]. The results show that it is a kinetically favorable process for two hydrogen atoms from the sp^3 carbon atoms of a pentacene adsorbate in the IB configuration to be abstracted by two silicon atoms of two dimers bearing unpaired electrons.

7.2.4 Perylene on Si(100)-(2×1)

Perylene constitutes the core structure of the well-known organic semiconductor, 3, 4, 9, 10-perylene-tetracarboxylic dianhydride (PTCDA), which has been studied as the active material in organic solar cells. It is interesting to note that perylene, in contrast to other studied PAHs, was found to adsorb in an upright position on Si (100)-(2×1) [14, 25]. The STM image of chemisorbed perylene on Si(100)-(2×1) is shown in Fig. 7.8a [14]. Chain-like structures that are aligned perpendicular to the Si dimer rows are apparent. Furthermore, each block making up the chain is comprised of five-units of bright features, with individual features being assigned to a single perylene molecule. The length of the brightest feature is 11.35 Å, which is comparable to the longer axis of a perylene molecule of 11.25 Å. Rada et al. proposed that C^8 and C^{11} of perylene form C–Si bonds with the substrate, via C–H bond cleavage and Si–H bond formation. With C^8 and C^{11} remaining in the sp^2 hybridization, perylene chemisorbs in an upright position and on top of two Si atoms from two pairs of adjacent Si dimers, as depicted in Fig. 7.8b.

The binding configuration of perylene on Si(100)-(2×1) as proposed by Rada et al. is consistent with the experimental and theoretical high-resolution electron

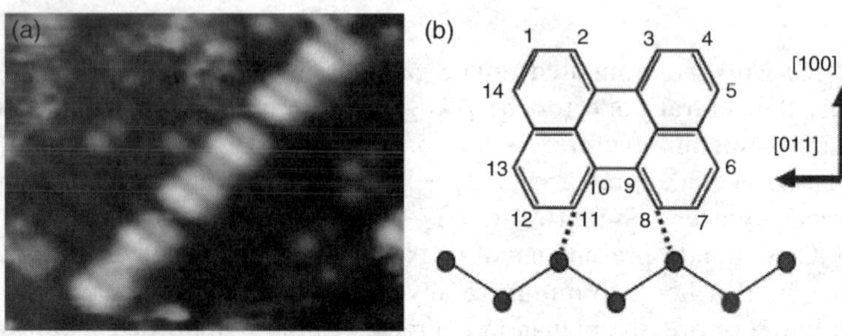

FIGURE 7.8 (a) Filled state STM image and (b) proposed binding configuration of perylene on Si(100)-(2×1) (from Ref. 14).

energy loss spectroscopy (HREELS) data from Preuss et al. [25]. Based on DFT method, Preuss et al. simulated the HREELS spectra of various binding configurations of perylene on Si(100)-(2×1) and found that only the spectrum that was simulated based on the binding model from Rada et al. is in agreement with the experimental HREELS data.

7.2.5 Coronene on Si(100)-(2×1)

Coronene is comprised of six peri-fused benzene rings and can be described by a set of three mobile Clar's aromatic sextets. The adsorption of this molecule on Si(100)-(2×1) has been investigated using STM combined with DFT calculations [26]. Three binding configurations have been identified, which include a major on-top binding mode (~80% of the total population) and two minor substrate defect-related binding modes.

Figure 7.9a(i) shows the filled state STM image (white markers indicate positions of the Si dimers) for the on-top binding mode, which is in good agreement with the simulated image based on the configuration as shown in Fig. 7.9a(ii) and (iii). This configuration involves the formation of four C–Si bonds between the four outer carbon atoms from two opposite rings of coronene (in lighter shade) with two

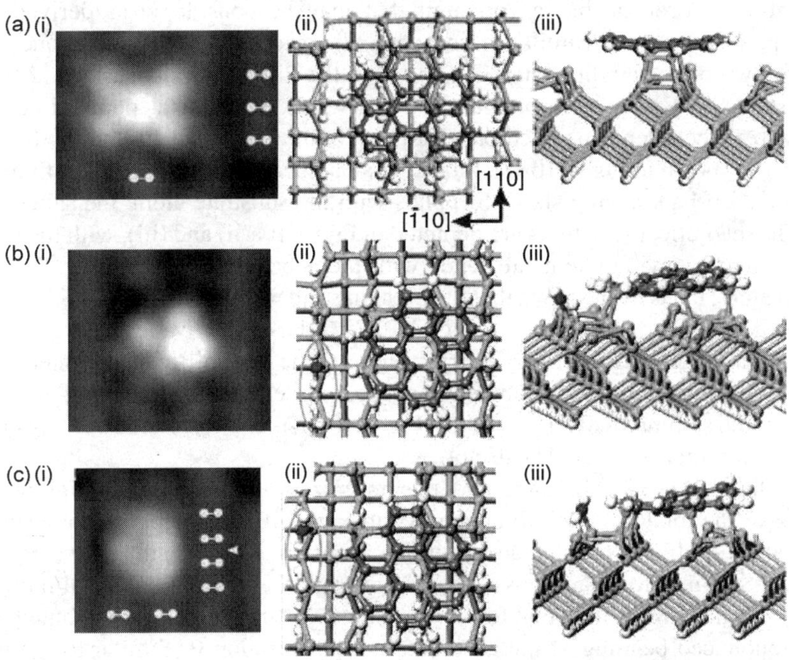

FIGURE 7.9 (i) Experimental STM images and (ii) top and (iii) side views of coronene on Si(100)-(2×1) for (a) major on-top binding mode and (b, c) two minor substrate defect-related binding modes (from Ref. 26).

pairs of Si dimers (with an unreacted Si dimer in-between) along the same row. Such a configuration is found to be stable with a calculated binding energy of 24.6 kcal/mol.

The other two minor binding configurations are located between dimer rows and near to a type C defect with a dissociated chemisorbed water molecule. Figure 7.9b shows the filled state STM image and the optimized configuration of one of the binding modes. The ellipse in Fig. 7.9b(ii) marks the dissociated water molecule with the darkly shaded ball indicating the oxygen atom. Similar to the on-top binding mode, coronene forms four C–Si bonds with the substrate, but at different dimer rows as marked by the lightly shaded balls, and also possesses a comparable binding energy of 24.2 kcal/mol. The other minor binding species of coronene occupies the same position with respect to the Si dimer rows as the first minor species but differs in the relative position with reference to the dissociated water molecule. This is depicted by the optimized configuration in Fig. 7.9c(ii) and (iii), with the filled state STM image shown in Fig. 7.9c(i). However, the calculated binding energy of 32 kcal/mol is much higher than the previous two configurations, which is attributed to the metallic characteristics of the electronic structure for the type C defect.

7.2.6 Dibenzo[*a, j*]coronene on Si(100)-(2×1)

Dibenzo[*a, j*]coronene has a core unit that may be considered as derived from coronene or pentacene. Similar to coronene, dibenzo[*a, j*]coronene has one major (A) and two minor binding sites (B and C) on the Si(100)-(2×1) surface [27]. The filled state STM image of configuration A (~75% of the total population), with dibenzo[*a, j*]coronene absorbed on top of the Si dimer rows (indicated by the white makers), is shown in Fig. 7.10a(i). Suzuki et al. associated this binding configuration with the adsorbate having six C–Si bonds with the substrate along the same dimer row. The theoretical structures are depicted in Fig. 7.10a(ii) and (iii), with the carbon atoms that are having chemical bonds with the substrate circled in Fig. 7.10a(ii). Furthermore, it has been shown that the breaking and formation of the C–Si bonds, as indicated by the dotted arrows in Fig. 7.10a(iii), has an activation energy of only 2.3 kcal/mol. Such a low energy barrier suggests the possibility of the interconversion of the two structures at room temperature, whose average structure gives rise to the observed symmetrical STM feature (Fig. 7.10a(i)). The calculated binding energy of this configuration is 26.2 kcal/mol.

The filled state STM image and the optimized structures of configuration B (~15% of the total population) are shown in Fig. 7.10b. Apparently, the adsorbate forms six C–Si bonds (circled atoms in Fig. 7.10b(ii)) with the substrate across two adjacent Si dimer rows and possesses an adsorption energy of 37.5 kcal/mol. This energy is much higher than that for configuration A, which could be attributed to the less pronounced bending of the adsorbate in configuration B. Similar to the major binding mode, dibenzo[*a,j*] coronene in configuration B could undergo interconversion between two structures via the breaking and formation of C–Si bonds indicated by the dotted arrows in Fig. 7.10b(iii). This is due to the low 0.8 kcal/mol energy

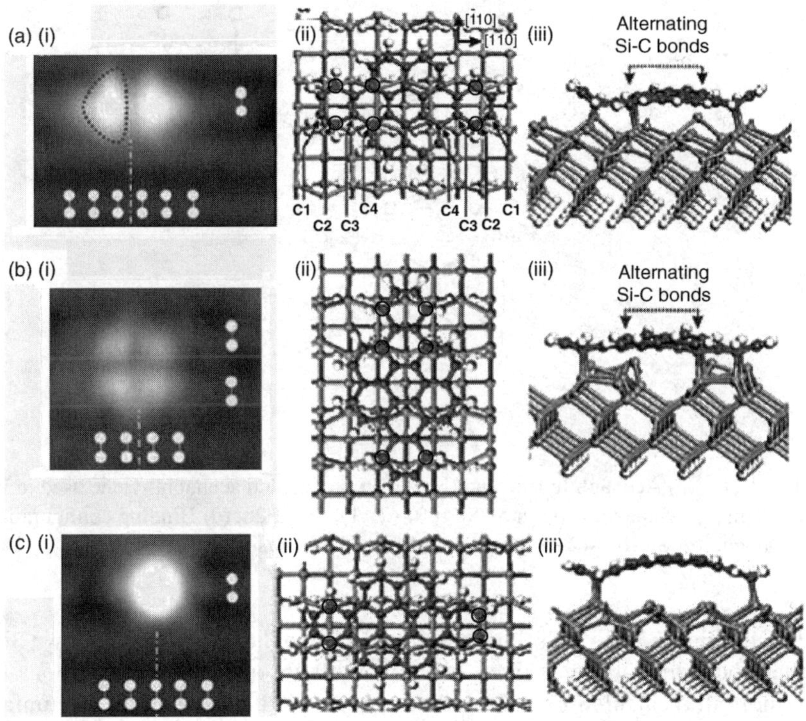

FIGURE 7.10 (i) Experimental STM images and (ii) top and (iii) side views of dibenzo[*a*, *j*]coronene on Si(100)-(2×1) for configurations (a) A, (b) B, and (c) C (from Ref. 27).

barrier for the isomerization to take place. As for configuration C, whose STM image is shown in Fig. 7.10c(i), the adsorbate forms four C–Si bonds with the surface along a Si dimer row as depicted in Fig. 7.10c(ii) and (iii). Due to the lower number of such bonds as compared to the previous two configurations, as well as the pronounced bending of the adsorbate as depicted in Fig. 7.10c(iii), dibenzo[*a*, *j*] coronene in configuration C possesses a much lower binding energy of 17.7 kcal/mol.

7.2.7 Acenaphthylene on Si(100)-(2×1)

As shown in Fig. 7.11a, acenaphthylene is made up of an aromatic naphthalene unit that is conjugated to a C=C bond (hereafter referred to as 1,2-alkene). STM study of the adsorption of acenaphthylene on Si(100)-(2×1) indicates that more than 90% of the adsorbates bind through one type of configuration that appears as a bright protrusion centered on a Si dimer row [28]. The exact binding configuration of the molecule was elucidated via FTIR measurement of Si(100)-(2×1) surfaces that were exposed to acenaphthylene or acenaphthylene molecules that were selectively isotopically labeled with deuterium (Fig. 7.11a–c) [28]. The FTIR data indicate that the 1,2-alkene unit of acenaphthylene reacts with a Si dimer via a [2+2]-like cycloaddition with the unreacted naphthalene unit protruding from the surface, as

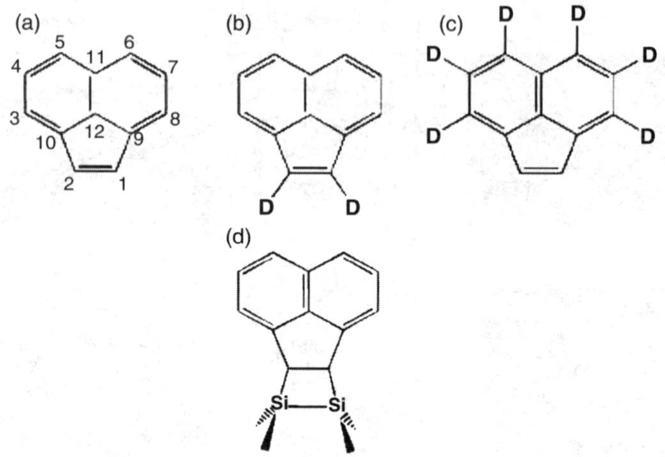

FIGURE 7.11 (a) Acenaphthylene and (b, c) isotopic labeled acenaphthylene used in FTIR study of acenaphthylene adsorption on Si(100)-(2×1) in Ref 28. (d) Binding configuration of acenaphthylene on Si(100)-(2×1) as obtained from FTIR results.

depicted in Fig. 7.11d. This binding configuration has also been shown to be both thermodynamically and kinetically favorable through DFT calculations [29]. Furthermore, the calculation result indicates that it is thermodynamically unfavorable for acenaphthylene to react with the Si dimer via C^3 and C^4 as well as C^{11} and C^{12} at the fusion positions.

It is interesting to note that the naphthalene unit within the adsorbed acenaphthylene retains its aromaticity since only the 1,2-alkene unit is involved in binding to the substrate. In contrast, a large degree of aromaticity is lost for most PAHs when they chemisorb on the Si(100)-(2×1) surface due to the transformations of sp^2 carbon atoms within the aromatic unit to sp^3 hybridization when they react with the substrate. Hence, the adsorption of acenaphthylene on Si(100)-(2×1) provides a strategy for functionalization or modification of the inorganic silicon surface with the aromatic organic unit.

7.3 PAHs on Si(111)-(7×7)

7.3.1 Naphthalene on Si(111)-(7×7)

Studies done using STM and DFT calculations [30] suggest that naphthalene chemisorbs on Si(111)-(7×7) mainly through the formation of two covalent bonds between C^1 and C^6 of ring I (or the equivalent C^2 and C^5 of ring II) and a neighboring adatom and rest atom pair on the substrate. Such a binding mode is denoted as configuration A, as shown in Fig. 7.12b. An STM image of Si(111)-(7×7) with chemisorbed naphthalene is shown in Fig. 7.13. The characteristic feature of at most three adatoms missing within a half unit cell with the three missing adatoms not

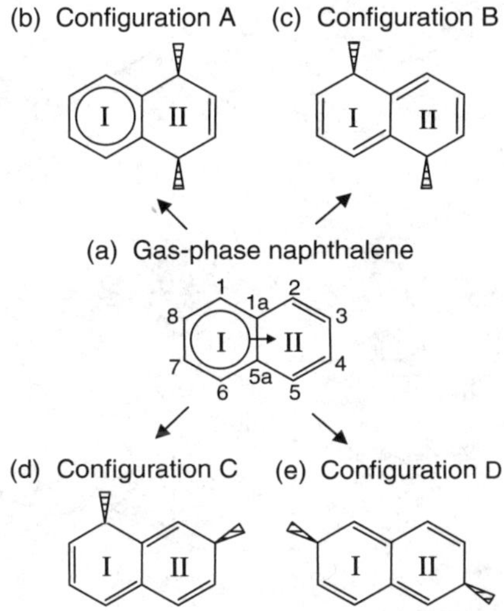

FIGURE 7.12 (a) Sharing of an aromatic sextet within naphthalene based on Clar's sextet concept. (b) Di-σ reactions at ring II of naphthalene produce one sextet in the resulting structure. Reactions at both rings I and II result in configurations (c) B, (d) C, and (e) D of naphthalene on Si(111)-(7×7) that contain conjugated double bonds but no sextet (from Ref. 30).

aligned in a straight line indicates the cooperative binding of naphthalene from an adatom–rest atom pair. The DFT optimized structures of the four possible type A configurations (denoted as configurations A1–A4) are shown in Fig. 7.14a–d. The notations of the clusters are defined in the figure caption.

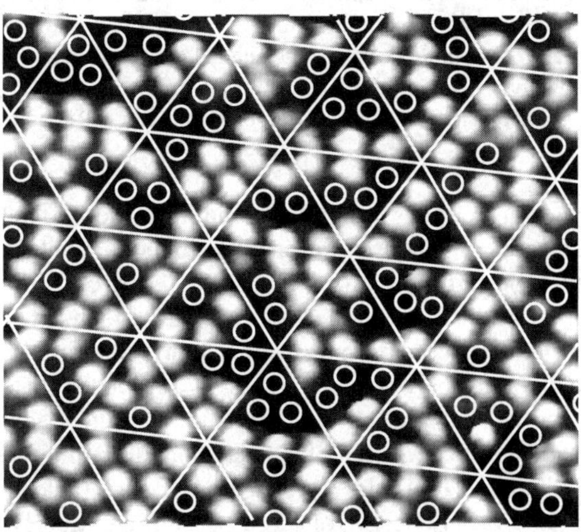

FIGURE 7.13 STM image of Si(111)-(7×7) with adsorbed naphthalene. The superimposed white lines mark out individual half unit cells and each circle indicates a disappeared adatom (from Ref. 30).

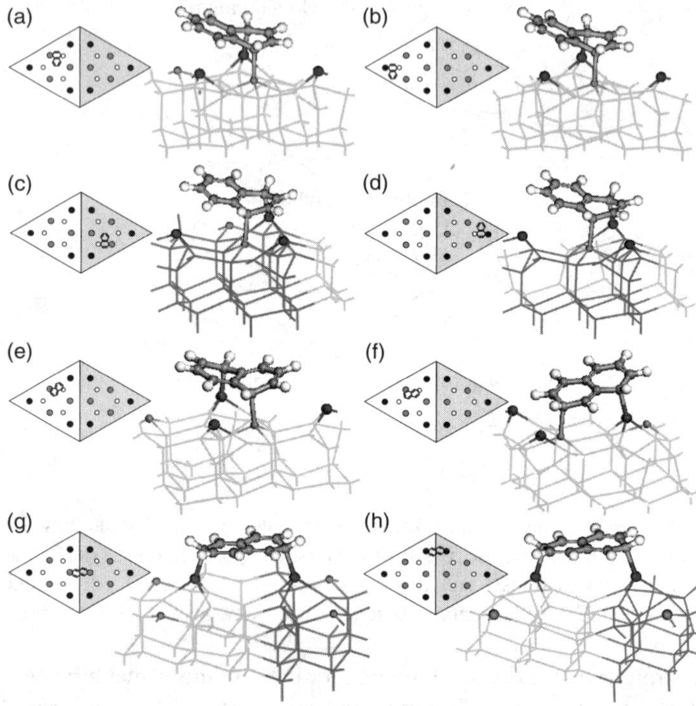

FIGURE 7.14 DFT optimized structures for type (a) A1, (b) A2, (c) A3, (d) A4, (e) B, (f) C, (g) D1, and (h) D2 configurations of naphthalene on Si(111)-(7×7). Notations for each cluster: ball-and-stick, adsorbate; lightly shaded lines, unfaulted half; darkly shaded lines, faulted half; larger and darker shaded spheres, adatoms; smaller and lighter shaded spheres, rest atoms. The schematic diagram on the left of each cluster shows the attachment position of the adsorbate within a (7×7) unit cell, where the shaded and unshaded triangles represent the faulted and unfaulted halves respectively. The center adatoms, corner adatoms, and rest atoms are denoted by the gray, black, and unfilled circles (from Ref. 30).

Besides configuration A, other minor di-σ binding modes of naphthalene with the Si(111)-(7×7) surface include configurations B (with bonding at C^1 and C^5), C (with bonding at C^1 and C^3), and D (with bonding at C^4 and C^8). The DFT optimized structures for configurations B–D are shown in Fig. 7.14e–h. In contrast to configurations A–C that bind to an adatom–rest atom pair, configuration D involves the formation of covalent bonds between the farthest two carbon atoms (C^4 and C^8 which are 5.07 Å apart) with two adjacent faulted and unfaulted adatoms, which are separated by a distance of 6.65 Å. The DFT calculated binding energies of configurations A–D are tabulated in Table 7.1. Apparently, naphthalene in configuration A has the highest adsorption energy followed by configuration D and then configurations B and C. A qualitative approach to comparison of these different binding energies is to utilize Clar's sextet concept [1]. Figure 7.12 shows the aromatic π-conjugation in a free naphthalene molecule and the bonded adsorbates in different

TABLE 7.1 DFT Calculated Binding Energy for Each of the Naphthalene Configuration on Si(111)-(7×7) with the Types of Surface Dangling Bonds That are Involved in the Binding Interaction

Configuration	Surface Dangling Bonds Attached to	Binding Energies (kcal/mol)
A1	UCe, URe	28.2
A2	UCo, URe	27.0
A3	FCe, FRe	27.4
A4	FCo, FRe	29.0
B	UCe, URe	17.4
C	UCe, URe	17.4
D1	UCe, FCe	24.5
D2	UCo, FCo	18.1

From Ref. 30. U: unfaulted half; F: faulted half; Ce: center adatom; Co: corner adatom; Re: rest atom.

configurations. Evidently, only configuration A maintains an aromatic sextet, whose resonance energy contributes to its extra stability as compared to the other configurations. As for configuration D, the planar structure (Fig. 7.14g and h) that allows better π-electron conjugation might account for its higher binding energy as compared to configurations B and C, which suffer greater strain.

Despite the higher stability of configuration A, the existence of one or more binding modes for naphthalene on Si(111)-(7×7) depends on their relative activation barriers to formation. Although no kinetic data concerning the formation of the various configurations are available, it is expected that the activation barriers for the production of type B and C species are relatively high. This can be attributed to the energy required for the more extensive breaking of the aromatic π-bonds within naphthalene to form the twisted structure in these binding modes. Therefore, it is proposed that configuration A is the major structure for chemisorbed naphthalene on Si(111)-(7×7). Furthermore, the possibility of naphthalene binding to an adatom–rest atom pair through the carbon atoms at the fusion positions (C^{1a} and C^{5a}) has been shown, via theoretical calculation, to be thermodynamically unfavorable. A possible explanation for the inertness of these ring fused carbon atoms could be the high steric rigidity due to their strong covalent bonding with three adjacent carbon atoms. In contrast, the nonfused carbon atoms bind to two neighboring carbon atoms and one hydrogen atom.

7.3.2 Tetracene on Si(111)-(7×7)

Both STM and DFT have been employed to study the chemisorption of tetracene on the Si(111)-(7×7) surface [31]. Figure 7.15a and b (with corresponding high-resolution images shown in Fig. 7.15c and d) displays two STM images of the same area of Si(111)-(7×7) that had been exposed to tetracene at sample bias of 1.6 and −2.5 V, respectively. Interestingly, changing of the sample bias resulted in transformations of the surface features from adatoms in (a) and (c) to rest atoms in (b) and (d). Comparison of the missing adatoms (labeled "A") in Fig. 7.15c and the

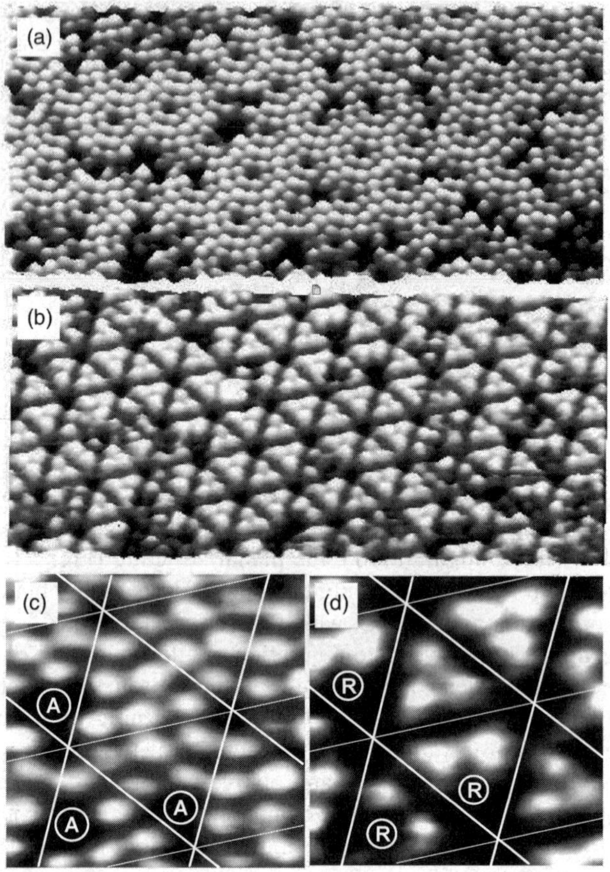

FIGURE 7.15 STM images of tetracene on Si(111)-(7×7) with adatoms visible in the empty state images (a, c) and rest atoms visible in the filled state images (b, d). "A" and "R" indicate the positions of the missing adatoms and rest atoms, respectively (from Ref. 31).

missing rest atoms (labeled "R") in Fig. 7.15d found that the "A" and "R" sites correspond to pairs of adatoms and rest atoms that are adjacent to each other. This result provides convincing evidence for the involvement of neighboring adatom–rest atom pairs on Si(111)-(7×7) in binding tetracene, which saturates the Si dangling bonds and causes them to disappear from the STM images.

Both experimental [32, 33] and theoretical [34, 35] studies have shown that the most reactive sites in tetracene for [4+2]-like addition with Si(111)-(7×7) are the *meso* carbons of the inner rings (namely C^2 and C^9 or the equivalent C^3 and C^8) in Fig. 7.16a. These results may be explained qualitatively by Clar's sextet concept [1], as illustrated in Fig. 7.16. Apparently, reactions at the inner rings create species with two sextets located on both sides of the reacted ring (Fig. 7.16b), whereas interactions at the outer rings result in only one sextet that is shared among the remaining three unreacted rings (Fig. 7.16c). Hence, the interaction of tetracene with the

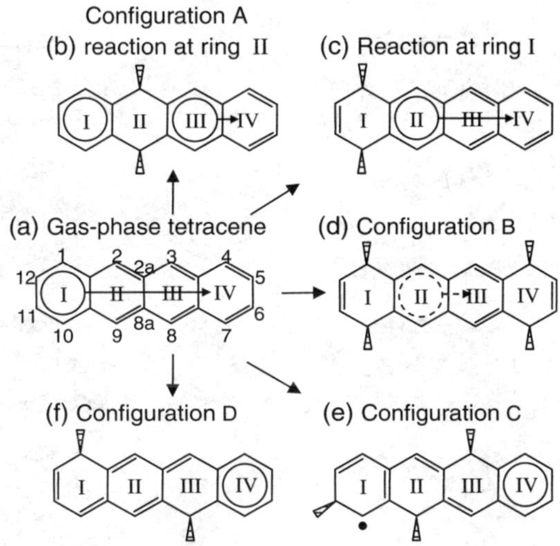

FIGURE 7.16 (a) Sharing of an aromatic sextet within tetracene based on Clar's sextet concept. Di-σ reactions at (b) ring II and (c) ring I of tetracene produce, respectively, two and one sextet in the resulting structure. (d) Loss of aromatic sextet due to nonplanar structure in configuration B of tetracene on Si(111)-(7×7). (e) Tetracene in configuration C contains a radical site and possibly a sextet. (f) Reactions at rings I and III in configuration D result in isolation of a sextet to ring IV (from Ref. 31).

Si(111)-(7×7) adatom–rest atom pair is expected to take place at the inner rings II or III to produce adsorbate that is referred to as type "A" configuration.

The DFT optimized structures of the four possible types of binding modes (referred to as configurations A1–A4) for configurations A are shown in Fig. 7.17a–d, which have the same notation as that used in Fig. 7.14. Apparently, tetracene in configuration A possesses an unsymmetrical butterfly structure that enhances the overlapping of the π-orbitals and in turn the aromaticity on both sides of the reacted ring.

These structures differ from the configuration B of tetracene on Si(100)-(2×1) as proposed by Rada et al. (Section 7.2.2), which involves di-σ reactions between the fused carbon atoms at positions 2a and 8a with the substrate Si dimer. However, DFT calculation indicates that the binding of tetracene to an adatom–rest atom pair via C^{2a} and C^{8a} is thermodynamically unfavorable, in accordance to the theoretical result and explanation as described for naphthalene adsorption on Si(111)-(7×7) in Section 7.3.1.

It is noted that no more than one tetracene in configuration A can bind simultaneously to two or three neighboring center adatoms due to steric hindrance, as illustrated in Fig. 7.18. This figure shows the plan view of three tetracene molecules (one real and two imaginary) adsorbed with configuration A on the unfaulted half of Si(111)-(7×7). The congestion at the central region shows that the dimension of the Si(111)-(7×7) unit cell does not allow the occupations of tetracene adsorbates with type A configuration at two neighboring center adatoms at

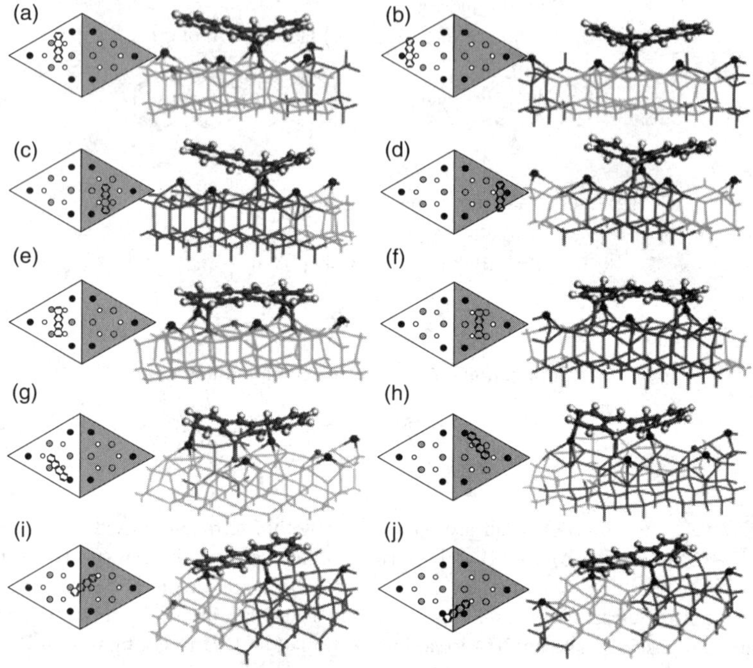

FIGURE 7.17 DFT optimized structures for tetracene binding configurations (a) A1, (b) A2, (c) A3, (d) A4, (e) B1, (f) B2, (g) C1, (h) C2, (i) D1, and (j) D2 on Si(111)-(7×7). The notations are defined the same way as in Fig. 7.14 (from Ref. 31).

the same instant. Nevertheless, the disappearance of two center adatoms within a half unit cell has been observed in the STM results, as shown in Fig. 7.19b. It is proposed that such observation is due to tetracene bonded to two neighboring pairs of center adatom–rest atoms via C^1 and C^{10} of ring I and C^4 and C^7 of ring IV (Fig. 7.19a). Such a binding configuration is feasible as the separation between C^1 and C^4 (7.42 Å from DFT calculation) is comparable to the distance of 7.68 Å between the two center adatoms. This adsorbate species is referred to as type "B" configuration and consists of B1 and B2 binding modes due to reactions at the unfaulted and faulted halves of the (7×7) unit cell, respectively.

The participation of the rest atoms in binding tetracene in configuration B is evidenced from the filled state STM image as shown in Fig. 7.19b(ii). The four circles in the figure denote the adatoms (larger circles) and rest atom (smaller circles) that interacted with tetracene to form covalent bonds. Furthermore, it is apparent that the three neighboring adatoms (marked by the crosses) on an unfaulted half (U) unit cell have greater brightness as compared to the adatoms in the other unfaulted half unit cells that have no reaction. This increase in charge density indicates a transfer of charges from their neighboring rest atoms (the two small circles) so as to facilitate the reaction of these rest atoms, via removal of their excess charges, with tetracene.

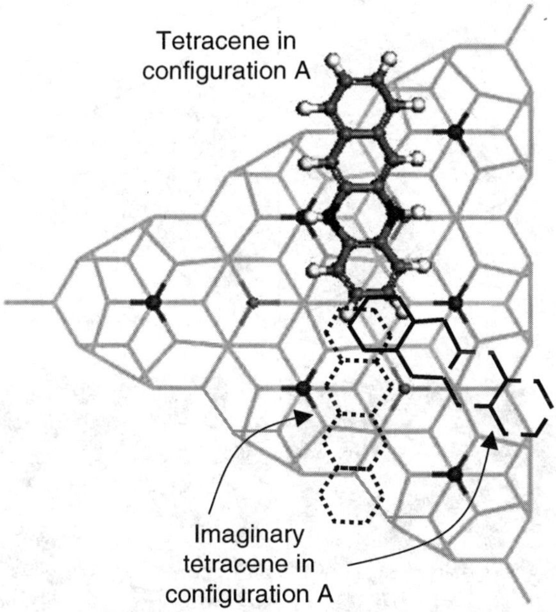

FIGURE 7.18 Cluster model illustrating the adsorption of three tetracene molecules (one real and two imaginary) with configuration A on the unfaulted half of Si(111)-(7×7). The congestion at the central region illustrates the impossibility for two neighboring center adatoms to bind two type A tetracene simultaneously (from Ref. 31).

The DFT optimized clusters for type B1 and B2 configurations are shown, respectively, in Fig. 7.17e and f. Tetracene in these configurations is almost parallel to the surface except for the slight tilt toward the rest atoms due to the approximate 1 Å height difference between the adatom and the rest atom. Furthermore, a sextet that is shared between rings II and III can be identified in these configurations, as shown in Fig. 7.16d. However, it is evident from Fig. 7.17e and f that rings II and III deviate slightly from planarity due to the conversions of the hybridizations of C^1, C^4, C^7, and C^{10} from sp^2 to sp^3. The resultant structure would impede the efficient overlapping of the π-orbitals within the sextet and thereby cause a loss in aromaticity.

Overall, configurations A and B adopt almost flat-lying orientations on the substrate, which agree with the conclusion obtained from a NEXAFS study of Si(111)-(7×7) with low coverage of tetracene [36]. Other binding modes such as configurations C and D were proposed as shown in Fig. 7.17g–j. Configuration C involves the formation of three covalent bonds between C^3, C^9, and C^{11} of tetracene and the Si dangling bonds at one half of the (7×7) unit cell while configuration D has C^1 and C^8 bonded to two adjacent adatoms at the border of a faulted and an unfaulted half unit cell. However, the STM results cannot exclusively assign a particular feature to the two binding modes since other configurations could possibly produce a similar feature.

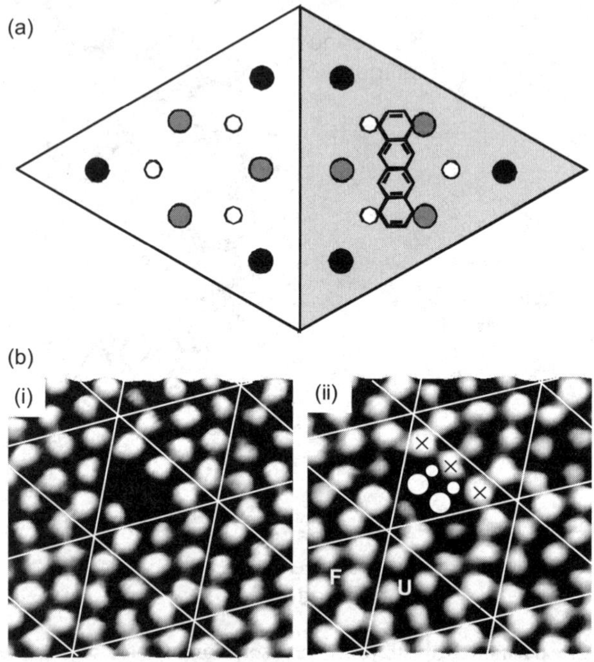

FIGURE 7.19 (a) Schematic diagram and (b) STM images of tetracene with type B binding configuration on Si(111)-(7×7). (b) (i) and (ii) were obtained at $V_s = 1.5$ and -1.5 V, respectively, with F and U denoting the faulted and unfaulted half unit cells (from Ref. 31).

The DFT calculated binding energies for the various configurations are tabulated in Table 7.2. The presence of four C–Si covalent bonds in configuration B may account for its higher adsorption energy as compared to configuration A, which comprises of only two such bonds. Furthermore, the B1 configuration that is located at the unfaulted half unit cell is slightly more stable than B2 at the faulted half. This is in accordance to Brommer's concept [37], which states that electron-donating species would prefer to react with the center adatoms on the faulted as compared to the unfaulted half. Mulliken charge difference calculation performed on a tetracene adsorbate in B1 configuration points to a charge transfer of ~0.79*e* from the substrate to the adsorbate. This indicates that tetracene is an electron acceptor with respect to Si(111)-(7×7) and hence is expected to be more stable on the unfaulted center adatoms.

7.3.3 Pentacene on Si(111)-(7×7)

Due to its larger size as compared to naphthalene and tetracene, pentacene exhibits more interesting adsorption configurations on Si(111)-(7×7), as revealed by a combined STM and DFT study [38]. Three types of binding configurations, referred to as configurations A, B, and C, were identified from the STM images. These

TABLE 7.2 DFT Calculated Binding Energy for Various Tetracene Binding Configuration on Si(111)-(7×7) with the Types of Surface Dangling Bonds That are Involved in the Interaction

Configuration	Surface Dangling Bonds Attached to	Binding Energies (kcal/mol)
A1	UCe, URe	42.8
A2	UCo, URe	43.1
A3	FCe, FRe	42.8
A4	FCo, FRe	42.1
B1	2UCe, 2URe	56.5
B2	2FCe, 2FRe	52.1
C1	UCe, UCo, URe	45.0
C2	FCe, FCo, FRe	45.7
D1	UCe, FCe	35.4
D2	UCo, FCo	37.5

From Ref. 31. U: unfaulted half; F: faulted half; Ce: center adatom; Co: corner adatom; Re: rest atom.

configurations comprise four C–Si covalent bonds that involve the surface adatoms and rest atoms, in accordance to the results from an XPS study [39]. Figure 7.20 shows the schematic diagram of each of these configurations with their corresponding STM images.

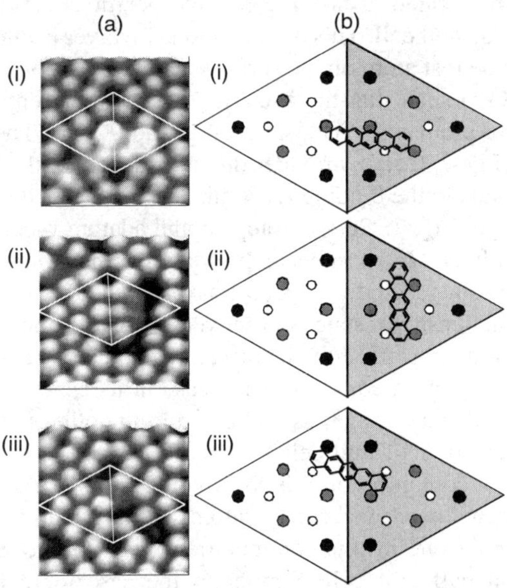

FIGURE 7.20 (a) Empty state STM images and (b) schematic diagrams showing the configurations of pentacene on a Si(111)-(7×7) unit cell for type (i) A, (ii) B, and (iii) C binding modes (from Ref. 38).

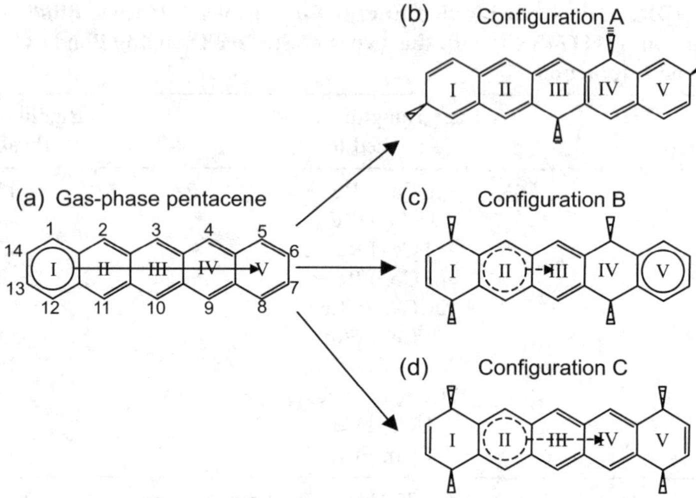

FIGURE 7.21 (a) Sharing of an aromatic sextet within pentacene based on Clar's sextet concept. (b) Pentacene in configuration A contains conjugated double bonds but no sextet. (c) Presence of a sextet at ring V of pentacene in configuration B. (d) Aromaticity loss at rings II, III, and IV due to twisted structure of pentacene in configuration C (from Ref. 38).

From the STM image shown in Fig. 7.20a(i), configuration A appears as two bright spots that are located at specific relative positions across the faulted and unfaulted halves of the unit cell. This binding mode involves a unique engagement of three adatoms and one rest atom on the surface with four carbon atoms of pentacene (C^4, C^6, C^{10}, and C^{13}) such that the π-conjugation within pentacene is preserved despite a loss in the aromaticity (Figs. 7.20b(i) and 7.21b). The filled state STM image, as shown in Fig. 7.22a(ii), provides direct evidence for the involvement in the surface dangling bonds in the binding reaction. The two spots that are observable in the empty state image (Fig. 7.22a(i)) disappear and adatom vacancies are also seen. The four circles in Fig. 7.22a(ii) indicate the adatoms (three bigger circles) and rest atom (smaller circle) that reacted with pentacene and hence become invisible due to a reduction in the local density of states. Moreover, the involvement of the rest atom is evidenced by the charge transfer from the rest atom to the neighboring corner adatom (marked by the cross) with a consequent increase in its charge density (and hence brightness in the filled state image) as compared to the other two unreacted corner adatoms within the same half unit cell.

The STM image of configuration B is shown in Fig. 7.20a(ii), which is comprised of a dark feature located over two center adatoms within a half unit cell. As depicted in Fig. 7.20b(ii), pentacene in this configuration binds to two center adatom–rest atom pairs within a half unit cell. Similar to the case for configuration A, the participation of the rest atom in the binding reaction can be observed from the filled state STM image as shown in Fig. 7.22b(ii). The two small circles in the figure denote the rest atoms that reacted with pentacene, with consequent charge transfer to their

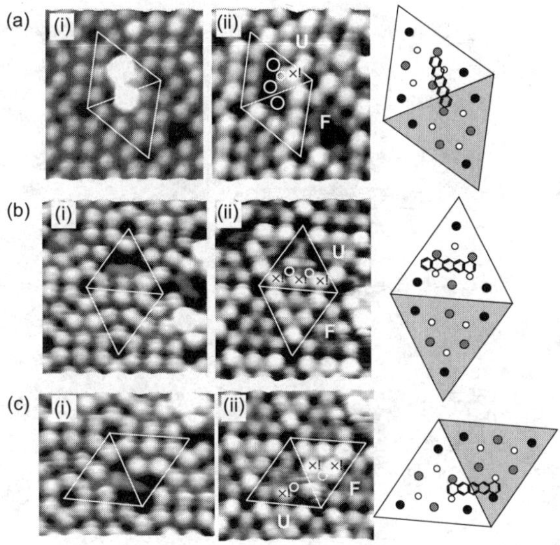

FIGURE 7.22 STM images of pentacene on Si(111)-(7×7) in binding configurations (a) A, (b) B, and (c) C acquired at sample biases of (i) 1.8 V and (ii) −1.8 V. F and U denote faulted and unfaulted half of the unit cell, respectively, and the corresponding schematic diagrams show the attachment positions of pentacene (from Ref. 38).

neighboring unfaulted adatoms (marked by the crosses) that achieve the same brightness as the corner adatoms in the faulted half.

Figure 7.20a(iii) shows the STM image of pentacene in configuration C, which also appears as a dark feature, but straddles the border between the faulted and unfaulted half unit cell. As depicted in Fig. 7.20b(iii), the two terminal rings of pentacene in this configuration form di-σ-bonds with two oppositely oriented adatom−rest atom pairs located on two different halves of a unit cell. Likewise, the involvement of the rest atoms in the binding reaction can be verified by the filled state STM image (Fig. 7.22c(ii)). The center adatoms (marked by the crosses) appear unusually bright, which points to an incoming charge transfer from their neighboring rest atoms (marked by the circles) following their reactions with pentacene.

The DFT optimized structures for configurations A, B, and C are shown in Fig. 7.23. Four types of binding modes are possible for configuration A and they are referred to as configurations A1–A4. As depicted in Fig. 7.23a–d, either a combination of three center adatoms plus one rest atom (for A1 and A3 configurations) or one center and two corner adatoms plus one rest atom (for A2 and A4 configurations) take part in binding pentacene in configuration A. Furthermore, the pentacene adsorbate adopts an almost flat-lying orientation on the surface in this configuration.

Figure 7.23e and f shows the structures for configurations B1 and B2, which are the two possible binding modes for configuration B due to reaction at the unfaulted and faulted halves, respectively. Due to the ~1 Å height difference between the

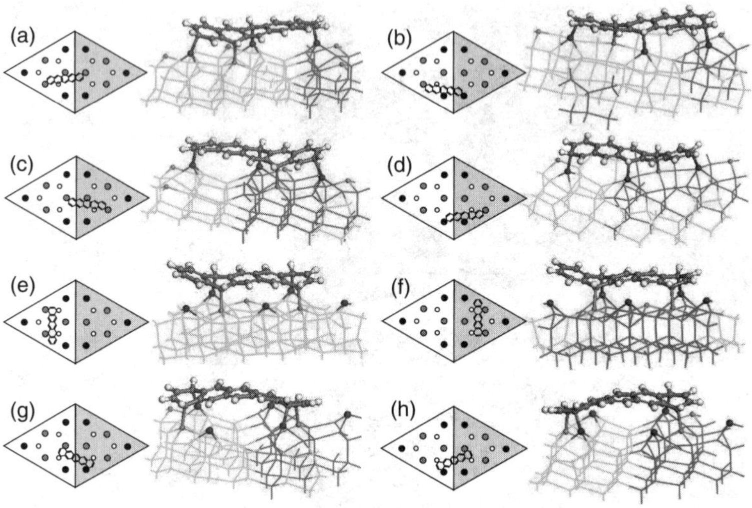

FIGURE 7.23 DFT optimized structures for pentacene binding configurations (a) A1, (b) A2, (c) A3, (d) A4, (e) B1, (f) B2, (g) C1, and (h) C2 on Si(111)-(7×7). The notations are defined the same way as in Fig. 7.14 (from Ref. 38).

adatom and rest atom, pentacene in configuration B is slightly titled toward the rest atoms. This binding mode is the same as that for configuration B of tetracene on Si(111)-(7×7) (Section 7.3.2) since pentacene is made up of a tetracene unit with an additional fused benzene ring. As for configuration C, the two possible binding modes (C1 and C2) are shown in Fig. 7.23g and h. Apparently, pentacene in configuration C is twisted due to the opposite orientation of the two reacted adatom–rest atom pairs as well as the height difference between the adatom and the rest atom.

TABLE 7.3 DFT Calculated Binding Energy for Various Pentacene Binding Configuration on Si(111)-(7×7) with the Types of Surface Dangling Bonds That are Involved in the Interaction

Configurations	Surface Dangling Bonds Attached to	Binding Energies (kcal/mol)
A1	2UCe, URe, FCe	56.8
A2	UCe, UCo, URe, FCo	56.7
A3	UCe, 2FCe, FRe	56.8
A4	UCo, FCe, FCo, FRe	56.1
B1	2UCe, 2URe	64.5
B2	2FCe, 2FRe	72.7
C1	UCe, URe, FCo, FRe	49.9
C2	UCo, URe, FCe, FRe	48.9

From Ref. 38. U: unfaulted half; F: faulted half; Ce: center adatom; Co: corner adatom; Re: rest atom.

The DFT calculated binding energies for the various configurations are tabulated in Table 7.3, which can again be qualitatively compared using Clar's sextet concept [1]. As illustrated in Fig. 7.21b, pentacene in configuration A experiences a complete loss of the aromatic sextet though π-conjugation can still be observed. In contrast, the di-σ reactions at rings I and IV that lead to formation of configuration B would isolate the sextet to ring V (Fig. 7.21c). The possible sextet shared between rings II and III could not be realized due to the deviations of these two rings from planarity (Fig. 7.23e and f), which impede the efficient overlapping of the π-orbitals. Since both configurations A and B possess four C–Si bonds, the extra stability of configuration B could be accounted for by the resonance energy from the aromatic sextet in ring V. As for configuration C, the twisted structures of the adsorbates, as shown in Fig. 7.23g and h, prevent the proper overlapping of the π-orbitals from taking place and hence destroy the aromatic sextet that could be formed and shared among rings II, III, and IV (Fig. 7.21d). Together with the strain induced within the pentacene adsorbate as caused by the twist, it is understandable that pentacene in configuration C possesses the lowest binding energy among the three configurations.

7.4 SUMMARY

Through the use of various experimental techniques combined with theoretical calculations, understanding has been gained on the reactions of PAHs with the Si(100)-(2×1) and Si(111)-(7×7) surfaces. The PAH carbon atoms react with the Si dangling bonds and undergo transformations from sp^2 to sp^3 hybridization to form C–Si covalent bonds with the substrates. The adsorbates take up configurations that are dependent on their relative dimensions with respect to the positions of the dangling bonds. On the Si(100)-(2×1) surface, the dangling bonds from each dimer pair cooperatively react with the adsorbates, which then adopt orientations that are parallel or perpendicular to the dimer rows. As for Si(111)-(7×7), the adatom–rest atom pair reacts and forms σ-bonds with two opposite carbon atoms from a benzene ring within the PAHs. Additional configurations exist for PAHs with larger dimensions, such as coronene and pentacene, which are able to react with individual dangling bonds from different Si dimer (on Si(100)-(2×1)) or adatom–rest atom (on Si(111)-(7×7)) pairs.

The binding energies of the various configurations differ and depend on the number of C–Si covalent bonds as well as the degree of π-conjugation or aromaticity in the resultant structures. In general, adsorbates with larger extent of bending or twisting have a reduced amount of π-orbital overlap and hence lower binding energies. The strong interactions between PAHs and the silicon surfaces lead to the almost flat-lying orientations of the adsorbates typically observed, though perylene was found to adopt a standing orientation on Si(100)-(2×1). It is also interesting to note that acenaphthylene chemisorbs on Si(100)-(2×1) via the 1,2-alkene unit and leaves the aromatic naphthalene unit unperturbed. Such atomistic understanding of the reaction mechanisms and binding configurations of PAHs on the silicon substrates contribute to knowledge of semiconductor surface science as well as research in the areas of organic–inorganic hybrid devices, organic electronics, and molecular electronics.

REFERENCES

1. (a) Clar, E. *Polycyclic Hydrocarbons*, Vols. 1 and 2. Academic Press, London, 1964. (b) Clar, E. *The Aromatic Sextet*. Wiley, London, 1972.
2. Shaw, J. M.; Seidler, P. F. *IBM J. Res. Dev.* **2001**, *45*, 3.
3. Bendikov, M.; Wudl, F.; Perepichka, D. F. *Chem. Rev.* **2004**, *104*, 4891.
4. Mabrook, M. F.; Yun, Y.; Pearson, C.; Zeze, D. A.; Petty, M. C. *Appl. Phys. Lett.* **2009**, *94*, 173302.
5. Pandey, A. K.; Nunzi, J.-M. *Appl. Phys. Lett.* **2006**, *89*, 213506.
6. Muccini, M. *Nat. Mater.* **2006**, *5*, 605.
7. Choi, J.-M.; Lee, J.; Hwang, D. K.; Kim, J. H., Im, S. *Appl. Phys. Lett.* **2006**, *88*, 043508.
8. Shao, Y.; Sista, S.; Chu, C.-W.; Sievers, D., Yang, Y. *Appl. Phys. Lett.* **2007**, *90*, 103501.
9. Filler, M. A.; Bent, S. F. *Prog. Surf. Sci.* **2003**, *73*, 1.
10. Cornil, J.; Beljonne, D.; Calbert, J.-P.; Brédas, J.-L. *Adv. Mater.* **2001**, *13*, 1053.
11. Okamura, K.; Ishii, H.; Kimura, Y.; Niwano, M. *Surf. Sci.* **2005**, *576*, 45.
12. Okamura, K.; Hosoi, Y.; Kimura, Y.; Ishii, H.; Niwano, M. *Appl. Surf. Sci.* **2004**, *237*, 439.
13. Rada, T.; Chen, Q.; Richardson, N. V. *J. Phys.* **2003**, *15*, S2749.
14. Rada, T.; Chen, Q.; Richardson, N. V. *Phys. Status Solidi b* **2004**, *241*, 2353.
15. Mao, H.; Guan, D.; Chen, M.; Dou, W.; Song, F.; Zhang, H.; Li, H.; He, P.; Bao, S. *J. Chem. Phys.* **2009**, *131*, 044703.
16. Kasaya, M.; Tabata, H.; Kawai, T. *Surf. Sci.* **1998**, *400*, 367.
17. Hughes, G.; Roche, J.; Carty, D.; Cafolla, T.; Smith, K. E. *J. Vac. Sci. Technol. B* **2002**, *20*, 1620.
18. Weidkamp, K. P.; Hacker, C. A.; Schwartz, M. P.; Cao, X.; Tromp, R. M.; Hamers, R. J. *J. Phys. Chem. B* **2003**, *107*, 11142.
19. Okamura, K.; Hosoi, Y.; Kimura, Y.; Ishii, H.; Niwano, M. *Appl. Surf. Sci.* **2004**, *237*, 439.
20. Suzuki, T.; Sorescu, D. C.; Yates, J. T., Jr. *Surf. Sci.* **2006**, *600*, 5092.
21. Lee, H.-K.; Han, J.-H.; Kim, K.-J.; Kang, T.-H.; Kim, B. *Surf. Sci.* **2007**, *601*, 1456.
22. Yamaguchi, T. *J. Phys. Soc. Jpn.* **1999**, *68*, 1321.
23. Choudhary, D.; Clancy, P.; Bowler, D. R. *Surf. Sci.* **2005**, *578*, 20.
24. Nunzi, F.; Sgamellotti, A.; Coletti, C.; Re, N. *J. Phys. Chem. C* **2008**, *112*, 6033.
25. Preuss, M.; Miotto, R.; Bechstedt, F.; Rada, T.; Richardson, N. V.; Schmidt, W. G. *Phys. Rev. B* **2006**, *74*, 115402.
26. Suzuki, T.; Sorescu, D. C.; Jordan, K. D.; Levy, J.; Yates, J. T., Jr. *J. Chem. Phys.* **2006**, *124*, 054701.
27. Suzuki, T.; Sorescu, D. C.; Jordan, K. D.; Yates, J. T., Jr. *J. Chem. Phys.* **2006**, *124*, 224708.

28. Schwartz, M. P.; Halter, R. J.; McMahon, R. J.; Hamers, R. J. *J. Phys. Chem. B* **2003**, *107*, 224.
29. Nunzi, F.; Sgamellotti, A.; Re, N. *J. Phys. Chem. B* **2004**, *108*, 10881.
30. Yong, K. S.; Zhang, Y. P.; Yang, S.-W.; Xu, G. Q. *Surf. Sci.* **2008**, *602*, 1921.
31. Yong, K. S.; Zhang, Y. P.; Yang, S.-W.; Wu, P.; Xu, G. Q. *J. Phys. Chem. A* **2007**, *111*, 12266.
32. Biermann, D.; Schmidt, W. *J. Am. Chem. Soc.* **1980**, *102*, 3163.
33. Murata, Y.; Kato, N.; Fujiwara, K.; Komatsu, K. *J. Org. Chem.* **1999**, *64*, 3483.
34. Schleyer, P. v. R.; Manoharan, M.; Jiao, H.; Stahl, F. *Org. Lett.* **2001**, *3*, 3643.
35. Cheng, M. F.; Li, W.-K. *Chem. Phys. Lett.* **2003**, *368*, 630.
36. Schedel, Th. ; Frank, K.-H.; Karisson, U.; Koch, E. E. *Vacuum* **1990**, *41*, 652.
37. Brommer, K. D.; Galván, M.; Dal Pino, Jr., A.; Joannopoulos, J. D. *Surf. Sci.* **1994**, *314*, 57.
38. Yong, K. S.; Zhang, Y. P.; Yang, S. W.; Wu, P.; Xu, G. Q. *J. Phys. Chem. C* **2007**, *111*, 4285.
39. Hughes, G; Carty, D.; Cafolla, A. A. *Surf. Sci.* **2005**, *582*, 90.

CHAPTER 8

Dative Bonding of Organic Molecules

YOUNG HWAN MIN, HANGIL LEE, DO HWAN KIM, AND SEHUN KIM

8.1 INTRODUCTION

8.1.1 What is Dative Bonding?

A dative bond is a coordinate covalent bond between atoms such that both bonding electrons are contributed by only one of the atoms involved in the bond. In other words, dative bonding occurs when one atom donates an electron pair and the other atom accepts the electron pair, and the atoms share the electrons equally. This bonding mode is distinct from ionic bonding modes. The donating atom is called a Lewis base (an electron pair donor) and the accepting atom is called a Lewis acid (an electron pair acceptor). Formation of a dative bond is, therefore, a Lewis acid/Lewis base reaction or a nucleophilic/electrophilic reaction. One of the most well-known examples of dative bonding is the bond between ammonia (NH_3) and boron trifluoride (BF_3). The boron of BF_3 has only six valence electrons and requires two more electrons to satisfy the octet rule. Due to this deficiency, BF_3 can act as a Lewis acid that accepts lone pair electrons. NH_3, on the other hand, can act as a Lewis base by donating its two lone pair electrons. Thus, NH_3 donates its two lone pair electrons to BF_3 to create a dative bond, NH_3:BF_3.

Interestingly, dative bonding occurs between Lewis bases (or Lewis acids) and reconstructed semiconductor surfaces such as Si(100)-(2×1), Ge(100)-(2×1), and Si(111)-(7×7). Semiconductor surfaces are prepared by cleavage and are reconstructed for stabilization by reducing the number of exposed unstable dangling bonds. In (2×1) reconstructed Si and Ge(100) surfaces, pairs of surface atoms form dimers held together by a strong σ-bond and a weak π-bond, similar to a C=C double bond. The dimers are tilted such that one of the two atoms is tilted up (elevated out of the plane of the substrate surface) and the other is tilted down (recessed below the plane of the substrate surface); the asymmetric dimer is more stable than the symmetric dimer [1]. The tilting of the dimer leads to partial electron transfer from

Functionalization of Semiconductor Surfaces, First Edition.
Edited by Franklin (Feng) Tao and Steven L. Bernasek.

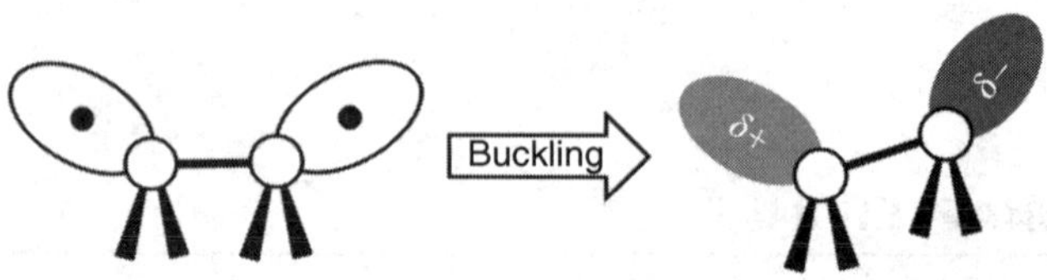

FIGURE 8.1 The tilted dimers of Si(100) and Ge(100) surfaces. The tilting of dimers leads to a partial electron transfer from the buckled-down atom to the buckled-up atom. The dark gray circle indicates the electron-rich orbital, and the light gray circle indicates the electron-deficient orbital.

the down atom to the up atom. As a result, the up atom is electron rich and the down atom is electron deficient (see Fig. 8.1) [2]. Because the down atom is electron deficient, it can react with a Lewis base to form a dative bond (see Fig. 8.2a) [3]. In contrast, the electron-rich up atom can donate its electron density to a Lewis acid to form a dative bond (see Fig. 8.2b).

In the case of Si(111)-(7×7), surface reconstruction reduces the number of unstable dangling bonds from 49 to 19 in each unit cell [4, 5]. Upon reconstruction, the rest atom and corner atom have a formal charge of −1, whereas each adatom has a formal charge of +7/12. The amount of charge transferred from a center adatom to a rest atom is twice that transferred from a corner adatom because each center adatom has two neighboring rest atoms and each corner adatom has only one neighboring rest atom. Due to this difference, the corner adatom has a higher electron density than the center adatom, and as a result, they have different reactivities. The couples formed by an adatom and a neighboring rest atom are zwitterionic in character, similar to the buckled dimers of Si(100)-(2×1) and Ge(100)-(2×1) surfaces.

8.1.2 Periodic Trends in Dative Bond Strength

Two factors determine the dative bond strength between a Lewis base and the electron-deficient down atom of a (100) surface dimer: the size of the HOMO on the Lewis base and the electronegativity (EN) of the electron-donating atom. The strength of a dative bond involving an atom from group V or VI decreases as one moves down the group because the HOMO size increases with increasing atomic mass, reducing the orbital overlap between the HOMO of the Lewis base and the electron-deficient down atom. However, the strength of a dative bond increases as the electronegativity decreases, that is, as one moves down the group, because a less electronegative atom donates electrons more easily to an electron-deficient down atom. Therefore, the strength of a dative bond, as one moves down a group in the periodic table, is determined by the net balance between orbital overlap and electronegativity (see Fig. 8.3).

Previously reported DFT calculations have compared the strengths of dative bonds between the group V Lewis bases dimethylamine (N), dimethylphosphine (P), and dimethylarsine (As) and the down Ge atom of a Ge(100) surface [6]. The calculations indicate that the P−Ge dative bond is slightly stronger than the N−Ge dative bond by 1.1 kcal/mol, whereas the As−Ge dative bond is weaker than the N−Ge

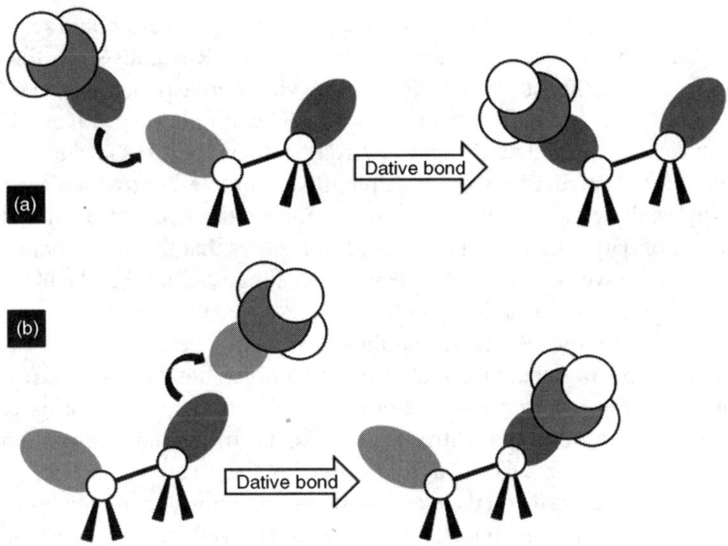

FIGURE 8.2 (a) The electron-deficient down atom can react with a Lewis base to form a dative bond. (b) The electron-rich up atom can react with a Lewis acid to form a dative bond. The dark gray circles indicate electron-rich orbitals, and the light gray circles indicate electron-deficient orbitals.

dative bond by 4.5 kcal/mol. In the case of P and N, the smaller orbital overlap in P–Ge than in N–Ge is compensated by the lower electronegativity of P (2.19) than of N (3.04). This compensation results in approximately equal N–Ge and P–Ge dative bond strengths. However, the electronegativities of P (2.19) and As (2.18) are similar. Therefore, orbital overlap dominates the formation of a dative bond, and the smaller orbital overlap in As–Ge results in a weaker dative bond than is observed in P–Ge.

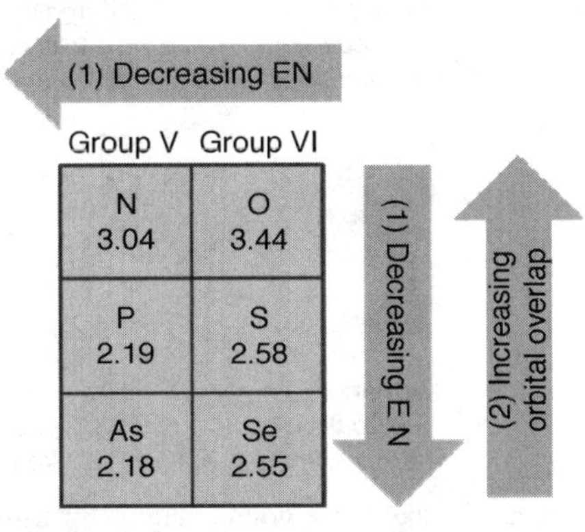

FIGURE 8.3 The overall effects of (1) electronegativity and (2) orbital overlap determine the dative bond strength. As the electronegativity decreases and the orbital overlap increases, the dative bond strength increases. The numbers under the elements indicate the Pauling electronegativity of each element.

In the case of group VI, the electronegativities of oxygen, sulfur, and selenium are 3.44, 2.58, and 2.55, respectively. Previously reported IR studies show that diethyl ether (O–Ge) and diethyl sulfide (S–Ge) adsorb via a dative bond on Ge(100)-(2×1) surfaces with bond strengths of 17.9 and 23.8 kcal/mol, respectively [7]. This indicates that the S–Ge dative bond is stronger than the O–Ge dative bond by 5.9 kcal/mol. DFT calculations of the strengths of dative bonds involving diethyl ether, diethyl sulfide, and diethyl selenide support these experimental data [6]. A comparison of dative bond energies (O, S, Se) shows that the dative bond strength increases as one moves down the group VI. This suggests that the electronegativity effects dominate group VI dative bonds.

The electronegativity effects are applicable along a period as well as down a group. Comparison of methanol and trimethylamine, both of which feature sp^3 hybridization on the electron-donor atom, indicates that oxygen forms a weaker dative bond (18.5 kcal/mol) than nitrogen (23.4 kcal/mol) with the down atom of a Si dimer (as shown in Fig. 8.4), in agreement with the relative electronegativities. Hybridization of the HOMO on the electron-donating atom also influences the dative bond strength. As the p character of the hybridized HOMO increases, orbital overlap between the HOMO and the down atom of the surface dimer increases. As a result, the dative bond energy increases [12]. Thus, sp^3 hybridized atoms can donate more

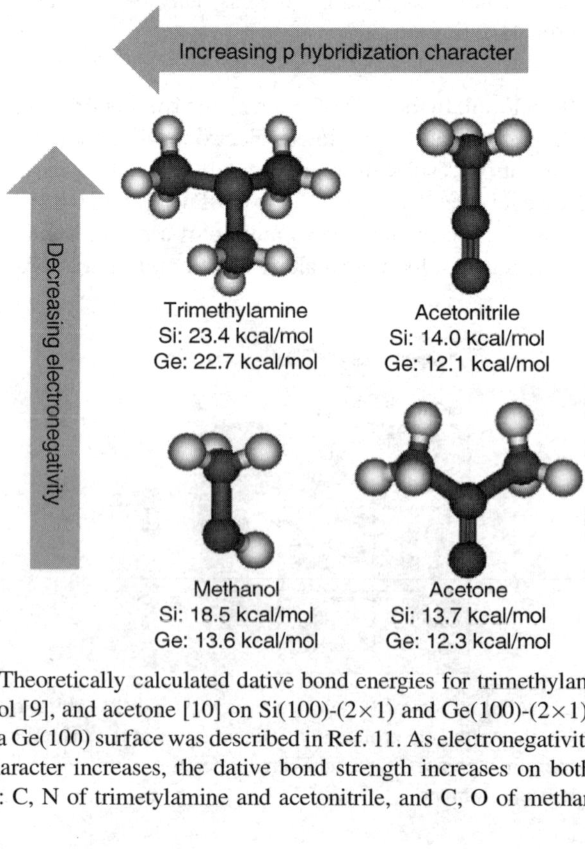

FIGURE 8.4 Theoretically calculated dative bond energies for trimethylamine [3], acetonitrile [8], methanol [9], and acetone [10] on Si(100)-(2×1) and Ge(100)-(2×1). The calculation for methanol on a Ge(100) surface was described in Ref. 11. As electronegativity decreases and p hybridization character increases, the dative bond strength increases on both surfaces. Gray: hydrogen; black: C, N of trimetylamine and acetonitrile, and C, O of methanol and acetone.

charge to form a more stable dative bond than sp hybridized atoms. As shown in Fig. 8.4, methanol (sp^3) and trimethylamine (sp^3) form stronger dative bonds than acetone (sp^2) and acetonitrile (sp), respectively.

8.1.3 Examples of Dative Bonding: Ammonia and Phosphine on Si(100) and Ge(100)

Although group V hydrides are not organic molecules, reactions of these compounds with Si(100) and Ge(100) surfaces are good examples of dative bond formation as well as proton transfer. Two examples are ammonia and phosphine, which may act as Lewis bases. Adsorption of these molecules on group IV semiconductor surfaces has been extensively studied.

Musgrave and coworkers suggested, based on DFT calculation results, that molecular adsorption of ammonia on Si(100)-(2×1) or Ge(100)-(2×1) occurs by formation of a dative bond with the down atom of the surface dimer [13, 14]. Formation of the molecularly adsorbed state is barrierless. The barrier separating the adsorbed state from the dissociated state was found to be 24 kcal/mol for Si (100) [13]. The transition state energy is lower than the energy of the reactants. As a result, dissociative adsorption is not activated with respect to the reactants. In summary, ammonia predominantly dissociates on Si(100) to form NH_2 and H species at 300 K. In addition to the dissociated species, a fraction of molecular NH_3 species may be present on Si(100). The fraction of dative-bonded species increases at lower temperatures. Increasing ammonia coverage favors dissociative adsorption. In contrast, the calculated activation barrier for the dissociation of NH_3 on Ge(100)-(2×1) is higher than the barrier for reversible desorption of the dative-bonded state [14]. The transition state energy is higher than the reactant energies. Although N–H dissociation is thermodynamically favored on Ge(100), the dissociation process is slow compared with the reversible desorption of NH_3 from the adsorbed state; the dissociation barrier is higher than the reactant energies, whereas molecular adsorption is barrierless. Thus, ammonia adsorbs molecularly and does not undergo dissociation on Ge(100) surfaces. Ammonia adsorbed onto Ge(100) forms a stable molecularly adsorbed species.

Phosphine (PH_3) behaves similar to ammonia and prefers dissociative adsorption on Si(100) surfaces. Radny and coworkers investigated the adsorption of phosphine on Si(100) and found that most adsorbed phosphine molecules dissociate to form PH_2+H at room temperature [15, 16]. The dative bond between phosphine and Si(100) has a binding energy of 14 kcal/mol, 12 kcal/mol less stable than that formed by ammonia (26 kcal/mol). The transition state associated with proton transfer from phosphine is approximately 16 kcal/mol higher than the energy of the dative-bonded state, a smaller barrier than the barrier to proton transfer from ammonia (24 kcal/mol). The thermodynamic stability of the dissociative structure with respect to the dative-bonded state is higher for phosphine adsorbed on Si(100) (32 kcal/mol) than for ammonia adsorbed on Si(100) (22 kcal/mol). Meanwhile, molecular adsorption on Ge(100) was found to prevail at room temperature [17] but comprehensive theoretical results are not yet available.

8.2 DATIVE BONDING OF LEWIS BASES (NUCLEOPHILIC)

8.2.1 Aliphatic Amines

8.2.1.1 Primary, Secondary, and Tertiary Amines on Si(100) and Ge(100) The reaction of aliphatic amines with Si(100)-(2×1) surfaces involves dative bonding, as shown in Fig. 8.5. Dative bond formation occurs without an energy barrier through donation of the lone pair electrons from the nitrogen atom to the electron-deficient down Si atom. Quaternary ammonium ions may be stabilized by the electron-donating alkyl groups on the amine, unlike ammonia. On the other hand, the up Si atom of the dimer is electron rich and acts a nucleophilic surface site. Nucleophilic attack by the up Si atom may occur, resulting in a N–H dissociative product.

The most important factor that influences the surface reactions of aliphatic amines on Si(100) is the presence of a N–H bond in the reacting amine molecule. Primary and secondary amines, both of which contain at least one N–H bond, can undergo proton transfer whereas tertiary amines cannot. The reactivity of cyclic aliphatic amines is similar to that of noncyclic amines.

Mui et al. investigated the bonding of aliphatic amines on Si(100)-(2×1) via multiple internal reflection Fourier transform infrared (MIR-FTIR) spectroscopy and density functional theory (DFT) calculations [3, 18]. Two simple primary and secondary amines, methylamine and dimethylamine, were found to undergo N–H dissociative adsorption on Si(100)-(2×1) via dative-bonded precursor states in a manner similar to that observed for ammonia on Si(100)-(2×1). The presence of a ν(Si–H) stretching mode near 2070 cm^{-1} in the MIR-FTIR spectrum provides strong evidence for the presence of the dissociated states of methylamine and dimethylamine (see Fig. 8.6b and d). In addition, strong ν(–C–H) stretching modes, observed at 2803 and 2787 cm^{-1}, indicate that a lone pair is present in the adsorption product. These modes, known as Bohlmann bands, originate from interactions between the nitrogen lone pair and the C–H σ-orbitals positioned *trans*-periplanar to the lone pair. Moreover, only a weak or no N–H stretching mode is observed at 3043 cm^{-1} for methylamine and dimethylamine, respectively. The spectral data support the presence of a N–H dissociative product.

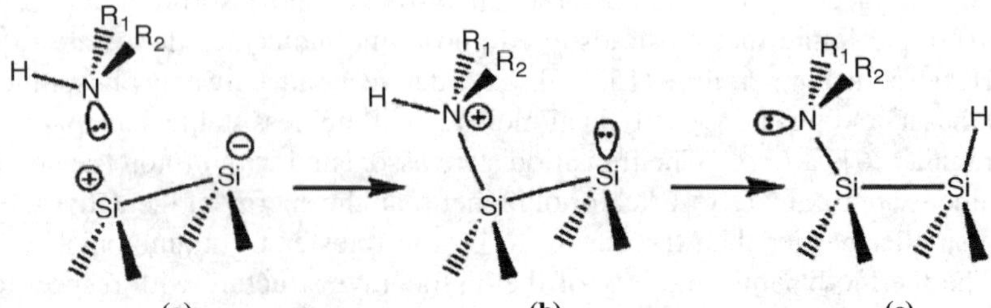

FIGURE 8.5 An illustration of the reaction of an amine with Si(100)-(2×1) surface (from Ref. 3). Reprinted with permission from Mui, C. et al., J. Am. Chem. Soc., **124**, 4027. Copyright (2002) American Chemical Society.

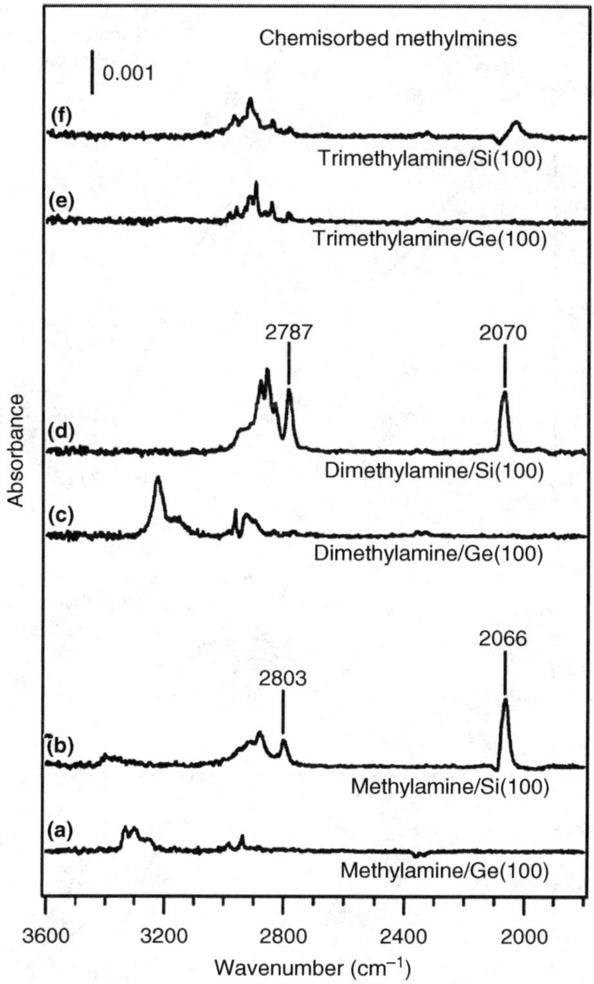

FIGURE 8.6 Infrared spectra of saturated coverage of methylamine, dimethylamine, or trimethylamine on Si(100)-(2×1) and Ge(100)-(2×1) surfaces at room temperature (from Ref. 3). Reprinted with permission from Mui, C. et al., J. Am. Chem. Soc., **124**, 4027. Copyright (2002) American Chemical Society.

The reaction pathway for dimethylamine on Si(100)-(2×1) (see Fig. 8.7) shows that this system is under kinetic control [3, 19]. Although the binding energy for the methyl dissociative product is higher than that for the hydrogen dissociative product, the barrier for N−C bond cleavage is high, 19.3 kcal/mol above the vacuum level. In contrast, the transition state for N−H cleavage is 9.3 kcal/mol below the entry level. The barrier for N−C cleavage is too high to be overcome at room temperature, explaining the experimental result that methylamine and dimethylamine undergo only N−H dissociation.

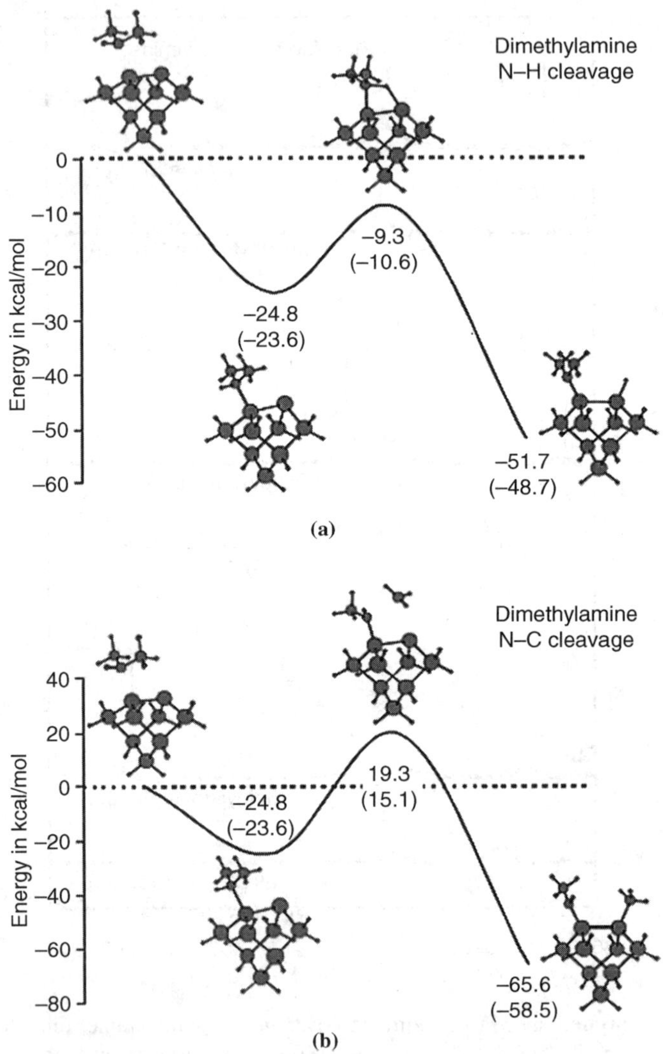

FIGURE 8.7 Theoretically calculated potential energy surfaces for (a) N–H dissociation and (b) N–C dissociation of dimethylamine on Si(100)-(2×1). All energies are given with respect to the vacuum level in kcal/mol (from Ref. 18). Reprinted with permission from Mui, C. et al., J. Chem. Phys. **114**, 10170. Copyright (2001) American Institute of Physics.

The difference between activation barriers of the two pathways can be understood by considering the transition state geometries [18]. The N–C bond must be stretched to accommodate front-side nucleophilic attack by the electron-rich up Si atom. The N–C and the Si–C bond lengths in this transition state are stretched by 36% and 63%, respectively, relative to the corresponding N–C and Si–C bonds in the reactant and the product. As a result, the energy of the transition state is high and the N–CH_3 dissociation process has a relatively high activation barrier [18]. These results indicate that the selectivity of N–H cleavage over N–CH_3 dissociation for aliphatic amines on Si(100)-(2×1) surfaces is due to kinetic effects rather than thermodynamic effects. Similar kinetic selectivity holds for other aliphatic amines, such as methylamine [18] and pyrrolidine [19], on Si.

Trimethylamine is a typical tertiary aliphatic amine and contains no N–H bonds to assist with dissociative adsorption. Figure 8.8 shows the N 1s XPS spectrum obtained after exposing a Si(100) surface to tetramethylamine at 300 K, followed by annealing at higher temperatures [20]. These data show that two N 1s photoelectron peaks are present near 402.2 and 398.9 eV at room temperature. The peak of N 1s for typical amines is around 398.5 eV. The peak at 402.2 eV represents a dative-bonded trimethylamine and contributes 85% of the total peak area. The lower energy peak at 398.9 eV is from a minor surface species, possibly a N–C dissociative product. When trimethylamine adsorbs onto Si(100)-(2×1) at 190 K, only one N 1s peak at 402.2 eV is observed, suggesting that the reaction pathway to N–C dissociation must overcome a high activation barrier. Thus, most of the trimethylamine molecules adsorb molecularly via dative bonding.

Dative bonding of trimethylamine on Si(100) was confirmed through MIR-FTIR analysis by Mui et al. [3]. Trimethylamine does not exhibit a significant peak in the ν(Si–H) stretching region. For multilayer adsorption of trimethylamine on Si(100) at 100 K, the observed C–H stretching vibrations below 2800 cm^{-1} are assigned to the vibrations of CH bonds positioned *trans*-periplanar to the nitrogen lone pair of the amine. These bands are known as Bohlmann bands and indicate the presence of a nitrogen lone pair [21]. This peak was very weak for trimethylamine adsorbed onto

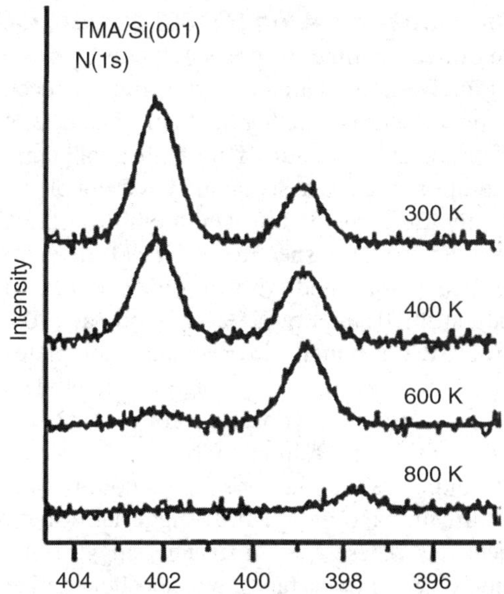

FIGURE 8.8 XPS spectral data for Si(100)-(2×1) surfaces exposed to trimethylamine, at room temperature and after annealing. Both spectra show two distinct binding energies for the N 1s core electrons (from Ref. 20). Reprinted with permission from Cao, X. et al., J. Am. Chem. Soc. **123**, 10988. Copyright (2001) American Chemical Society.

Si(100) at 300 K, suggesting loss of the lone pair and formation of a dative-bonded product. The formation of a dative bond between the lone pair of the nitrogen atom and the electron-deficient down atom of a Si dimer is exothermic with an adsorption energy of 23.4 kcal/mol. Cleavage of the N–C bond from the dative-bonded state requires crossing the transition state barrier, located 21.7 kcal/mol above the vacuum level. The N–C dissociative product energy is 62.3 kcal/mol below the vacuum level. Although the N–C dissociative product is thermodynamically stable, the high activation barrier makes the dissociation process kinetically unfavorable.

Mui et al. studied a series of alkylamines on Ge(100)-(2×1) surfaces and found that N–H cleavage does not occur for any of the alkylamines [3]. A distinct absorption peak in the N–H stretching region indicates the retention of N–H bonds. The difference between the chemical reactivities of the Ge(100)-(2×1) and Si (100)-(2×1) surfaces toward N–H dissociation is related to a decrease in the electron density as one moves down a group in the periodic table. The proton transfer process is facilitated by electron donation from the electron-rich up atom of a dimer. Thus, the dimer atom with largest electron density will most easily attack the proton of a N–H bond. Because the electron density on the nucleophilic Ge dimer atom is less than its Si counterpart, the N–H bond must be stretched to acquire sufficient electron density to form a Ge–H bond. The additional stretching would induce additional strain, increasing the activation energy. The adsorption of trimethylamine on either Ge(100) or Si(100) results in a datively bonded product. Trimethylamine has no N–H bonds but contains nitrogen lone pair electrons that can interact with the electrophilic down atom of a tilted dimer to form a dative bond.

8.2.1.2 Cyclic Aliphatic Amines on Si(100) and Ge(100) Pyrrolidine is a secondary amine like dimethylamine, whereas *N*-methylpyrrolidine is analogous to trimethylamine (see Fig. 8.9). The main concern about the reaction of these amines on (100) semiconductor surfaces is whether the final product is N dative bonded or if it proceeds to a N–H dissociated structure. For comparison, pyrroline and *N*-methyl-3-pyrroline have an additional reactive site at the C=C double bond. Thus, reactions may occur competitively either at the nitrogen atom or at the C=C double bond.

Figure 8.10 shows the infrared spectra of Si(100)-(2×1) and Ge(100)-(2×1) surfaces after the surface is saturated with pyrrolidine [19]. The appearance of a Si–H stretching mode at 2070 cm^{-1} or a Ge–H mode at 1904 cm^{-1} suggests that N–H or C–H dissociative products are formed. Complete loss of the N–H stretching mode at 3190 cm^{-1} indicates cleavage of the N–H bond on Si(100). In contrast, a significant ν(N–H) mode is observed after saturating doses of pyrrolidine on Ge(100). The presence of a Bohlmann band at 2780 cm^{-1} for pyrrolidine adsorbed onto Si(100) indicates that the lone pair is not involved in dative bonding. The absence of a Bohlmann band on Ge(100) suggests formation of a dative bond with the lone pair electrons. These spectra strongly suggest that pyrrolidine undergoes N–H dissociation on Si(100) surfaces, whereas both dative bonding and N–H dissociative products form on Ge(100).

The reaction product of pyrrolidine adsorbed onto Ge(100) depends on the adsorbate coverage. The infrared spectra shown in Fig. 8.11 reveal that molecular

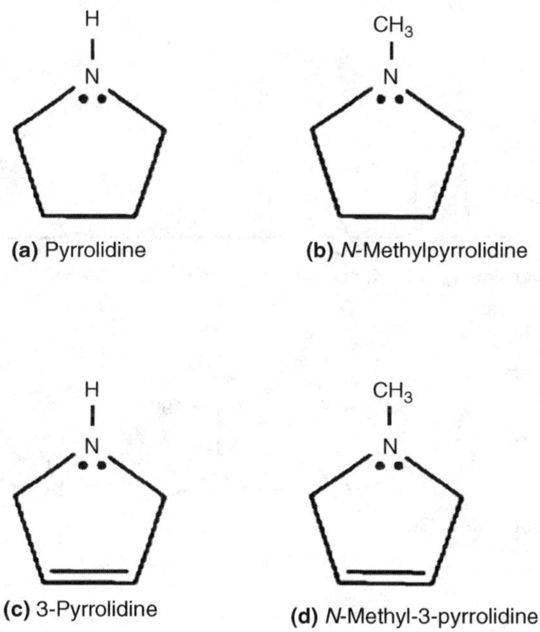

FIGURE 8.9 Examples of five-membered cyclic aliphatic amines.

adsorption through dative bonding is the only product at low coverage, whereas N–H dissociative adsorption also occurs at high coverage [19]. The different behavior of pyrrolidine on Si or Ge surfaces can be explained in terms of kinetic effects. The dative-bonded products form without crossing an activation barrier in each case. On Ge (100), the barrier to N–H dissociative adsorption is 21.9 kcal/mol and the transition state is only 1 kcal/mol below the energy of the reactants. On Si(100) surfaces, the barrier is 13.3 kcal/mol and the transition state is 11.9 kcal below the reactant energy. These differences explain why some pyrrolidine molecules are trapped in a dative-bonded state on Ge(100). The presence of neighboring dative-bonded species lowers the barrier for N–H dissociation and favors the dissociative products at high coverage. The chemistry of pyrrolidine binding to Si(111)-(7×7) surfaces is characterized by two types of dative bonding (β_1 and β_2) at 85 K [22]. In the β_1 state, one pyrrolidine molecule engages in both electron acceptance by the adatom and electron donation by the rest atom simultaneously through dative bonding and hydrogen bonding, respectively. In the second dative-bonded state (β_2), the adsorbed pyrrolidine is directly bonded to the silicon surface adatom via a N–Si (adatom) dative bond. The presence of a N–H dissociative product at higher temperatures could not be clearly determined.

N-Methylpyrrolidine forms a stable dative bond at room temperature on both Si(100) and Ge(100) surfaces [19]. The saturation coverage is significantly higher on the Ge(100) than on the Si(100) surface, probably due to steric effects. Ge(100) has a slightly larger spacing between adjacent dimers in each row than does Si(100): 4.00 Å versus 3.84 Å. Thus, the steric repulsion between adjacent adsorbates may be weaker, and a higher coverage may be attained for Ge(100). 3-Pyrroline

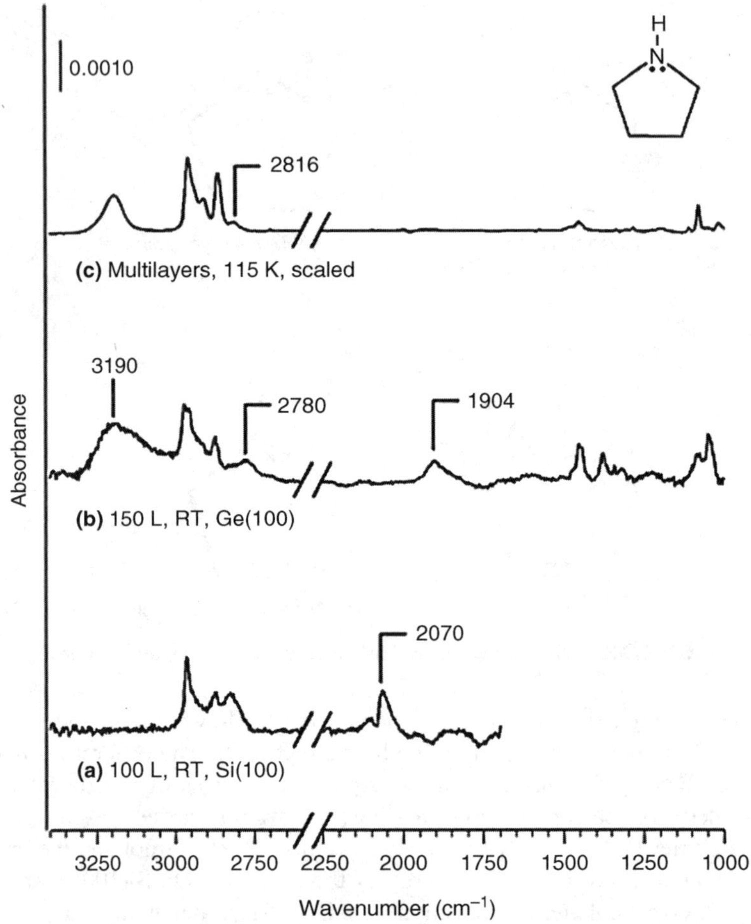

FIGURE 8.10 Infrared spectra of pyrrolidine on Si(100)-(2×1) and Ge(100)-(2×1) surfaces: (a) 100 L at 298 K on Si(100); (b) 150 L at 298 K on Ge(100); (c) multilayers (scaled to fit the figure) at 117 K (from Ref. 19). Reprinted with permission from Wang, G. T. et al., J. Phys. Chem. B, **107**, 4982. Copyright (2003) American Chemical Society.

primarily undergoes molecular adsorption through Ge–N dative bonding on Ge (100) [19], whereas N–H dissociation is favored on Si(100) along with [2+2] cycloaddition as a minor product [4]. 3-Pyrroline exhibits coverage-dependent adsorption on Ge(100), similar to pyrrolidine. Some of the dative-bonded species undergo N–H dissociation at higher coverage. *N*-Methyl-3-pyrroline adsorbs through Ge–N dative bonding on Ge(100) surfaces, as expected from the high barrier to cleavage of the N–CH_3 bond [22].

8.2.1.3 Ethylenediamine on Ge(100) Ethylenediamine is a bifunctional molecule with two amino end groups. Either one or two of the amino groups may be involved in ethylenediamine surface reactions. The reaction can be

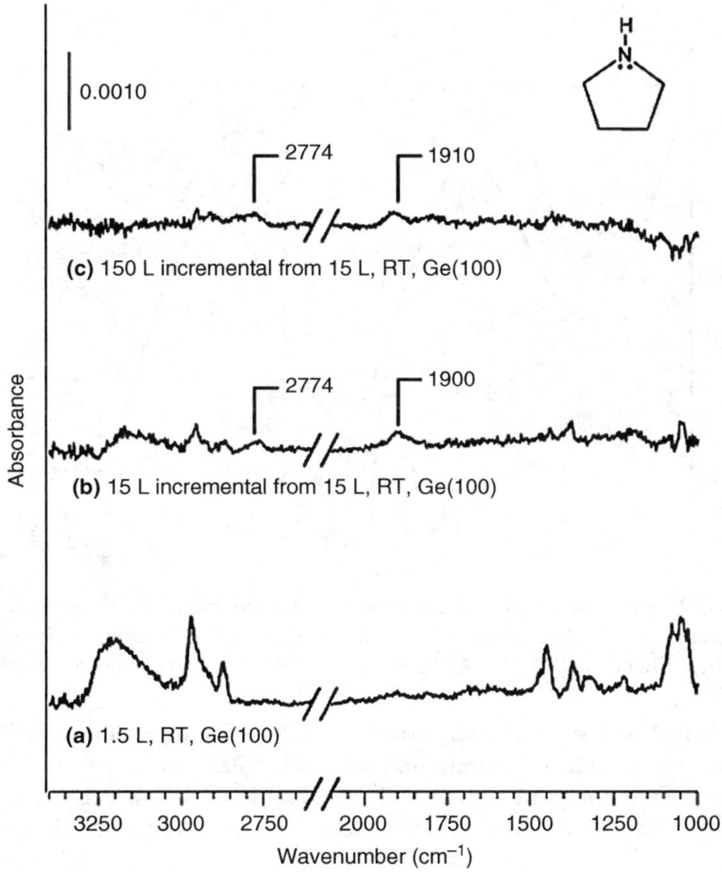

FIGURE 8.11 Infrared spectra of pyrrolidine on a Ge(100)-(2×1) surface as a function of exposure: (a) 1.5 L at 298 K; (b) 15 L, in ratio with 1.5 L at 298 K; (c) 150 L, in ratio with 15 L at 298 K (from Ref. 19). Reprinted with permission from Wang, G. T. et al., J. Phys. Chem. B, **107**, 4982. Copyright (2003) American Chemical Society.

performed through N dative bonding or N–H dissociation. The selectivity for forming each of the possible products is important for exerting appropriate control during multilayer organic functionalization.

Both amino end groups were found to undergo dissociative adsorption on Ge(100) surfaces at low coverage, and two N–Ge covalent bonds form with two down Ge atoms in adjacent dimer rows [23]. Although ethylenediamine is a primary amine, it behaves differently from other primary amines. The formation of an interdimer row bridging structure affects the reaction pathway, leading to dual N–H dissociative products. The calculated potential energy surfaces (see Fig. 8.12) support the experimental results that indicate dual N–H dissociation reactions. Initially, ethylenediamine adsorbs in a dual dative-bonded precursor state, in which both amine lone pairs are donated to the electrophilic dimer atoms of neighboring dimers, a state which lies 43.7 kcal/mol below the energy of the reactants. From the dual dative bond, it is now possible for

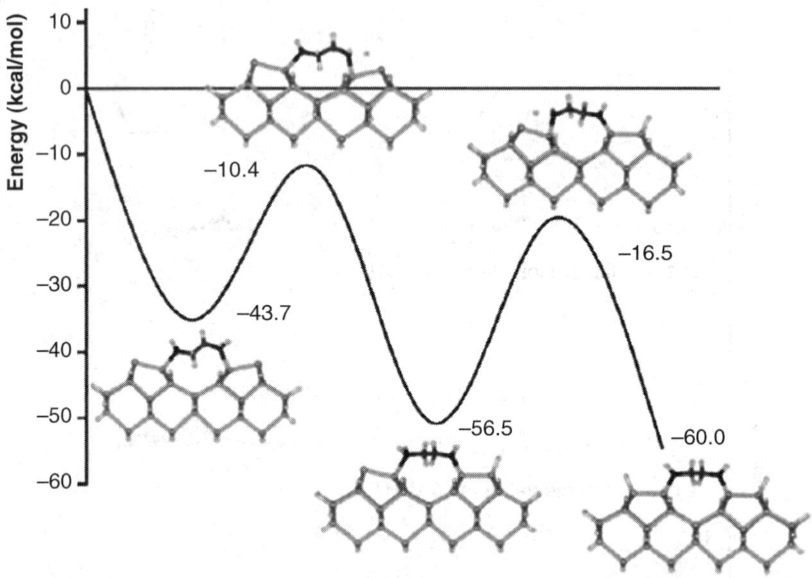

FIGURE 8.12 Calculated potential energy surface for the interdimer N−H dissociation of ethylenediamine on a Ge(100) surface (from Ref. 23). Reprinted with permission from Kim, A. et al., J. Phys. Chem. B, **109**, 19817. Copyright (2005) American Chemical Society.

ethylenediamine to transfer a hydrogen atom to each surface dimer, thus creating a dual N−H dissociation product. The transition states associated with transfer of the first and second hydrogen atoms are −10.4 and −16.5 kcal/mol below the energy of the reactants, respectively. As the coverage increases, N−H dissociation is inhibited, and formation of a single Ge−N dative-bonded structure dominates. Changes in the electronic structure at the surface due to neighboring adsorbates could account for the inhibition of the N−H dissociation reaction at higher coverage.

The adsorption product at high coverage, a single dative-bonded ethylenediamine on Ge(100), was utilized for the controlled formation of multiple organic layers on a Ge(100) surface. One of the ultimate goals of organic functionalization is the appropriate control of the deposition of multiple organic layers so that precisely tailored surfaces may be prepared with a variety of useful functionalities. A multilayer organic functionalization system was built on Ge(100)-(2×1) using a combination of ethylenediamine and 1,4-phenylene diisocyanate under vacuum conditions to form urea linkages [24]. The idealized scheme for producing the multilayer film is illustrated in Fig. 8.13.

8.2.2 Aromatic Amines

Aromatic amines may be classified into two different categories. In the first group, nitrogen atoms are external to the aromatic phenyl ring. In this case, the presence of the phenyl ring may affect the reactivity of the amine on semiconductor surfaces. Reaction through the phenyl ring via [4+2] or [2+2] cycloaddition is not competitive

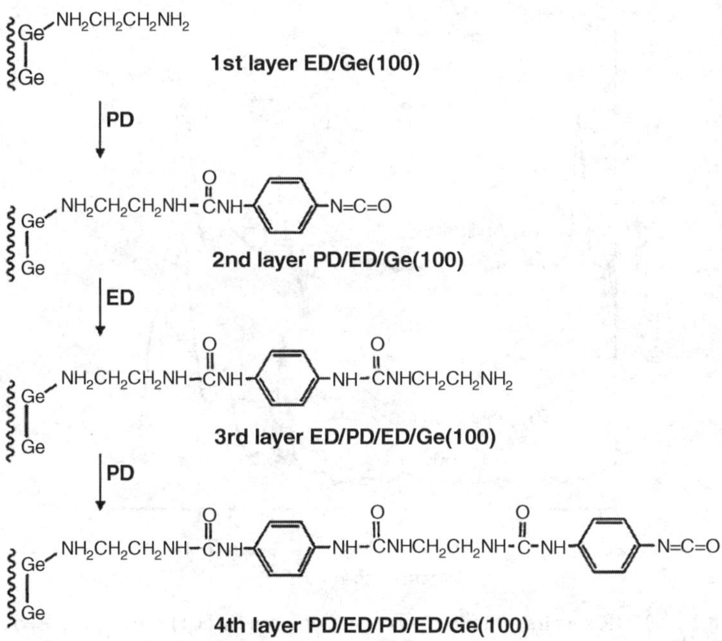

FIGURE 8.13 Schematic illustration of the synthetic route for the layer-by-layer growth of polyurea films on a Ge(100)-(2×1) surface (from Ref. 24). Reprinted with permission from Kim, A. et al., J. Am. Chem. Soc., **127**, 6123. Copyright (2005) American Chemical Society.

with the reaction through the amine group, mainly due to the aromaticity of the phenyl ring. The second group includes heteroaromatic amines, in which some of the carbon atoms in the aromatic ring are substituted with nitrogen atoms. The concepts of aromaticity and resonance stabilization are applied to the bonding of heteroaromatic molecules on Si(100)-(2×1) or Ge(100)-(2×1). Pyrrole, pyridine, pyrazine, and *s*-triazine have all been investigated experimentally and/or theoretically on semiconductor surfaces. Beyond possessing a large resonance stabilization energy and nonbonding electrons, many heteroaromatic molecules have significant dipole moments due to resonance and induction effects. This leads to an asymmetric charge distribution in the cyclic π-system and can activate new reaction pathways and produce new surface species not observed for benzene. The participation of the lone pair electrons of the nitrogen atom in an aromatic π-electron system is an important factor for determining whether the dative-bonded state is the major product.

8.2.2.1 Aniline on Si(100) and Ge(100) The reaction of aniline ($C_6H_5NH_2$) with semiconductor surfaces may occur either at the amine group or at the phenyl ring. Figure 8.14 shows an FTIR spectrum of a Si(001) surface exposed to 10 L aniline at 300 K [4]. A comparison of the spectra of surface-adsorbed aniline (Fig. 8.14a) and liquid aniline (Fig. 8.14b) reveals the spectral changes resulting from adsorption of aniline onto Si(001). The presence of a strong Si−H stretching vibration at 2072 cm^{-1} shows that adsorption is accompanied by cleavage of either

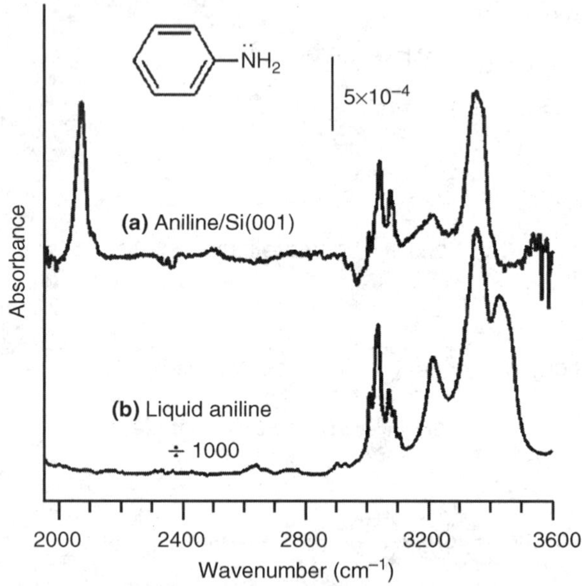

FIGURE 8.14 FTIR spectra of aniline. (a) Spectrum of a Si(001) surface exposed to 10 L of aniline at 300 K. (b) Spectrum of neat liquid aniline (from Ref. 4). Reprinted with permission from Cao, X. et al., J. Phys. Chem. B, **105**, 3759. Copyright (2001) American Chemical Society.

N–H or C–H bonds. Meanwhile, the C–H spectral region and the ring-stretching combination band at 3217 cm^{-1} remain nearly unchanged. No significant absorption is detected in the 2600–3000 cm^{-1} region, in which alkane-like C–H vibrations are typically observed. These observations suggest that the benzene ring in aniline remains unperturbed. The two N–H modes at 3356 and 3431 cm^{-1} (Fig. 8.14b) of liquid aniline collapse into a single N–H mode at 3356 cm^{-1} (Fig. 8.14a) for the surface-adsorbed aniline, which strongly suggests cleavage of one of the N–H bonds.

The N 1s XPS data in Fig. 8.15 show a single peak at a binding energy of 398.9 eV [4]. This energy is almost identical with the energy associated with binding of primary amines on Si(100) surfaces, which was demonstrated to undergo N–H dissociation as discussed above. The similarity in N 1s binding energy between aniline and primary amine compounds suggests that aniline also binds to Si(100) surfaces via N–H dissociative adsorption.

Thus, it was concluded that aniline exclusively undergoes dissociative adsorption to form Si–N–C and Si–H bonds, as depicted in Fig. 8.16. The reaction is similar to the adsorption of the primary aliphatic amine, methylamine, onto Si(100).

It is notable that the reaction between aniline and Si(100) proceeds exclusively along one of the two or more adsorption modes available: Si–N–C linkage or a benzene-like direct ring interaction. Reaction through the π-electrons of the phenyl ring results in loss of aromaticity and is not thermodynamically favorable. No reports have yet described the adsorption of aniline on Ge(100); however, considering the dative bonding of aliphatic primary amines on Ge(100) and the reactivity of the

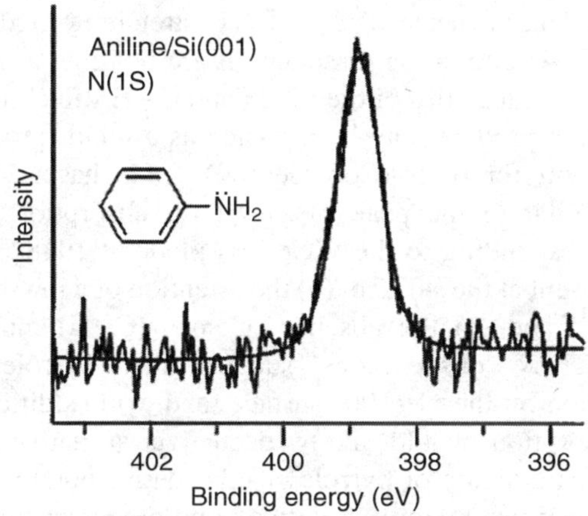

FIGURE 8.15 XPS N 1s spectrum of a Si(001) sample exposed to 5 L of aniline at 300 K (from Ref. 4). Reprinted with permission from Cao, X. et al., J. Phys. Chem. B, **105**, 3759. Copyright (2001) American Chemical Society.

phenyl ring of aniline, aniline is predicted to adsorb onto Ge(100) through N–Ge dative bonding.

8.2.2.2 Five-Membered Heteroaromatic Amines: Pyrrole on Si(100) and Ge(100) Pyrrole is a five-membered cyclic aromatic compound with one nitrogen atom. In this molecule, the two lone pair electrons of the nitrogen atom and the four π-electrons form a conjugated π-electron system. High-resolution electron energy loss spectroscopy (HREELS) and IR spectroscopic studies reveal that pyrrole on Si(100) and Ge(100) undergoes dissociative adsorption through either N–H cleavage (major product) or C–H cleavage (minor product) [4, 19, 25]. The retention of aromatic ν(C–H) stretching modes near 3100 cm^{-1} and the growth of ν(Si–H) stretching modes at 2107 cm^{-1} in HREELS support the notion that the dominant surface species is a N–H dissociative product on Si (100) [25]. The adsorption peak corresponding to ν(Si–H) at 2093 cm^{-1} in the IR spectrum also supports either N–H or C–H cleavage [4]. Few or no absorption bands are observed in the 2800–3000 cm^{-1} region normally associated with saturated C–H vibrational modes, excluding the possible cycloaddition through a C=C double bond

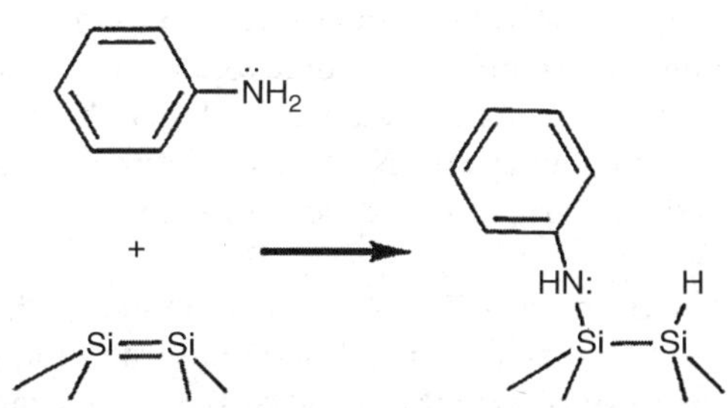

FIGURE 8.16 Bonding configuration of aniline on the Si(100) surface (from Ref. 4).

in the pyrrole ring. The presence of a ν(N−H) stretching mode near 3400 cm^{-1} suggests that both C−H and N−H dissociation are possible at room temperature. Comparing the relative intensities of the Si−H and C−H stretching peaks, the C−H dissociative adsorption product may be assigned as a minor product. On the other hand, the IR spectrum for pyrrole on Ge(100)-(2×1) has a significant peak at 1988 cm^{-1} [19], similar to the peak observed for absorption on Si(100): (a) a significant peak corresponding to the ν(Ge−H) mode at 1988 cm^{-1} indicates that Ge−H bonds are present at the surface; (b) the retention of a low-amplitude ν(N−H) mode at 3403 cm^{-1}; and (c) the absence of sp^3 ν(C−H) modes in the region 2800–3000 cm^{-1}. These observations suggest that pyrrole also undergoes dissociative adsorption on the Ge(100) surface, and cycloaddition does not occur.

Dissociative adsorption would likely occur via a dative-bonded precursor state [19]. The aromatic nature of pyrrole enables dative bonding at either the ring carbons or the nitrogen heteroatom. A dative bond precursor through the nitrogen lone pair of the pyrrole accompanies loss of aromaticity, and the transition state is unstable relative to the energy of reactants. Thus, this pathway must overcome a high activation barrier for N−H dissociation. The other type of dative bonding, via a ring carbon, is possible because of resonance stabilization by the ring. The precursor state may give way to N−H dissociation to form a pyrrolyl radical, followed by isomerization to produce a N−H dissociated state. Otherwise, a C−H dissociation product may form from the latter type of dative bonding, analogous to electrophilic aromatic substitution. This reaction is specific to aromatic systems and occurs when an electrophilic down atom of a dimer attacks a ring carbon to cleave the aromatic C−H bond, and the resultant carbocation is stabilized through resonance. In the reaction of pyrrole with Si(100)-(2×1), the N−H dissociative product dominates. This product supposedly forms via the lower energy α-C−H dative bonding.

The adsorption of pyrrole on Ge(100) results in a mixture of dissociation products, whereas other heterocyclic aromatic compounds, such as pyridine and pyrimidine, adsorb selectively through nitrogen dative bonds to the Ge(100)-(2×1) surface. The main product of adsorption of pyrrole on Ge(100) is a N−H dissociation product [19], the same adsorption product on Si(100) surfaces. Evidence for the minor electrophilic aromatic substitution product of C−H bond cleavage is also observed. The major factor that distinguishes these bonding species from those of other aromatic heterocyclic compounds adsorbed onto Ge(100) is the delocalization of the nitrogen lone pair in the pyrrole over the ring as part of the aromatic π-system. Consequently, the N dative-bonded state is relatively unstable due to loss of resonance energy. Thus, initial adsorption through dative bonding is unfavorable, and electrophilic aromatic substitution may compete as a reaction pathway such that the nitrogen atom is not a reaction site. The reaction pathways for N−H dissociation in Fig. 8.17 compare the two possible intermediate structures, which involve either Ge−N or Ge−C dative bonding [19].

The adsorption of pyrrole on Si(111)-(7×7) is similar to that on Si(100). The disappearance of the N−H stretching band upon adsorption at room temperature, along with the appearance of Si−H vibrational peaks, clearly indicates that pyrrole chemisorbs dissociatively on Si(111)-(7×7) [26]. The N−H dissociative product is

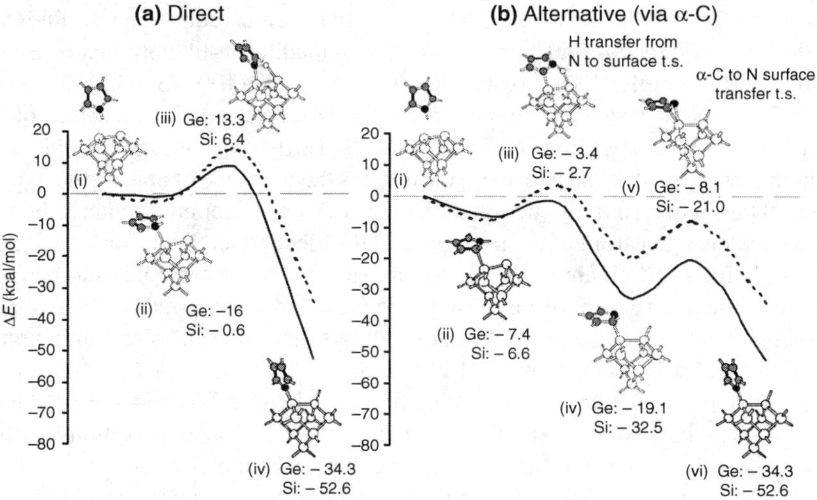

FIGURE 8.17 Reaction pathways for N−H dissociation in pyrrole adsorbed onto Si(100)-(2 × 1) and Ge(100)-(2 × 1) surfaces: (a) direct pathway; (b) alternative pathway (from Ref. 19). Reprinted with permission from Wang, G. T. et al., J. Phys. Chem. B, **107**, 4982. Copyright (2003) American Chemical Society.

calculated to be 24–31 kcal/mol more stable than the 2,5-dihydropyrrole-like adduct formed via a [4+2]-like addition, implying that the retention of aromaticity is an important factor in the reaction pathway. Furthermore, pyrrolyl binding at an adatom is sterically more favorable than attachment to a lower lying rest atom, in which the ring may experience repulsive interactions with the neighboring adatoms.

Protection of the pyrrole nitrogen with a methyl group reverses the reactivity [27]. Because C−N bonds have a much higher dissociation barrier than N−H bonds, *N*-methylpyrrole reacts exclusively through both [2+2]-like and [4+2]-like additions.

8.2.2.3 Six-Membered Heteroaromatic Amines Pyridine, pyrazine, and *s*-triazine are six-membered heterocyclic amines with different number of nitrogen atoms in the ring. Two reaction modes compete during adsorption of these molecules on Si(100)-(2×1): adsorption may occur through the lone pairs of the nitrogen atom or through the π-electrons of the double bond [28]. DFT studies suggest that all three molecules adsorb through N dative bonding with the buckled-down silicon atom of the silicon surface dimer [29]. In addition, [4+2] cycloaddition occurs, similar to the adsorption of benzene. Reaction between the remaining double bond and the adjacent Si dimer may follow via cycloaddition, leading to a tetra-σ-bonded state called a "tight-bridged product."

Pyridine on Si(100) and Ge(100) The lone pair electrons on the N atom of pyridine are not directly involved in aromatic π-conjugation, and instead are donated to form a dative bond with the electron-deficient atom of a surface dimer. Two [4+2] cycloaddition configurations may be attained from pyridine. Adducts may have either N–Si/C–Si bonds or C–Si/C–Si bonds. In fact, it is found that pyridine forms both the dative-bonded product and the [4+2] cycloaddition-like product on Si(100) and the nearly completely dative-bonded product on Ge(100) [30–32].

The N 1s XPS spectrum shown in Fig. 8.18 indicates the presence of two adsorption states for pyridine adsorbed onto Si(100) [30]. Observation of two N 1s binding energies at 398.8 and 401.8 eV implies the presence of at least two types of species. The higher binding energy state (401.8 eV) corresponds to dative bonding between the nitrogen atom of the pyridine and the electron-deficient surface Si atom. Dative bonding is dominant at 110 K. An increase in temperature leads to a gradual decrease in the intensity of the dative bonding peak and growth of the 398.8 eV peak. Increasing the temperature to 350 K causes all pyridine molecules remaining on the surface to yield a N 1s peak at 398.8 eV.

The other peak at 398.8 eV was identified as a butterfly-like [4+2] product by HREELS (see Fig. 8.19) and DFT calculations [30]. The observation of Si–C (615 cm^{-1}) and Si–N (495 cm^{-1}) peaks, together with the coexistence of C(sp^2)–H (3045 cm^{-1}) and C(sp^3)–H (2895 cm^{-1}) stretching modes, suggests formation of a di-σ-bonded pyridine on Si(100) at 350 K. The suggested configuration was a butterfly-like [4+2] product with di-σ-bonding between the N/C atoms of the pyridine molecule and Si dimer atoms, similarly to the bonding of benzene on the same surface.

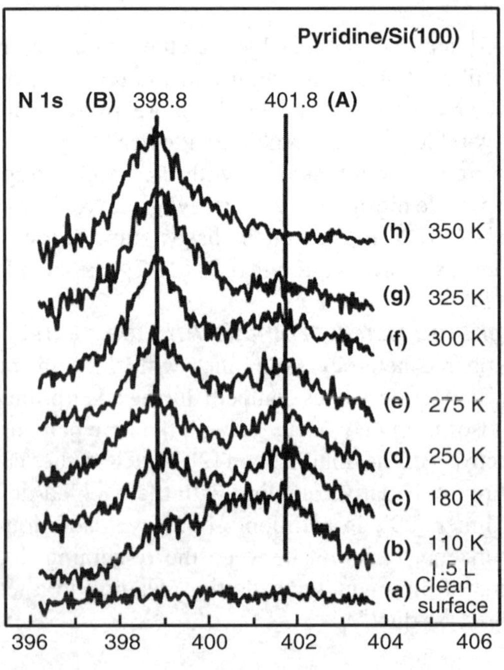

FIGURE 8.18 The N 1s core level spectra for pyridine adsorbed onto Si(100) as a function of temperature (from Ref. 30). Reprinted with permission from Tao, F. et al., J. Phys. Chem. B, **107**, 6384. Copyright (2003) American Chemical Society.

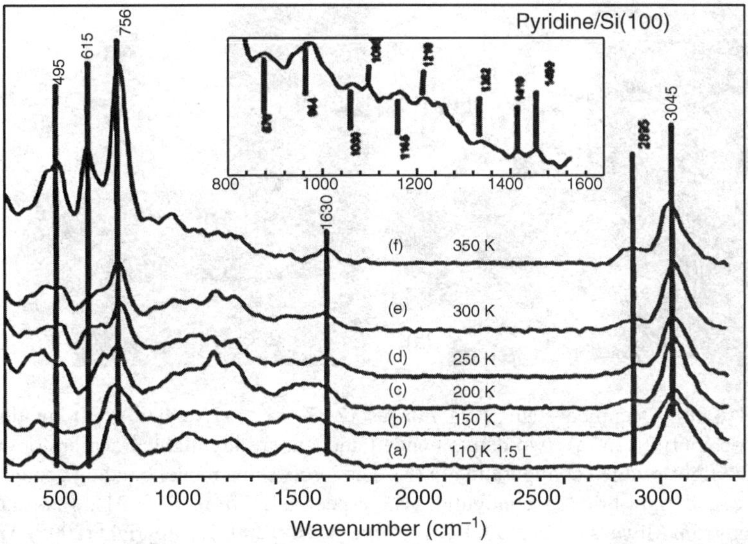

FIGURE 8.19 HREELS spectra for pyridine adsorbed onto Si(100) as a function of temperature (110–350 K). The inset shows an enlarged view of spectrum (f) in the region 830–1530 cm^{-1} (from Ref. 30). Reprinted with permission from Tao, F. et al., J. Phys. Chem. B, **107**, 6384. Copyright (2003) American Chemical Society.

STM measurements identified two interesting features [31]. The feature shown in Fig. 8.20b is a two-dimer footprint in which a bright feature, centered over one of the dimers, represents a tight-bridged configuration. Gradual conversion of the dative-bonded configuration (see Fig. 8.20a) into a tight-bridged species (see Fig. 8.20b) is observed at low coverage. The barrier for conversion is estimated from the observed rate of conversion to be 0.9 eV. The di-σ-bonded adduct may further proceed to a tight-bridged species via appropriate choice of adsorbate coverage and temperature, among other factors. The formation of a tight-bridged configuration requires two surface dimers. Thus, they may be invisible under certain coverage conditions.

An XPS study of pyridine on Si(111)-(7×7) at 110 K also shows two features in the N 1s edge: one at 401.8 eV (a dative-bonded state) and another at 398.8 eV (the monolayer coverage) [33]. Based on the vibrational studies, the second feature is assigned to the [4+2] adduct with Si−N and Si−C bonds. The dative-bonded state is converted into a cycloadduct with increasing temperature, similar to the behavior observed on Si(100).

In contrast, pyridine adsorbs selectively through a Ge−N dative bond on the Ge(100) surface. Furthermore, STM images show formation of a highly ordered monolayer at the surface with coverage of 0.25 ML, as shown in Fig. 8.21 [32]. The pyridine overlayer forms a $c(4\times2)$ structure in which the molecules bind to the down atoms of every other dimer to minimize repulsive interactions between pyridine molecules.

Theoretical calculations show that the dative-bonded adduct is more stable than the other possible reaction products on Ge. The calculations for the most stable

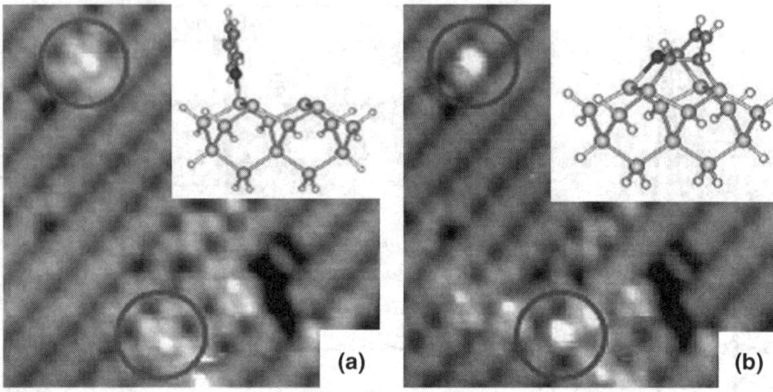

FIGURE 8.20 Occupied state STM images (7×7 nm^2, −2 V, 40 pA) during dosing of pyridine on Si(100). In (a), two dative-bonded adducts are highlighted and are observed to relax to the tight-bridged configuration in (b). The insets show model bonding geometries for the dative and tight-bridged configurations, respectively (from Ref. 31). Reprinted with permission from Miwa, J. A. et al., J. Phys. Chem. B, **109**, 20055. Copyright (2005) American Chemical Society.

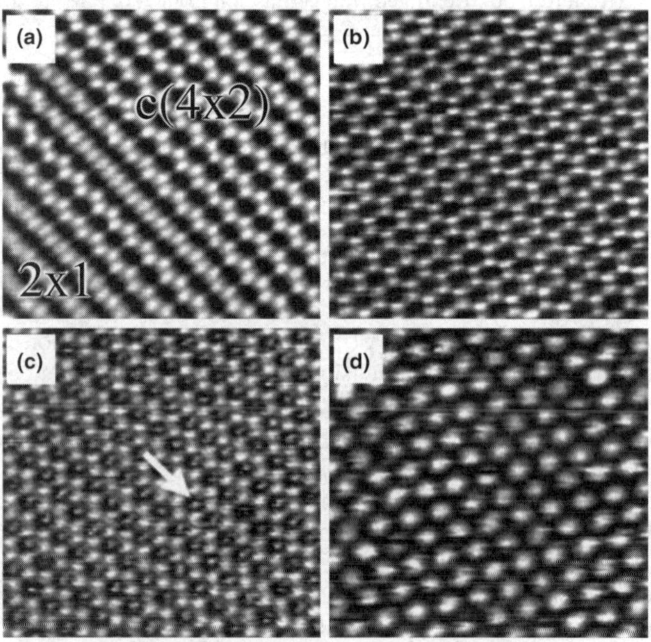

FIGURE 8.21 Sequence of filled state STM images (10×10 nm^2, 0.1 nA) of (a) a clean Ge (100) surface (−1.8 V) and 0.25 ML pyridine adsorbed onto the surface at bias voltages (V_s) of (b) −1.0 V, (c) −1.8 V, and (d) −2.4 V. Arrow indicates a pyridine molecule (from Ref. 32). Reprinted with permission from Cho, Y. E. et al., J. Am. Chem. Soc., **125**, 7514. Copyright (2003) American Chemical Society.

dative-bonded structure indicate that (1) the Ge–N bond length (2.04 Å) is significantly longer than the covalent bond length (1.70 Å) such that the pyridine molecule datively binds to the down Ge atom; (2) the Ge dimers are rearranged into a $c(4\times2)$ structure; (3) the adsorbed pyridine molecules retain their aromaticity; and (4) the plane of the aromatic ring is tilted with respect to the surface (see Fig. 8.22).

The electronic character of the dative-bonded structure can be clearly described through the charge density difference $\Delta\rho(r)$ before and after adsorption [34]. The $\Delta\rho(r)$ contours corresponding to the clean Ge(100) surface show a pronounced maximum (red color) in the buckled dimer region, indicating bonding between the two Ge atoms. Above the up Ge atoms, the existence of an electron-rich region may be explained by charge transfer from the buckled-down atom to the buckled-up atom. Figure 8.22b shows a pronounced maximum (indicated in red) between the down Ge atom and the pyridine molecule. The maximum resides closer to the pyridine molecule, represents a nitrogen lone pair, and, thus, indicates a dative bond.

Comparison of pyridine adsorption on Ge(100) with that on Si(100) reveals significant differences. The tight-bridged-type configuration readily available on Si(100)-(2×1) cannot form on Ge(100)-(2×1) because of the apparent negative adsorption energy predicted by DFT calculations [35]. This energetic difference causes pyridine to attach to a single Ge atom by dative bonding to form a very stable $c(4\times2)$ pattern observed by STM [32]. The difference between the Ge(100) and the Si(100) surfaces mainly arises from the fact that the strength of the covalent bond between C and Ge, formed by a cycloaddition-type reaction, is weaker than that between C and Si. Indeed, the Ge–C bond strength was found to be 8.9 kcal/mol less than its silicon analogue. Thus, on Ge(100), the dative-bonded product is favored such that the aromaticity of the pyridine ring is retained after adsorption onto the surface.

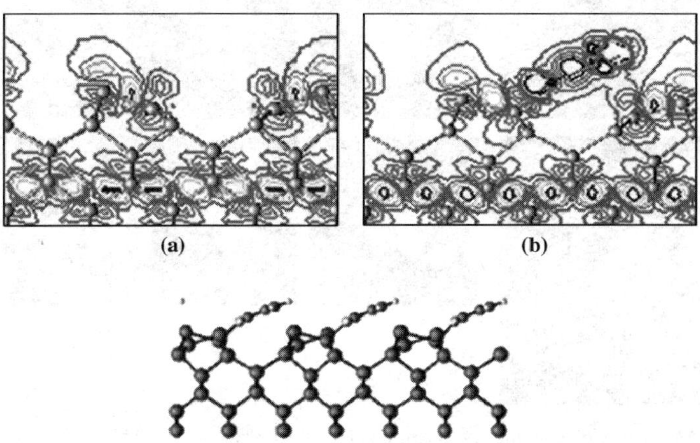

FIGURE 8.22 Plot of the charge density difference between the total valence charge density and the superposition of atomic valence charge densities for (a) clean Ge(100)-$c(4\times2)$ and (b) pyridine/Ge(100)-$c(4\times2)$. Side view of the optimized dative-bonded structure (from Ref. 34). Reprinted with permission from Hong, S. et al., J. Phys. Chem. B, **108**, 15229. Copyright (2004) American Chemical Society.

Pyrimidine on Ge(100) Pyrimidine (1,3-diazine, $C_4H_4N_2$) is a six-membered aromatic molecule with four carbon atoms and two nitrogen atoms, each with a lone pair of electrons. Pyrimidine is one of the important parent molecules for synthesizing nucleic acid bases. The structure of pyrimidine-adsorbed semiconductor (100) surfaces can assist an understanding of the interaction between biomolecules, such as amino acids, nucleic acids, or DNA, and semiconductor (100) surfaces.

Lee et al. studied the adsorption of pyrimidine on Ge(100) using STM and found that pyrimidine undergoes molecular adsorption through dative bonding on Ge(100) [36]. For coverage of pyrimidine up to 0.25 ML, a well-ordered $c(4\times2)$ structure results from states that appear as oval-shaped protrusions in the STM micrographs (see Fig. 8.23). The oval-shaped protrusions, surrounded by six neighboring up atoms of the buckled-up Ge dimers, are located between the Ge dimer rows. The STM image indicates that pyrimidine molecules are located between adjacent dimer rows.

Through DTF calculations, Lee et al. found that the adsorbed pyrimidine molecules are tilted by 40° with respect to the Ge surface and form bridges between the down Ge atoms of neighboring Ge dimer rows via double Ge–N dative bonding without loss of aromaticity (see Fig. 8.24) [36]. At higher coverage (0.5 ML), the pyrimidine overlayer converts to a $p(2\times2)$ structure. This high-coverage overlayer is attributed to pyrimidine in a single dative-bonded configuration, a state that is stable only in the presence of a continuous pyrimidine flux.

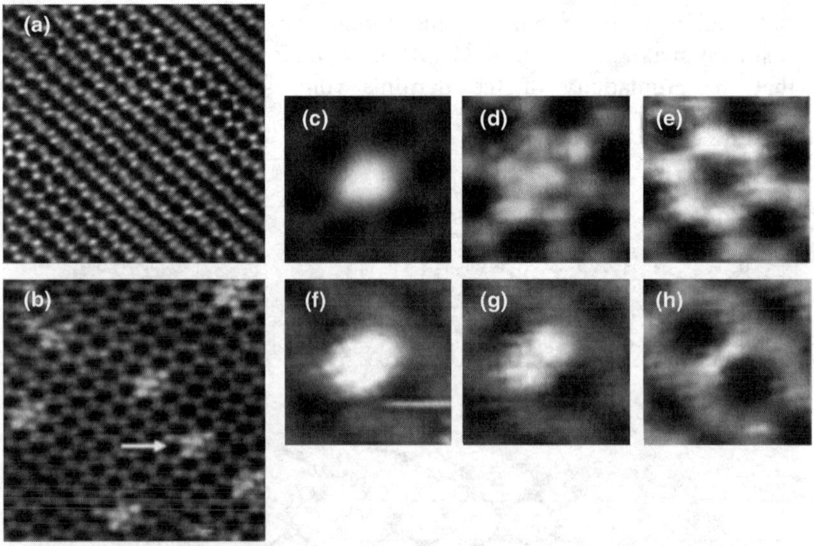

FIGURE 8.23 (a) A filled state STM image (15×15 nm^2, $V_s = -1.8$ V, $I_t = 0.1$ nA) of a clean Ge(100) surface. (b) A filled state STM image ($V_s = -1.6$ V) of pyrimidine on Ge(100) at a coverage of 0.01 ML. The arrow indicates an adsorbed pyrimidine molecule. STM images of pyrimidine on Ge(100) at various bias voltages: (c) −2.0 V, (d) −1.6 V, (e) −1.0 V, (f) +1.4 V, (g) +1.0 V, and (h) +0.8 V (from Ref. 36). Reprinted with permission from Lee, J. Y. et al., J. Phys. Chem. B, **109**, 348. Copyright (2005) American Chemical Society.

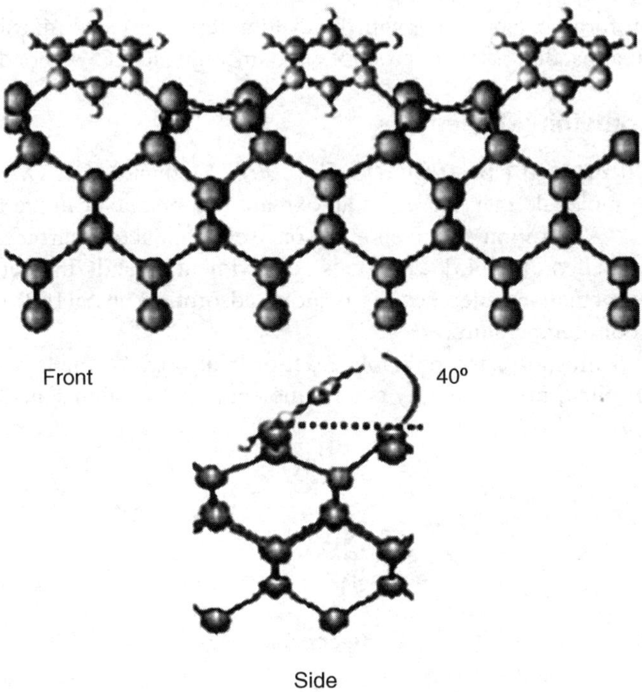

FIGURE 8.24 Front and side views of the adsorption structure of pyrimidine on Ge(100) (from Ref. 36). Reprinted with permission from Lee, J. Y. et al., J. Phys. Chem. B, **109**, 348. Copyright (2005) American Chemical Society.

Pyrazine on Si(100) and Si(111) Pyrazine is expected to adsorb through either N dative bonding or [4+2] cycloaddition on Si(100) and Si(111). The [4+2] product may have either two C–Si bonds or two N–Si bonds. The formation of N dative bonding is favored only at low temperatures [27]. The binding energy for the dative-bonded pyrazine (−18.6 kcal/mol) is lower than that of pyridine (−27.0 kcal/mol), reflecting the smaller electrostatic interactions between nonpolar pyrazine and electron-deficient down Si atoms. More polar pyridine molecules have stronger attractions with Si atoms. Moreover, the electron density at the nitrogen atom in pyridine is higher than that in pyrazine. Thus, pyrazine is a poor electron donor, and dative bonding through the nitrogen atom is unfavorable at room temperature. At elevated temperatures, the thermodynamically most favored [4+2] cycloaddition may occur to form a 1,4-*N,N*-dihydropyrazine-like state [37, 38]. Chemisorption does not cause significant shifts in the stretching frequency of C–H, indicating that all four C atoms of the molecule are not rehybridized. In addition, a new peak at 1617 cm^{-1}, attributed to the unconjugated C=C double bond, is observed. Furthermore, the characteristic vibrational modes of the aromatic ring at 1330–1547 cm^{-1} are absent in the HREELS spectra of chemisorbed pyrazine, indicating the disruption of aromaticity in the pyrazine. The formation of a [4+2] product is accompanied by loss of aromaticity. The reaction of pyrazine with the Si (111)-(7×7) surface results in formation of a [4+2] product in which both nitrogen atoms of the pyrazine form covalent bonds with the adatom–rest atom pair on the Si(111)-(7×7) surface. Such a reaction pathway produces two unconjugated C=C bonds [39].

8.2.3 O-Containing Molecules

8.2.3.1 Alcohols on Si(100) and Ge(100) Methanol (CH_3OH) is a simple O-containing molecule that is used as an oxidant and precursor in the formation of oxide layers. Adsorption of methanol on semiconductor surfaces has been extensively studied [40–45]. Previous experimental and theoretical results revealed that methanol undergoes dissociative adsorption on Si(100) and Ge(100) surfaces at room temperature.

Figure 8.25 illustrates the probable reaction pathways of methanol adsorption onto Ge(100). First, methanol may enter a molecular adsorption state as a result of

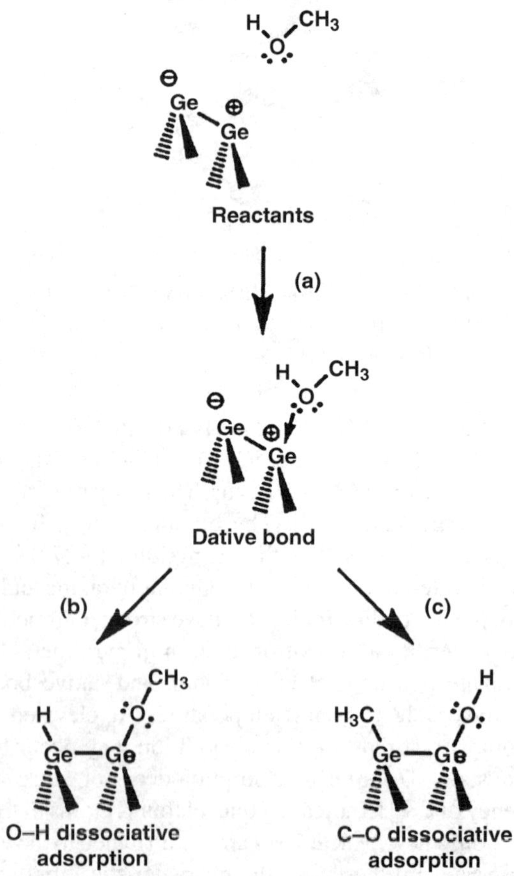

FIGURE 8.25 Schematic illustration of the possible reactions of methanol on Ge(100) surfaces: (a) Lewis acid–base interaction via Ge–O dative bonding; (b) proton transfer via O–H bond dissociation; and (c) methyl group transfer via C–O bond dissociation (from Ref. 44). Reprinted with permission from Bae, S.-S. et al., J. Phys. Chem. C, **111**, 15013. Copyright (2007) American Chemical Society.

Ge–O dative bonding between the O lone pair electrons and the electrophilic down Ge atom of a dimer (see Fig. 8.25a). This surface reaction occurs via a nucleophilic/electrophilic reaction. If the molecular adsorption state is unstable, the molecule subsequently undergoes proton transfer via O–H bond dissociation (see Fig. 8.25b) or methyl group transfer via C–O bond dissociation (see Fig. 8.25c).

Kim et al. studied the adsorption of methanol on Ge(100) using STM and DFT calculations [11, 44]. At low coverage, high-resolution experimental STM (see Fig. 8.26a) shows that methanol undergoes O–H bond dissociative adsorption on a single Ge–Ge dimer. As the methanol coverage increases to saturation, dissociative adsorption of methanol results in formation of dimer row-based chain-like arrays. DFT calculations (see Fig. 8.26b) show that at room temperature, O–H bond dissociative adsorption is kinetically more favorable than C–O bond dissociative adsorption, although the final product of C–O bond dissociative adsorption is more stable geometrically and thermodynamically. STM experiments and DFT calculations together suggested that the adsorption structure of methanol on Ge(100) at room temperature has a H–Ge–Ge–OCH_3 geometry as a result of O–H dissociative adsorption onto a single dimer.

Kachian and Bent described the adsorption structure of ethanol on Ge(100) [7]. Figure 8.27 displays the IR spectra for the interdimer O–H dissociated product and the intradimer O–H dissociative product for ethanol on a Ge(100) surface. As shown in this figure, it is difficult to identify the difference between inter- and intradimer O–H dissociated products. Thus, DFT calculations are needed to determine the most favorable adsorption structure.

8.2.3.2 Ketones on Si(100) and Ge(100)

Armstrong et al. reported that the majority of surface adducts formed by acetone, acetaldehyde, and biacetyl at low temperatures are [2+2] C=O cycloaddition products, as illustrated in Fig. 8.28a for the case of acetone [46, 47]. Barriocanal and Doren identified a barrierless pathway that passes through a dative-bonded precursor state for the [2+2] C=O cycloaddition

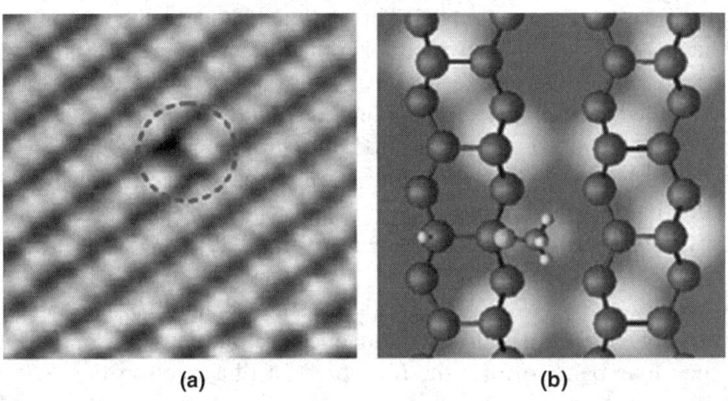

FIGURE 8.26 (a) The experimental filled state STM image ($V_s = -1.6$ V); (b) the corresponding theoretical schematic illustration of OH dissociative adsorption, with the OH group parallel to the Ge dimer row (from Ref. 44). Reprinted with permission from Bae, S.-S. et al., J. Phys. Chem. C, **111**, 15013. Copyright (2007) American Chemical Society.

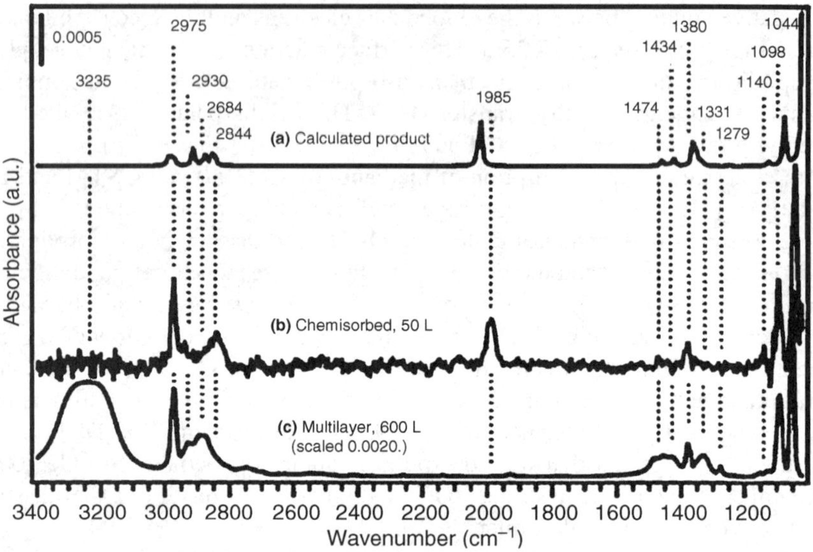

FIGURE 8.27 (a) Calculated IR spectrum of the intradimer O–H dissociation product of ethanol across a Ge dimer on the two-dimer intrarow $Ge_{15}H_{16}$ cluster. (b) The average of two IR spectra taken following saturation exposure of ethanol to Ge(100)-(2×1) at 310 K. (c) An IR spectrum of a multilayer of ethanol on Ge(100)-(2×1), taken at 140 K (intensity scaled). The peak labels correspond to chemisorbed and multilayer spectra. The product used to calculate the IR spectrum displayed in part (a) was geometrically optimized without geometric constraints (from Ref. 7). Reprinted with permission from Kachian, J. S. et al., J. Am. Chem. Soc., **131**, 7005. Copyright (2009) American Chemical Society.

product of glyoxal [48]. On the other hand, Wang et al. were not able to find a barrierless pathway for the reaction of acetone on Si(100)-(2×1), and they propose that steric interactions of the methyl groups with the surface likely play a role in hindering the reaction of acetone [10]. The conjugated double bonds in glyoxal may weaken the C=O bond and reduce or remove any activation barrier. Using IR spectroscopic results, Wang et al. [10] showed that acetone becomes trapped in a dative-bonded well at low temperatures on Ge(100)-(2×1), as evidenced by a 70–90 cm^{-1} red shift observed in the m(C=O) stretching mode.

We also consider the α-CH dissociation product, as depicted in Fig. 8.28b, which was first identified on a semiconductor surface by Wang et al. in their study of acetone on Ge(100)-(2×1) [10]. As shown in Fig. 8.29, the IR spectrum revealed absorption peaks near 1638 and 1967 cm^{-1} corresponding to ν(C=C) and ν(Ge–H) stretching modes, respectively. Wang et al. postulated that the nucleophilic dimer atom more easily abstracts a hydrogen atom once acetone datively bonds through one of the oxygen lone pairs to form the final product illustrated in Fig. 8.28b.

8.2.3.3 Carboxyl Acids on Si(100) and Ge(100) Several research groups have previously investigated carboxylic acids on Si(100)-(2×1) and have provided convincing evidence that the major reaction pathway proceeds via O–H

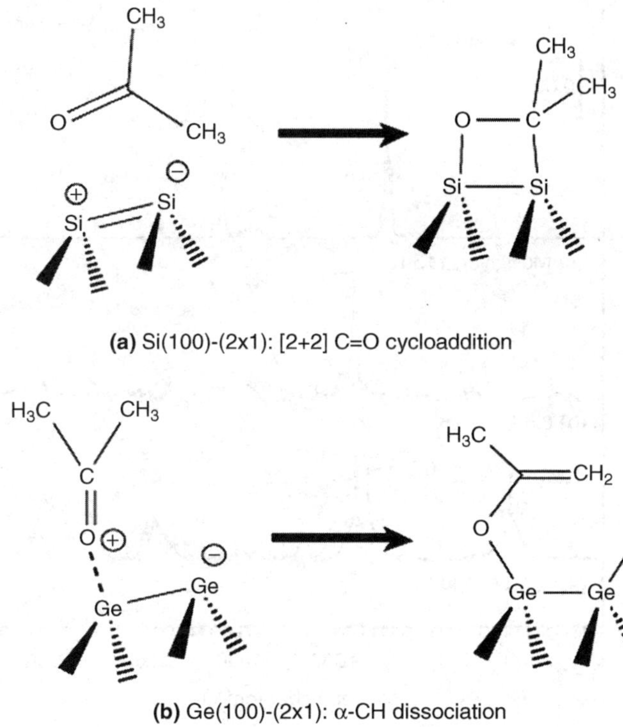

FIGURE 8.28 The major products of acetone adsorption onto Si(100)-(2×1) or Ge(100)-(2×1) are different. (a) [2+2] C=O cycloaddition is kinetically favored and is observed on Si(100)-(2×1) at low temperatures. (b) An α-CH dissociation product is thermodynamically favored and is observed on Ge(100)-(2×1) at room temperature.

dissociation [49, 50]. On the basis of the observed energy difference between the ν(C=O) and ν(C−O) peaks in the HREELS spectra, Tanaka et al. concluded that formic acid forms a monodentate O−H dissociation product on Si(100)-(2×1) at room temperature. More recently, Lopez et al. [51] reported an O−H dissociation product on Si(100)-(2×1), following a reaction with benzoic acid or the multifunctional molecules glycine and 4-aminobenzoic acid, respectively. Hwang et al. [52] investigated the reaction of vinyl acetic acid on Si(100)-(2×1) using low-energy electron diffraction (LEED), X-ray photoelectron spectroscopy (XPS), and synchrotron radiation photoemission spectroscopy (SRPES).

Bent and coworkers reported that the major reaction pathway of carboxylic acids on Si(100) was via O−H dissociation [53]. Although O−H dissociation is reported to be the major pathway on Si, different or additional surface reactions may potentially be observed for acetic acid adsorption.

Figure 8.30 shows the IR spectra corresponding to the reaction of acetic acid on Ge(100)-(2×1) as a function of coverage. Infrared results from a coverage-dependent study of acetic acid adsorbed onto Ge(100)-(2×1) at room temperature

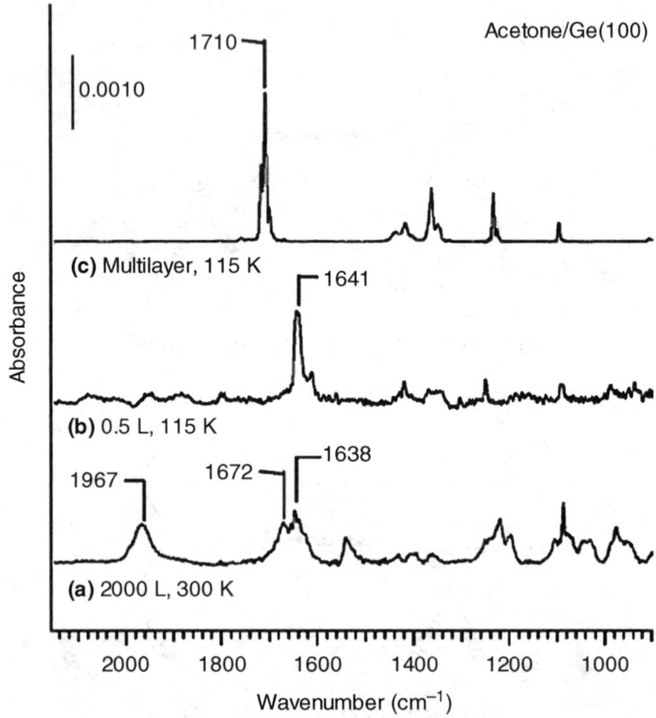

FIGURE 8.29 Infrared spectra of acetone adsorbed onto a Ge(100)-(2×1) surface: (a) 2000 L at 300 K, (b) 0.5 L at 115 K, and (c) multilayers (scaled) at 115 K. Stretching modes at 1967 and 1672/1638 cm^{-1} are evidence of ν(Ge–H) and ν(C=C) stretching modes, respectively, and indicate an α-CH dissociation product. At low temperatures, the presence of a red-shifted ν(C=O) mode at 1641 cm^{-1} indicates a dative bond species. Figure reprinted with permission from Ref. 10, copyright 2001 American Chemical Society. Reprinted with permission from Wang, G. T. et al., J. Phys. Chem. B, **105**, 12559. Copyright (2001) American Chemical Society.

or multilayers adsorbed at low temperatures are shown in Fig. 8.30a and b, respectively. The key peaks of the multilayer spectrum, appearing at 1697, 1440, 1414, 1360, and 1302 cm^{-1}, are in excellent agreement with previously published low-temperature spectra of molecular acetic acid [54]. The strong peak at 1697 cm^{-1} is attributed to a ν(C=O) stretching mode. A comparison with Haurie and Novak [55] suggests that the peaks at 1440 and 1360 cm^{-1} may be assigned to asymmetric and symmetric δ(CH_3) bending modes, respectively, whereas the mode at 1414 cm^{-1} may be assigned to a δ(OH) bending mode. The intense absorption feature at 1302 cm^{-1} may be attributed to a skeletal stretching vibration with the strongest contribution attributed to the C–O single bond [54, 56]. They conclusively confirmed, based on IR spectral studies, that carboxylic acids undergo O–H dissociation at 310 K, and the product is both kinetically and thermodynamically favored.

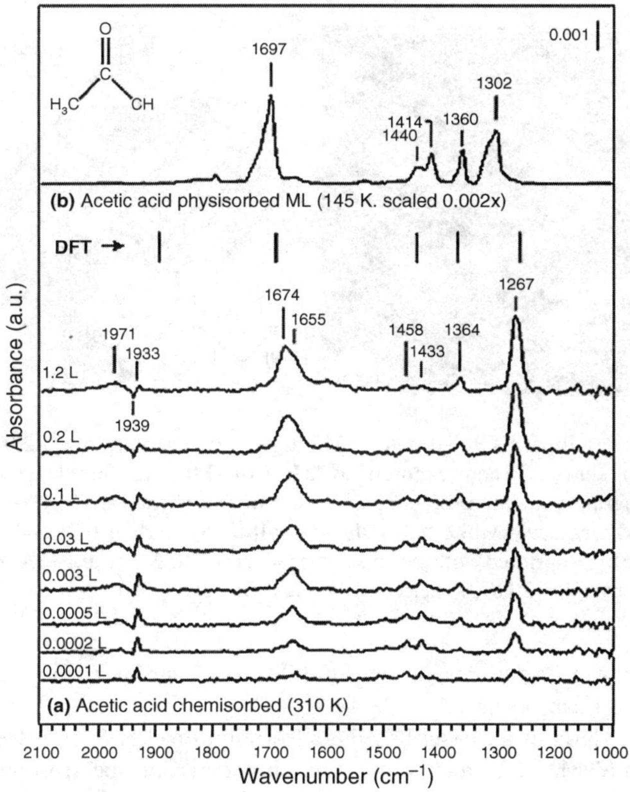

FIGURE 8.30 (a) Coverage-dependent infrared spectra for the reaction of acetic acid on Ge(100)-(2×1) at 310 K, and (b) acetic acid multilayers adsorbed at 145 K (scaled). The vertical lines mark the theoretically calculated vibrational frequencies of the O–H dissociation product (scaled by 0.96 for the B3LYP/6-31G* level of theory [53]. Reprinted with permission from Filler, M. A. et al., J. Am. Chem. Soc., **128**, 770. Copyright (2006) American Chemical Society.

8.2.4 S-Containing Molecules

8.2.4.1 Thiophene on Si(100) and Ge(100) A heteroaromatic thiophene molecule contains a sulfur atom and an aromatic ring. Thus, it can adsorb on Si and Ge(100) surfaces in a variety of ways, such as via [4+2] cycloaddition, [2+2] cycloaddition, or Lewis acid–base reaction. Lu et al. reported the adsorption of thiophene (C_4H_4S) on a reconstructed Si(100)-(2×1) surface using a hybrid density functional B3LYP method in combination with the cluster model approach [57]. Two chemisorption mechanisms, [4+2] or [2+2] cycloaddition of C_4H_4S onto a surface dimer site, have been examined. The calculations revealed that the former process is barrierless and favored over the latter process, which requires an activation energy of 2.6 kcal/mol. The di-σ-bonded surface species formed by the [4+2] cycloaddition-type chemisorption can undergo further [2+2] cycloaddition with a neighboring

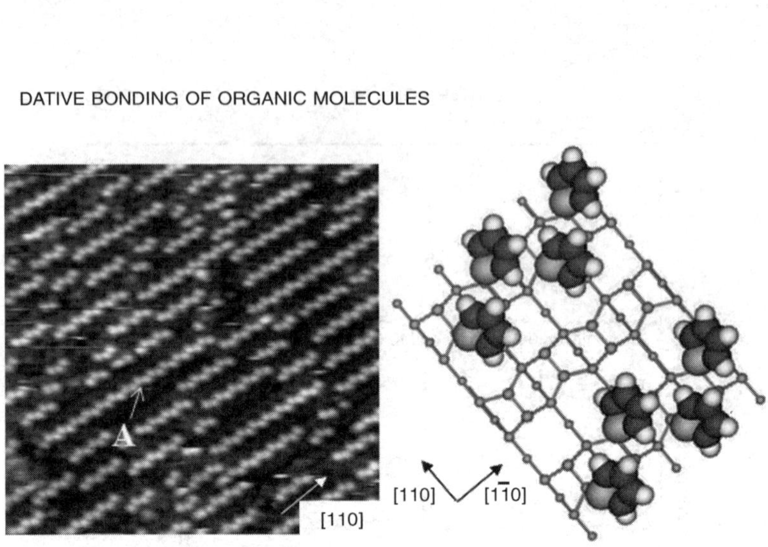

FIGURE 8.31 (left panel) Filled state STM images ($20 \times 20\,nm^2$, $V_s = -2.0\,V$, $I_t = 0.1\,nA$) of a Ge(100) surface when exposed to 2 L (0.13 ML) of thiophene. (right panel) Schematic diagram for dative bonding structure of thiophene on Ge(100) system. Each light gray, dark gray, and white ball indicates sulfur (S), carbon (C), and hydrogen (H), respectively [58]. Reprinted with permission from Jeon, S. M. et al., J. Am. Chem. Soc., **128**, 6296. Copyright (2006) American Chemical Society.

Si=Si dimer site. As a result, on Si(100)-(2×1) surfaces, only cycloaddition (chemisorption) can occur.

In contrast, Jeon et al. described the adsorption of C_4H_4S on a Ge(100)-(2×1) surface using STM, DFT calculations, and photoemission spectroscopy. They conclude that C_4H_4S on Ge(100) can be concurrently adsorbed through dative bonding and [4+2] cycloaddition at a coverage of <0.25 ML [58]. The initial deposition of thiophene on Ge(100)-(2×1) proceeds via dative bonding between thiophene and the Ge(100)-(2×1) surface through the sulfur atom. Figure 8.31 shows that the thiophene molecules preferentially form one-dimensional molecular chains (feature A: dative bonding feature) if a clean Ge(100) surface is exposed to 2 L of thiophene.

DFT calculations of the C_4H_4S/Ge_9H_{12} system, shown in Fig. 8.32, agree with the theoretical results for thiophene on Si(100) [57] in that the [4+2] cycloaddition

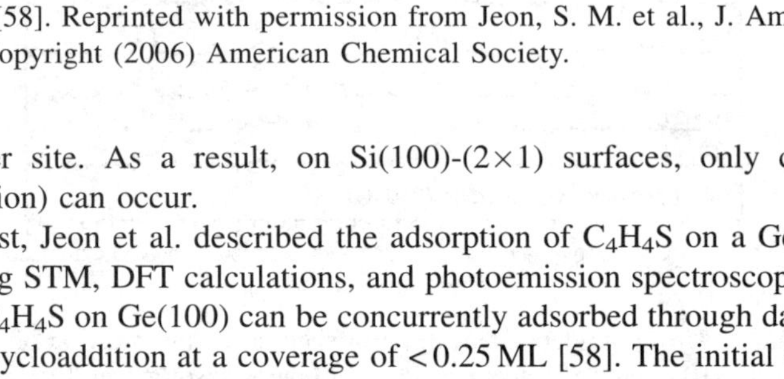

FIGURE 8.32 Local minima for the C_4H_4S/Ge_9H_{12} system calculated at the B3LYP/LACVP** level of theory. The values indicate the adsorption energies with respect to free thiophene and Ge_9H_{12}.

reaction product is predicted to be the most thermodynamically favorable product. The Ge–S dative bonding configuration was found to be the least stable structure from a thermodynamic perspective, although this reaction may have the lowest activation barrier, making it the most favorable from a kinetic perspective. The sulfur atom in thiophene is sp^3-hybridized, and therefore, the lone pair electrons are tilted away from the aromatic ring plane. As a consequence, the ring plane of the adsorbed thiophene is tilted with respect to the surface when it is bound to Ge(100) via Ge–S dative bonding.

For thiophene adsorbed onto Ge(100)-(2×1), Ge–S dative bonding can occur, although [4+2] cycloaddition is more energetically favorable.

8.3 DATIVE BONDING OF LEWIS ACIDS (ELECTROPHILIC)

As mentioned in previous sections, the formation of datively bonded Lewis bases with electron-deficient atoms on semiconductor surfaces has been intensively studied. In contrast, relatively few examples of the formation of dative bonds between Lewis acids and electron-rich atoms on semiconductor surfaces are available.

Konecny and Doren investigated the adsorption of BH_3 on Si(100) using nonlocal DFT with a cluster model for the surface. They predicted that a Si–B bond would form via nucleophilic attack on the boron, leaving BH_2 and H fragments bound to the surface [59]. Yoshinobu and coworkers studied the adsorption of BF_3 (a Lewis acid) on Si(100)-*c*(4×2) using high-resolution Si 2p photoelectron spectroscopy [60]. They found that BF_3 dissociates to form Si–F and Si–BF_2 species on Si(100), and the dissociated species (BF_2 and F) adsorb predominantly on the up atoms of Si(100) dimers. Cao and Hamers reported that the adsorption of trimethylamine (a Lewis base) and BF_3 (a Lewis acid) on a Si(100) surface leads to formation of a novel surface-mediated donor–acceptor complex with the structure TMA–Si–Si–BF_3 [61]. They found that the sequential adsorption of TMA and BF_3 leads to a dative bond between BF_3 (a Lewis acid) and the electron-rich up atom of a single Si(100) dimer.

Aluminum trichloride ($AlCl_3$) is a Lewis acid that forms donor–acceptor complexes with electron-rich molecules. Interestingly, Jung et al. studied the adsorption structures of $AlCl_3$ on Ge(100) using STM and high-resolution photoelectron spectroscopy. They found that $AlCl_3$ molecules adsorb onto Ge(100) surfaces via a cycloaddition-like reaction to form di-σ-bonds in which the $AlCl_3$ molecules interact with the surface dimers of Ge(100) [62]. These results are quite different from those of NH_3 adsorbed onto Ge(100), in which a single dative bond forms. Recently, an STM investigation by Kim et al. revealed that vinylferrocene, an organometallic molecule, adsorbs onto Ge(100) surfaces through Ge–Fe dative bonding between the electron-deficient Fe atom of the vinylferrocene and the electron-rich up atoms of Ge(100) dimers (see Fig. 8.33) [63]. They also predicted that there would be an interaction between the electron-rich cyclopentadiene (Cp) rings and the electron-deficient down Ge atoms at both ends of the adsorbate. An STM image of vinylferrocene adsorbed onto Ge(100) displays dumbbell-shaped

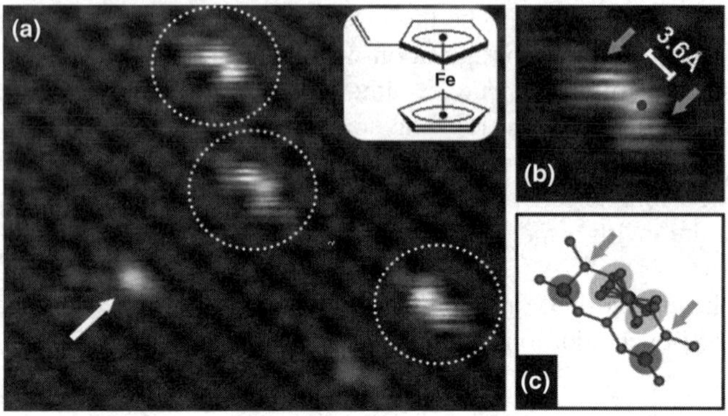

FIGURE 8.33 (a) A filled state STM image of a vinylferrocene adsorbed onto a Ge(100) surface, acquired at room temperature ($V_s = -1.1$ V, $I_t = 0.1$ nA). The dotted circles and arrow indicate parallel vinylferrocenes and tilted vinylferrocenes, respectively. (b, c) An enlarged image and a proposed schematic representation of the adsorption structure of the parallel vinylferrocene. Arrows indicate the positions of down Ge atoms.

fuzzy features (dotted circles), consisting of two coupled bright protrusions on one side of a dimer row, attributed to adsorbed vinylferrocene molecules. As indicated in an enlarged STM image of this feature (see Fig. 8.33b), the distance between the top positions of coupled protrusions is 3.6 Å, in agreement with both experimental and theoretical values of 3.3 Å for the distance between the Cp rings in a ferrocene molecule [64]. Therefore, it is reasonable that the bright dumbbell-shaped protrusions are associated with the coupled Cp rings of a vinylferrocene molecule. An atom-resolved STM investigation indicated that the center of the dumbbell-shaped protrusion, the position of the iron atom, is exactly located on top of an up Ge atom (see Fig. 8.33c). Thus, formation of a dative bond between the Fe and the up Ge atom is expected. The formal charge of the iron atom in the vinylferrocene molecule is considered to be positive [65], and the electron-rich buckled-up Ge atom is predicted to donate electrons to an iron atom.

8.4 SUMMARY

The chemical reactions of organic compounds with semiconductor surfaces such as Si(100)-(2×1), Si(111)-(7×7), and Ge(100)-(2×1) are summarized in Table 8.1. Additional surface reconstructions may occur after the initial reactions. Cycloaddition and dissociative adsorption compete with dative bonding. With respect to aliphatic amines, both primary and secondary amine molecules dissociatively adsorb, whereas tertiary amines molecularly adsorb onto Si(100). In contrast, most amines absorb mainly through dative bonding on Ge(100) surfaces, except for ethylenediamine and pyrrole. O-containing molecules such as alcohol, ketone, and carboxylic acids on Si(100)-(2×1) and Ge(100)-(2×1) mainly undergo dissociative

TABLE 8.1 Reactions Between Organic Molecules and Si(100), Si(111), and Ge(100) Surfaces

Reactant	Si(100)-(2×1)	Si(111)-(7×7)	Ge(100)-(2×1)
Aliphatic amines			
Methylamine	N–H dissociative [66]		N dative [3]
Dimethylamine	N–H dissociative [3, 18, 20, 67]	N–H dissociative [74]	N dative [3]
Trimethylamine	N dative [3, 18, 20, 66, 67]	N dative/N–C dissociative [20]	N dative [3]
N,N-Dimethylbutylamine	N dative/N–C dissociative [66]		
N-Methylpentylamine	N–H dissociative [66]		
Hexylamine	N–H dissociative [66]		
Ethylenediamine			N–H dissociative/N dative [23]
Pyrrolidine	N–H dissociative [4, 19, 70, 71]		N dative/N–H dissociative [19]
N-Methylpyrrolidine	N dative [20]		N dative [19]
3-Pyrroline	N–H dissociative/C=C [2+2] [4, 19, 71, 72]		N dative/N–H dissociative [19]
1-Amino-3-cyclopentene	C=C [2+2] [73]		
Aromatic amines			
Aniline	N–H dissociative [4, 68]	C=C–C=C [4+2] [75]	
1,4-Phenylenediamine	N–H dissociative [69]		
Pyrrole	N–H dissociative [4, 19, 25]	N–H dissociative [27]	N–H dissociative/C–H dissociative [19, 76]
N-Methylpyrrole	C–H dissociative/C=C–C=C [4+2] [19]	C=C–C=C [4+2] [27]	
Pyridine	N=C–C=C [4+2] [28–31, 35]	N dative/C=N–C=C [4+2] [33]	N dative [32, 34, 35]
Pyrimidine			N dative [36]
Pyrazine	1,4- or 2,5-[4+2] cycloaddition [28, 29, 37]	C=N–C=N [4+2] [39]	
s-Triazine	N=C–N=C [4+2] cycloaddition [28]		

(Continued)

TABLE 8.1 (*Continued*)

Reactant	Si(100)-(2×1)	Si(111)-(7×7)	Ge(100)-(2×1)
Alcohols			
Methanol	O–H dissociative [9, 68, 69]		O–H dissociative [11, 44]
Ethanol	O–H dissociative [69, 77]		O–H dissociative [7]
1-Propanol	O–H dissociative [69, 78]		
2-Propanol	O–H dissociative [79]		
tert-Butyl alcohol	O–H dissociative [79]		
2,3-Butanediol	O–H dissociative [80]		
Phenol	O–H dissociative [81]		
Acids			
Carboxylic acid	O–H dissociative [49, 50, 53]		O–H dissociative [54–56]
Formic acid	O–H dissociative [10, 75]		
Benzoic acid	O–H dissociative [42, 51, 68]		
Ketones/aldehydes			
Acetone	[2+2] cycloaddition [48]		α-C–H dissociative [10]
Sulfur thiophene	Cycloaddition [57]		S dative and cycloaddition [58]

reaction through a dative-bonded intermediate state. S-containing molecules such as thiophene adsorb on Si(100)-(2×1) and Ge(100)-(2×1) through dative bonding and cycloaddition reaction.

REFERENCES

1. Kang, H. C. *Surf. Interface Anal.* **1999**, *28*, 92.
2. Lui, Q.; Hoffmann, R. *J. Am. Chem. Soc.* **1995**, *117*, 4082.
3. Mui, C.; Han, J. H.; Wang, G. T.; Musgrave, C. B.; Bent, S. F. *J. Am. Chem. Soc.* **2002**, *124*, 4027.
4. Cao, X.; Coulter, S. K.; Ellison, M. D.; Liu, H.; Hamers, R. J. *J. Phys. Chem. B* **2001**, *105*, 3759.
5. Waltenburg, H. N.; Yates, J. T., Jr., *Chem. Rev.* **1995**, *95*, 1589.
6. Kachian, J. S.; Wong, K. T.; Bent, S. F. *Acc. Chem. Res.* **2010**, *43*, 346.
7. Kachian, J. S.; Bent, S. F. *J. Am. Chem. Soc.* **2009**, *131*, 7005.
8. Filler, M. A.; Mui, C.; Musgrave, C. B.; Bent, S. F. *J. Am. Chem. Soc.* **2003**, *125*, 4928.
9. Lu, X.; Zhang, Q.; Lin, M. C. *Phys. Chem. Chem. Phys.* **2001**, *3*, 2156.
10. Wang, G. T.; Mui, C.; Musgrave, C. B.; Bent, S. F. *J. Phys. Chem. B* **2001**, *105*, 12559.
11. Kim, D. H.; Bae, S.-S.; Hong, S.; Kim, S. *Surf. Sci.* **2010**, *604*, 129.
12. Bent, H. A. *Chem. Rev.* **1961**, *61*, 275.
13. Widjaja, Y.; Mysinger, M. M.; Musgrave, C. B. *J. Phys. Chem. B* **2000**, *104*, 2527.
14. Mui, C.; Musgrave, C. B. *Langmuir* **2005**, *21*, 5230.
15. Warschkow, O.; Wilson, H. F.; Marks, N. A; Schofield, S. R.; Curson, N. J.; Smith, P. V.; Radny, M. W.; McKenzie, D. R.; Simmons, M. Y. *Phys. Rev. B* **2005**, *72*, 125328.
16. Schofield, S. R.; Curson, N. J.; Warschkow, O.; Marks, N. A; Wilson, H. F.; Simmons, M. Y.; Smith, P. V.; Radny, M. W.; McKenzie, D. R.; Clark, R. G. *J. Phys. Chem. B* **2006**, *110*, 3173.
17. Tsai, H.-W.; Lin, D.-S. *Surf. Sci.* **2001**, *482*, 654.
18. Mui, C.; Wang, G. T.; Bent, S. F.; Musgrave, C. B. *J. Chem. Phys.* **2001**, *114*, 10170.
19. Wang, G. T.; Mui, C.; Tannaci, J. F.; Filler, M. A.; Musgrave, C. B.; Bent, S. F. *J. Phys. Chem. B* **2003**, *107*, 4982.
20. Cao, X.; Hamers, R. J. *J. Am. Chem. Soc.* **2001**, *123*, 10988.
21. McKean, D. C.; Ellis, I. A. *J. Mol. Struct.* **1975**, *29*, 81.
22. Tao, F.; Cai, Y.; Ning, Y.; Xu, G.; Bernasek, S. L. *J. Phys. Chem. C* **2008**, *112*, 15474.
23. Kim, A.; Filler, M. A.; Kim, S.; Bent, S. F. *J. Phys. Chem. B* **2005**, *109*, 19817.
24. Kim, A.; Filler, M. A.; Kim, S.; Bent, S. F. *J. Am. Chem. Soc.* **2005**, *127*, 6123.
25. Qiao, M. H.; Cao, Y.; Deng, J. F.; Xu, G. Q. *Chem. Phys. Lett.* **2000**, *325*, 508.
26. Yuan, Z. L.; Chen, X. F.; Wang, Z. H.; Yong, K. S.; Cao, Y.; Xu, G. Q. *J. Chem. Phys.* **2003**, *119*, 10389.
27. Tao, F.; Yuan, Z. L.; Chen, X. F.; Qiao, M. H.; Wang, Z. H.; Dai, Y. J.; Huang, H. G.; Cao, Y.; Xu, G. Q. *Phys. Rev. B* **2003**, *67*, 245406.
28. Lu, X.; Lin, M. C. *Int. Rev. Phys. Chem.* **2002**, *21*, 137.

29. Lu, X.; Xu, X.; Wu, J.; Wang, N.; Zhang, Q. *New J. Chem.* **2002**, *26*, 160.
30. Tao, F.; Qiao, M. H.; Wang, Z. H.; Xu, G. Q. *J. Phys. Chem. B* **2003**, *107*, 6384.
31. Miwa, J. A.; Eves, B. J.; Rosei, F.; Lopinski, G. P. *J. Phys. Chem. B* **2005**, *109*, 20055.
32. Cho, Y. E.; Maeng, J. Y.; Kim, S.; Hong, S. *J. Am. Chem. Soc.* **2003**, *125*, 7514.
33. Tao, F.; Lai, Y. H.; Xu, G. Q. *Langmuir* **2004**, *20*, 366.
34. Hong, S.; Cho, Y. E.; Maeng, J. Y.; Kim, S. *J. Phys. Chem. B* **2004**, *108*, 15229.
35. Kim, H.-J.; Cho, J.-H. *J. Chem. Phys.* **2004**, *120*, 8222.
36. Lee, J. Y.; Jung, S. J.; Hong, S.; Kim, S. *J. Phys. Chem. B* **2005**, *109*, 348.
37. Huang, H. G.; Huang, J. Y.; Ning, Y. S.; Xu, G. Q. *J. Chem. Phys.* **2004**, *121*, 4820.
38. Shimomura, M.; Ichikawa, D.; Fukuda, Y.; Abukawa, T.; Aoyama, T.; Kono, S. *Phys. Rev. B* **2005**, *72*, 033303.
39. Huang, H. G.; Wang, Z. H.; Xu, G. Q. *J. Phys. Chem. B* **2004**, *108*, 12560.
40. Kanaya, H.; Usuda, K.; Yamada, K. *Appl. Phys. Lett.* **1995**, *67*, 682.
41. Horie, T.; Takakuwa, Y.; Miyamoto, N. *Jpn. J. Appl. Phys.* **1994**, *33*, 4684.
42. Kato, T.; Kang, S.-Y.; Xu, X.; Yamabe, T. *J. Phys. Chem. B* **2001**, *105*, 10340.
43. Zhang, L.; Carman, A. J.; Casey, S. M. *J. Phys. Chem. B* **2003**, *107*, 8424.
44. Bae, S.-S.; Kim, D. H.; Kim, A.; Jung, S. J.; Hong, S.; Kim, S. *J. Phys. Chem. C* **2007**, *111*, 15013.
45. Carbone, M.; Meloni, S.; Caminiti, R. *Phys. Rev. B* **2007**, *76*, 085332.
46. Armstrong, J. L.; Pylant, E. D.; White, J. M. *J. Vac. Sci. Technol. A* **1998**, *16*, 123.
47. Armstrong, J. L.; White, J. M.; Langell, M. *J. Vac. Sci. Technol. A* **1997**, *15*, 1146.
48. Barriocanal, J. A.; Doren, D. J. *J. Am. Chem. Soc.* **2001**, *123*, 7340.
49. Tanaka, S.; Onchi, M.; Nishijima, M. *J. Chem. Phys.* **1989**, *91*, 2712.
50. Lopez, A.; Heller, T.; Bitzer, T.; Richardson, N. V. *Chem. Phys.* **2002**, *277*, 1.
51. Lopez, A.; Bitzer, T.; Heller, T.; Richardson, N. V. *Surf. Sci.* **2001**, *480*, 65.
52. Hwang, H.-N.; Baik, J. Y.; An, K.-S.; Lee, S. S.; Kim, Y.; Hwang, C. C.; Kim, B. *J. Phys. Chem. B* **2004**, *108*, 8379.
53. Filler, M. A.; Deventer, J. A. V.; Keung, A. J.; Bent, S. F. *J. Am. Chem. Soc.* **2006**, *128*, 770.
54. Teragni, P.; Masetti, G.; Zerbi, G. *Chem. Phys.* **1978**, *28*, 55.
55. Haurie, M.; Novak, A. *Spectrochim. Acta* **1965**, *21*, 1217.
56. Srivastava, T. N.; Onyszchuk, M. *Can. J. Chem.* **1963**, *41*, 1244.
57. Lu, X.; Xu, X.; Wang, N.; Zhang, Q.; Lin, M. C. *J. Phys. Chem. B* **2001**, *105*, 10069.
58. Jeon, S. M.; Jung, S. J.; Lim, D. K.; Kim, H.-D.; Lee, H.; Kim, S. *J. Am. Chem. Soc.* **2006**, *128*, 6296.
59. Konecny, R.; Doren, D. J. *J. Phys. Chem. B* **1997**, *101*, 10983.
60. Machida, S.; Nagao, M.; Yamamoto, S.; Kakefuda, Y.; Mukai, K.; Yamashita, Y.; Yoshinobu, J. *Surf. Sci.* **2003**, 532–535, 716.
61. Cao, X.; Hamers, R. J. *J. Phys. Chem. B* **2002**, *106*, 1840.
62. Jung, S. J.; Youn, Y.-S.; Lee, H.; Kim, K.-J.; Kim, B. S; Kim, S. *J. Am. Chem. Soc.* **2008**, *130*, 3288.
63. Kim, Y. B; Jung, S. J.; Min, Y. H.; Kim, S. *J. Am. Chem. Soc.* **2010**, *132*, 12782.

64. Koch, H.; Jørgensen, P.; Helgaker, T. *J. Chem. Phys.* **1996**, *104*, 9528.
65. Haaland, A. *Acc. Chem. Res.* **1979**, *12*, 415.
66. Cao, X.; Hamers, R. J. *J. Vac. Sci. Technol. B* **2002**, *20*, 1614.
67. Carman, A. J.; Zhang, L.; Liswood, J. L.; Casey, S. M. *J. Phys. Chem. B* **2003**, *107*, 5491.
68. Bitzer, T.; Alkunshalie, T.; Richardson, N. V. *Surf. Sci.* **1996**, *368*, 202.
69. Bitzer, T.; Richardson, N. V. *Appl. Phys. Lett.* **1997**, *71*, 662.
70. Wang, G. T.; Mui, C.; Musgrave, C. B.; Bent, S. F. *J. Phys. Chem. B* **2001**, *105*, 3295.
71. Liu, H.; Hamers, R. J. *Surf. Sci.* **1998**, *416*, 354.
72. Romero, A. H.; Sbraccia, C.; Silvestrelli, P. L. *J. Chem. Phys.* **2004**, *120*, 9745.
73. Cantele, G.; Trani, F.; Ninno, D.; Cossi, M.; Barone, V. *J. Phys.: Condens. Matter* **2006**, *18*, 2349.
74. Cao, X.; Hamers, R. J. *Surf. Sci.* **2003**, *523*, 241.
75. Tomimoto, H.; Sumii, R.; Shirota, N.; Yagi, S.; Taniguchi, M.; Sekitani, T.; Tanaka, K. *J. Vac. Sci. Technol. B* **2000**, *18*, 2335.
76. Kim, D. H.; Choi, D. S.; Kim, A.; Bae, S.-S.; Hong, S.; Kim, S. *J. Phys. Chem. B* **2006**, *110*, 7938.
77. Silvestrelli, P. L. *Surf. Sci.* **2004**, *552*, 17.
78. Zhou, J.-G.; Hagelberg, F. *Int. J. Quantum Chem.* **2005**, *105*, 359.
79. Kim, J.; Kim, K.; Yong, K. *J. Vac. Sci. Technol. A* **2002**, *20*, 1582.
80. Kim, J. W.; Carbone, M.; Tallarida, M.; Dil, J. H.; Horn, K.; Casaletto, M. P.; Flammini, R.; Piancastelli, M. N. *Surf. Sci.* **2004**, *559*, 179.
81. Casaletto, M. P.; Carbone, M.; Piancastelli, M. N.; Horn, K.; Weiss, K.; Zanoni, R. *Surf. Sci.* **2005**, *582*, 42.

CHAPTER 9

Ab Initio Molecular Dynamics Studies of Conjugated Dienes on Semiconductor Surfaces

MARK E. TUCKERMAN AND YANLI ZHANG

9.1 INTRODUCTION

The chemistry of hybrid structures composed of organic molecules and semiconductor surfaces is opening up exciting new areas of development in molecular electronics, nanoscale sensing devices, and surface lithography [1,2]. Covalent attachment of organic molecules to a semiconducting surface can yield active devices, such as molecular switches and sensors or passivating insulating layers. Moreover, it is intriguing to consider how the reactions might be controlled by "engineering" specific modifications to organic molecules and/or surfaces, suggesting possible new lithographic techniques.

One of the goals of controlling the surface chemistry is the creation of ordered nanostructures on semiconducting surfaces. Indeed, there has been some success in obtaining locally ordered structures on the hydrogen-terminated Si(100) surface [3–7]. These methods require a dangling Si bond without a hydrogen to initialize the self-replicating reaction. Another popular approach eliminates the initialization step by exploiting the reactivity between surface dimers on certain reconstructed surfaces with the π bonds in many organic molecules. The challenge with this approach lies in designing the surface and/or the molecule so as to eliminate all but one desired reaction channel. Charge asymmetries, such as occur on the Si(100)-2 $\times$ 1 surface, lead to a violation of the usual Woodward-Hoffman selection rules, which govern many purely organic reactions, allowing a variety of possible [4 + 2] and [2 + 2] surface adducts. Multiple reactive sites, such as occur on some of the SiC surfaces, also allow for a variety of possible adducts.

Functionalization of Semiconductor Surfaces, First Edition.
Edited by Franklin (Feng) Tao and Steven L. Bernasek.

The exploration of hybrid organic-semiconductor materials and the reactions associated with them is an area in which theoretical and computational tools can play an important role. Indeed, modern theoretical methods combined with high-performance computing, have advanced to a level such that the thermodynamics and reaction mechanisms of a variety of chemical processes can now be routinely studied. These studies can aid in the interpretation of experimental results and can leverage theoretical mechanisms to predict the outcomes of new experiments. This chapter will focus on a description of one set of such techniques, namely, those based on density functional theory and first-principles or *ab initio* molecular dynamics (AIMD) [8–11]. As these methods employ an explicit representation of the electronic structure, electron localization techniques can be used to follow local electronic rearrangements during a reaction and, therefore, generate a clear picture of the reaction mechanism. In addition, statistical mechanical tools can be employed to obtain thermodynamic properties of the reaction products, including relative free energies and populations of the various products. Following a detailed description of the computational approaches, we will present a series of applications to conjugated dienes reacting with different semiconductor surfaces [12–17]. We will explore the role of surface thermal motions on the reaction mechanisms, and we will demonstrate how the predicted mechanisms can be used to rationalize product distributions [18]. We will investigate how the surface structure influences the thermodynamics of the reaction products and how these thermodynamic properties can be used to "reverse engineer" the molecule and/or the surface in order to control the product distribution and associated free energies. Finally, we will describe the problem of computing theoretical scanning tunneling microscopy (STM) images and the challenges inherent in simple perturbative schemes. Unless otherwise stated, all of the calculations to be presented in this chapter were carried out using the implementation of plane-wave based AIMD in the PINY_MD package [19]. Details of this methodology are presented in the next section.

9.2 COMPUTATIONAL METHODS

Because the problem of covalently attaching an organic molecule to a semiconductor surface requires the formation of chemical bonds, a theoretical treatment of this problem must be able to describe this bond formation process, which generally requires an *ab initio* approach in which the electronic structure is accounted for explicitly. Assuming the validity of the Born–Oppenheimer approximation, the goal of any *ab initio* approach is to approximate the ground-state solution of the electronic Schrödinger equation $\hat{H}_{\text{elec}}(\mathbf{R})|\Psi_0(\mathbf{R})\rangle = E_0(\mathbf{R})|\Psi_0(\mathbf{R})\rangle$, where $\hat{H}_{\text{elec}}$ is the electronic Hamiltonian, and $\mathbf{R}$ denotes a classical configuration of the chemical nuclei in the system. Ultimately, in order to predict reaction mechanisms and thermodynamics of a large, condensed-phase system, one needs to use approximate solutions of the Schrödinger equation and based on these solutions, propagate the nuclei dynamically using Newton's laws of motion. This is the essence of the method known as *ab initio* molecular dynamics (AIMD) [8–11].

In this section, we will briefly review density functional theory (DFT) as the electronic structure method of choice for the studies to be described in this chapter.

DFT represents a compromise between accuracy and computational efficiency. This is an important consideration, as a typical AIMD calculation requires that the electronic structure problem be solved tens of thousands to hundreds of thousands of time in order to generate one or more trajectories of sufficient length to extract dynamic and thermodynamic properties. We will then describe the AIMD approach, including several technical considerations such as basis sets, boundary conditions, and electron localization schemes.

9.2.1 Density Functional Theory

As previously above, we seek approximate solutions to the electronic Schrödinger equation. To this end, we begin by considering a system of N nuclei at positions $\mathbf{R}_1, \ldots, \mathbf{R}_N \equiv \mathbf{R}$ and M electrons with coordinate labels $\mathbf{r}_1, \ldots, \mathbf{r}_M$ and spin states $s_1, \ldots, s_M$. The fixed nuclear positions allow us to define the electronic Hamiltonian (in atomic units) as

$$\hat{H}_{\text{elec}}(\mathbf{R}) = -\frac{1}{2}\sum_{i=1}^{M} \nabla_i^2 + \sum_{i>j} \frac{1}{|\mathbf{r}_i - \mathbf{r}_j|} - \sum_{I=1}^{N}\sum_{i=1}^{M} \frac{Z_I}{|\mathbf{R}_I - \mathbf{r}_i|} \tag{9.1}$$

where Z_I is the charge on the Ith nucleus. The first term in Equation 9.1 represents the electron kinetic energy, the second represents the electron–electron Coulomb repulsion, and the third represents the electron–nuclear Coulomb attraction. The time-independent electronic Schrödinger equation or electronic eigenvalue problem

$$\left[-\frac{1}{2}\sum_{i=1}^{M} \nabla_i^2 + \sum_{i>j} \frac{1}{|\mathbf{r}_i - \mathbf{r}_j|} - \sum_{I=1}^{N}\sum_{i=1}^{M} \frac{Z_I}{|\mathbf{R}_I - \mathbf{r}_i|}\right] \Psi(\mathbf{x}_1, \ldots, \mathbf{x}_M, \mathbf{R}) = E(\mathbf{R})\Psi(\mathbf{x}_1, \ldots, \mathbf{x}_M, \mathbf{R}) \tag{9.2}$$

must be solved in order to generate all of the electronic energy levels and eigenfunctions at the given nuclear configuration $\mathbf{R}$. Here $\mathbf{x}_i = \mathbf{r}_i, s_i$ is a combination of coordinate and spin variables. Unfortunately, for large condensed-phase problems of the type to be considered here, an exact solution of the electronic eigenvalue problem is computationally intractable.

The Kohn–Sham (KS) [20] formulation of density functional theory (DFT) [21] replaces the fully interacting electronic problem in Equation 9.2 by an equivalent noninteracting system that is required to yield the same *ground-state* energy and ground state electronic density as the original interacting system. As the name implies, the central quantity in DFT is this ground-state electron density $n_0(\mathbf{r})$ generated from the ground-state electronic wave function Ψ_0 via

$$n_0(\mathbf{r}) = M \sum_{s_1=-1/2}^{1/2} \cdots \sum_{s_M=-1/2}^{1/2} \int d\mathbf{r}_2, \ldots, d\mathbf{r}_M |\Psi_0(\mathbf{r}, s_1, \ldots, \mathbf{x}_M)|^2 \tag{9.3}$$

where, for notational convenience, the dependence on $\mathbf{R}$ is left off. The central theorem of DFT is the Hohenberg–Kohn theorem, which states that there exists an exact energy $E[n]$ that is a functional of electronic densities $n(\mathbf{r})$ such that when $E[n]$ is minimized with respect to $n(\mathbf{r})$ subject to the constraint that $\int d\mathbf{r}\, n(\mathbf{r}) = M$ (each $n(\mathbf{r})$ must yield the correct number of electrons), the true ground-state density $n_0(\mathbf{r})$ is obtained. The true ground-state energy is then given by $E_0 = E[n_0]$. (Remember that E_0 is a function of nuclear positions: $E_0 = E_0(\mathbf{R})$.) The KS noninteracting system is constructed in terms of a set of mutually orthogonal single-particle orbitals $\psi_i(\mathbf{r})$ in terms of which the density $n(\mathbf{r})$ is given by

$$n(\mathbf{r}) = \sum_{i=1}^{N_s} f_i |\psi_i(\mathbf{r})|^2 \tag{9.4}$$

where f_i are the occupation numbers of a set of N_s such orbitals, where $\sum_i f_i = M$. In closed-shell systems, the orbitals are all doubly occupied so that $N_s = M/2$, and $f_i = 2$. In open-shell systems, we treat all of the electrons in double and singly occupied orbitals explicitly and take $N_s = M$. When virtual or unoccupied orbitals are needed, we can take $N_s > M/2$ or $N_s > M$ for closed-and open-shell systems, respectively, and take $f_i = 0$ for the virtual orbitals.

In KS theory, the energy functional is taken to be

$$\mathscr{E}[\{\psi\}] = -\frac{1}{2}\sum_{i=1}^{N_s} f_i \langle \psi_i | \nabla^2 | \psi_i \rangle + \frac{1}{2}\int d\mathbf{r}\, d\mathbf{r}' \frac{n(\mathbf{r})n(\mathbf{r}')}{|\mathbf{r}-\mathbf{r}'|} + E_{\mathrm{xc}}[n] + \int d\mathbf{r}\, n(\mathbf{r}) V_{\mathrm{ext}}(\mathbf{r}, \mathbf{R}) \tag{9.5}$$

The first term in the functional represents the noninteracting quantum kinetic energy of the electrons, the second term is the direct Coulomb interaction between two charge distributions, the third term is the exchange-correlation energy, whose exact form is unknown, and the fourth represents the "external" Coulomb potential on the electrons due to the fixed nuclei, $V_{\mathrm{ext}}(\mathbf{r}, \mathbf{R}) = -\sum_I Z_I / |\mathbf{r} - \mathbf{R}_I|$. Minimization of Equation 9.5 with respect to the orbitals subject to the orthogonality constraint leads to a set of coupled self-consistent field equations of the form

$$\left[-\frac{1}{2}\nabla^2 + V_{\mathrm{KS}}(\mathbf{r})\right]\psi_i(\mathbf{r}) = \sum_j \lambda_{ij}\psi_j(\mathbf{r}) \tag{9.6}$$

where the KS potential $V_{KS}(\mathbf{r})$ is given by

$$V_{\mathrm{KS}}(\mathbf{r}) = \int d\mathbf{r}' \frac{n(\mathbf{r}')}{|\mathbf{r}-\mathbf{r}'|} + \frac{\delta E_{\mathrm{xc}}}{\delta n(\mathbf{r})} + V_{\mathrm{ext}}(\mathbf{r}, \mathbf{R}) \tag{9.7}$$

and λ_{ij} is a set of Lagrange multipliers used to enforce the orthogonality constraint $\langle \psi_i | \psi_j \rangle = \delta_{ij}$. If we introduce into Equation 9.6 a unitary transformation U that diagonalizes the matrix λ_{ij}, then we obtain the Kohn–Sham equations in the form

$$\left[-\frac{1}{2}\nabla^2 + V_{\mathrm{KS}}(\mathbf{r})\right]\phi_i(\mathbf{r}) = \varepsilon_i \phi_i(\mathbf{r}) \tag{9.8}$$

where $\phi_i(\mathbf{r}) = \sum_j U_{ij}\psi_j(\mathbf{r})$ are the KS orbitals and ε_i are the KS energy levels,that is, the eigenvalues of the matrix λ_{ij}. If the exact exchange-correlation functional were known, the KS theory would be exact. However, because $E_{\mathrm{xc}}[n]$ is unknown, approximations must be introduced for this term in practice. The accuracy of DFT results depends critically on the quality of the approximation. One of the most widely used forms for $E_{\mathrm{xc}}[n]$ is known as the generalized-gradient approximation (GGA), where in $E_{\mathrm{xc}}[n]$ is approximated as a local functional of the form

$$E_{\mathrm{xc}}[n] \approx \int d\mathbf{r}\, f_{\mathrm{GGA}}(n(\mathbf{r}), |\nabla n(\mathbf{r})|) \tag{9.9}$$

where the form of the function f_{GGA} determines the specific GGA approximation. Commonly used GGA functionals are the Becke–Lee–Yang–Parr (BLYP) [22,23] and Perdew–Burke–Ernzerhof (PBE) [24] functionals.

9.2.2 Ab Initio Molecular Dynamics

Solution of the KS equations yields the electronic structure at a set of fixed nuclear positions $\mathbf{R}_1, \ldots, \mathbf{R}_N \equiv \mathbf{R}$. Thus, in order to follow the progress of a chemical reaction, we need an approach that allows us to propagate the nuclei in time. If we assume the nuclei can be treated as classical point particles, then we seek the nuclear positions $\mathbf{R}_1(t), \ldots, \mathbf{R}_N(t)$ as functions of time, which are given by Newton's second law

$$M_I \ddot{\mathbf{R}}_I = \mathbf{F}_I \tag{9.10}$$

where M_I and $\mathbf{F}_I$ are the mass and total force on the Ith nucleus. If the exact ground-state wave function $\Psi_0(\mathbf{R})$ were known, then the forces would be given by the Hellman–Feynman theorem

$$\mathbf{F}_I = -\langle \Psi_0(\mathbf{R}) | \nabla_I \hat{H}_{\mathrm{elec}}(\mathbf{R}) | \Psi_0(\mathbf{R}) \rangle - \nabla_I U_{\mathrm{NN}}(\mathbf{R}) \tag{9.11}$$

where we have introduced the nuclear–nuclear Coulomb repulsion potential

$$U_{\mathrm{NN}}(\mathbf{R}) = \sum_{I>J} \frac{Z_I Z_J}{|\mathbf{R}_I - \mathbf{R}_J|} \tag{9.12}$$

Within the framework of KS DFT, the force expression becomes

$$\mathbf{F}_I = -\int d\mathbf{r}\, n_0(\mathbf{r})\nabla_I V_{\text{ext}}(\mathbf{r},\mathbf{R}) - \nabla_I U_{\text{NN}}(\mathbf{R}) \tag{9.13}$$

The equations of motion, Equations 9.10, are integrated numerically using a set of discrete times $t = 0, \Delta t, 2\,\Delta t, \ldots, P\Delta t$ subject to a set of initial coordinates $\mathbf{R}_1(0), \ldots, \mathbf{R}_N(0)$ and velocities $\dot{\mathbf{R}}_1(0), \ldots, \dot{\mathbf{R}}_N(0)$ for a solver such as the velocity Verlet algorithm:

$$\begin{aligned}\mathbf{R}_I(\Delta t) &= \mathbf{R}_I(0) + \Delta t\dot{\mathbf{R}}_I(0) + \frac{\Delta t^2}{2M_I\mathbf{F}_I(0)}\\ \dot{\mathbf{R}}_I(\Delta t) &= \dot{\mathbf{R}}_I(0) + \frac{\Delta t}{2M_I}[\mathbf{F}_I(0) + \mathbf{F}_I(\Delta t)]\end{aligned} \tag{9.14}$$

where $\mathbf{F}_I(0)$ and $\mathbf{F}_I(\Delta t)$ are the forces at t = 0 and $t = \Delta t$, respectively. Iteration of Equation 9.14 yields a full trajectory of P steps. Equations 9.13 and 9.14 suggest an algorithm for generating the finite-temperature dynamics of a system using forces generated from electronic structure calculations performed "on the fly" as the simulation proceeds: Starting with the initial nuclear configuration, one minimizes the KS energy functional to obtain the ground-state density, and Equation 9.13 are used to obtain the initial forces. These forces are then used to propagate the nuclear positions to the next time step using the first of Equation 9.14. At this new nuclear configuration, the KS functional is minimized again to obtain the new ground-state density and forces using Equation 9.13, and these forces are used to propagate the velocities to time $t = \Delta t$. These forces can also be used again to propagate the positions to time $t = 2\,\Delta t$. The procedure is iterated until a full trajectory is generated. This approach is known as "Born–Oppenheimer" dynamics because it employs, at each step, an electronic configuration that is fully quenched to the ground-state Born–Oppenheimer surface.

An alternative to Born–Oppenheimer dynamics is the Car–Parrinello (CP) method [8–10]. In this approach, an initially minimized electronic configuration is subsequently "propagated" from one nuclear configuration to the next using a fictitious Newtonian dynamics for the orbitals. In this "dynamics", the orbitals are given a small amount of fictitious thermal kinetic energy and are made "light" compared to the nuclei. Under these conditions, the orbitals actually generate a potential of mean force surface that is very close to the true Born–Oppenheimer surface. The equations of motion of the CP method are

$$\begin{aligned}M_I\ddot{\mathbf{R}}_I &= -\nabla_I\mathscr{E}[\{\psi\},\mathbf{R}] - \nabla_I U_{\text{NN}}(\mathbf{R})\\ \mu|\ddot{\psi}_i\rangle &= -\frac{\partial}{\partial\langle\psi_i|}\mathscr{E}[\{\psi\},\mathbf{R}] + \sum_j \lambda_{ij}|\psi_j\rangle\end{aligned} \tag{9.15}$$

where μ is a mass-like parameter for the orbitals (which actually has units of energy $\times$ time^2), and λ_{ij} is the Lagrange multiplier matrix that enforces the orthogonality of the orbitals as a holonomic constraint on the fictitious orbital dynamics. Choosing μ small ensures that the orbital dynamics is adiabatically decoupled from the true nuclear dynamics, thereby allowing the orbitals to generate the aforementioned potential of mean force surface. For a detailed analysis of the CP dynamics, see Refs. [10,25]. As an illustration of the CP dynamics, Fig. 1 of Ref. [26] shows the temperature profile for a short CPAIMD simulation of bulk silicon together with the kinetic energy profile from the fictitious orbital dynamics. The figure demonstrates that the orbital dynamics is essentially a "slave" to the nuclear dynamics, which shows that the electronic configuration closely follows the dynamics of the nuclei in the spirit of the Born–Oppenheimer approximation.

9.2.3 Plane Wave Bases and Surface Boundary Conditions

In AIMD calculations, the most commonly employed boundary conditions are periodic boundary conditions, in which the system is replicated infinitely in all three spatial directions. This is clearly a natural choice for solids and is particularly convenient for liquids. In an infinite periodic system, the KS orbitals become Bloch functions of the form

$$\psi_{i\mathbf{k}}(\mathbf{r}) = e^{i\mathbf{k}\cdot\mathbf{r}}u_{i\mathbf{k}}(\mathbf{r}) \tag{9.16}$$

where $\mathbf{k}$ is a vector in the first Brioullin zone and $u_{i\mathbf{k}}(\mathbf{r})$ is a periodic function. A natural basis set for expanding a periodic function is the Fourier or plane-wave basis set, in which $u_{i\mathbf{k}}(\mathbf{r})$ is expanded according to

$$u_{i\mathbf{k}}(\mathbf{r}) = \frac{1}{\sqrt{V}}\sum_{\mathbf{g}} c_{i,\mathbf{g}}^{\mathbf{k}}\, e^{i\mathbf{g}\cdot\mathbf{r}} \tag{9.17}$$

where V is the volume of the cell, $\mathbf{g} = 2\pi\mathbf{h}^{-1}\hat{\mathbf{g}}$ is a reciprocal lattice vector, $\mathbf{h}$ is the cell matrix, whose columns are the cell vectors ($V = \det(\mathbf{h})$), $\hat{\mathbf{g}}$ is a vector of integers, and $\{c_{i,\mathbf{g}}^{\mathbf{k}}\}$ are the expansion coefficients. An advantage of plane waves is that the sums needed to go back and forth between reciprocal space and real space can be performed efficiently using fast Fourier transforms (FFTs) if the orbitals are represented on a regular lattice. In general, the properties of a periodic system are only correctly described if a sufficient number of $\mathbf{k}$-vectors are sampled from the Brioullin zone. However, for the applications we will consider, we are able to choose sufficiently large system sizes that we can restrict our $\mathbf{k}$-point sampling to the single point, $\mathbf{k} = (0, 0, 0)$, known as the Γ-point. At the Γ-point, the plane-wave expansion reduces to

$$\psi_i(\mathbf{r}) = \frac{1}{\sqrt{V}}\sum_{\mathbf{g}} c_{i,\mathbf{g}}\, e^{i\mathbf{g}\cdot\mathbf{r}} \tag{9.18}$$

At the Γ-point, the orbitals can always be chosen to be real functions. Therefore, the plane-wave expansion coefficients satisfy the following property

$$c^*_{i,\mathbf{g}} = c_{i,-\mathbf{g}} \tag{9.19}$$

which requires keeping only half of the full set of plane-wave expansion coefficients. In actual applications, plane waves up to a given cutoff $|\mathbf{g}|^2/2 < E_{\mathrm{cut}}$ only are retained. Similarly, the density $n(\mathbf{r})$ given by Equation 9.4 can also be expanded in a plane-wave basis:

$$n(\mathbf{r}) = \frac{1}{V}\sum_{\mathbf{g}} n_{\mathbf{g}}\, e^{i\mathbf{g}\cdot\mathbf{r}} \tag{9.20}$$

However, since $n(\mathbf{r})$ is obtained as a square of the KS orbitals, the cutoff needed for this expansion is $4E_{\mathrm{cut}}$ for consistency with the orbital expansion.

At first glance, it might seem that plane waves are ill-suited to treat surfaces because surfaces are naturally periodic in only two dimensions. However, in a series of papers, [27–29] Martyna, Tuckerman, and coworkers showed that clusters (systems with no periodicity), wires (systems with one periodic dimension), and surfaces (systems with two periodic dimensions) could all be treated using a plane-wave basis within a single unified formalism as follows. Let $n(\mathbf{r})$ be a particle density with a Fourier expansion given by Equation 9.20, and let $\phi(\mathbf{r}-\mathbf{r}')$ denote an interaction potential. In a fully periodic system, the energy of a system described by $n(\mathbf{r})$ and $\phi(\mathbf{r}-\mathbf{r}')$ is given by

$$E = \frac{1}{2}\int d\mathbf{r}d\mathbf{r}'n(\mathbf{r})\,\phi(\mathbf{r}-\mathbf{r}')n(\mathbf{r}') = \frac{1}{2V}\sum_{\mathbf{g}} |n_{\mathrm{g}}|^2\tilde{\phi}_{-\mathrm{g}} \tag{9.21}$$

where $\tilde{\phi}_{\mathrm{g}}$ is the Fourier transform of the potential. For systems with fewer than three periodic dimensions, the idea is to replace Equation 9.21 with its first-image approximation

$$E \approx E^{(1)} \equiv \frac{1}{2V}\sum_{\mathbf{g}} |n_{\mathrm{g}}|^2\bar{\phi}_{-\mathbf{g}} \tag{9.22}$$

where $\bar{\phi}_g$ denotes a Fourier expansion coefficient of the potential in the nonperiodic dimensions and a Fourier transform along the periodic dimensions. For clusters, $\bar{\phi}_g$ is given by

$$\bar{\phi}_{\mathrm{g}} = \int_{-L_z/2}^{L_z/2} dz \int_{-L_y/2}^{L_y/2} dy \int_{-L_x/2}^{L_x/2} dx\, \phi(\mathbf{r})e^{-i\mathrm{g}\cdot\mathrm{r}} \tag{9.23}$$

for wires, it becomes

$$\bar{\phi}_{\mathrm{g}} = \int_{-L_z/2}^{L_z/2} dz \int_{-L_y/2}^{L_y/2} dy \int_{-\infty}^{\infty} dx\, \phi(\mathbf{r})e^{-i\mathrm{g}\cdot\mathrm{r}} \tag{9.24}$$

and for surfaces, we obtain

$$\bar{\phi}_g = \int_{-L_z/2}^{L_z/2} dz \int_{-\infty}^{\infty} dy \int_{-\infty}^{\infty} dx\, \phi(\mathbf{r}) e^{-i\mathbf{g}\cdot\mathbf{r}} \tag{9.25}$$

The error in the first-image approximation drops off as a function of the volume, area, or length in the nonperiodic directions, as analyzed in Refs. [27–29].

In order to have an expression that is easily computed within the plane-wave description, consider two functions $\phi^{\text{long}}(\mathbf{r})$ and $\phi^{\text{short}}(\mathbf{r})$, which are assumed to be the long- and short-range contributions to the total potential, that is,

$$\phi(\mathbf{r}) = \phi^{\text{long}}(\mathbf{r}) + \phi^{\text{short}}(\mathbf{r}) \tag{9.26}$$

The Fourier expansion coefficients, therefore, also satisfy this decomposition:

$$\bar{\phi}(\mathbf{g}) = \bar{\phi}^{\text{long}}(\mathbf{g}) + \bar{\phi}^{\text{short}}(\mathbf{g}) \tag{9.27}$$

We require that $\phi^{\text{short}}(\mathbf{r})$ vanish exponentially quickly at large distances from the center of the parallelepiped and that $\phi^{\text{long}}(\mathbf{r})$ contain the long-range dependence of the full potential, $\phi(\mathbf{r})$. With these two requirements, it is possible to write

$$\begin{aligned} \bar{\phi}^{\text{short}}(\mathbf{g}) &= \int_{D(V)} d\mathbf{r}\, e^{-i\mathbf{g}\cdot\mathbf{r}} \phi^{\text{short}}(\mathbf{r}) \\ &= \int_{\text{all space}} d\mathbf{r} e^{-i\mathbf{g}\cdot\mathbf{r}} \phi^{\text{short}}(\mathbf{r}) + \varepsilon(\mathbf{g}) \\ &= \tilde{\phi}^{\text{short}}(\mathbf{g}) + \varepsilon(\mathbf{g}) \end{aligned} \tag{9.28}$$

with exponentially small error, $\varepsilon(\mathbf{g})$, provided the range of $\phi^{\text{short}}(\mathbf{r})$ is small compared to the size of the parallelepiped. In order to ensure that Equation 9.28 is satisfied, a convergence parameter, α, is introduced that can be used to adjust the range of $\phi^{\text{short}}(\mathbf{r})$ such that $\varepsilon(\mathbf{g}) \sim 0$ and the error, $\varepsilon(\mathbf{g})$, will be neglected in the following.

The function, $\tilde{\phi}^{\text{short}}(\mathbf{g})$, is the Fourier transform of $\phi^{\text{short}}(\mathbf{r})$. Therefore,

$$\begin{aligned} \bar{\phi}(\mathbf{g}) &= \bar{\phi}^{\text{long}}(\mathbf{g}) + \tilde{\phi}^{\text{short}}(\mathbf{g}) \\ &= \bar{\phi}^{\text{long}}(\mathbf{g}) - \tilde{\phi}^{\text{long}}(\mathbf{g}) + \tilde{\phi}^{\text{short}}(\mathbf{g}) + \tilde{\phi}^{\text{long}}(\mathbf{g}) \\ &= \hat{\phi}^{\text{screen}}(\mathbf{g}) + \tilde{\phi}(\mathbf{g}) \end{aligned} \tag{9.29}$$

where $\tilde{\phi}(\mathbf{g}) = \tilde{\phi}^{\text{short}}(\mathbf{g}) + \tilde{\phi}^{\text{long}}(\mathbf{g})$ is the Fourier transform of the full potential, $\phi(\mathbf{r}) = \phi^{\text{short}}(\mathbf{r}) + \phi^{\text{long}}(\mathbf{r})$ and

$$\hat{\phi}^{\text{screen}}(\mathbf{g}) = \bar{\phi}^{\text{long}}(\mathbf{g}) - \tilde{\phi}^{\text{long}}(\mathbf{g}). \tag{9.30}$$

Thus, Equation 9.30 becomes

$$\langle \phi \rangle = \frac{1}{2V} \sum_{\hat{g}} |\bar{n}(\mathbf{g})|^2 \left[\tilde{\phi}(-\mathbf{g}) + \hat{\phi}^{\text{screen}}(-\mathbf{g}) \right] \tag{9.31}$$

The new function appearing in the average potential energy, Equation 9.31, is the difference between the Fourier series and Fourier transform form of the long-range part of the potential energy and is referred to as a screening function because it is constructed in order to "screen" the interaction of the system with an infinite array of periodic images. The specific case of the Coulomb potential,

$$\phi(\mathbf{r}) = \frac{1}{r} \tag{9.32}$$

can be separated into short- and long-range components via

$$\frac{1}{r} = \frac{\text{erf}(\alpha r)}{r} + \frac{\text{erfc}(\alpha r)}{r} \tag{9.33}$$

where erf(x) and erfc(x) are the error function and its complement, respectively:

$$\text{erf}(x) = \frac{2}{\sqrt{\pi}} \int_0^x e^{-t^2} dt$$
$$\text{erfc}(x) = 1 - \text{erf}(x)$$

The first term in Equation 9.33 is long range. The parameter α determines the specific ranges of these terms. The screening function for the cluster case is easily computed by introducing an FFT grid and performing the integration numerically [27]. For the wire [28] and surface [29] cases, analytical expressions can be worked out. In particular, for surfaces, the screening function is

$$\begin{aligned}\bar{\phi}^{\text{screen}}(\mathbf{g}) = -\frac{4\pi}{g^2} \Bigg\{ \cos\left(\frac{g_c L_c}{2}\right) \\ \times \Bigg[\exp\left(-\frac{g_s L_c}{2}\right) - \frac{1}{2}\exp\left(-\frac{g_s L_c}{2}\right) \text{erfc}\left(\frac{\alpha^2 L_c - g_s}{2\alpha}\right) \\ - \frac{1}{2}\exp\left(\frac{g_s L_c}{2}\right) \text{erfc}\left(\frac{\alpha^2 L_c + g_s}{2\alpha}\right) \Bigg] \\ + \exp\left(-\frac{g^2}{4\alpha^2}\right) \text{Re}\left[\text{erfc}\left(\frac{\alpha^2 L_c + i g_c}{2\alpha}\right) \right] \Bigg\}\end{aligned} \tag{9.34}$$

When a plane-wave basis set is employed, the external energy is made somewhat complicated by the fact that very large basis sets are needed to treat the rapid spatial fluctuations of core electrons. Therefore, core electrons are often replaced by atomic pseudopotentials or augmented plane-wave techniques. Here, we shall discuss the former. In the atomic pseudopotential scheme, the nucleus plus the core electrons are treated in a frozen core type approximation as an "ion" carrying only the valence charge. In order to make this approximation, the valence orbitals, which, in principle must be orthogonal to the core orbitals, must see a different pseudopotential for each angular momentum component in the core, which means that the pseudopotential must generally be nonlocal. In order to see this, we consider a potential operator of the form

$$\hat{V}_{\text{pseud}} = \sum_{l=0}^{\infty}\sum_{m=-l}^{l} v_l(r)|lm\rangle\langle lm| \tag{9.35}$$

where r is the distance from the ion, and $|lm\rangle\langle lm|$ is a projection operator onto each angular momentum component. In order to truncate the infinite sum over l in Equation 9.35, we assume that for l greater than or equal to some $\bar{l}$ some, $v_l(r) \approx v_{\bar{l}}(r)$ and add and subtract the function $v_{\bar{l}}(r)$ in Equation 9.35:

$$\begin{aligned}\hat{V}_{\text{pseud}} &= \sum_{l=0}^{\infty}\sum_{m=-l}^{l}(v_l(r)-v_{\bar{l}}(r))|lm\rangle\langle lm| + v_{\bar{l}}(r)\sum_{l=0}^{\infty}\sum_{m=-l}^{l}|lm\rangle\langle lm| \\ &= \sum_{l=0}^{\infty}\sum_{m=-l}^{l}(v_l(r)-v_{\bar{l}}(r))|lm\rangle\langle lm| + v_{\bar{l}}(r) \\ &\approx \sum_{l=0}^{\bar{l}-1}\sum_{m=-l}^{l}\Delta v_l(r)|lm\rangle\langle lm| + v_{\bar{l}}(r)\end{aligned} \tag{9.36}$$

where the second line follows from the fact that the sum of the projection operators is unity, $\Delta v_l(r) = v_l(r)-v_{\bar{l}}(r)$, and the sum in the third line is truncated before $\Delta v_l(r) = 0$. The complete pseudopotential operator is

$$\hat{V}_{\text{pseud}}(\mathbf{r};\mathbf{R}_1,\ldots,\mathbf{R}_N) = \sum_{I=1}^{N}\left[v_{\text{loc}}(|\mathbf{r}-\mathbf{R}_I|) + \sum_{l=0}^{\bar{l}-1}\Delta v_l(|\mathbf{r}-\mathbf{R}_I|)|lm\rangle\langle lm|\right] \tag{9.37}$$

where $v_{\text{loc}}(r) \equiv v_{\bar{l}}(r)$ is known as the local part of the pseudopotential (having no projection operator attached to it). Now, the external energy, being derived from the ground-state expectation value of a one-body operator, is given by

$$\varepsilon_{\text{ext}} = \sum_i f_i\langle\psi_i|\hat{V}_{\text{pseud}}|\psi_i\rangle \tag{9.38}$$

The first (local) term gives simply a local energy of the form

$$\varepsilon_{\text{loc}} = \sum_{I=1}^{N}\int d\mathbf{r}\, n(\mathbf{r})v_{\text{loc}}(|\mathbf{r}-\mathbf{R}_I|) \tag{9.39}$$

which can be evaluated in reciprocal space as

$$\varepsilon_{\rm loc} = \frac{1}{\Omega}\sum_{I=1}^{N}\sum_{\mathbf{g}} n_{\mathbf{g}}^{*}\tilde{v}_{\rm loc}(\mathbf{g})e^{-i\mathbf{g}\cdot\mathbf{R}_I} \tag{9.40}$$

where $\tilde{V}_{\rm loc}(g)$ is the Fourier transform of the local potential. Note that at $\mathbf{g} = (0,0,0)$, only the nonsingular part of $\tilde{v}_{\rm loc}(\mathbf{g})$ contributes. In the evaluation of the local term, it is often convenient to add and subtract a long-range term of the form $Z_I \mathrm{erf}(\alpha_I r)/r$, where $\mathrm{erf}(x)$ is the error function, for each ion in order to obtain the nonsingular part explicitly and a residual short-range function $\bar{v}_{\rm loc}(|\mathbf{r}-\mathbf{R}_I|) = v_{\rm loc}(|\mathbf{r}-\mathbf{R}_I|) - Z_I \mathrm{erf}(\alpha_I|\mathbf{r}-\mathbf{R}_I|)/|\mathbf{r}-\mathbf{R}_I|$ for each ionic core.

9.2.4 Electron Localization Methods

An important feature of the KS energy functional is the fact that the total energy $E[\{\psi\},\mathbf{R}]$ is invariant with respect to a unitary transformation within the space of occupied orbitals. That is, if we introduce a new set of orbitals $\psi'_i(\mathbf{r})$ related to the $\psi_i(\mathbf{r})$ by

$$\psi'_i(\mathbf{r}) = \sum_{j=1}^{N_s} U_{ij}\psi_j(\mathbf{r}) \tag{9.41}$$

where U_{ij} is a $N_s \times N_s$ unitary matrix, then the energy $E[\{\psi'\},\mathbf{R}] = E[\{\psi\},\mathbf{R}]$. We say that the energy is invariant with respect to the group $SU(N_s)$, that is, the group of all $N_s \times N_s$ unitary matrices with unit determinant. This invariance is a type of gauge invariance, specifically that in the occupied orbital subspace. The fictitious orbital dynamics of the AIMD scheme as written in Equation 9.15 does not preserve any particular unitary representation or gauge of the orbitals but allows the orbitals to mix arbitrarily according to Equation 9.41. This mixing happens intrinsically as part of the dynamics rather than by explicit application of the unitary transformation.

Although this arbitrariness has no effect on the nuclear dynamics, it is often desirable for the orbitals to be in a particular unitary representation or gauge. For example, we might wish to have the true Kohn–Sham orbitals $\phi_i(\mathbf{r})$ (see Equation 9.8) at each step in an AIMD simulation in order to calculate the Kohn–Sham eigenvalues and generate the corresponding density of states from a histogram of these eigenvalues. This would require choosing U_{ij} to be the unitary transformation that diagonalizes the matrix of Lagrange multipliers in Eq. 9.6. Another important representation is that in which the orbitals are maximally localized in real space. In this representation, the orbitals are closest to the classic "textbook" molecular orbital picture.

In order to obtain the unitary transformation U_{ij} that generates maximally localized orbitals, we seek a functional that measures the total spatial spread of the orbitals. One possibility for this functional is simply to use the variance of the position operator $\hat{\mathbf{r}}$ with respect to each orbital and sum these variances:

$$\Omega[\{\psi\}] = \sum_{i=1}^{N_s} \left[\langle \psi_i | \hat{r}^2 | \psi_i \rangle - \langle \psi_i | \hat{\mathbf{r}} | \psi_i \rangle^2 \right] \tag{9.42}$$

The procedure for obtaining the maximally localized orbitals is to introduce the transformation in Equation 9.41 into Equation 9.42 and then to minimize the spread functional with respect to U_{ij}:

$$\frac{\partial}{\partial U_{ij}} \Omega[\{\psi'\}] = 0 \tag{9.43}$$

The minimization must be carried out subject to the constraint that U_{ij} be an element of $SU(N_s)$. This constraint condition can be eliminated if we choose U to have the form

$$\mathrm{U} = e^{i\mathrm{A}} \tag{9.44}$$

where A is an $N_s \times N_s$ Hermitian matrix, and performing the minimization of Ω with respect to A.

A little reflection reveals that the spread functional in Equation 9.42 is actually not suitable for periodic systems. The reason for this is that the position operator $\hat{\mathbf{r}}$ lacks the translational invariance of the underlying periodic supercell. A generalization of the spread functional that does not suffer from this deficiency is [30,31]

$$\Omega[\{\psi\}] = \frac{1}{(2\pi)^2} \sum_{i=1}^{N_s} \sum_{I} \omega_I f(|z_{I,ii}|^2) + \mathcal{O}((\sigma/L)^2) \tag{9.45}$$

where σ and L denote the typical spatial extent of a localized orbital and box length, respectively, and

$$z_{I,ii} = \int d\mathbf{r} \psi_i^*(\mathbf{r}) e^{i\mathbf{G}_I \cdot \mathbf{r}} \psi_j(\mathbf{r}) \equiv \langle \psi_i | \hat{O}^I | \psi_j \rangle \tag{9.46}$$

Here $G_I = 2\pi(\mathbf{h}^{-1})^T \hat{\mathbf{g}}_I$, where $\hat{\mathbf{g}}_I = (l_I, m_I, n_I)$ is the Ith Miller index and ω_I is a weight having dimensions of (length)^2. The function $f(|z|^2)$ is often taken to be $1 - |z|^2$, although several choices are possible. The orbitals that result from minimizing Equation 9.45 are known as *Wannier orbitals* $|w_i\rangle$. If $z_{I,ii}$ is evaluated with respect to these orbitals, then the orbital centers, known as *Wannier centers*, can be computed according to

$$w_\alpha = -\sum_\beta \frac{h_{\alpha\beta}}{2\pi} \operatorname{Im} \ln z_{\beta,ii} \tag{9.47}$$

Wannier orbitals and their centers are useful in analyzing chemically reactive systems and will be employed in the present surface chemistry studies.

Like the KS energy, the Lagrangian that generates the fictitious, CP dynamics is invariant with respect to gauge transformations of the form given in Eq. 9.41. It is not, however, invariant under time-dependent unitary transformations of the form

$$\psi'_i(\mathbf{r}, t) = \sum_{j=1}^{N_s} U_{ij}(t)\psi_j(\mathbf{r}, t) \tag{9.48}$$

and consequently, the orbital gauge changes at each step of an AIMD simulation. If, however, we impose the requirement of invariance under Equation 9.48 on the CP Lagrangian, then not only would we obtain a gauge-covariant version of the AIMD algorithm, but we could also then fix a particular orbital gauge and have this gauge be preserved under the CP evolution. Using techniques for gauge field theory, it is possible to devise such a AIMD algorithm [32]. Introducing orbital momenta $|\pi_i\rangle$ conjugate to the orbital degrees of freedom, the gauge-covariant AIMD equations of motion have the basic structure

$$\begin{aligned} M_I\ddot{\mathbf{R}}_I &= -\nabla_I(\mathscr{E}[\{\psi\}, \mathbf{R}] + U_{\mathrm{NN}}(\mathbf{R})) \\ |\dot{\psi}_i\rangle &= |\pi_i\rangle + \sum_j B_{ij}(t)|\psi_j\rangle \\ |\dot{\pi}_i\rangle &= -\frac{1}{\mu}\frac{\partial}{\partial\langle\psi_i\rangle|}\mathscr{E}[\psi\}, \mathbf{R}] + \sum_j \lambda_{ij}|\psi_j\rangle + \sum_j B_{ij}(t)|\pi_j\rangle \end{aligned} \tag{9.49}$$

where

$$B_{ij}(t) = \sum_k U_{ki}\frac{d}{dt}U_{kj} \tag{9.50}$$

Here, the terms involving the matrix $B_{ij}(t)$ are gauge-fixing terms that preserve a desired orbital gauge. If we choose the unitary transformation $U_{ij}(t)$ to be the matrix that satisfies Equation 9.43, then Equation 9.49 will propagate maximally localized orbitals [33]. As was shown in Refs. [32,33], if it possible to evaluate the gauge-fixing terms in a way that does not require explicit minimization of the spread functional [34]. In this way, if the orbitals are initially localized, they remain localized throughout the trajectory. Explicit expressions for B_{ij} are given in Ref. [32].

While the Wannier orbitals and Wannier centers are useful concepts, it is also useful to have a measure of electron localization that does not depend on a specific orbital representation, as the latter does have some arbitrariness associated with it. An alternative measure of electron localization that involves only the electron density $n(\mathbf{r})$ and the so-called kinetic energy density

$$\tau(\mathbf{r}) = \sum_{i=1}^{N_s} f_i|\nabla\psi_i(\mathbf{r})|^2 \tag{9.51}$$

was introduced by Becke and Edgecombe [35]. Defining the ratio

$$\chi(\mathbf{r}) = \frac{D(\mathbf{r})}{D_0(\mathbf{r})} \tag{9.52}$$

where

$$\begin{aligned} D(\mathbf{r}) &= \tau(\mathbf{r}) - \frac{1}{4}\frac{|\nabla n(\mathbf{r})|^2}{n(\mathbf{r})} \\ D_0(\mathbf{r}) &= \frac{3}{4}\left(6\pi^2\right)^{2/3} n^{5/3}(\mathbf{r}) \end{aligned} \tag{9.53}$$

the function

$$f(\mathbf{r}) = \frac{1}{1+\chi^2(\mathbf{r})} \tag{9.54}$$

can be shown to lie in the interval $f(\mathbf{r}) \in [0, 1]$, where $f(\mathbf{r}) = 1$ corresponds to perfect localization, and $f(\mathbf{r}) = 1/2$ corresponds to a gas-like localization. The function $f(\mathbf{r})$ is known as the *electron localization function* or ELF. In the studies to be presented below, we will make use both of the ELF and the Wannier orbitals and centers to quantify electron localization.

9.3 REACTIONS ON THE SI(100)-(2 × 1) SURFACE

When solid silicon is cleaved along its (100) plane, the dangling atoms reconstruct in such a way that parallel rows of silicon dimers are formed. This is the well-known 2×1 reconstruction. The Si–Si bond in these surface dimers consists of a σ bond and a weak, partial π bond, so that the bond order is less than 2. Nevertheless, there is sufficient π character in the dimers for them to act as dieneophiles in reactions with conjugated dienes of the type that we will consider in this section.

Experiments [36,37] have shown that the surface dimers do not lie in a single plane but preferentially form a buckled $c(4 \times 2)$ structure, in which neighboring dimers in each row buckle in opposite directions and in which neighboring rows are mirror images of each other (see Figure 9.1 (top) and (bottom)). Low-temperature scanning tunneling microscopy has identified the persistence of this buckled structure down to temperatures below 10 K [37]. The buckling gives rise to a polarization of the Si–Si bond, with the lower silicon atom acquiring a net positive charge and the upper silicon acquiring a net negative charge (see Figure 9.2). At room temperature, the buckling pattern is dynamic with the dimers executing a "rocking" motion or oscillation about the idealized pattern in Fig. 9.1(top). Electron energy loss spectroscopy (EELS) [38] gives a dimer rocking frequency of 20 meV or ~200 fs at 300 K but does not reveal how frequently the dimer flips to the opposite orientation. *Ab initio* [39] and force

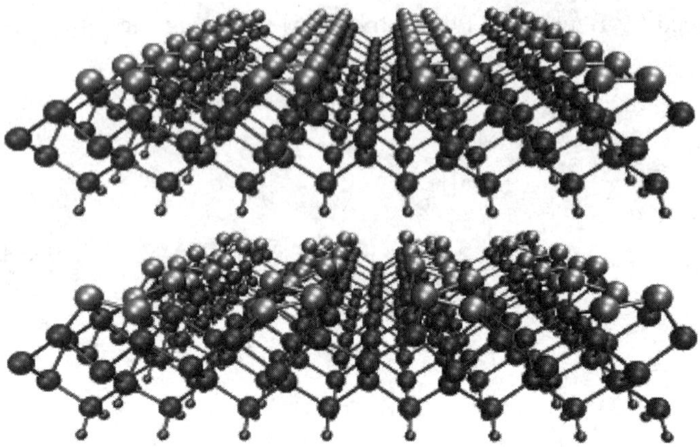

FIGURE 9.1 (Top) Symmetric dimers on the Si(100)-(2 × 1) surface. (Bottom) Buckled dimers on the Si(100)-(2 × 1) surface.

field-based [40,41] molecular dynamics simulations confirm the rocking period is between 200 and 300 fs, while the flipping period ranges from 200 fs to more than 1.5 ps. We will see that the charge asymmetry in the dimers and the surface dynamics influence both the reaction mechanisms and the reaction kinetics.

As a tangential point, we briefly comment on the zero-temperature surface structure. The point is tangential because all of the calculations to be presented are at 300 K where the buckling and dimer flipping are well-established surface characteristics. At $T = 0$, however, there is a discrepancy between DFT and high-level quantum chemical approaches that include multireference methods. The former can be carried out in a well-converged plane-wave basis with the periodicity of the surface properly accounted for. The latter, while more accurate in their treatment of the electgronic structure, can only be carried out on small clusters that contain one or two Si–Si dimers, which is a crude representation of the surface. Interestingly, the DFT calculations yield the buckled configuration as the zero-temperature structure [15], in agreement with the low-temperature STM [37], while the cluster calculations yield the symmetric unbuckled structure [42]. It is not obvious whether the source of the discrepancy is due primarily to the level of accuracy of the two approaches and possible fortuitous cancellation of errors in the DFT calculations, or whether the dimer-dimer interactions and surface periodicity are critical in stabilizing the buckled structure.

It is also worth mentioning that the lattice constant of bulk silicon and Si–Si dimer length on the Si(100)-(2 × 1) surface in DFT depend somewhat on the exchange-correlation functional employed. Two popular choices are the BLYP [22,23] and PBE [24] functionals. BLYP gives a bulk lattice constant of 5.52 Å and a surface dimer bond length of 2.3 Å at 300K. PBE gives a bulk lattice constant of 5.47 Å, in better agreement with the experimental value of 5.43Å, [43] and a dimer bond length of 2.3 Å at 300K. Thus, overall, the PBE functional is somewhat better for treating the silicon surface, but even BLYP provides a reasonable description. In what follows, we will present results employing both functionals.

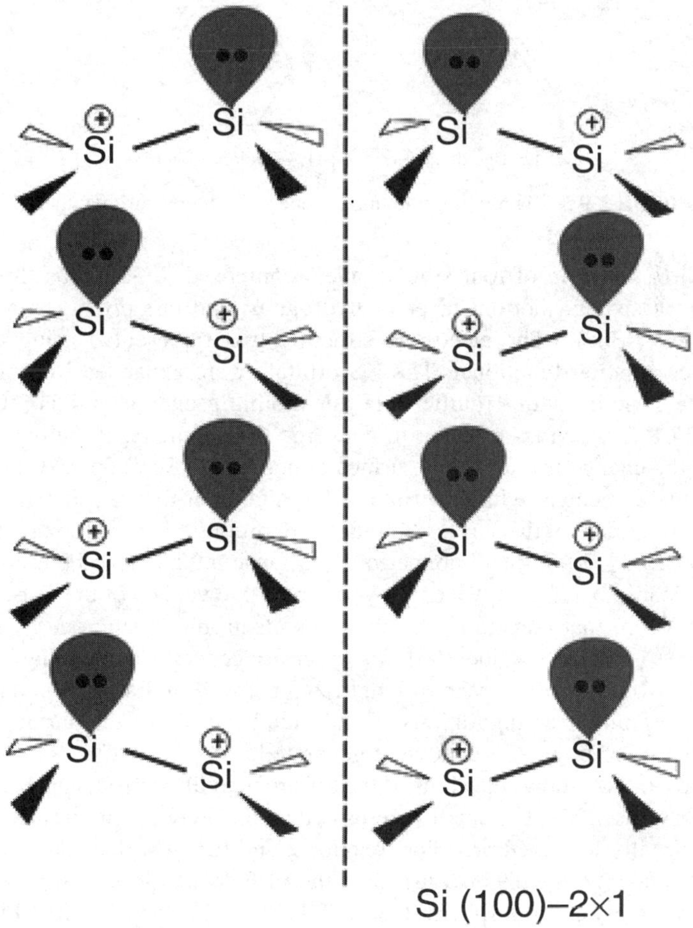

FIGURE 9.2 Diagram of the $c(4 \times 2)$ structure of the Si(100)-(2 × 1) surface showing the charge asymmetry within the surface dimers.

9.3.1 Attachment of 1,3-Butadiene to the Si(100)-(2 × 1) Surface

The first reaction we will discuss is the covalent attachment of 1,3-butadiene to the Si(100)-(2 × 1) surface. This benchmark system has been studied both experimentally and theoretically [12,13,44–48]. In addition, it is electronically similar to the conjugated region of the 1,3-cyclohexadiene molecule, which has been investigated using STM, [18,49] thus making comparisons with these measurements possible (see Figure 9.3). Because experiments and static *ab initio* calculations cannot identify specific mechanisms by which these addition products form, this system is ideal to study using AIMD.

We begin by describing a calculation aimed at probing the distribution of products formed when 1,3-butadiene reacts with the surface. AIMD calculations were

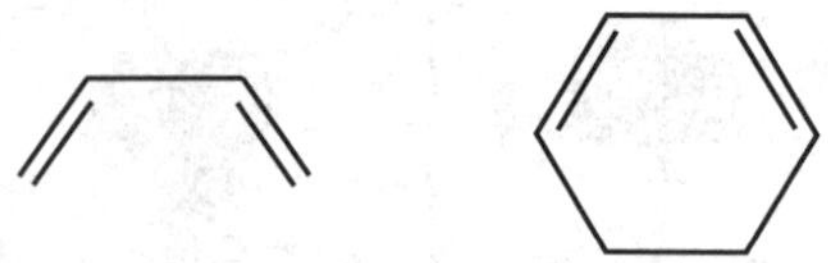

1,3-Butadiene 1,3-Cyclohexadiene

FIGURE 9.3 The two conjugated dienes considered in this chapter.

performed on a system of four silicon layers composed of 32 atoms (four surface dimers), a passivating bottom layer of hydrogens, and one *cis*-1,3-butadiene at a temperature of 300K. The electronic structure was represented using the BLYP exchange-correlation functional. The KS orbitals were expanded in a plane-wave basis at the Γ-point, with Troullier–Martins atomic pseudopotentials [50] up to a cutoff of 35 Ry, which is sufficient to converge the geometry of the butadiene and reproduce the change in energy per surface dimer (~ -1.5 eV) upon reconstruction, in reasonable agreement with experiment [51] and other theoretical estimates [52]. Our proper treatment of the surface boundary conditions allowed a box with periodic dimensions 15.34, 7.67 Å, and nonperiodic dimension 22.53 Å to be used. Because the system is highly reactive, we employed a modified version of the Car--Parrinello AIMD algorithm that constrains the fictitious electronic kinetic energy in the CP Lagrangian to a preset value [53]. In order to generate a meaningful product distribution, 40 trajectories, each of length 2–3 ps, were initiated from an unbiased distribution of initial configurations of the butadiene above the surface. In total, roughly 110 ps of trajectory data were generated.

In Figure 9.4, we show snapshots of the different addition products obtained from the trajectory ensemble. The small spheres in the bond between atoms are the centers of the maximally localized Wannier functions, and the isosurface shown in panels (a), (c), (d), and (f) is the 0.95 isosurface of the ELF. In addition, each panel contains one or more darkly colored spheres (see Ref. [13] for a color version of Figure 9.4); these designate local fully or partially positively charged sites. Product A (shown in panel (a)) is a [4+2] product in which the butadiene attaches to a single Si–Si dimer. The two Wannier centers between Carbons C_2 and C_3 indicate a double bond between these two carbons, and the single Wannier center between the two silicon atoms in the Si–Si dimer to which the butadiene is attached indicates a single bond. Hence, product A is essentially the analog of a Diels–Alder adduct in which the Si–Si dimer plays the role of the ethylene reactant (see Figure 9.5). Product B (shown in panel (b)) is an intrarow, interdimer [4+2] product, in which the butadiene attaches to two dimers within a single row. Once again, the presence of two Wannier centers between carbons C_2 and C_3 indicates the conversion of the single C–C bond to a double bond in the final product. Product C (shown in panel (c)) is an interrow, interdimer [4+2] product, in which the butadiene bridges two dimers in neighboring rows. Product D (shown in panel (d)) is a [2+2] product in which carbons C_1 and C_2 bridge two dimers in a single row. Notice, in this, case that there are two Wannier centers between Carbons C_3 and C_4, indicating that this double bond, which normally exists in 1,3-butadiene, remains in tact. Panel (e) shows an intermediate “fluxional”

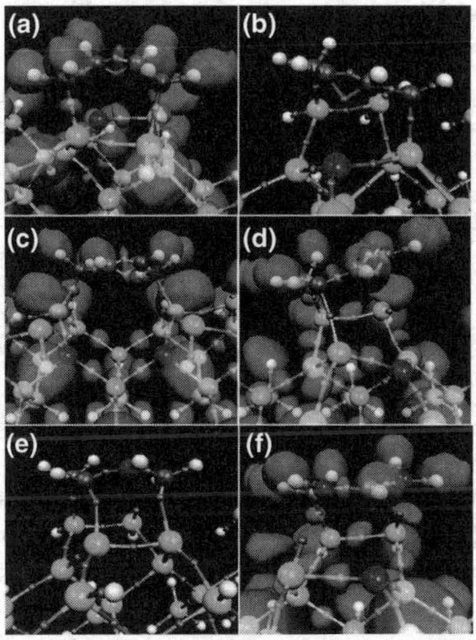

FIGURE 9.4 Snapshots of the addition products obtained. Large, medium, and small spheres denote Si, C, and H atoms, respectively, and very small spheres in bonds between atoms indicate the location of Wannier centers. Darkly shaded spheres locate atoms with positive full or partial charge. Full positive charge is defined to be an atom surrounded by three Wannier centers. The surface is the ELF 0.95 isosurface. A full color version of this figure can be found in Ref. [13].

species that can occur after the formation of a [2+2] adduct when the $C_3 = C_4$ double bond is positioned for a second nucleophilic attack. This species rapidly converts to a stable [4+2]-like interdimer adduct via an electron pair reorganization. Finally, panel (f) shows a stable intermediate carbocation state that forms in most of the reactions observed. Such carbocation intermediates have been observed STM measurements, [54] and we will have more to say about this species below.

From the ensemble of trajectories, we can predict the product distribution for comparison with STM measurements on 1,3-cyclohexadiene [18]. This distribution, reported in Table 9.1, shows the percentages of the addition products together with the experimental values for 1,3-cyclohexadiene. In Table 9.1, the "H" population is the fraction of the ensemble that did not form a definite product on the time scale of the simulations, while in the STM measurement, it designates all adducts that could not be identified as one of A–E. Generally, it can be seen that the yields of all products

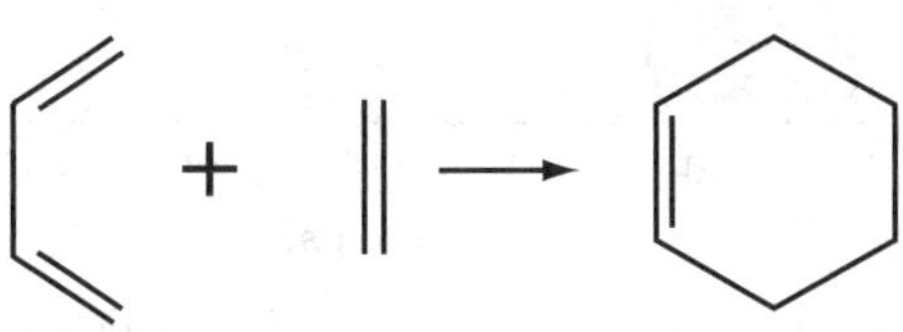

FIGURE 9.5 Classic Diels–Alder reaction between 1,3-butadiene and ethylene yielding cyclohexene.

TABLE 9.1 Final Product Distribution

Adduct	Theory(%)	Expt[a](%)	Adduct	Theory(%)	Expt[a](%)
A	15	11 ± 3	D(+E)	30	22 ± 15
B	30	31 ± 6	H	10	21 ± 5
C	15	16 ± 7			

[a]Ref. [18].

analogous to those of Ref. [18], that is, A–D(+E), agree with experiment within the error bars. Some slight differences between the 1,3-butadiene and 1,3-cyclohexadiene products are (1) for the former, the B adduct forms diagonally across two intrarow dimers and (2) the [2+2] adduct is actually an interdimer [2+2]-like adduct within a row.

In order to rationalize the distribution observed in the STM measurement and in the AIMD calculations, we next focus on the mechanism of these cycloaddition reactions. Figure 9.6 shows the carbon–carbon and carbon–silicon bond distances as functions of time over a typical trajectory that leads to the Diels–Alder [4+2] intradimer product. It can be seen that, of the two C = C double bonds, one reverts to a single bond before the other. This is consistent with the formation of the carbocation intermediate shown in panel (f) of Figure 9.4. Similarly, one of the Si–C bonds forms before the other. These bond-length trajectories, which are similar to those seen in many of the ensemble members, suggest an asymmetric, non-concerted mechanism that proceeds via the carbocation intermediate.

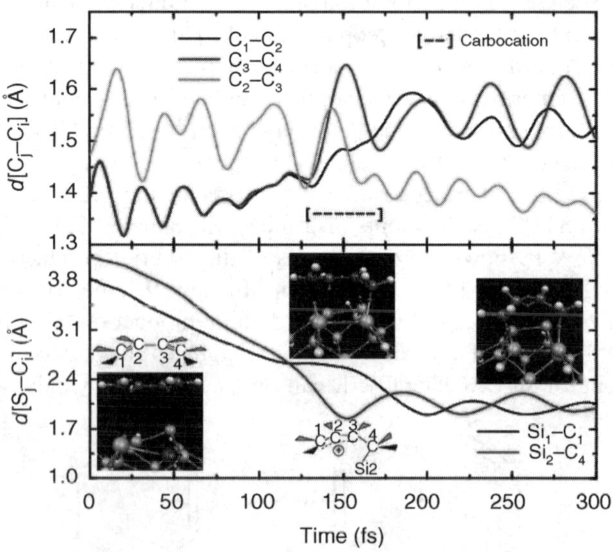

FIGURE 9.6 Si–C and C–C bond lengths versus simulation time for a representative trajectory leading to the [4+2] Diels--Alder adduct. From left to right the three snapshots depict: (i) butadiene above the Si(100)-(2 × 1) surface, (ii) carbocation formation, (iii) final [4+2] DA adduct. Sphere coloring is the same as in Figure 9.4. A full color version of this figure can be found in Ref. [13].

To this end, we have proposed the following mechanism [13] that serves to rationalize the observed product distribution both in the AIMD trajectory ensemble and in the STM measurements. This mechanism is illustrated in the scheme of Figure 9.7.

The mechanism, which is common to all of the reactions simulated, begins with a nucleophilic attack of the C = C double bond on the positively charge atom in one of the Si–Si dimers. The subsequent migration of the local positive charge from the Si atom, now neutralized by formation of the Si–C bond, dictates the outcome of the reaction. In all cases, the next step involves the migration of the positive charge into the organic molecule, leading to a carbocation intermediate (Figure 9.7R). The carbocation, which can exist for up to 1–2 ps, is stabilized by resonance, as illustrated by the three ELF lobes and the three Wannier centers around the positive carbon atoms. The Wannier center of the delocalized orbital is located below the middle carbon atom, indicating that it is shared by two bonds. Since the carbocation bonds to an Si atom of a surface dimer and the other dimer member has a net negative charge, an intermediate zwitterionic state is formed. In one of the resonant structures, the positive charge is "localized" on the end of the butadiene (Figure 9.7R-right), which allows the carbon to attack the negatively charged Si atom. This leads to the formation of the Diels–Alder [4+2] adduct in panel (a) of Figure 9.4.

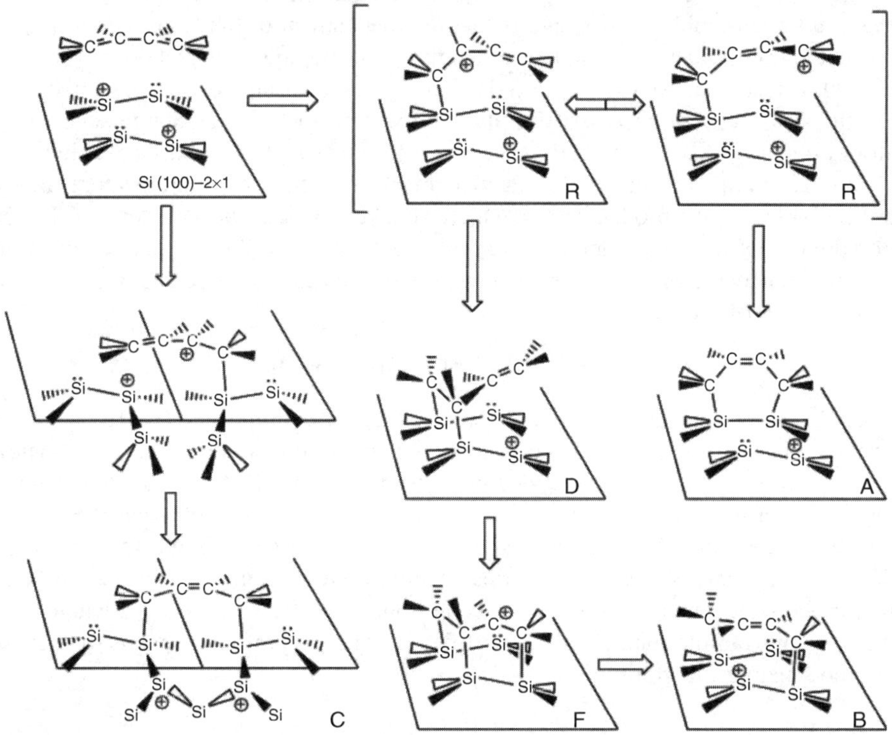

FIGURE 9.7 Scheme depicting the proposed reaction mechanism of 1,3-butadiene with the Si(100)-(2 × 1) surface.

If we consider the resonant structure in which the double bond is localized on the end of the butadiene (Figure 9.7R-right), then the positive carbon can attack the negatively charged Si of a neighboring dimer. This leads to the interrow [2+2]-like adduct in panel (d) of Figure 9.4. If the remaining double bond in the butadiene is positioned for a second nucleophilic attack, then the fluxional intermediate (panel (e) of Figure 9.4) is formed, and this intermediate is seen to convert rapidly to the final [4+2] product. The fluxional intermediate was also suggested in the STM measurements of Ref. [18]. The [2+2]-like product is stabilized if the double bond at the end of the butadiene is oriented between the rows. Finally, the inter-row adduct formation is initiated by a second nucleophilic attack of the C=C double bond in the zwitterion on the positively charged Si atom in a neighboring row. This product constitutes a nine-membered ring, which is accompanied by migration of the local positive charge into the bulk layer as seen in the ELF isosurface in panel (c) of Figure 9.4.

The preceding discussion highlights the predominance of a stepwise zwitterionic mechanism governing addition product formation on the Si(100)-(2 $\times$ 1) surface. Although we cannot rule out a concerted mechanism, one might expect on statistical grounds that the initially asymmetric charge distribution in the surface dimers allows only very special initial conditions to lead to a concerted reaction path. In fact, one of the trajectories in the ensemble exhibited a concerted path, however, as we will see in the free energy analysis to follow, this type of mechanism is not a highly probability event in the ensemble. This suggests that the mechanism of the Diels–Alder adduct formation is dominated by a nonconcerted mechanism, indicating that on the Si(100)-(2 $\times$ 1) surface, the usual Woodward–Hoffman rules governing organic cycloaddition reactions [55] do not necessarily apply [13,15]. Specifically, the charge asymmetry allows for the violation and leads to the observed distribution of surface adducts.

We can explore the reaction mechanism in more depth by computing a free energy profile for one of the product states, which we take to be the Diels–Alder-type [4+2] intradimer product. In order to define a free energy profile, we need a reaction coordinate capable of following the progress of the reaction. For this, we choose a coordinate ξ of the form

$$\xi = \frac{1}{2}|(\mathbf{r}_{\mathrm{Si}_1} + \mathbf{r}_{\mathrm{Si}_2}) - (\mathbf{r}_{\mathrm{C}_1} + \mathbf{r}_{\mathrm{C}_4})| \tag{9.55}$$

where Si_1 and Si_2 are the two silicon atoms in the surface dimer, and C_1 and C_4 are the two outer carbons in the butadiene (see Fig. 9.8 for an illustration of this coordinate). Over the course of the reaction, ξ decreases from 3.90 to 1.96 Å. In order to calculate the free energy profile, we employ the blue moon ensemble approach [56,57], in which ξ is constrained at 13 equally spaced points (with a spacing of 0.15 Å) between the two endpoints. At each constrained value, a simulation of length 9.0 ps was performed over which we compute a conditional average $\langle \partial H/\partial \xi \rangle_{\mathrm{cond}}$, where H is the nuclear Hamiltonian. Finally, the full free energy profile is reconstructed via thermodynamic integration:

$$\Delta G(\xi) = \int_{\xi_0}^{\xi} d\xi' \left\langle \frac{\partial H}{\partial \xi'} \right\rangle_{\mathrm{cond}} \tag{9.56}$$

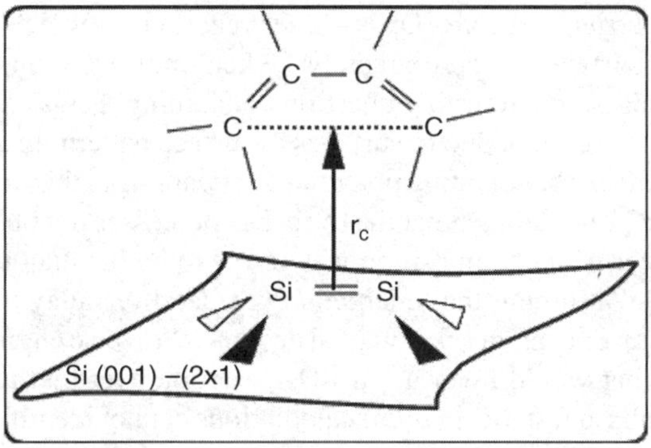

FIGURE 9.8 Illustration of the reaction coordinate in Equation 9.55.

The resulting free energy profile, shown in Figure 9.9, reveals a deep minimum corresponding to the final DA product and a plateau at $\xi = 2.75$ Å, implying a short-lived intermediate state. If the mechanism were concerted, one would expect a profile with a single minimum, corresponding simply to the final product state. We will see an example of such a profile later when we investigate reactions on the SiC(100)-(2 × 2) surface. The profile in Figure 9.9 is clearly more characteristic of a nonconcerted mechanism involving a well-defined reaction intermediate. That the initial free energy barrier at $\xi \approx 3.25$ Å is relatively low (3–4 kcal/mol) suggests that the reaction can occur at room temperature.

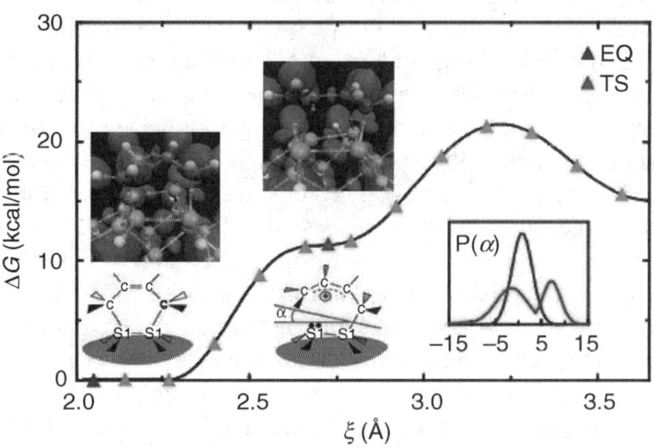

FIGURE 9.9 Free energy along the reaction pathway leading to a Diels–Alder [4+2] adduct. Blue and red triangles indicate the product (EQ) and intermediate states (IS), respectively. Inset shows the buckling angle (α) distribution of the Si dimer for both the IS and the EQ configurations. The snapshots include configurations representing the IS and EQ geometries. Large, medium, and small spheres denote Si, C, and H atoms, respectively, and very small spheres indicate the location of Wannier centers. Darkly shaded spheres locate positively charged atoms. The purple surface is the ELF 0.95 isosurface. A full color version of this figure can be found in Ref. [12].

Note that the barrier to a retro-Diels–Alder reaction, in which the 1,3-butadiene desorbs from the surface is approximately 23 kcal/mol according to the profile in Figure 9.9. This raises an interesting question concerning the use of the chemistry of conjugated dienes on semiconductor surfaces in surface patterning and lithography. In principle, this chemistry has exciting potential applications in this area, however, it has been shown for the [4+2] adduct that the retro-DA process is not observed on Si(100)-(2 × 1). Instead, thermal decomposition was found to be the major reaction pathway upon heating [45]. Assuming that energetic considerations play a major role in the reaction mechanism, one might ask if we can design or "reverse engineer" a chemically substituted diene that would favor a retro-DA reaction over thermal decomposition. Because experiments and static *ab initio* calculations cannot identify specific mechanisms by which these addition products form, this system is ideal to study using AIMD. Moreover, we will show that AIMD can actually aid in the design strategy.

It is known in organic chemistry that DA reactions increase their rates with increasing strength of the electron-donating groups on the diene. Therefore, it would seem natural to assume that electron-withdrawing groups on the diene have an opposite effect. Furthermore, it can be expected that such groups would destabilize the DA [4+2] adduct, so that it is more likely to undergo a heating induced retro-DA reaction. Based on these ideas, we suggest a butadiene derivative that leads to a lower free energy barrier for the retro-DA reaction. To be considered useful for lithographic purposes, a substituted diene candidate should have the following properties: (i) it should participate in a spontaneous DA [4+2] reaction on the room temperature surface and (ii) the [4+2] adduct should undergo a retro-DA reaction upon heating. In order to account for both requirements, the candidate substituent must have moderate electron-withdrawing ability to favor the retro-DA pathway, while it should also spontaneously react producing the DA [4+2] adduct at a sufficient rate. Based on qualitative arguments, fluorine was chosen as a test substituent, because it is electronegative yet is not considered to be a highly effective electron-withdrawing substituent such as CF_3 or SO_3H. Specifically, one of the hydrogens on the C=C double bond of the isolated 1,3-butadiene was replaced by F[14]. The substitution is illustrated in Figure 9.10.

The utility of the modified butadiene shown in Figure 9.10 for the retro-DA reaction can be demonstrated by computing the free energy profile for its reaction with the surface and comparing it to that in Figure 9.9. For this calculation, we employ the same protocol as above except that the plane-wave cutoff is increased to 80 Ry because of the highly repulsive pseudopotential needed to describe the fluorine atom. The resulting profile comparison is shown in Figure 9.11. Both profiles show a plateau at the intermediate carbocation state and, therefore, further support the

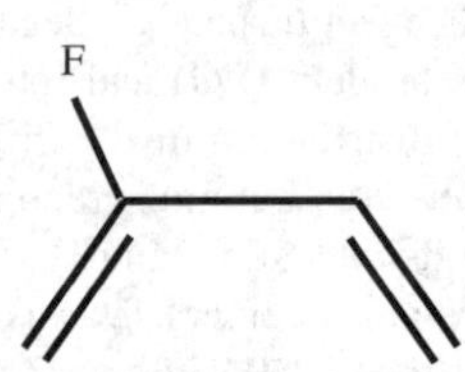

FIGURE 9.10 The 1,3-butadiene derivative 2-F-1,3,butadiene.

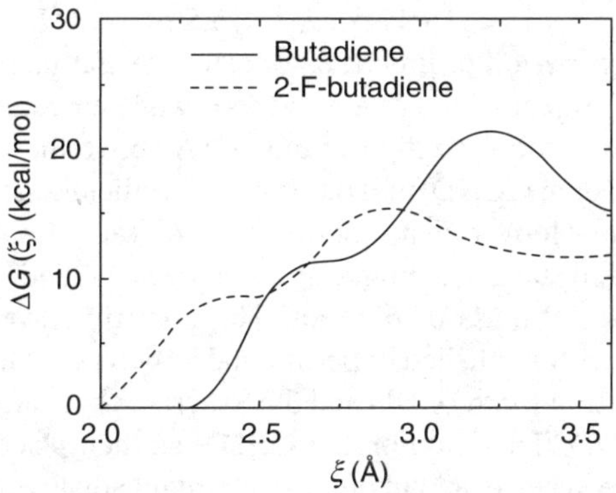

FIGURE 9.11 Free energy along the reaction pathway leading to a Diels--Alder [4+2] adduct. The solid line corresponds to the unmodified 1,3-butadiene, and the dashed line corresponds to the 2-F-1,3-butadiene of Figure 9.10.

mechanistic picture of Figure 9.7. The figure also shows that, like 1,3-butadiene, 2-F-1,3-butadiene has negligible activation barrier (4–5 kcal/mol) toward the DA [4+2] adduct, so that a spontaneous reaction is expected at room temperature. Although similar in many qualitative aspects, there is a 6–8 kcal/mol decrease in the activation free energy for the retro-DA pathway. The activation energy is still too high for a spontaneous retro-DA reaction upon heating, but this result serves as a proof of concept and illustrates the power of a potential computational engineering principle toward designing functionalizing agents with required chemical characteristics.

9.3.2 Attachment of 1,3-Cyclohexadiene to the Si(100)-(2 × 1) Surface

Since the STM measurements of Ref. [18] were based on 1,3-cyclohexadiene rather than 1,3-butadiene, is it instructive to carry out an analogous set of calculations using the former molecule in order to contrast with the results described in the previous subsection. This system has also been the subject of numerous experimental [1,18,49,58,59] and static *ab initio* studies [60,63]. Again, however, static calculations cannot reveal mechanistic details or elucidate the role of surface dimer dynamics in the reaction process.

For these calculations, we chose to improve the descriptions of the electronic structure and of the Si(100)-(2 × 1) surface by increasing the size of the periodic slab and by employing the PBE functional rather than the BLYP functional. In these calculations [15], the Kohn–Sham orbitals were expanded in a plane-wave basis with a kinetic energy cutoff of 35 Ry, and Troullier–Martins norm-conserving pseudopotentials [64], with S, P, and D treated as local for H, C, and Si, respectively, were employed. The Si(100)-(2×1) system was comprised of 16 2×1 units each five layers deep, with the bottom layer fixed at the bulk lattice positions and terminated with H. The top surface forms two rows of four buckled dimers each. The larger surface dimensions prevents periodic images from interacting with themselves, which eliminates the need for **k**-point sampling beyond the gamma point. The simulation box dimensions were $L_x = 15.47$ Å, $L_y = 15.5$ Å, and $L_z = 31.8$ Å, where the z-direction is the non periodic direction. Because of the increased system size, a smaller ensemble of trajectories, eleven in total, was generated with the cyclohexadiene initiated from random configurations 3.0 Å above the surface.

First the cyclohexadiene (CHD) and reconstructed silicon surface were equilibrated separately using Nosé–Hoover chain thermostats [65] at 300K for over 1 ps each to obtain well-thermalized starting configurations. These calculations used a time step of 0.1 fs and a fictitious CP mass of 670 a.u. The final time averaged configurations compared favorably with published experimental structures. All trajectories used the same initial configuration for the CHD and the Si surface but varied in the position and orientation of the CHD. The equilibrated CHD was then placed randomly at 3.0 Å above the equilibrated surface, and in order to maintain adiabaticity for longer periods of time in the combined CHD + Si system, the time step was dropped to 0.05 fs and the fictitious CP mass lowered to 400 a.u. The combined system was then annealed from 0 to 300K under NVE conditions. Next, the center of mass of the CHD was fixed while the system equilibrated for 1 ps using Nosé–Hoover chain thermostats [65] under NVT conditions. Finally, the thermostat and center of mass constraint were removed, and trajectories were run until a reaction product formed or a maximum of 12 ps was reached, after which the trajectory was considered unreactive.

In Figure 9.12, we show the adducts obtained. In particular, we observe four (A, B, C_t, and D) out of the five STM-identified cycloaddition products [18,49] as well as four additional adducts. In contrast to the 1,3-butadiene case, the most probable adduct (C), the [4+2] interdimer, intrarow product, now splits into two isomers that differ in the orientation of the CH_2 group. In the previously predicted case, (C_t), the CH_2 groups are over the trough, while for the new adduct, (C_r), the CH_2 groups are situated over the dimer row. Previous static DFT calculations found that C_r is slightly favored over C_t [62], consistent with the present findings. Experiment is,

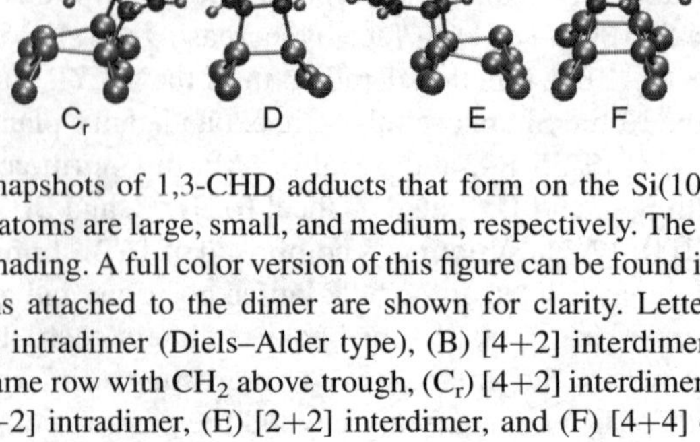

FIGURE 9.12 Snapshots of 1,3-CHD adducts that form on the Si(100)–(2×1) surface at 300K. Si, H, and C atoms are large, small, and medium, respectively. The remaining C π-bond are given a darker shading. A full color version of this figure can be found in Ref. [15]. Only the Si dimer and atoms attached to the dimer are shown for clarity. Letters label the various adducts: (A) [4+2] intradimer (Diels–Alder type), (B) [4+2] interdimer across trough, (C_t) [4+2] interdimer same row with CH_2 above trough, (C_r) [4+2] interdimer same row with CH_2 above row, (D) [2+2] intradimer, (E) [2+2] interdimer, and (F) [4+4] 4-bond.

most likely, unable to distinguish between the two isomers due to the symmetry of the adduct and strong unpaired Si-adduct orbital interactions. Likewise, the [2 + 2] interdimer adduct (E) is oriented such that the π-bond is over the trough, not the dimer row as identified in the STM experiment. However, this adduct may be able to flip orientation to match the experimentally derived structure since the barrier to move the H nearest the Si from the dimer to the trough side of the CHD should be small. In another case, both π-bonds in the CHD reacted to form a 4-bond adduct (F) as was predicted by Lee et al. [66] (labeled 5 in their notation), although the actual mechanism is somewhat different. It should be noted that an H-abstraction from a partially reacted CHD occurred in one of the trajectories. H-abstraction has been observed experimentally between 400 and 700 K [67]. Its presence in our simulation either indicates (1) H-abstraction does occur at lower temperatures, but in such small quantities that it is not detectable experimentally, (2) the extra energy released from the first bond formation created a local hot-spot, or (3) the barriers for H-abstraction are too low within the chosen DFT framework. The percentages of product A, B, C_t, C_r, D, E, and other obtained are 7.1, 7.1, 7.1, 21.4, 28.6, 7.1, and 21.4, respectively [15]. These are to be compared to the experimental populations reported in Table 9.1, however, it must be kept in mind that the ensemble used here is considerably smaller than that used in the butadiene study.

The reaction mechanism of 1,3-cyclohexadiene is very similar to that observed for 1,3-butadiene. We illustrate this by showing trajectories of the carbon–carbon and carbon–silicon bonds as functions of time for typical trajectories of each of the product states (see Figure 9.13).

As in Figure 9.6, Figure 9.13 shows that the C=C double bonds revert to single bonds at different times during each of the reactions and that the C–Si bonds also

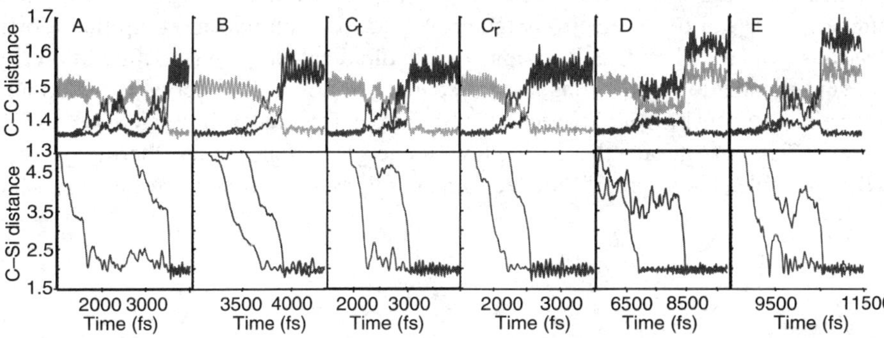

FIGURE 9.13 The top row shows the three possible π-bond lengths averaged over 25 fs for the adducts (see Fig. 9.12) in the reactive region. The bottom row displays the corresponding first and second C–Si bond lengths in light and dark lines, respectively. A full color version of this figure can be found in Ref. [15]. The small figure in A labels the three C–C bonds and the first C–Si bond. The second C–Si would form at positions 1 or 2. All distances given in Å. During the reaction, the C–C and C = C bond lengths fluctuate significantly. After the adduct forms, the remaining π-bond reverts to the original length, while the single C–C bonds are expanded. Change in the π bond lengths coincides with C–Si bond formation.

form asymmetrically. Thus, we can conclude that both 1,3-cyclohexadiene and 1,3-butadiene react with the Si(100)-(2×1) surface via a nonconcerted, asymmetric reaction pathway. This is illustrated in Figure 9.14 for the [4 + 2] Diels–Alder-like intradimer product. An analysis of the role played by the dimer tilt angle in determining the reaction kinetics reveals that as the molecule approaches the lower, positively charged silicon, the dimer tilt angle decreases somewhat. This is illustrated in Figure 9.15, which shows a plot of tilt angle (ϕ) versus. the carbon–silicon distance (d) for the complete set of trajectories. This figure helps us understand why the C_t and C_r interdimer products are favored: The second attack is more likely to occur on a neighboring dimer because it will generally have a larger tilt angle than the dimer on which the first C–Si bond has already formed and, therefore, its upper silicon will have a larger negative charge (see Figure 9.7 and the ELF plot in Figure 9.16).

The longer trajectories and improvement in the electronic structure representation for the 1,3-CHD system allow us to investigate other aspects of the reaction mechanism as well. An important question concerns the possibility of a radical mechanism. Multireference cluster based calculations of the minimum energy pathways for the cycloaddition of CHD to the Si(100)-(2×1) surface employing second order perturbation theory [66] suggested that diradical mechanisms should play a major role. Despite experimental evidence to the contrary, [36,68] multi-reference cluster calculations predict that Si surface dimers are symmetric, [42] not tilted, which could alter reaction pathways. As previously noted, periodic DFT calculations performed here capture the correct dimer tilt. At the same time, these calculations neglect surface crossing events since they are single reference.

In order to estimate how important radical mechanisms and surface crossings might be during adduct formation, we have calculated single point energies along four representative trajectories for three spin states: spin restricted (SR), which assumes that all electrons are paired (singlet), spin unrestricted (SU) with the same number of up and down electrons (singlet), and SU with two more up than down electrons (triplet). The SU calculations mimic diradical electronic configurations by allowing up and down electrons to vary spatially to lower the energy. If the up and down electron densities are identical, the SU singlet functional simply reduces to SR. Since DFT is a variational theory, the lowest energy configuration is thermodynamically favorable. Therefore, if the electronic configuration of the lowest energy state

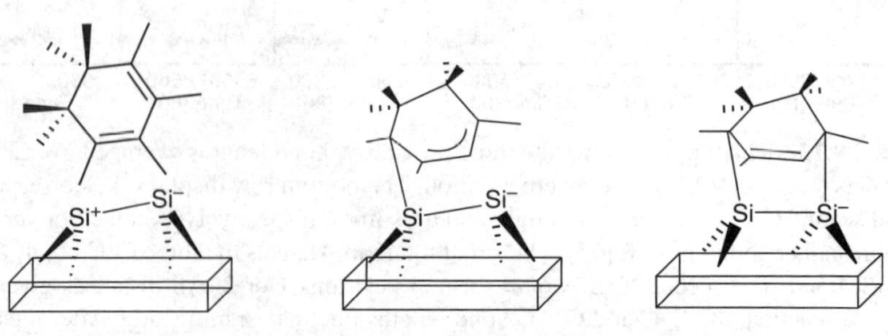

FIGURE 9.14 Schematic representation of the asymmetric, nonconcerted mechanism for the addition of 1,3-cyclohexadiene on the Si(100)-(2 × 1) surface.

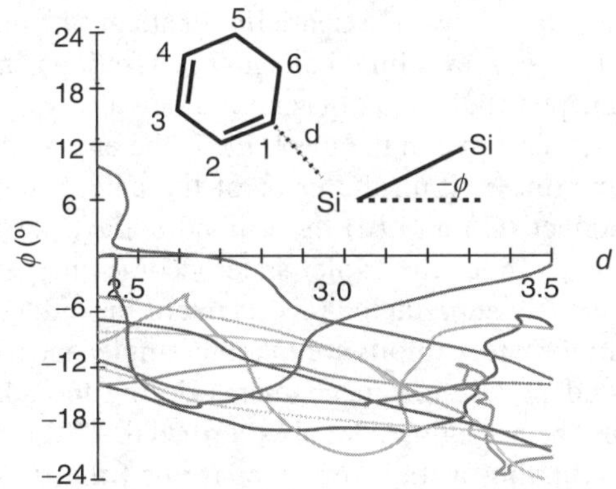

FIGURE 9.15 $d_{C\text{–}Si}$ versus. the dimer tilt angle (ϕ). All the dimers are oriented down relative to CHD. Properties in the reactive region immediately prior to the first bond formation. The different shades denote different trajectories. $d_{C\text{–}Si}$ is the distance between the C and Si in the first bond. Stars appear at $d_{C\text{–}Si} = 3.5$ Å. A full color version of this figure can be found in Ref. [15].

changes, the difference between the SR or SU (singlet) and SU (triplet) energies becomes small, and a surface crossing might occur even though it is not allowed within standard DFT.

Figure 9.17 compares the three energies for four representative trajectories. Single point energy calculations were carried out at 0K using configurations taken from the 300K trajectory. In all cases, the SR (black down triangles) and SU singlet (up, light

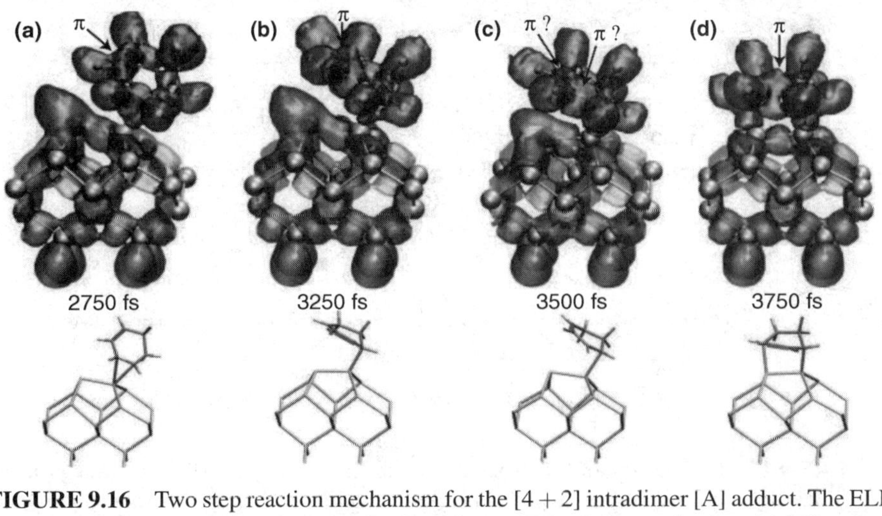

FIGURE 9.16 Two step reaction mechanism for the [4 + 2] intradimer [A] adduct. The ELF is approximately divided between contributions due to CHD (darker surface) and the Si surface (lighter surface). A full color version of this figure can be found in Ref. [15]. Stick figures at the bottom show all bonds, with the π-bonds highlighted in red. The first bond shifts between a "down" Si and (a) the middle of a π-bond or (b) the C closest to the CH_2 groups. In (c), the π-bond delocalizes between three adjacent C. Finally (d), the negative "up" Si reacts with the positive C next to the CH_2 groups to form the adduct.

triangles) single point energies were essentially identical (the upper line in the inset shows the energy difference) revealing that spin polarization is not necessary for the singlet case. The SU triplet (light squares) energies are always higher in energy than the SR energies (the lower curve in the inset line is the energy difference), although the difference is sometimes within the error of the calculation as seen in (b) the [4 + 2] interdimer adduct (C_t) and (d) the 4-bond adduct. At no point in (a), the [4 + 2] intradimer adduct, does the triplet state become energetically accessible at 300 K. In (c), the previously unpredicted [4 + 2] interdimer adduct (C_r), only the final adduct, not the transition state, exhibits comparable singlet and triplet energies. Since we are most interested in reaction mechanisms during the adduct formation, the triplet influence should be negligible. We also verified that SR is sufficiently accurate for our system by rerunning a trajectory that formed the (C_r) [4 + 2] interdimer adduct shown in Figure 9.17c using singlet SU. Although there were some variations, the same final adduct formed. Due to the difference between periodic DFT and multireference cluster systems, a definitive statement on the reaction mechanism cannot be made. These calculations do reveal, however, that within periodic GGA DFT, the spin restricted reaction mechanisms studied are favored over the equivalent triplet mechanisms, and hence should play the major role.

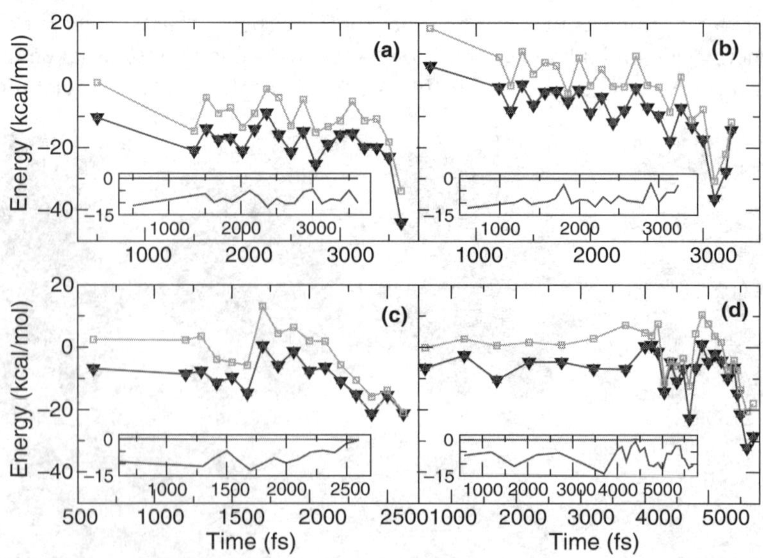

FIGURE 9.17 Comparison of SR (black down triangle), singlet SU (light up triangle), and triplet SU (light squares) 0K single point energies (kcal/mol) relative to the isolated CHD and Si(100) surface. Insets show the energy difference (kcal/mol) between SR-singlet SU (upper line) and SR-triplet SU (lower line). A full color version of this figure can be found in Ref. [15]. Representative configurations taken along 300 K trajectories for the (a) [4 + 2] intradimer [A], (b) [4 + 2] interdimer (same row) with CH_2 over the trough [C_t], (c) interdimer (same row) with CH_2 over the dimer row [C_r], and (d) 4 bond adducts [F]. A singlet radical mechanism is never favorable, although a triplet diradical mechanism may be possible during (b) and (d).

9.4 REACTIONS ON THE SIC(100)-(3×2) SURFACE

Silicon carbide (SiC) and its associated reactions with a conjugated diene are interesting surface systems to study and compare to the pure silicon surface case discussed in the preceding section. As we have seen, the Si(100)-(2×1) surface allows for a relatively broad distribution of products because the surface dimers are relatively closely spaced. Because of this, creating ordered organic layers on this surface using conjugated dienes seems unlikely unless some method can be found to enhance the population of one of the adducts, rendering the remaining adducts negligible. SiC exhibits a number of complicated surface reconstructions depending on the surface orientation and growth conditions. Some of these reconstructions offer the intriguing possibility of restricting the product distribution due to the fact that carbon–carbon or silicon–silicon dimer spacings are considerably larger.

SiC is often the material of choice for electronic and sensor applications under extreme conditions [69–71] or subject to biocompatibility constraints [72]. Although most reconstructions are still being debated both experimentally and theoretically [73,74], there is widespread agreement on the structure of the 3C-SiC(001)-(3×2) surface [75,76] (see Figure 9.18), which will be studied in this section. SiC(001) shares the same zinc blend structure as pure Si(001), but with alternating layers of Si and C. The top three layers are Si, the bottom in bulk-like positions and the top decomposed into an open 2/3 + 1/3 adlayer structure. Si atoms in the bottom two-thirds layers are four-fold coordinated dimers while those Si atoms in the top one-third are asymmetric tilted dimers with dangling bonds. Given the Si-rich surface environment and presence of asymmetric surface dimers, one might expect much of the same Si-based chemistry to occur with two significant differences: (1) altered reactivity due to the surface strain (the SiC lattice constant is ~20% smaller than Si) and (2) suppression of interdimer adducts due to the larger dimer spacing compared to Si (~60% along a dimer row, ~20% across dimer rows). Previous theoretical studies used either static (0 K) DFT calculations of hydrogen, [77–81] a carbon nanotube, [82] or ethylene/acetylene [83,84] adsorbed on SiC(001)-(3×2) or employed molecular dynamics of water [85] or small molecules of the CH_3-X

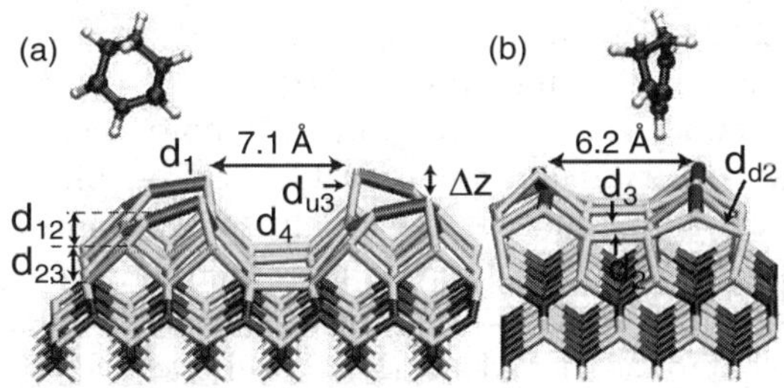

FIGURE 9.18 View of 1,3-CHD + 3C-SiC(001)-(3×2) system (a) along dimer rows and (b) between dimers in a row. Si, C, H, and the top Si surface dimers are represented by darker or lighter shading. A full color version of this figure can be found in Ref. [16]. The dimers are spaced farther apart by ~60% along a dimer row and ~20% across dimer rows relative to Si(100)-(2×1).

family [86] on the less thermodynamically stable SiC(001)-(2×1) surface. Here, we consider cycloaddition reactions on the SiC-(3×2) surface that include dynamic and thermal effects. A primary goal for considering this surface is to determine whether 3C-SiC(001)-(3×2) is a promising candidate for creating ordered semiconductor-organic interfaces via cycloaddition reactions.

In this study, [16] the KS orbitals were expanded in a plane-wave basis setup to a kinetic energy cutoff of 40 Ry. As in the 1,3-CHD studies described above, exchange and correlation are treated with the spin restricted form of the PBE functional, [24] and core electrons were replaced by Troullier–Martins pseudopotentials [50] with S, P, and D treated as local for H, C, and Si, respectively. The resulting SiC theoretical lattice constant, 4.39 Å, agrees well with the experimental value of 4.36 Å. [76] The full system is shown in Figure 9.18. The 3×2 unit cell is doubled in both directions to include four surface dimers to allow the possibility of all interdimer adducts. Again, the resulting large surface area, (18.6 Å×12.4 Å), allows the Γ-point approximation to be used in lieu of explicit k-point sampling. Two bulk layers of Si and C, terminated by H on the bottom surface, provide a reconstructed (1/3 + 2/3) Si surface in reasonable agreement with experiment (see below). The final system has 182 atoms [24 atoms/layer * (1 Si adlayer + 4 atomic layers) + 2*24 terminating H]. The simulation cell employed lengths of 18.6 Å and 12.4 Å along the periodic directions and 31.2 Å along the nonperiodic z direction.

Both the CHD and SiC(001) surface were equilibrated separately under NVT conditions using Nosé–Hoover chain thermostats [65] at 300 K with a timestep of 0.1 fs for 1 and 3 ps, respectively. When the equilibrated CHD was allowed to react with the equilibrated surface, the time step was reduced to 0.05 fs in order to ensure adiabaticity. The CHD was placed 3 Å above the surface, as defined by the lowest point on the CHD and the highest point on the surface. Each of 12 trajectories was initiated from the same CHD and SiC structures but with the CHD placed at a different orientations and/or locations over the surface. The subsequent initialization procedure was identical to the CHD–Si(100) system: First the system was annealed from 0 to 300 K in the NVE ensemble. Following this, it was equilibrated with Nosé–Hoover chain thermostats for 1 ps at 300 K under NVT conditions, keeping the center of mass of the CHD fixed. Finally, the CHD center of mass constraint was removed and the system was allowed to evolve under the NVE ensemble until an adduct formed or 20 ps elapsed.

The reactions that occur on this surface all take place on or in the vicinity of a single surface Si–Si dimer. However, as Figure 9.19 shows, there is not one but rather four adducts that are observed to form. Adduct labels from the Si + CHD study are used for consistency. As postulated, the widely spaced dimers successfully suppressed the interdimer adducts that formed on the Si(100)-(2×1) surface [15]. From the 12 trajectories, 3 formed the [4 + 2] Diels–Alder type intradimer adduct (A), 1 produced the [2 + 2] intradimer adduct (D), 5 exhibited hydrogen abstraction (H), and 1 resulted in a novel [4 + 2] subdimer adduct between Si in d_1 and d_2 (G) (see Figure 9.18). The remaining trajectories only formed 1 C–Si bond within 20 ps. Although the statistics are limited, these results suggest that H abstraction is favorable, consistent with the high reactivity of atomic H observed in experimental

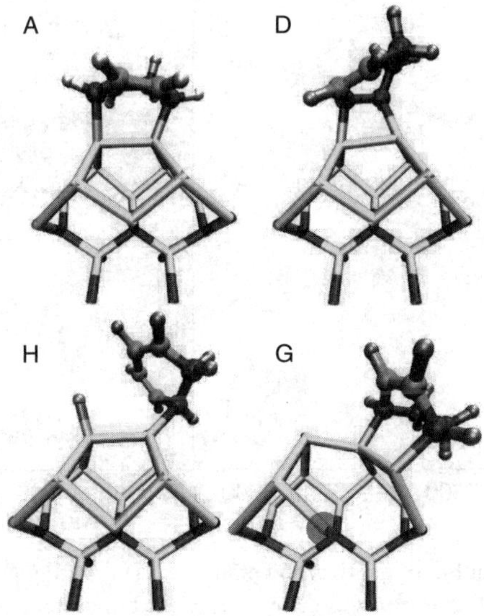

FIGURE 9.19 Snapshots of the four adducts that formed on the SiC surface: (A) [4 + 2] intradimer adduct, (D) [2 + 2] intradimer adduct, (H) hydrogen abstraction, and (G) [4 + 2] subsurface dimer adduct. Si, C, and H are represented by lighter or darker shading, respectively. The remaining C = C bond(s) is more darkly shaded. A full color version of this figure can be found in Ref. [16]. The larger spacing between dimers suppresses interdimer adducts. However, adduct (G) destroys the surface, rendering this system inappropriate for applications requiring well-defined organic-semiconducting interfaces.

studies on this system. [87,88] What is somewhat more troublesome, from the point of view of creating well-ordered organic-semiconducting interfaces is the presence of the subdimer adduct G. All the surface bonds directly connected to the adduct slightly expand to 2.42–2.47 Å, with the exception of one bond to a Si in the third layer, which disappears entirely. The energetic gain of the additional strong C–Si bond outweighs the loss of a strained Si–Si bond. The end effect is the destruction of the perfect surface and the creation of an unsaturated Si in the bulk. One adduct is noticeably missing: the [2 + 2] subdimer adduct. At several points during the simulation this adduct was poised to form but quickly left the vicinity. Most likely, the strain caused by the four-membered ring combined with the two energetically less stable unsaturated Si prevented this adduct from forming, even though the [2 + 2] intradimer and [4 + 2] subdimer adducts are stable.

In Figure 9.20, we show the carbon–carbon and CHD–Si distances as functions of time for the different adducts observed. This figure reveals that the mechanism of the reactions proceeds in a manner very similar to that of CHD and 1,3-butadiene on the Si(100)-(2×1) surface: It is an asymmetric, nonconcerted mechanism that involves a carbocation intermediate. What differs from Si(100) is the time elapsed before the first bond forms and the intermediate lifetime. On the Si(100)-(2×1) surface the CHD always found an available "down" Si to form the first bond within less than 10 ps or 40 Å of wandering over the surface. On the SiC(001)-(3×2) surface the

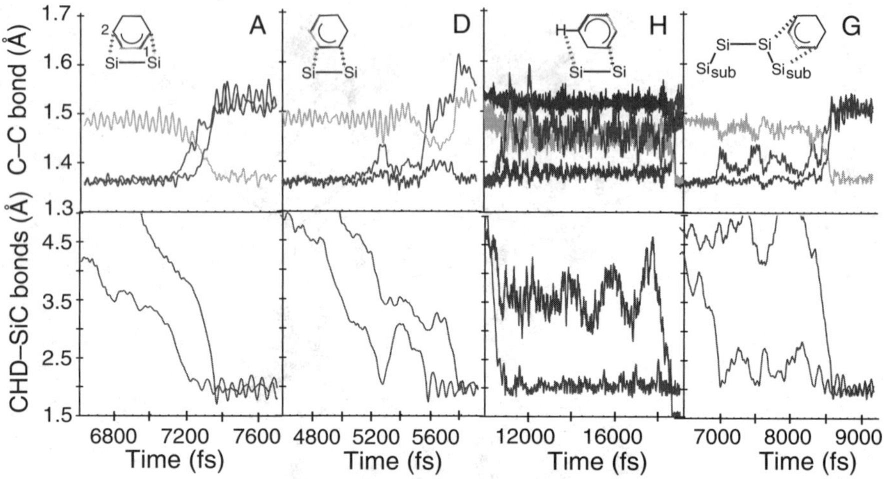

FIGURE 9.20 Relevant bond lengths (Å) versus time (fs) during product formation for four representative adducts. The top row displays the C–C bonds lengths (moving average over 25 fs) while the bottom row plots the first and second CHD – SiC surface bond. The shading follows the same pattern as that of Figure 9.13. A full color version of this figure can be found in Ref. [16]. Change in the C–C bond length closely correlates with surface-adduct bond formation. Intermediate lifetimes over all trajectories range from 0.05 to 18^+ ps.

exploration process sometimes required up to 20 ps and over 100 Å. While the exact numbers are only qualitative, the trend is significant. The Si(100) dimers are more tilted on average, and hence expected to be slightly more reactive. However, the dominant contribution is likely the density of tilted dimers: Si has 0.033 dimers/Å^2, but SiC only has 0.017 dimers/Å^2. Regardless of whether dimer flipping occurs, it is simply more difficult to find a dimer on the SiC surface.

9.5 REACTIONS ON THE SIC(100)-(2×2) SURFACE

There is considerable interest in the growth of molecular lines or wires on semiconductor surfaces. Such structures allow molecular scale devices to be constructed using semiconductors such as H-terminated Si(111) and Si(100) or Si(100)-2×1 as the preferred substrates. Various molecules can be grown into lines on the H-terminated surfaces, [89] and on the Si(100)-2×1 surface, styrene and derivatives such as 2,4-dimethylstyrene or longer chain alkenes can be used to grow wires along the dimer rows. [90–97]. More recently, allylic mercaptan and acetophenone have been shown to grow across dimer rows on the H:Si(100)-(2×1) surface. [91,97–100] Other semiconductor surfaces can be considered for such applications, however, these have not received as much attention. An intriguing possible alternate in the silicon–carbide family is the SiC(100)-(2×2) surface.

The SiC(100)-(2×2) surface exhibits a crucial difference from the SiC(100)-(3×2) in that it is characterized by C≡C triple bonds, which bridge Si–Si single bonds. These triple bonds are well separated and reactive, suggesting the possibility of restricting the product distribution for the addition of conjugated dienes on this surface. Figure 9.21 shows a snapshot of this surface with a 1,3-cyclohexadiene above it. Previous *ab initio* calculations suggest that these dimers react favorably with 1,4-cyclohexadiene [101]. Here, we present preliminary results on the free energy profile at 300 K for the reaction of this surface with 1,3-cyclohexadiene, also shown in Figure 9.21. A more detailed study is now available in Ref. [102].

In this study, the KS orbitals were expanded in a plane-wave basis setup to a kinetic energy cutoff of 65 Ry. As in the 1,3-CHD studies described above, exchange and correlation are treated with the spin restricted form of the PBE functional, [24] and core electrons were replaced by Troullier-Martins pseudopotentials [50] with S, P, and D treated as local for H, C, and Si, respectively. The periodic slab contains 128 atoms arranged in six layers (including a bottom passivating hydrogen layer). Proper treatment of surface boundary conditions allowed for a simulation cell with dimensions $L_x = 17.56$Å, $L_y = 8.78$ Å, and $L_z = 31$ Å along the nonperiodic dimension. The surface contains eight C≡C dimers. This setup is capable of reproducing the experimentally observed dimer buckling [103] that static *ab initio* calculations using cluster models are unable to describe [101].

In Figure 9.22, we show the free energy profile for the [4+2] cycloaddition reaction of 1,3-cyclohexadiene with one of the C≡C surface dimers. The free energy profile is calculated by dividing the ξ interval $\xi \in [1.59, 3.69]$ into 15 equally spaced intervals, and each constrained simulation was equilibrated for 1.0 ps followed by 3.0 ps of averaging using a time step of 0.025 fs. All calculations are carried out in the NVT ensemble at 300 K using Nosé–Hoover chain thermostats [65] In contrast to the free energy profile of Fig. 9.9, the profile in Figure 9.22 shows no evidence of a stable intermediate. Rather, apart from an initial barrier of approximately 8 kcal/mol,

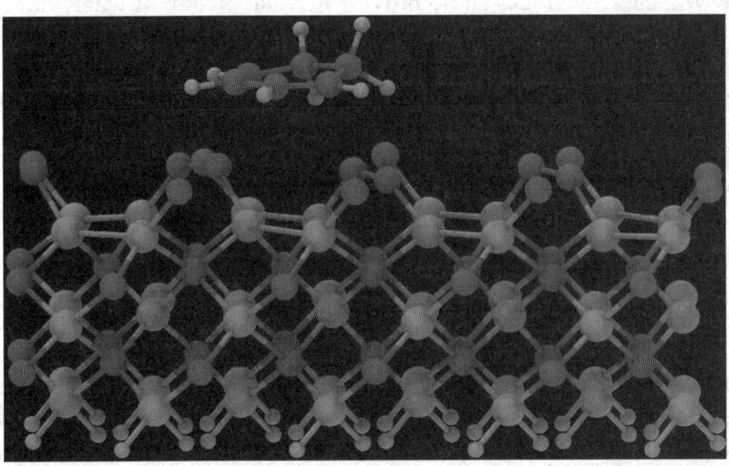

FIGURE 9.21 Snapshot of the SiC-(2×2) surface and 1,3-cyclohexadiene molecule. Dark, medium, and lightly shaded spheres represent carbon, silicon, and hydrogen atoms, respectively. A full color version of this figure appears in Ref. [102].

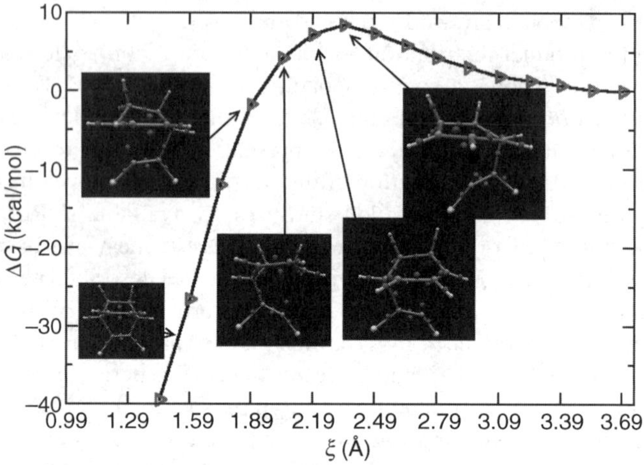

FIGURE 9.22 Free energy profile for the formation of the [4 + 2] Diels–Alder-like adduct between 1,3-cyclohexadiene a C≡C dimer on the SiC-(2 × 2) surface. Dark, light, and medium shaded spheres represent C, H, and Si, respectively. Small spheres in the bonds between atoms are the centers of maximally localized Wannier functions. A full color and slightly expanded version of this figure can be found in Ref. [102].

the free energy is strictly downhill. The reaction is thermodynamically favored by approximately 48 kcal/mol. The suggestion from Figure 9.22 is that the reaction is symmetric and concerted in contrast to the reactions on the other surfaces we have considered thus far. Figure 9.22 also shows snapshots of the molecule and the surface atoms with which it interacts at various points along the free energy profile. In these snapshots, spheres in the bonds between atoms represent the centers of maximally localized Wannier functions. These provide a visual picture of where new covalent bonds are forming as the reaction coordinate ξ is decreased. By following these, we clearly see that one C–C bond forms before the other, demonstrating the asymmetry of the reaction, which is a result of the buckling of the surface dimers. The buckling gives rise to a charge asymmetry in the C≡C surface dimer, and as a result, the first step in the reaction is a nucleophilic attack of one of the C≡C bonds in the cyclohexadiene on the positively charged carbon in the surface dimer, this carbon being the lower of the two. Once this first CC bond forms, the second CC bond follows after a change of approximately 0.3 Å in the reaction coordinate with no stable intermediate along the way toward the final [4 + 2] cycloaddition product. In addition, the Wannier centers show the conversion of the triple bond on the surface to a double bond in the final product state.

Further evidence for the concerted nature of the reaction is provided in Figure 9.23, which shows the average carbon–carbon lengths computed over the constrained trajectories at each point of the free energy profile. It can be seen that there is a slight tendency for an asymmetric reaction, despite its being concerted.

In order to demonstrate that the [4 + 2] Diels–Alder type cycloaddition product is highly favored over other reaction products on this surface, we show one additional example of a free energy profile, specifically, that for the formation of

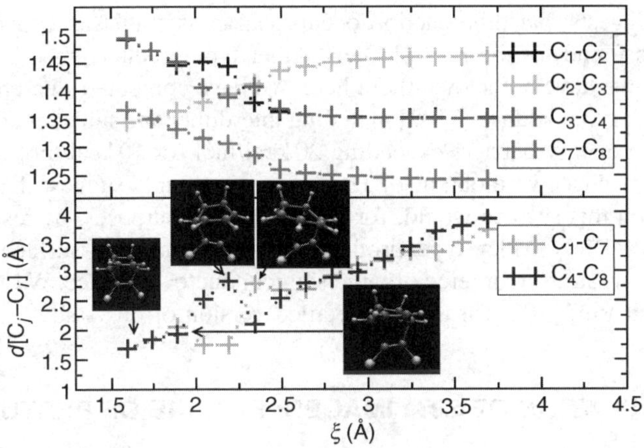

FIGURE 9.23 Average carbon–carbon bond lengths obtained from each constrained simulation. A full color version of this figure can be found in the Supporting Information for Ref. [102].

a [2 + 2] cycloaddition reaction with a single surface C≡C dimer. This profile is shown in Figure 9.24. In contrast to the [4 + 2] Diels–Alder type adduct, the barrier to formation of this adduct is roughly 27 kcal/mol (compared to 8 kcal/mol for the Diels–Alder product). Thus, although the [2 + 2] reaction is thermodynamically favorable, this barrier is sufficiently high that we would expect this particular reaction channel to be highly suppressed compared to one with a 19 kcal/mol lower barrier. The free energy profile, together with the snapshots taken along the reaction

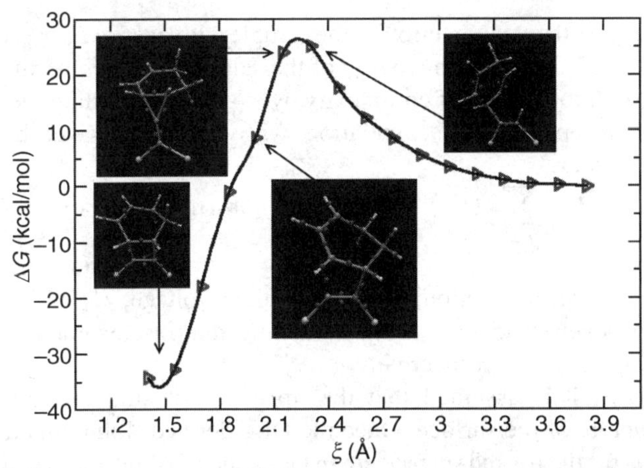

FIGURE 9.24 Free energy profile for the formation of the [2 + 2] adduct between 1,3-cyclohexadiene a C≡C dimer on the SiC-(2 × 2) surface. Dark, light, and medium shaded spheres represent C, H, and Si, respectively. Small spheres in bonds between atoms are the centers of maximally localized Wannier functions. A full color version of this figure can be found in Ref. [102].

path, also suggests that this reaction occurs via an asymmetric, concerted mechanism, as was found for the [4 + 2] Diels–Alder type product.

Although we have not shown them here, we have computed free energy profiles for a variety of other adducts, [102] including interdimer and sublayer adducts, and in all cases, free energy barriers exceeding 20 kcal/mol (or 40 kcal/mol in the case of the sublayer adduct) were obtained. These results strongly suggest that the product distribution on this surface would, for all intents and purposes, be restricted to the single [4 + 2] Diels–Alder type product, implying that this surface might be a candidate for creating an ordered organic/semiconductor interface. We encourage the reader to visit Ref. [102] for a more detailed version of this study.

9.6 CALCULATION OF STM IMAGES: FAILURE OF PERTURBATIVE TECHNIQUES

Because of the importance of STM techniques in imaging neat as well as functionalized surfaces, it is instructive to comment briefly on the problem of computing such images via density functional approaches. DFT derived STM images have traditionally been calculated by two methods: Tersoff–Hamann [104] and Bardeen perturbation theory [105]. Several excellent reviews of DFT-derived STM methods exist in the literature [106,107]. Tersoff–Hamann, the simplest approximation, assumes that the tunneling current is proportional to the electron density above the surface:

$$I \propto \sum_{\nu} |\psi_{\nu}(\mathbf{r}_{\text{tip}})|^2 \delta(E_{\nu} - E_{\text{F}}) \tag{9.57}$$

where $\psi_{\nu}(\mathbf{r})$ is the νth wave function of the sample with energy E_{ν}, $\mathbf{r}_{\text{tip}}$ is the location of the probe, and E_{F} is the Fermi energy of the sample. Bardeen perturbation theory [105] adds an additional level of complexity by explicitly including the contributions of the tip from a separate DFT calculation, with the current given by

$$I \propto \sum_{\mu,\nu} f(E_{\mu})[1 - f(E_{\nu} + eV)]|M_{\mu\nu}|^2 \delta(E_{\mu} - E_{\nu}) \tag{9.58}$$

Here $f(E)$ is the Fermi function, V is the applied voltage, E_{μ} and E_{ν} are tip and sample energy levels, and $M_{\mu\nu}$ is the tunneling matrix element between tip and surface states ψ_{μ} and ψ_{ν}, respectively.

In either case, it is assumed that the tip does not alter the geometry or the electronic structure of the surface. Once the DFT-derived Kohn–Sham energies and wave functions for the tip and surface are in hand, the STM image over a range of bias and isocurrents can be calculated using the same data set. In this section, we employ an STM code developed by Kováĉik and Marx [108,109] based on the Bardeen perturbation theory implementation given by Hofer et al. [106]. The spatial resolution of the calculated STM images is 0.14 Å.

In order to investigate the applicability of perturbative schemes, we return to the 1,3-CHD/Si(100)-(2×1) system considered earlier. The DFT implementation employed in this section is that in the Quantum Espresso code [110]. The KS orbitals are expanded in a plane waves are expanded up to 36 Ry and exchange correlation is treated at the LDA level [111]. The Kleinman–Bylander [64] pseudopotential forms are employed. During 0 K geometry optimizations, the total force on each atom is reduced to less than 1.0×10^{-3} a.u.

The tip is taken to be a tungsten tip, for which separate DFT calculations are required in order to apply the Bardeen perturbation theory. Six electrons are included for each W atom. The W(110) bcc tip is comprised of three pyramidal layers on two (2×3) bulk layers fixed at the theoretically derived lattice constant (3.12 Å) with $L_z = 26.9$ Å. The three pyramidal layers and any adatom on the apex of the tip (Si or O) are allowed to fully relax. Combined tip–surface calculations fix the lateral dimensions at the bulk Si value (15.5 Å × 15.5 Å) and expand L_z to 34.8 Å. The expanded lateral dimensions require (3×5) W-bulk layers (30 atoms/layer), resulting in a minimum bulk W strain of (0.9%×16.8%). In order to maintain the integrity of the tip crystalline structure, all of the tip must be fixed at scaled ideal coordinates except for the z-component of the three pyramidal layers. Representative finite temperature and ideal clusters are extracted from the corresponding full surface configurations by taking the positions of the CHD and nearest 10 Si and terminating all 12 broken bonds with H. In all cases, the large unit cell allowed the gamma point approximation to be employed.

Teague and Boland [18,49] characterized at least five types of CHD adducts on the Si(100)-(2×1) surface by identifying unpaired Si dimers and π^* orbitals on the C=C bond through empty-state STM imaging. We focus here on the [4+2] intradimer CHD adduct for simplicity, but similar results hold for the other adducts. Figure 9.25 shows a representative set of images computed over an AIMD trajectory of 3.5 ps. The images appear to match closely those reported by Teague and Boland, [18] but they match for the wrong reasons. Superimposing the Si surface dimers and CHD adduct on top of the isosurface reveals that the even through the Si-dimers are correctly described, the dumbbell-like protrusions correspond to the CH_2 groups, not the C=C bond! Adding impurities to the tip is not sufficient to resolve this problem [17].

However, if we compute forces in all three spatial directions (two parallel to the surface and one perpendicular) duc to the STM tip (see Figures 9.26 and 9.27 for an illustration of the system setup and the forces in the three spatial directions exerted on

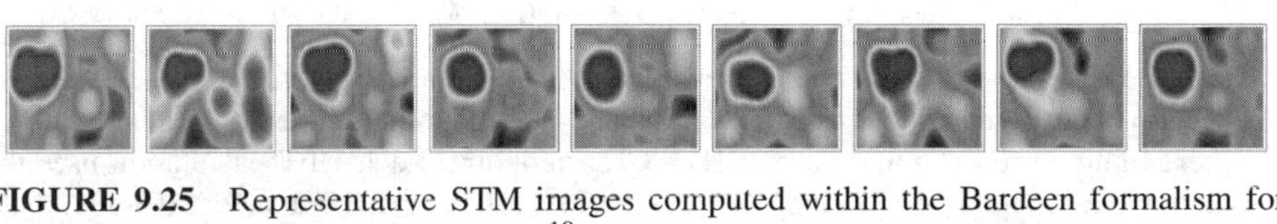

FIGURE 9.25 Representative STM images computed within the Bardeen formalism for $V_{surf} = 0.2$ V and isocurrent $I = 1.0\times10^{-10}$ A at different time points along a 3.5 ps trajectory. The last panel on the right is the thermal averaged image over the trajectory. A full color version of these images can be found in Ref. [17].

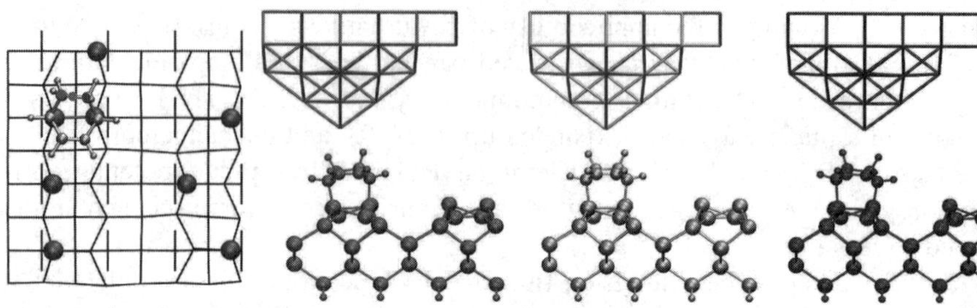

FIGURE 9.26 W (110) tip over the [4 + 2] intradimer cyclohexadiene (CHD) adduct adsorbed on the Si(100) reconstructed surface. The geometry optimized tip is initially placed 3.0 Å vertically (relative to the C = C midpoint) above the independently optimized CHD at the left C in CH_2 (left), midpoint of CH_2–CH_2 bond (middle), and right C in CH_2 (right) as marked by the red, orange, and blue dots, respectively, in the far left sketch. The dark spheres highlight which Si are at least the same height as the Si connected to CHD. A full color version of this figure can be found in Ref. [17].

the molecule by the tip), we find that the tip exerts a strong attractive force on the C = C bond in agreement with recent experiments, [59] but it also exerts a strong repulsive force on the CH_2 groups, and these various perturbations change the molecular geometry as well as the energetic ordering of the virtual orbitals [17]. These forces are large enough that perturbative techniques cannot be reliably applied to compute images of covalently bound floppy molecules.

Given the failures of perturbative methods, we suggest [17] employing the NEGF formalism [112–114] to compute STM images. NEGF methods, which include tip–surface interactions in the calculated *I–V* curve, can circumvent the problems with perturbative methods encountered above. They do so, however, at a substantial increase in computational overhead, particularly for empty-state imaging. In a NEGF

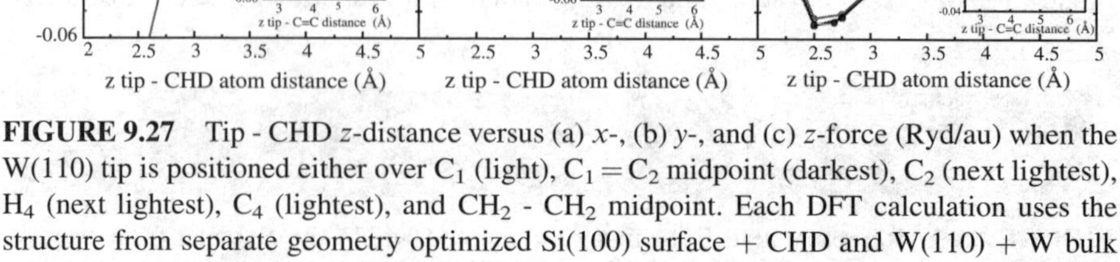

FIGURE 9.27 Tip - CHD *z*-distance versus (a) *x*-, (b) *y*-, and (c) *z*-force (Ryd/au) when the W(110) tip is positioned either over C_1 (light), $C_1 = C_2$ midpoint (darkest), C_2 (next lightest), H_4 (next lightest), C_4 (lightest), and CH_2 - CH_2 midpoint. Each DFT calculation uses the structure from separate geometry optimized Si(100) surface + CHD and W(110) + W bulk calculations. The insets show the same data, but with the corresponding tip–C = C midpoint *z*-distance. In (a) there is a strong drive to push the CH_2 groups away from the tip. A full color version of this figure can be found in Ref. [17].

calculation, the current is computed according to

$$I = \frac{e}{\pi\hbar}\int_{-\infty}^{\infty} \mathrm{Tr}\left[G^{\dagger}(E)\Gamma_2(E)G(E)\Gamma_1(E)\right][f_1(E) - f_2(E)]dE \qquad (9.59)$$

where $G = [E - H_0 + eV - \Sigma(E)]^{-1}$ is the Green's function at energy E, the Hamiltonian H_0 includes the molecule, the STM tip, and the surface region near the molecule treated atomistically. $f_1(E)$ and $f_2(E)$ are the Fermi functions of the tip and surface, respectively, and $\Gamma_1(E)$ and $\Gamma_2(E)$ are related to the imaginary part of the self-energy operators $(\Sigma(E))$ of the tip and surface, respectively. The latter treat the coupling of the atoms included explicitly in H_0 to the extended external environment. Since the tip perturbs both the wave function and geometry as a function of the relative tip–surface position, separate NEGF calculations must be done at each point instead of treating the tip and surface separately once and for all. If each NEGF calculation were inexpensive, then generating a 3D map in order to extract a fixed height or isocurrent image would be tedious, but not prohibitive. In practice, each individual calculation is costly using standard methods, and the development of more efficient self-consistent techniques based on the NEGF approach remains an outstanding challenge.

REFERENCES

1. Filler, M. A.; Bent, S. F. *Prog. Surf. Sci.* **2002**, *73*, 1–56.
2. Kachian, J. S.; Wong, K. T.; Bent, S. F. *Acc. Chem. Res.* **2010**, *43*, 346–355.
3. Basu, R.; Kinser, C. R.; Tovar, J. D.; Hersam, M. C. *Chem. Phys.* **2006**, *326*, 144–150.
4. Pitters, J. L.; Dogel, I.; DiLabio, G. A.; Wolkow, R. A. *J. Phys. Chem. B* **2006**, *110*, 2159–2163.
5. Hossain, M. Z.; Kato, H. S.; Kawai, M. *J. Am. Chem. Soc.* **2005**, *127*, 15030–15031.
6. Kirczenow, G.; Piva, P. G.; Wolkow, R. A. *Phys. Rev. B* **2005**, *72*, 245306.
7. Lopinski, G. P.; Wayner, D. D. M.; Wolkow, R. A. *Nature* **2000**, *406*, 48–51.
8. Car, R.; Parrinello, M. *Phys. Rev. Lett.* **1985**, *55*, 2471–2474.
9. Marx, D.; Hutter, J. Modern methods and algorithms of quantum chemistry. In Grotendorst, J., Ed. *Ab Initio Molecular Dynamics: Theory and Implementations.* Fz Juelich, NIC, **2000**, pp. 301–449.
10. Tuckerman, M. E. *J. Phys.:Condens. Matter* **2002**, *14*, R1297–R1355.
11. Marx, D.; Hutter, J. *Ab Initio Molecular Dynamics*. Cambridge University Press, Cambridge, **2009**.
12. Minary, P.; Tuckerman, M. E. *J. Am. Chem. Soc.* **2004**, *126*, 13920–13921.
13. Minary, P.; Tuckerman, M. E. *J. Am. Chem. Soc.* **2005**, *127*, 1110–1111.
14. Iftimie, R.; Minary, P.; Tuckerman, M. E. *P. Natl. Acad. Sci. U.S.A.* **2005**, *102*, 6654–6659.
15. Hayes, R. L.; Tuckerman, M. E. *J. Am. Chem. Soc.* **2007**, *129*, 12172–12180.
16. Hayes, R. L.; Tuckerman, M. E. *J. Phys. Chem. C* **2008**, *112*, 5880–5887.
17. Hayes, R. L.; Tuckerman, M. E. *J. Phys. Chem. C* **2010**, *114*, 15102.

18. Teague, L. C.; Boland, J. J. *J. Phys. Chem. B* **2003**, *107*, 3820–3823.
19. Tuckerman, M. E.; Yarne, D. A.; Samuelson, S. O.; Hughes, A. L.; Martyna, G. J. *Comput. Phys. Commun.* **2000**, *128*, 333–376.
20. Kohn, W.; Sham, L. J. *Phys. Rev. A* **1965**, *140*, 1133–1138.
21. Hohenberg, P.; Kohn, W. *Phys. Rev. B* **1964**, *136*, 864–871.
22. Becke, A. D. *Phys. Rev. A* **1988**, *38*, 3098–3100.
23. Lee, C.; Yang, W.; Parr, R. G. *Phys. Rev. B* **1988**, *37*, 785–789.
24. Perdew, J. P.; Burke, K.; Ernzerhof, M. *Phys. Rev. Lett.* **1996**, *77*, 3865–3868.
25. Marx, D.; Tuckerman, M.; Martyna, G. *Comp. Phys. Commun.* **1999**, *118*, 166–184.
26. Tuckerman, M. E.; Parrinello, M. *J. Chem. Phys.* **1994**, *101*, 1302–1315.
27. Tuckerman, M. E.; Martyna, G. J. *J. Chem. Phys.* **1999**, *110*, 2810–2821.
28. Minary, P.; Tuckerman, M. E.; Pihakari, K. A.; Martyna, G. J. *J. Chem. Phys.* **2002**, *116*, 5351–5362.
29. Minary, P.; Morrone, J. A.; Yarne, D. A.; Tuckerman, M. E.; Martyna, G. J. *J. Chem. Phys.* **2004**, *121*, 11949–11956.
30. Resta, R.; Sorella, S. *Phys. Rev. Lett.* **1999**, *82*, 370–373.
31. Berghold, G.; Mundy, C. J.; Romero, A. H.; Hutter, J.; Parrinello, M. *Phys. Rev. B* **2000**, *61*, 10040–10048.
32. Thomas, J. W.; Iftimie, R.; Tuckerman, M. E. *Phys. Rev. B* **2004**, *69*, 125105.
33. Iftimie, R.; Thomas, J. W.; Tuckerman, M. E. *J. Chem. Phys.* **2004**, *120*, 2169.
34. Sharma, M.; Wu, Y. D.; Car, R. *Intl. J. Quant. Chem.* **2003**, *95*, 821.
35. Becke, A. D.; Edgecombe, K. E. *J. Chem. Phys.* **1990**, *92*, 5397.
36. Over, H.; Wasserfall, J.; Ranke, W.; Ambiatello, C.; Sawitzki, R.; Wolf, D.; Moritz, W. *Phys. Rev. B* **1997**, *55*, 4731–4736.
37. Ono, M.; Kamoshida, A.; Matsuura, E.; Ishikawa, T. E.; Hasegawa, Y. *Phys. Rev. B* **2003**, *67*, 201306.
38. Takagi, N.; Shimonaka, S.; Aruga, T.; Nishijima, M. *Phys. Rev. B* **1999**, *60*, 10919–10925.
39. Stoll, E. P.; Baratoff, A.; Mangili, A.; Maric, D. *Mater. Sci. Eng.* **1996**, *B37*, 155–157.
40. Weakliem, P. C.; Smith, G. W.; Carter, E. A. *Surf. Sci.* **1990**, *232*, L219–L223.
41. Gryko, J.; Allen, R. E. *Ultramicroscopy* **1992**, *42–44*, 793–800.
42. Olson, R. M.; Gordon, M. S. *J. Chem. Phys.* **2006**, *124*, 081105.
43. Madelung, O. *Data in Science and Technology—Semiconductors: Group IV Elements and III–V Compounds. Springer-Verlag; **1991**.*
44. Teplyakov, A. V.; Kong, M. J.; Bent, S. F. *J. Am. Chem. Soc.* **1997**, *119*, 11100–11101.
45. Teplyakov, A. V.; Kong, M. J.; Bent, S. F. *J. Chem. Phys.* **1998**, *108*, 4599–4606.
46. Hovis, J. S.; Liu, J. S.; Hamers, R. J. *J. Phys. Chem. B* **1998**, *102*, 6873–6879.
47. Mui, C.; Bent, S. F.; Musgrave, C. B. *J. Phys. Chem. A* **2000**, *104*, 2457–2462.
48. Bent, S. F. *Surf. Sci.* **2003**, *500*, 879–903.
49. Teague, L. C.; Boland, J. J. *Thin Solid Films* **2004**, *464–465*, 1–4.
50. Troullier, N.; Martins, J. L. *Phys. Rev. B* **1991**, *43*, 1993–2006.
51. Brewer, L. *Lawrence Berkeley Laboratory Report* **1977**, LBL–3720.

52. Yin, M. T.; Cohen, M. L. *Phys. Rev. B* **1980**, *24*, 2303.
53. Minary, P.; Martyna, G. J.; Tuckerman, M. E. *J. Chem. Phys.* **2003**, *118*, 2527–2538.
54. Yoshinobu, J.; Yamashita, Y.; Yasui, F.; Mukai, K.; Tsuneyuki, S.; Akagi, K.; Hamaguchi, K.; Machida, S.; Nagao, M.; Sato, T.; Iwatsuki, M. *J. Electron. Spectrosc. Relat. Phenom.* **2001**, *114–116*, 383–387.
55. Jones, M., Jr. *Organic Chemistry. W. W. Norton and Co, New York, 2005.*
56. Carter, E. A.; Ciccotti, G.; Hynes, J. T.; Kapral, R. *Chem. Phys. Lett.* **1989**, *156*, 472.
57. Sprik, M.; Ciccotti, G. *J. Chem. Phys.* **1998**, *109*, 7737.
58. Naydenov, B.; Ryan, P.; Teague, L. C.; Boland, J. J. *Phys. Rev. Lett.* **2006**, *97*, 098304.
59. Naydenov, B.; Teague, L. C.; Ryan, P. M.; Boland, J. J. *Nano Lett.* **2006**, *6*, 1752–1756.
60. Konecny, R.; Doren, D. J. *J. Am. Chem. Soc.* **1997**, *119*, 11098–11099.
61. Choi, C. H.; Gordon, M. S. *J. Am. Chem. Soc.* **1999**, *121*, 11311–11317.
62. Teague, L. C.; Chen, D.; Boland, J. J. *J. Phys. Chem. B* **2004**, *108*, 7827–7830.
63. Galperin, M.; Beratan, D. N. *J. Phys. Chem. B* **2005**, *109*, 1473–1480.
64. Kleinman, L.; Bylander, D. M. *Phys. Rev. Lett.* **1982**, *48*, 1425.
65. Martyna, G. J.; Klein, M. L.; Tuckerman, M. E. *J. Chem. Phys.* **1992**, *97*, 2635–2643.
66. Lee, H. S.; Choi, C. H.; Gordon, M. S. *J. Am. Chem. Soc.* **2005**, *127*, 8485–8491.
67. Kong, M. J.; Teplyakov, A. V.; Jagmohan, J.; Lyubovitsky, J. G.; Mui, C.; Bent, S. F. *J. Phys. Chem. B* **2000**, *104*, 3000–30007.
68. Mizuno, S.; Shirasawa, T.; Shiraishi, Y.; Tochihara, H. *Phys. Rev. B* **2004**, *69*, 241306.
69. Capano, R. J.; Trew, R. J. *MRS Bull.* **1997**, *22*, 19.
70. Starke, U. *Silicon Carbide: Fundamental Questions and Applications to Current Device Technology*. Springer-Verlag, Berlin, Germany, **2004**.
71. Mélinon, P.; Masenelli, B.; Tournus, F.; Perez, A. *Nat. Mater.* **2007**, *6*, 479–490.
72. Stutzmann, M.; Garrido, J. A.; Eickhoff, M.; Brandt, M. S. *Phys. Stat. Sol. A* **2006**, *203*, 3424–3437.
73. Soukiassian, P. G.; Enriquez, H. B. *J. Phys.:Condens. Matter* **2004**, *16*, S1611–S1658.
74. Pollmann, J.; Krüger, P. *J. Phys.:Condens. Matter* **2004**, *16*, S1659–S1703.
75. D'angelo, M.; Enriquez, H.; Aristov, V. Y.; Soukiassian, P.; Renaud, G.; Barbier, A.; Noblet, M.; Chiang, S.; Semond, F. *Phys. Rev. B* **2003**, *68*, 165321.
76. Tejeda, A.; Dunham, D.; García de Abajo, F. J.; Denlinger, J. D.; Rotenberg, E.; Michel, E. G.; Soukiassian, P. *Phys. Rev. B* **2004**, *70*, 045317.
77. Di Felice, R.; Bertoni, C. M.; Pignedoli, C. A.; Catellani, A. *Phys. Rev. Lett.* **2005**, *94*, 116103.
78. Peng, X.; Krüger, P.; Pollmann, J. *Phys. Rev. B* **2005**, *72*, 245320.
79. Chang, H.; Wu, J.; Gu, B.-L.; Liu, F.; Duan, W. *Phys. Rev. Lett.* **2005**, *95*, 196803.
80. Peng, X.; Krüger, P.; Pollmann, J. *Phys. Rev. B* **2007**, *75*, 073409.
81. Peng, X.; Krüger, P.; Pollmann, *Phys. Rev. B* **2007**, *76*, 125303.
82. de Brito Mota, F.; de Castilho, C. M. C. *Phys. Rev. B* **2006**, *74*, 165408.
83. Wieferink, J.; Krüger, P.; Pollmann, J. *Phys. Rev. B* **2006**, *74*, 205311.
84. Wieferink, J.; Krüger, P.; Pollmann, J. *Phys. Rev. B* **2007**, *75*, 153305.
85. Cicero, G.; Catellani, A.; Galli, G. *Phys. Rev. Lett.* **2004**, *93*, 016102.

86. Cicero, G.; Catellani, A. *J. Chem. Phys.* **2005**, *122*, 214716.
87. Amy, F.; Chabal, Y. J. *J. Chem. Phys.* **2003**, *119*, 6201–6209.
88. Derycke, V.; Soukiassian, P. G.; Amy, F.; Chabal, Y.; D'Angelo, M. D.; Enriquez, H. B.; Silly, M. G. *Nat. Mater.* **2003**, *2*, 253–254.
89. McNab, I. R.; Polanyi, J. C. *Chem. Rev.* **2006**, *106*, 4321.
90. DiLabio, G. A.; Piva, P. G.; Kruse, P.; Wolkow, R. A. *J. Am. Chem. Soc.* **2004**, *126*, 16048.
91. Hossain, M. Z.; Kato, H. S.; Kawai, M. *J. Phys. Chem. B* **2005**, *109*, 23129.
92. Hossain, M. Z.; Kato, H. S.; Kawai, M. *J. Am. Chem. Soc.* **2005**, *127*, 15030.
93. DiLabio, G. A.; Dogel, S. A.; Anagaw, A.; Pitters, J. L.; Wolkow, R. A. *Phys. Chem. Chem. Phys.* **2007**, *9*, 1629.
94. Hossain, M. Z.; Kato, H. S.; Kawai, M. *J. Am. Chem. Soc.* **2007**, *129*, 3328.
95. Hossain, M. Z.; Kato, H. S.; Kawai, M. *J. Am. Chem. Soc.* **2007**, *129*, 12304.
96. Zikovsky, J.; Dogel, S. A.; Haider, A. B.; DiLabio, G. A.; Wolkow, R. A. *J. Phys. Chem. A* **2007**, *111*, 12257.
97. Hossain, M. Z.; Kato, H. S.; Kawai, M. *J. Am. Chem. Soc.* **2008**, *130*, 11518.
98. Hossain, M. Z.; Kato, H. S.; Kawai, M. *J. Phys. Chem. C* **2009**, *113*, 10751.
99. Ferguson, G. A.; Than, C. T. L.; Raghavachari, K. *J. Phys. Chem. C* **2009**, *113*, 11817.
100. Ferguson, G. A.; Than, C. T. L.; Raghavachari, K. *J. Phys. Chem. C* **2010**, *1*, 679.
101. Bermudez, V. M. *Surf. Sci.* **2003**, *540*, 255–264.
102. Zhang, Y. and Tuckerman, M. E. *J. Phys. Chem. Lett.* **2011**, *2*, 1814–1819.
103. Derycke, V.; Soukiassian, P.; Mayne, A.; Dujardin, G. *Surf. Sci.* **2000**, *446*, L101–L107.
104. Tersoff, J.; Hamann, D. R. *Phys. Rev. B* **1985**, *31*, 805–813.
105. Bardeen, J. *Phys. Rev. Lett.* **1961**, *6*, 57–59.
106. Hofer, W. A.; Foster, A. S.; Shluger, A. L. *Rev. Mod. Phys.* **2003**, *75*, 1287–1331.
107. Gottlieb, A. D.; Wesoloski, L. *Nanotechnology* **2006**, *17*, R57–R65.
108. Kovâĉik, R.; Meyer, B.; Marx, D. *Angew. Chem. Int. Ed.* **2007**, *46*, 4894–4897.
109. Kovâĉik, R. *Characterization of Catalyst Surfaces by Ab Initio Thermodynamics and STM Data Calculations*. Thesis, Ruhr-Universitat Bochum, **2007**.
110. Baroni, S.; Dal Corso, A.; de Gironcoli, S.; Giannozzi, P.; Cavazzoni, C.; Ballabio, G.; Scandolo, S.; Chiarotti, G.; Focher, P.; Pasquarello, A.; Laasonen, K.; Trave, A.; Car, R.; Marzari, N.; Kokalj, A. *Quantum Espresso Package*, http://www.pwscf.org/, **2008**.
111. Ceperley, D. M.; Alder, B. J. *Phys. Rev. Lett.* **1980**, *45*, 566–569.
112. Keldysh, L. V. *Sov. Phys. JETP* **1965**, *20*, 1018.
113. Calzolari, A.; Marzari, N.; Souza, I.; Nardelli, M. B. *Phys. Rev. B* **2004**, *69*, 035108.
114. Datta, S. *Nanotechnology*. **2004**, *15*, S433.

CHAPTER 10

Formation of Organic Nanostructures on Semiconductor Surfaces

MD. ZAKIR HOSSAIN AND MAKI KAWAI

10.1 INTRODUCTION

Single molecule or an assembly of organic molecules has the potential to be used as a basic component of future nanoscale electronic devices [1–3]. Indeed, a suitably functionalized ordered assembly of molecules on the surface can act as a wire, diode, transistor, and so on [4, 5]. One of the key challenges for realization of such a device is assembling the functional organic molecules in a desired fashion and connecting them into a circuit. Among the various approaches for fabricating nanostructures on the material surface, the atom-by-atom manipulation with scanning tunneling microscope and the self-assembly of molecules are widely used [6–9]. However, these approaches are not suitable for semiconductor surface because the desired molecules are strongly bonded to the surface, and the molecular diffusion on the surface is greatly inhibited. Hence, the formation of organic nanostructures on semiconductor surfaces remains a challenging task.

Among various semiconductor materials, silicon is considered to be the most promising alternative substrate for molecular electronics because of its compatibility with organic molecules, semiconductor band structure, and existing microelectronics [10, 11]. Recently, the H-terminated silicon surface appeared as an ideal template for fabricating organic nanostructures on the silicon surface. The key initiative of making organic nanostructures is to create a reactive site on the H-terminated surface. This reactive site can be created by removing a H atom from a surface silicon atom resulting in an unpaired dangling bond (DB) site either using a scanning tunneling microscopy (STM) tip or using thermal annealing [12]. Indeed, the H-terminated surface intrinsically contains a dilute concentration of DB sites due to incomplete H termination of the surface silicon atoms [13]. These DB sites

Functionalization of Semiconductor Surfaces, First Edition.
Edited by Franklin (Feng) Tao and Steven L. Bernasek.

are highly reactive toward various molecules and can act as a radical site to initiate a chain reaction when appropriate molecules, such as alkene, are dosed onto the surface. Linford et al. first suggested DB-initiated radical chain reaction mechanism for explaining the alkyl monolayer formation by 1-alkenes and 1-alkynes on the H-terminated Si(111) surface [14]. Taking advantage of the anisotropic nature of surface silicon atoms, Lopinski et al. demonstrate the one-dimensional (1D) organic nanostructure by reacting styrene molecules with the DB site on the H-terminated Si (100)-(2×1) surface [15].

In this chapter, we focus on the formation of 1D organic nanostructures on the H-terminated Si(100)-(2×1) surface through DB-initiated chain reaction, and the physical and electronic properties of these nanostructures. To date, a variety of molecules such as alkene, alkyne, aldehyde, ketone, and so on have been studied for the nanostructure formation on the Si(100)–H surface [15–31]. Though the basic reaction occurs between the DB of the substrate and the >C=C< or >C=O group of the molecules, many molecules containing those functional groups do not form any 1D nanostructure at room temperature. While most of the molecules form nanostructures along the dimer row, a few molecules show nanostructure across the dimer rows. Control over the directional selectivity allows for fabricating interconnected nanostructure junctions in two dimensions. Molecular composition of a nanostructure can be varied by the alternate dosing of different molecules sequentially. The valence state of 1D nanostructures formed by carbonyl molecules is discussed.

10.2 EXPERIMENTAL

Experimental works were performed in an ultrahigh vacuum (UHV) chamber with a base pressure of $\sim 5 \times 10^{-11}$ mbar. The silicon sample was cut into desired size from commercially available silicon wafer. The surface was cleaned by prolonged degassing at ~850 K (~8 h) followed by repeated flashing up to 1400 K, which results in the reconstructed bare (2×1) surface. The sample was heated by resistive heating, and the temperature of the surface was measured by an optical pyrometer.

The bare Si(100)-(2×1) surface was terminated with H by exposure to atomic H, generated by a hot W filament (~2100 K), at the surface temperature of ~625 K. STM measurements were performed using an electrochemically etched W tip. For low-temperature measurements, the sample was cooled down by liquid N_2. Ultraviolet photoelectron spectroscopy (UPS) measurements were done using a He discharge lamp as photon source and a hemispherical electron analyzer. The diameter of the ultraviolet (UV) light irradiation area is ~1.5 mm. The incident and emission angles for UV light and photoelectron were 45° and 0°, respectively, with respect to the surface normal.

All the chemicals used for the studies are commercially available, and loaded into the ampule without further processing. The sample ampule was purified by pumping out the vapor above the liquid or solid phases before dosing onto the silicon surface. Liquid chemicals such as styrene, allyl mercaptan (AM), acetone, and so on were dosed onto the silicon surface through electronically controlled pulse-valve dosing system, and the solids such as benzophenone were dosed by opening a gate valve between the benzophenone ampule and the silicon sample in a load-lock chamber. The number of molecules dosed onto the surface is expressed in langmuir ($1\,L = 1 \times 10^{-6}$ Torr · s), which is estimated from the background pressure reading by B. A. (Bayard-Alpert) gauge during the exposure.

10.3 RESULTS AND DISCUSSION

10.3.1 Individual 1D Nanostructures on Si(100)–H: STM Study

10.3.1.1 Styrene and Its Derivatives on Si(100)-(2×1)–H The bare Si (100)-(2×1) surface exposed to atomic H at a substrate temperature of ~625 K generally produces monohydrogenated surface, where each surface silicon atom is terminated by a H atom. The as-prepared H-terminated surface possesses a dilute concentration of DB sites due to incomplete H termination [13]. A typical STM image of a freshly prepared H-terminated Si(100) surface is shown in Fig. 10.1a. As

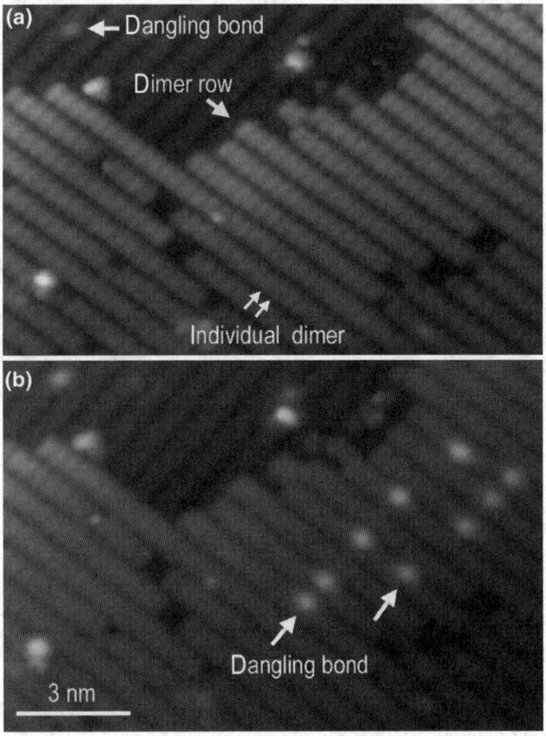

FIGURE 10.1 Selected area STM images of (a) as-prepared H-terminated Si(100)-(2×1) surface and (b) surface "a" after creation of few dangling bond sites by electron injection from STM tip. Scanning parameters: $V_{sample} = -1.45$ V (a) and -2.20 V (b); $I = 0.2$ nA. Electron pulse injection parameters: $V_{sample} = -3.8$ V; $I = 0.5$ nA, $\tau = 0.5$ s.

seen in the STM image, the H-terminated Si(100) surface retains the dimer structures, which are arranged in parallel rows on a terrace. A dimer row and two individual dimers in a row are indicated. Note that the surface periodicity is similar to that of the bare surface; that is, the interdimer distance in a row is 3.84 Å and the distance between two neighboring rows is 7.68 Å. At a typical scanning bias, a DB on freshly prepared Si(100)-(2×1)–H surface appeared as a bright protrusion, which is located slightly off the center of the dimer row, as seen in Fig. 10.1a. Besides these intrinsic DB sites on the as-prepared H-terminated surface, a DB at a desired location can also be created by injecting an electron pulse from the STM tip [12]. A number of such DB sites created by the STM tip are shown in Fig. 10.1b.

Ideally, a DB site contains an unpaired electron, which can act as a reactive radical site on the surface. When a molecule with suitable functionality is dosed on the surface, the DB site readily reacts with the molecule. It is observed that styrene and its derivatives such as methylstyrene, bromostyrene, or trifluoromethylstyrene undergo chain reaction with the DB site of the substrate and form a 1D nanostructure on the H-terminated Si(100)-(2×1) surface [15, 17, 19, 21, 22]. The typical STM images of such nanostructures formed through the reaction of styrene molecules with the dangling bond sites on the H-terminated surface are shown in Fig. 10.2 [15]. These sets of images reveal that the line-shaped nanostructure grown with increasing doses of styrene molecules is initiated at the DB sites indicated in Fig. 10.2a. The longest nanostructure in Fig. 10.2d is estimated to be ~130 Å long, corresponding to 34 dimers in a row. Sometimes growth of the nanostructures is observed to

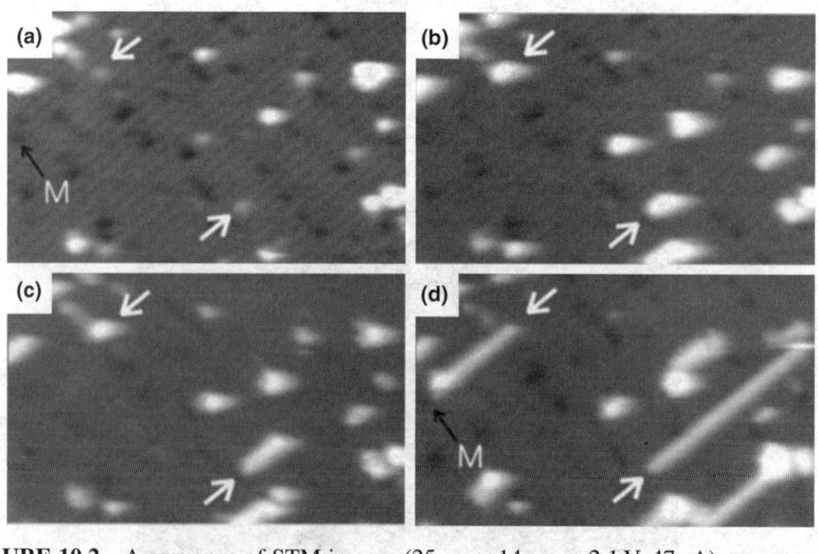

FIGURE 10.2 A sequence of STM images (25 nm × 14 nm, −2.1 V, 47 pA) corresponding to an increasing exposure to styrene: (a) 3 L; (b) 28 L; (c) 50 L; (d) 105 L. The white arrows denote two particular dangling bond sites that lead to the growth of long styrene nanostructures. The missing dimer defect (M) marked in the figure terminates the growth of the nanostructure in the top left-hand corner of the image. Reproduced from Ref. 15.

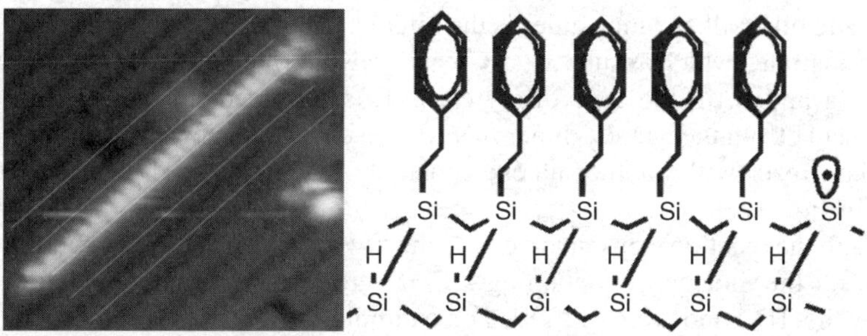

FIGURE 10.3 A high-resolution STM image of the nanostructure formed by styrene derivatives on the Si(100)-(2×1)–H surface revealing the individual adsorbed molecules. Reproduced from Ref. 21 by permission of the American Chemical Society. $V_{sample} = -2.5$ V; $I = 0.2$ nA. Scan area $= 8 \times 8\ \text{nm}^2$. Schematic of adsorbed styrene aligned one side of a dimer row is shown in the right panel.

be stopped at a preexisting defect site. The 1D nanostructure formed by styrene derivatives such as methylstyrene, bromostyrene, and so on appears similar to that in Fig. 10.2 [19, 21, 22]. In addition to single 1D nanostructures, some doubled nanostructures, where both of the silicon dimer atoms are occupied, are also formed.

A high-resolution STM image revealing the individual adsorbed molecules in a nanostructure is shown in Fig. 10.3. The 1D nanostructure is found to be located slightly off the center of the dimer row. The individual protrusions are spaced by 3.8 Å, corresponding to the distance between dimers in a row of the Si(100) surface. This image is consistent with molecules bound to adjacent dimers on the same side of a row, as shown in Fig. 10.3 (right panel). Note that the stacking of the phenyl rings in an assembly is important for determining the electronic properties of this 1D nanostructure. If the rings are parallel, some degree of overlap between π-states is expected, suggesting that the 1D nanostructure could function as a molecular wire [15]. The valence electronic states of such a 1D nanostructure are discussed in Section 10.3.3.

The reaction mechanism for the formation of a 1D nanostructure by styrene molecules is shown in Fig. 10.4 [15]. The initiation step of the chain reaction of

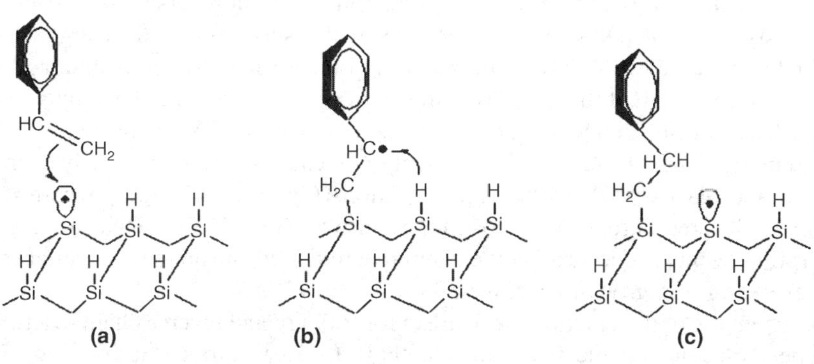

FIGURE 10.4 The chain reaction mechanism of styrene molecule with the dangling bond of the Si(100)-(2×1)–H surface.

styrene and other alkene molecules is the interaction of the >C=C< part with the DB of the substrate, which results in a C-centered intermediate radical, as shown in Fig. 10.4a and b (in the case of styrene). This intermediate radical (Fig. 10.4b) abstracts a H from the nearest dimer in a row resulting in a new DB site (Fig. 10.4c), which can react with another alkene molecule leading to a 1D nanostructure, as shown in Fig. 10.3.

Though the reaction steps involve only the C=C group and the DB bond site of the surface, all the molecules containing a C=C group do not undergo chain reaction forming the 1D nanostructure [28]. For example, allylbenzene, vinylcyclohexane, 4-vinylpyridene, propylene, and so on do not form nanostructures at room temperature [15, 18, 21, 28]. It is suggested that the stabilization of intermediate radical to abstract H from the neighboring site plays an important role in promoting the chain reaction forming the nanostructure [18, 32, 33].

Reverse Chain Reaction As seen in Figs. 10.3 and 10.4, the 1D nanostructure grown via the H abstraction mechanism should always be ended with a DB site unless the growth process encounters a defect site [15]. This end DB of a nanostructure can trigger a reverse chain reaction leading to the desorption of styrene molecule [15]. It is suggested that the observation of nanostructure on the surface by STM measurement depends on the relative rate of forward and reverse chain reaction at a given temperature. A controlled experiment with styrene molecules reveals that the reverse chain reaction at $\sim$100°C is fast enough to completely desorb a whole nanostructure in a few minutes [24].

Figure 10.5 shows a series of STM images demonstrating the reverse chain reaction at elevated temperature. When the surface "a" was exposed to $\sim$3 L of styrene at 300 K, four 1D nanostructures (i–iv) with different lengths (shown in Fig. 10.5b) were observed. The wider part of the line may be the doubled nanostructures. It was found that the length of the styrene line does not decrease even after several hours at 300 K. Since all of these growing 1D nanostructures (Fig. 10.5b) do not end at any preexisting defect, it is expected that there is a free DB at the end of each line, as shown in Fig. 10.3.

To see if the end DB of a nanostructure can initiate a reverse chain reaction to desorb styrene molecules, the temperature of the surface shown in Fig. 10.5b is gradually increased to 400 K while scanning of the same area is maintained. It is found that the length of the nanostructures does not decrease until the temperature of the surface reaches 400 K. Sequential measurements of STM images at 400 K (as shown in Fig. 10.5c–f) indicate that the length of a nanostructure gradually decreases to zero and leaves a DB at the same position (Fig. 10.5c–f and h) where it was originally located before the nanostructure growth (Fig. 10.5a). This suggests that desorption events occur through the chain reaction mechanism that is similar to, but the reverse of, the growth processes.

A schematic of the reaction mechanism for forward and reverse chain reactions of styrene molecules on the H-terminated Si(100)-(2$\times$1) surface at 300 and 400 K, respectively, is shown in Fig. 10.6. In the reverse reaction, H from the adsorbed styrene goes back to the DB site (as indicated in Fig. 10.6c) followed by breaking of

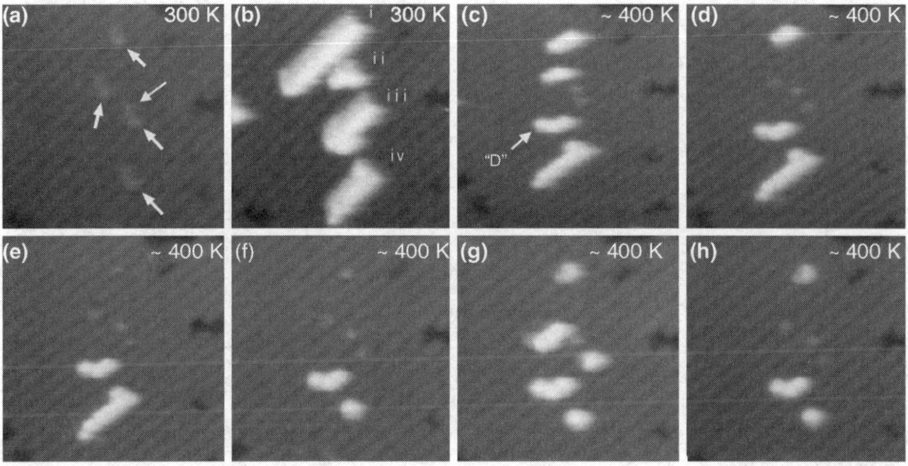

FIGURE 10.5 Series of STM images of the same area on the H-terminated Si(100)-(2×1) surface measured at 300 K (a, b) and 400 K (c, h). (a) Five dangling bonds indicated by arrows are created using high-voltage pulse injection from STM tip. The shadow feature of each DB is due to the tip artifact. (b) The surface "a" was exposed to ~3 L of styrene at 300 K. (c–f) Sequential STM images after the temperature of the surface "b" increased to 400 K. (g) The surface "f" was exposed to ~4 L of styrene at 400 K. (h) Three minutes after "g". For (g), $V_{sample} = -1.1$ V; $I = 0.2$ nA; scan area = 9.2 × 6.3 nm^2. For (h), $V_{sample} = -1.8$ V; $I = 0.2$ nA; scan area = 7.0 × 5.5 nm^2. Reproduced from Ref. 24 by permission of the American Chemical Society.

the Si–C bond (Fig. 10.6b). As mentioned earlier, the stabilization of the intermediate radical plays an important role in successful growth of a nanostructure. Computational studies have suggested that the stability of a C-centered radical and an adsorbed molecule adjacent to a DB is the key factor for the successful growth of nanostructures [18, 32, 33].

From the direct observation of reverse chain reaction, one can estimate the activation barrier for desorption. The change in the number of molecules (N) in the "styrene" nanostructure as a function of time (t) is observed to be linear, as shown for the two different "styrene" nanostructures in Fig. 10.6d. The N versus t plots clearly suggest that desorption events of styrene molecules through the chain reaction mechanism are zero order; that is, the rate of reverse reaction is independent of the number of molecules in a nanostructure. The observed zero-order reaction is consistent with the fact that the molecule at the next nearest neighbor site to the DB plays no role in the desorption process of the molecule at the nearest neighbor site to the DB. The slight slope difference between the two lines (gray and black) is perhaps due to the difference in substrate temperature (±20 K).

Employing the zero-order rate equation ($-dN/dt = k$) in Fig. 10.6d, the average rate constant k for the reverse reaction at 400 K is estimated to be 1.17×10^{-2} s^{-1}. If the preexponential factor (ν) is assumed to be 10^{14} s^{-1} (as in Ref. 15), E_d is estimated to be 1.26 ± 0.06 eV. This estimated activation barrier for the reverse reaction lies within the calculated values (0.90–1.43 eV) reported earlier [18, 32, 33]. Compared

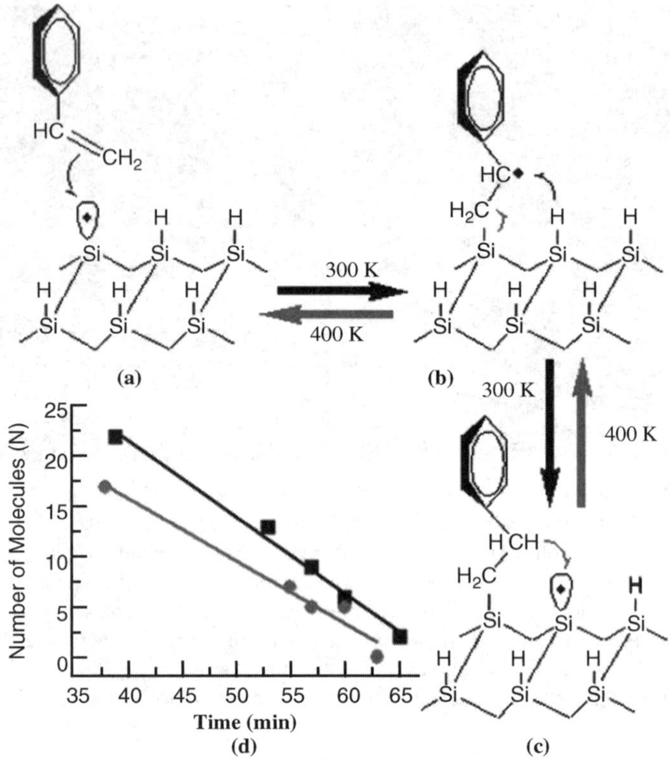

FIGURE 10.6 (a–c) Schematic of reaction mechanism for forward and reversed chain reactions of styrene molecules on H-terminated Si(100)-(2 × 1) surface at 300 and 400 K, respectively. (d) Change in number of molecules in two typical styrene nanostructures as a function of time at 400 K. The solid lines are the linear fit to the data points for two different styrene nanostructures. Reproduced from Ref. 24 by permission of the American Chemical Society.

with that in the "styrene" nanostructure, the activation barrier for the reverse reaction in the 2,4-dimethylstyrene (2,4-DMS) nanostructure seems to be lower, as indicated by the frequent reverse reaction seen at 300 K. Neglecting the number of molecules adsorbed from the background, the rate constant k for 2,4-DMS at 300 K is estimated to be $2.59 \times 10^{-2}\,s^{-1}$, which gives an E_d value of 0.93 eV [24]. This lower E_d for 2,4-DMS compared to that for the "styrene" nanostructure may be related to the steric crowding of two methyl groups in the phenyl ring. These relative activation barriers for the desorption of styrene and methylstyrene are in agreement with the recent theoretical calculation [34].

10.3.1.2 Long-Chain Alkenes on Si(100)-(2×1)–H A number of long-chain alkene molecules have been studied by DiLabio et al. [18]. They observed 1D nanostructure of only those alkene molecules whose chain length is higher than eight carbons (i.e., C_nH_{2n}; $n \geq 8$) such as 1-octene, 1-undecene, 1-tertadecacene, and so on [18]. As discussed above, the desorption rate at a certain substrate temperature is crucial for observing the 1D nanostructure in the STM image. If the desorption rate is so fast that all the molecules in a nanostructure desorb before the image is scanned, no 1D nanostructure will be seen in the STM image, even if the molecules might have undergone a chain reaction and formed a nanostructure during the exposure. Thus, the alkenes (i.e., C_nH_{2n}; $n \leq 8$) that show no nanostructure growth at 300 K might show a nanostructure in the STM image at lower temperatures (<300 K) because of the reduced desorption rate. Indeed, 1-hexene and 1-heptene molecules are also found to undergo a chain reaction initiated at unpaired DB sites on the Si(100)-(2×1)–H surface, and to form a stable nanostructure at 180 K, as shown in Fig. 10.7 [24]. These observations widen the scope of nanostructure formation by various molecules.

10.3.1.3 Cross-Row Nanostructure As discussed in previous sections, all the molecules such as styrene, styrene derivatives, alkenes, and so on form a 1D nanostructure only along the dimer row. However, to fabricate a nanostructure network on the surface, control over the directional selectivity of the nanostructure growth should be achieved. It is found that unlike other short-chain alkene molecules, allyl mercaptan readily undergoes chain reaction with the DB site and forms a 1D nanostructure even at room temperature [20]. In this particular case, the nanostructure is formed exclusively across the dimer row direction. Figure 10.8a shows the STM image of the Si(100)-(2×1)–H surface exposed to ~3 L of ALM molecules at 300 K. Four lines of small protrusions running across the dimer rows are 1D nanostructures formed by ALM molecules. These nanostructures are randomly distributed all over the surface, which suggests that the STM tip has no role in the growth of these nanostructures. All the observed nanostructures lie across the dimer rows. The high-resolution zoomed-in STM image (Fig. 10.8b) reveals that each protrusion in nanostructure has two parts, the "head" and "tail" as indicated in Fig. 10.8b. The "head" of a protrusion is located on the edge of a dimer row and its "tail" is on the preceding row (discussed later). The observed lines are stable even at 650 K (Fig. 10.8c) and for at least 10 days in vacuum at 300 K, which ensures that the molecules in the nanostructure are chemically bonded to the silicon substrate.

In accordance with the reported mechanism for the chain reaction of alkene molecules triggered at the DB site [14, 15], the proposed mechanism for the cross-row chain reaction by ALM molecules on Si(100)-(2×1)–H surface is shown in Fig. 10.9. Since propylene ($CH_2{=}CH{-}CH_3$) and allyl methyl sulfide ($CH_2{=}CH{-}CH_2{-}S{-}CH_3$) do not show any nanostructure growth, the –SH group in ALM must have an important role in stabilization of the intermediate radical (Fig. 10.9b), which is the key for the successful growth of a nanostructure [15, 18]. The chemistry of C-centered radical in the thiol molecule suggests that the radical centered at C can be transferred on S through tautomerism, and results in a thiyl radical (Fig. 10.9c), which is expected to be stable enough to abstract a H from the substrate.

The distance between the stable intermediate radical and the anchoring site of the molecule is one of the factors that determines the selectivity in the direction (across or along the dimer rows) of nanostructure growth. It is evident that the length of the

FIGURE 10.7 Selected area STM images of the Si(100)-(2×1)–H surface containing some nanostructures formed by DB-initiated chain reactions of (a) 1-hexene and (b) 1-heptene at 180 K. Typical single and doubled nanostructures are indicated. The doses of molecules for (a) and (b) are ~2 and ~8 L, respectively. (c) and (d) are STM images showing 1D nanostructure of adsorbed 1-hexene and 1-heptene after warming up the sample to 300 K. $V_{sample} = -2.5$ V; $I = 0.2$ nA. Scan areas: (a) 48 × 30 nm^2; (b) 54 × 34 nm^2. Reproduced from Ref. 24 by permission of the American Chemical Society.

ALM molecule makes it possible that the radical on the S atom (Fig. 10.9c) reaches to the nearest H on the next dimer row and abstracts it. To date, several computational studies have been reported regarding the detailed mechanism and energetics of the cross-row chain reaction of allyl mercaptan [35–37]. Choi and Cho suggested that instead of H abstraction by thiyl radical, the sulfur atom makes a bond with the silicon atom followed by desorption of H_2 from the dimer resulting in a DB site on the

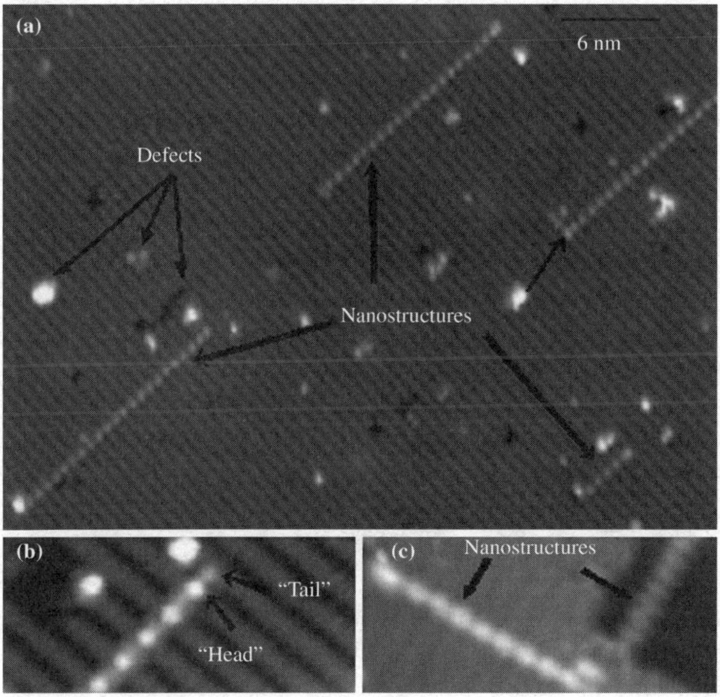

FIGURE 10.8 (a) STM image of Si(100)-(2×1)–H surface containing a dilute concentration of dangling bonds exposed to ~3 L of allyl mercaptan (CH_2=CH–CH_2–SH) molecules at 300 K. V_{sample} = –2.5 V; I = 0.2 nA. (b) High-resolution zoomed-in image of nanostructure formed by allyl mercaptan at 300 K. V_{sample} = –1.2 V; I = 0.2 nA. (c) STM image at 650 K. V_{sample} = –3.0 V; I = 0.2 nA. Reproduced from Ref. 20 by permission of the American Chemical Society.

next silicon atom [36]. Thus, allyl mercaptan binds through the Si–C and Si–S bond across the dimer, thereby yielding a higher thermodynamic stability compared to other nanostructures (containing a single Si–C bond per molecule) grown along the dimer rows [36]. This success in forming a 1D nanostructure across the dimer rows provides a means to make a predesigned interconnection of nanostructures on the Si(100)-(2×1)–H surface.

10.3.1.4 Aldehyde and Ketone: Acetophenone – A Unique Example In addition to the C=C group, the >C=O group also reacts with the DB site leading to the formation of a nanostructure on the surface. Both aldehyde and ketone are observed to form nanostructures at room temperature. To date, a number of simple aldehydes and ketones such as acetaldehyde, bezaldehyde, cyclopropyl methyl ketone, acetone, acetophenone, and benzophenone molecules have been studied [23, 28, 38, 39]. All these molecules except cyclopropyl methyl ketone undergo chain reaction with directional selectivity along the dimer row. Cyclopropyl methyl ketone shows poor directional selectivity resulting in

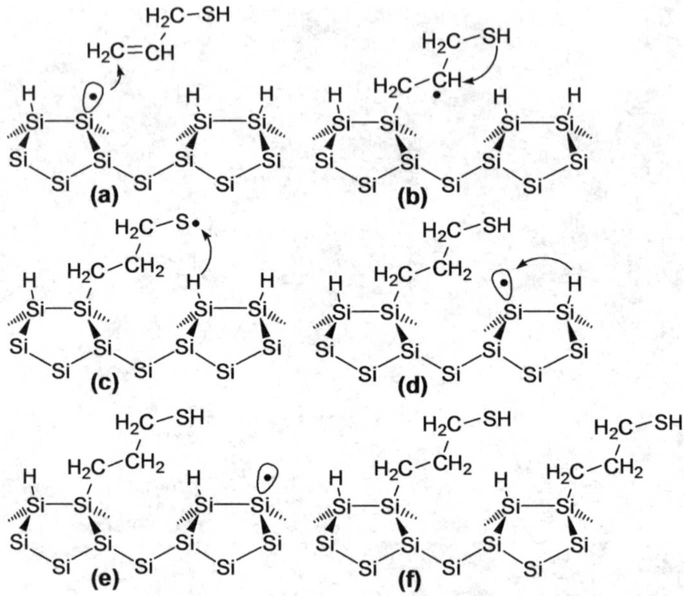

FIGURE 10.9 Proposed reaction mechanism for allyl mercaptan molecule with an unpaired dangling bond on H-terminated Si(100)-(2 × 1) surface.

random nanostructures on the surface [39]. It is found that acetone undergoes the most facile chain reaction among the molecules forming nanostructures along the dimer row on the surface.

A series of STM images demonstrating the formation of 1D nanostructures by dosing ~0.02 L of acetone molecules at 300 K are shown in Fig. 10.10. Acetone molecules always form a single nanostructure; that is, unlike styrene and other molecules that form nanostructures in the parallel-row direction [15, 18, 38], acetone molecules never form doubled nanostructures where both Si dimer atoms in a row are occupied by adsorbed molecules [23, 28]. In rare events, only one or two underlying Si dimers are observed to be doubly occupied by acetone molecules in a 1D structure.

It was observed that the nanostructures formed by adsorbed acetone molecules are normally stable during scanning in the occupied state at sample biases below ~4.0 V. At higher scanning biases (>−5 V), the nanostructures were decomposed (data not shown). However, when a surface containing several freshly grown nanostructures was scanned in the unoccupied state, even at lower biases (e.g., at +2 V), some of the molecular lines disappeared leaving a clean H-terminated dimer row, as observed in Fig. 10.10c. This disappearance of nanostructure is also ascribed to the reverse chain reaction [24]. It seems that the scan at a positive sample bias (scanning in the unoccupied state) facilitates the reverse chain reaction in the present case.

The contrast of nanostructure of adsorbed acetone molecules in an STM image strongly depends on the bias voltage applied during scanning. As shown in Fig. 10.10d, the nanostructure appeared as depressions at a lower sample bias of

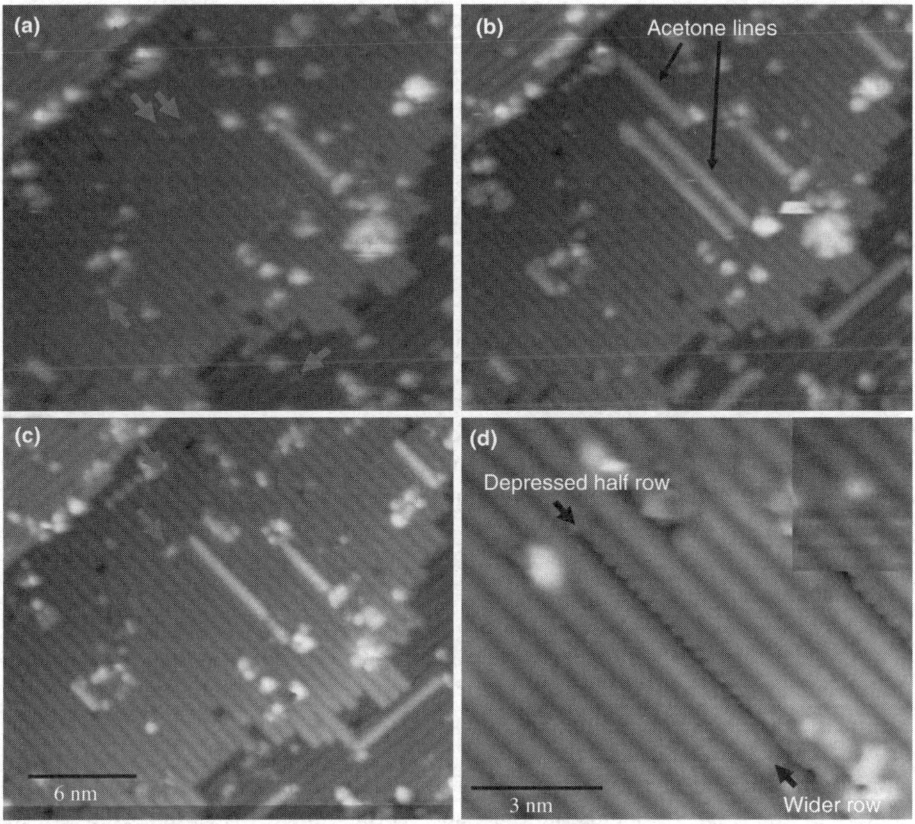

FIGURE 10.10 Selected area STM images of the Si(100)-(2×1)−H surface showing growth of nanostructures of adsorbed acetone molecules through chain reaction mechanism at 300 K. (a) Initial Si(100)-(2×1)−H surface containing several DB sites indicated by arrows. In addition to those DB sites, the surface also contains several defects and some preadsorbed molecules. (b) Surface "a" exposed to ∼0.02 L of acetone molecules at room temperature. (c) Surface "b" after scanning at positive sample bias. Arrows indicate the dimer rows from which the nanostructure disappeared. (d) Selected area of surface "c" scanned at −1.6 V sample bias. The inset shows the location of a DB site in a dimer row. $V_{sample} = -2.3$ V (a–c); $I = 0.2$ nA (a–d). Reproduced from Ref. 23 by permission of the American Chemical Society.

−1.6 V, whereas they appeared as bright protrusions at −2.3 V (Fig. 10.10a–c). Careful inspection of Fig. 10.10d reveals that one-half of the dimer row appeared as depression (indicated by depressed half-row), while the other half appeared similarly to that of a clean H-terminated dimer row. In addition, the dimer row adjacent to the depression side appears wider (indicated in Fig. 10.10d) than that of a row in clean terrace. The depressed half-row in Fig. 10.10d lies on the same side of the dimer row at which the initial DB site is located in Fig. 10.10a (as shown in the inset). This registry of acetone line on a dimer row gives the direct observation of undergoing reaction paths of adsorbing molecules.

Uniqueness of Acetophenone Acetone and benzophenone form similar 1D nanostructures along the dimer row, and the adsorbed states of all molecules in a nanostructure are identical. In contrast, the 1D nanostructure of acetophenone molecules grows both "along" and "across" the dimer row, as shown in Fig. 10.11 [28]. The nanostructures lying across the dimer rows are indicated by a black arrow (Fig. 10.11a). The contrasts of nanostructures along and across the dimer rows are similar in the STM image; this establishes that both cross-row and parallel-row nanostructures are formed by the same chemical species. The cross-row nanostructures of adsorbed acetophenone molecules are clearly distinguishable from those of allyl mercaptan molecules; a cross-row nanostructure formed by preexposure to a small amount ($\sim$0.02 L) of allyl mercaptan molecules is indicated by the gray arrow [20]. It is found that almost 30% of the 1D nanostructures formed by acetophenone molecules in Fig. 10.11a lie across the dimer rows. However, this percentage of cross-row nanostructures varies depending on the cleanliness of the surfaces. Apparently, the higher concentration of dark defect sites (missing dimer and/or dihydride) on the surface favors the higher number of cross-row nanostructures [20].

One can easily recognize that two different types of molecular arrangements exist in nanostructures along the dimer row (Fig. 10.11b). Some of the nanostructures resolved into many small protrusions, which are arranged in a zigzag pattern (indicated as "zigzag"), while others appear as uniform bar-like structure (indicated as "uniform"). These two different types of molecular arrangements can be seen together even within the same nanostructure; however, the boundary between the two arrangements cannot be clearly determined. It is found that almost 70% of nanostructures consist of zigzag arrangement of the small protrusions. In the case of nanostructures across the dimer rows, only the zigzag arrangements of small protrusions are observed, as shown in the lower right inset of Fig. 10.11b.

The CH_3COCH_3, PhCOPh, and $PhCOCH_3$ contain only the >C=O reactive group, and the nanostructures are formed through a common chain reaction mechanism involving the >C=O group and DB site of the substrate, as shown in Fig. 10.12 [28, 38]. Although the reactions leading to the formation of nanostructures by these molecules are basically similar, the directional selectivity of acetophenone molecules is different from that of acetone (CH_3COCH_3) and benzophenone (PhCOPh). Indeed, all the molecules that form well-defined nanostructures undergo chain reactions either along or across the dimer rows [28]. Hence, acetophenone is a unique example of an organic molecule that can lead to a DB-initiated chain reaction propagating both across and along the dimer rows on the Si(100)-(2×1)–H surface.

The unique behavior of the acetophenone is attributed to the formation of chiral carbon atom after adsorption (Fig. 10.12c), which did not happen for benzophenone and acetone molecules. Hence, depending on the spatial orientation of the CH_3 and Ph groups, the chiral carbon atoms may attain the absolute configuration of *R* or *S*. Thus, the creation of a chiral center allows the acetophenone molecules to attain one of two possible stacking sequences (*RSRS* and *RRR* or *SSS*) along the growth direction as shown schematically in Fig. 10.13.

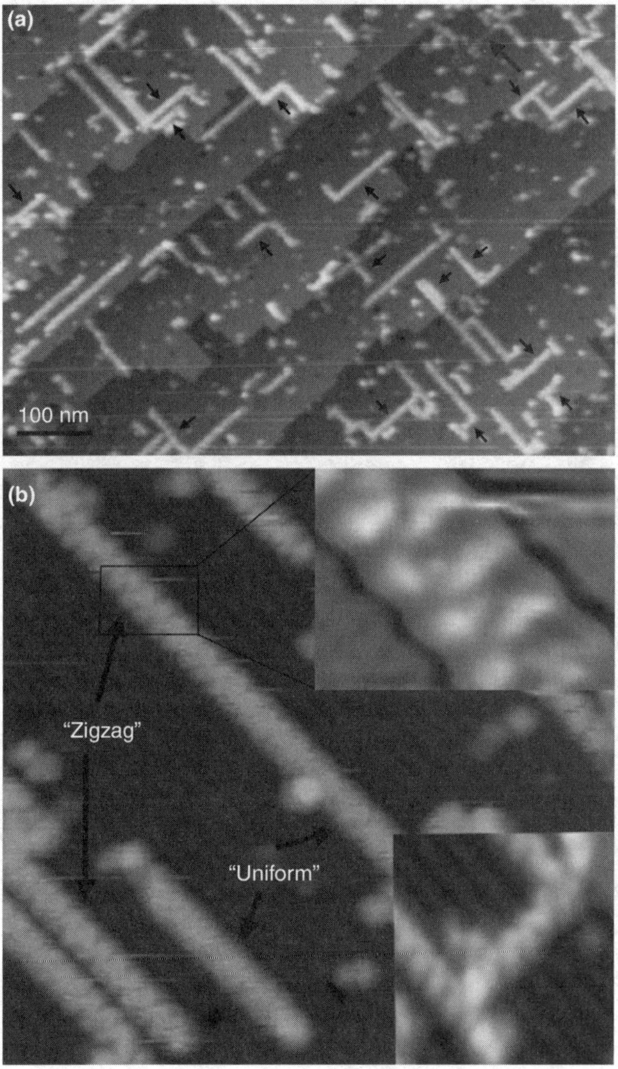

FIGURE 10.11 (a) STM images of the Si(100)-(2×1)–H surface exposed to ~10 L of acetophenone ($PhCOCH_3$) molecules at 300 K. Before acetophenone exposure, the surface was preexposed to a small amount (~0.02 L) of allyl mercaptan molecules. The black arrows indicate nanostructure grown across the dimer rows. The gray arrow indicates a cross-row nanostructure formed by allyl mercaptan molecules. (b) A high-resolution STM image of a few nanostructures of adsorbed acetophenone. Two different appearances of the nanostructures are indicated as "zigzag" and "uniform." Upper right inset: zoomed-in 3D view of a "zigzag" nanostructure. Lower right inset: high-resolution image of a cross-row nanostructure. Nanostructures along the dimer row are also seen. $V_{sample} = -2.1$ V; $I = 0.2$ nA. Reproduced from Ref. 28 by permission of the American Chemical Society.

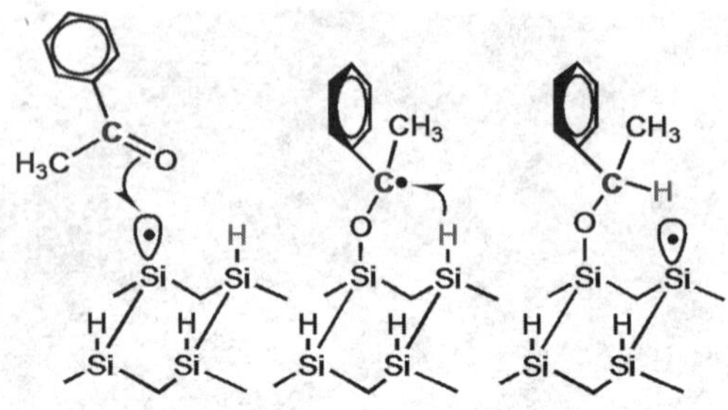

FIGURE 10.12 Schematic of the chain reaction mechanism of acetophenone and a DB site of the Si(100)-(2×1)–H surface along the dimer row.

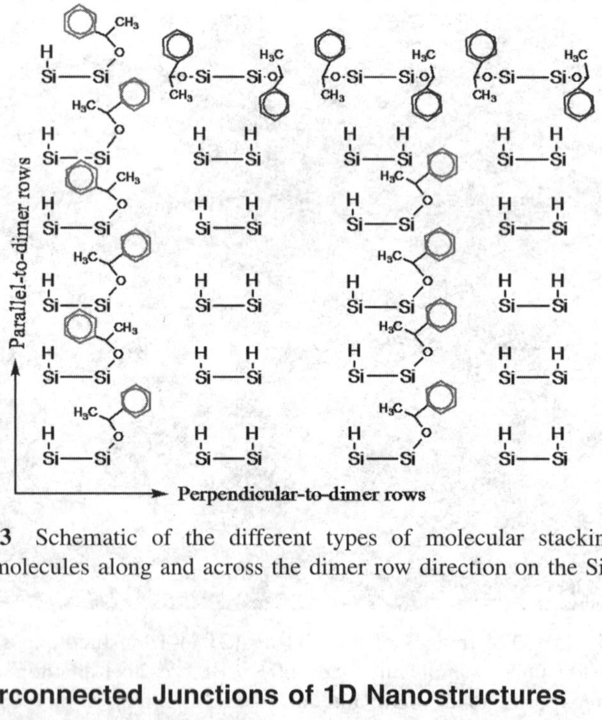

FIGURE 10.13 Schematic of the different types of molecular stacking of adsorbed acetophenone molecules along and across the dimer row direction on the Si(100)-(2×1)–H surface.

10.3.2 Interconnected Junctions of 1D Nanostructures

10.3.2.1 Perpendicular Junction As discussed above, the growth of 1D nanostructure can be directed not only along but also across the dimer rows on Si(100)-(2×1)–H by selecting the appropriate molecular systems [21]. Thus, control over the growth direction allows to fabricate the interconnected nanostructure network on the surface. Figure 10.14a shows an STM image of the Si(100)-(2×1)–H surface exposed to ~5 L of ALM at 300 K. Two nanostructure of unequal lengths, running perpendicular to the dimer rows, are formed by adsorbed ALM molecules through the chain reaction mechanism [20].

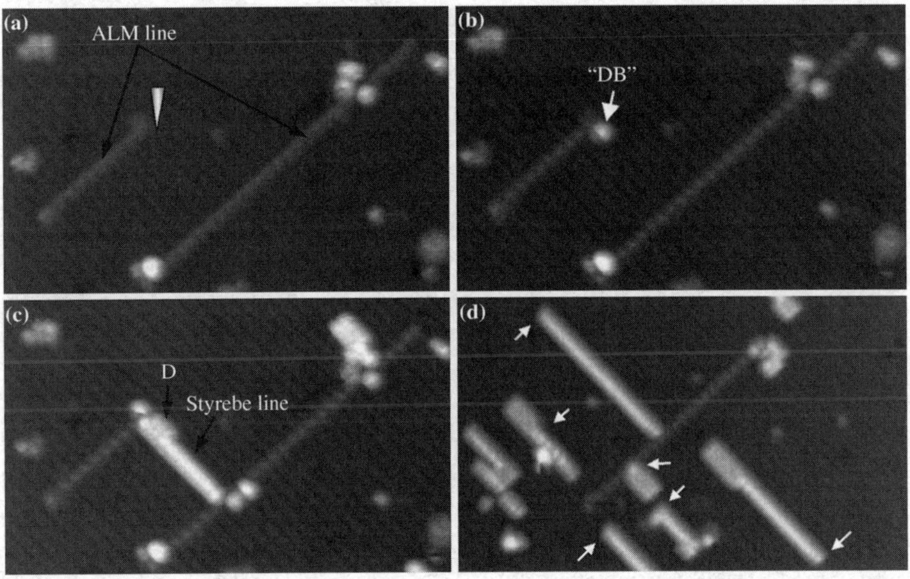

FIGURE 10.14 Selected area STM images of (a) the Si(100)-(2×1)−H surface containing two ALM nanostructures, (b) a dangling bond created using STM tip at the position shown in (a), and (c) the surface "b" exposed to ~5 L of styrene at room temperature. (d) The Si(100)-(2×1)−H surface containing an ALM nanostructure is exposed to 10 L of styrene after creating several DBs at the position indicated by arrows. $V_{sample} = -2.6$ V; $I = 0.2$ nA. Scan area: (a−c) 26 × 16 nm^2; (d) 32 × 20 nm^2. Reproduced from Ref. 21 by permission of the American Chemical Society.

Since there are several molecules that undergo the chain reaction along the dimer rows, it should be possible to connect those cross-row nanostructures by a parallel-row nanostructure. To achieve this, a single DB is created at the position shown in Fig. 10.14a using high-voltage pulse injection from the STM tip. The bright protrusion indicated by "DB" in Fig. 10.14b may be the unpaired DB or a styrene molecule (from the background) already reacted with the freshly created DB. When the surface shown by Fig. 10.14b is exposed to ~5 L of styrene, a bright nanostructure connecting both the ALM nanostructures appeared (Fig. 10.14c). This newly appeared bright nanostructure is formed by styrene molecules.

In order to grow a styrene nanostructure that perpendicularly connects a pre-existing ALM nanostructure, it is not essential to create DB adjacent to the ALM nanostructure as done in Fig. 10.14a. Since the styrene nanostructure can grow either side of the DB site along the dimer rows, a reaction initiated at a DB site located even a little away from the ALM nanostructure can eventually touch the ALM nanostructures. Figure 10.14d shows such growths of styrene nanostructures initiated by the DB sites located at different distances from an ALM nanostructure, in which arrows indicate the location of DB sites on the initial surface. It is observed that some of the styrene nanostructures have grown toward the ALM nanostructure and ultimately connected with it, while others have grown away from the ALM nanostructure.

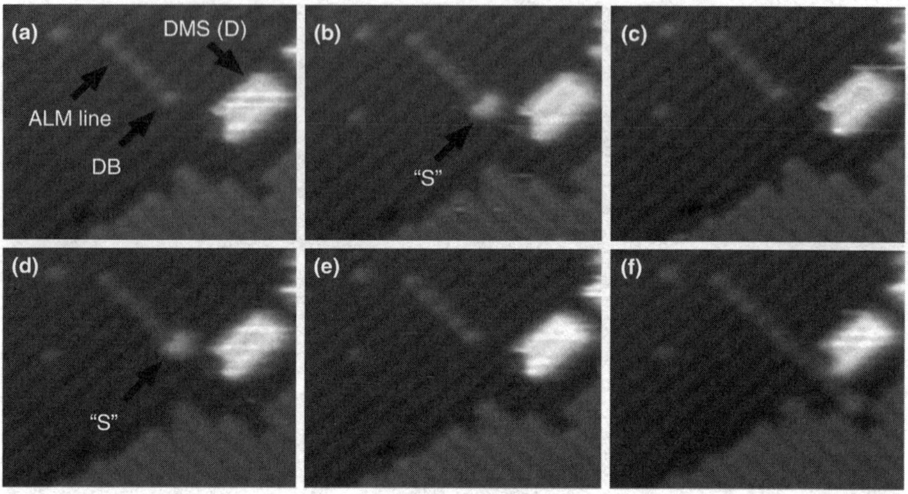

FIGURE 10.15 Selected area STM images of Si(100)-(2×1)–H surface showing (a) a growing ALM nanostructure with an end DB site and a nearby DMS doubled nanostructure, (b) the "a" exposed to 5 L of DMS, (c) 6 min after "b", (d) the "c" exposed to 7 L of DMS, (e) 6 min after "d", and (f) the "e" exposed to 0.5 L of ALM. $V_{sample} = -2.5$ V; $I = 0.2$ nA. Scan area: (a–f) $11.2 \times 9.6\,nm^2$. Reproduced from Ref. 21 by permission of the American Chemical Society.

Note that there exists a DB site at the end of the molecular nanostructure (growing nanostructure) where the growth has not yet been terminated by a preexisting DB or defect, among others [15, 21, 24]. Hence, instead of creating a new DB at the vicinity of the ALM nanostructure, the end DB site of a growing nanostructure should also be able to initiate a nanostructure growth along the rows if the surface is exposed to styrene or DMS. However, the repeated attempts of growing a 1D nanostructure from the end DB of a growing ALM nanostructure were not successful, as demonstrated in Fig. 10.15. Though a single styrene molecule (denoted as "s") is found to be adsorbed at the end DB site, no nanostructure is grown [21]. Instead, the adsorbed styrene at the end DB site desorbs in a few minutes as seen in Fig. 10.15c and e.

As noted above, acetone molecules undergo the most facile chain reaction with a DB site, which proceeds selectively on the Si(100)-(2×1)–H surface resulting in only a single 1D nanostructure along the dimer row. Indeed, the smaller size and higher reactivity of acetone molecules enable us to successively grow a parallel-row nanostructure from the end of a cross-row nanostructure simply by changing the feed of gas molecules into the reaction chamber [23]. The interconnected junction of 1D nanostructures fabricated from a single DB site is shown in Fig. 10.16. The fabrication of such a 1D nanostructure from the end of a freshly grown ALM nanostructure, without creating any additional DB site, is well reproduced.

An attempt to grow an ALM nanostructure from the end of a nanostructure grown by acetone dosing was also made but did not succeed [21]. This limits the formation of a more extended junction of interconnected nanostructures on the surface. The low probability of ALM molecules to undergo a cross-row chain

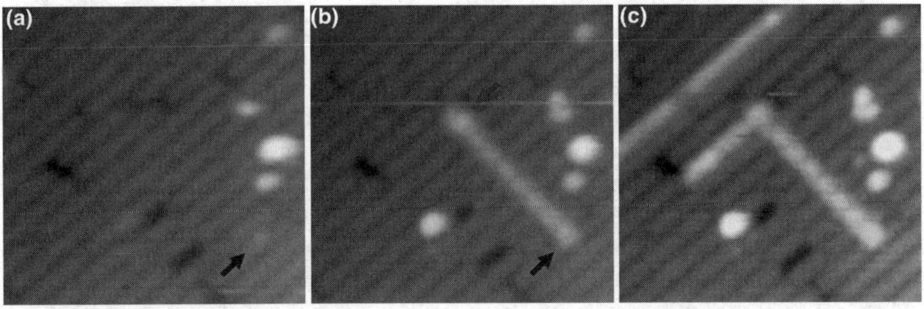

FIGURE 10.16 Successive growth of mutually perpendicular nanostructures by propagating a chain reaction from the cross-row direction to the parallel-row direction on the Si(100)-(2×1)–H surface at 300 K. (a) Initial Si(100)-(2×1)–H surface containing a DB site indicated by black arrow. (b) Surface "a" exposed to ~0.4 L of ALM molecules. The black and gray arrows indicate the initiation and end of ALM nanostructure growth, respectively. (c) Surface "b" exposed to ~0.2 L of acetone molecules. Scan area = 11 × 11 nm^2. $V_{sample} = -2.6$ V; $I = 0.2$ nA. Reproduced from Ref. 23 by permission of the American Chemical Society.

reaction or the steric hindrance induced by the nearby adsorbed acetone molecule may be the reason for this.

10.3.2.2 One-Dimensional Heterojunction In addition to growing of L-shaped junction of 1D nanostructure, a nanostructure consisting of different molecules can also be fabricated by alternate dosing of different molecules that undergo chain reaction in the same direction. The growth of 1D heteromolecular junctions consisting of styrene/CH_3–styrene and CF_3–styrene/CH_3O–styrene systems has been reported [19, 31]. Very recently, Walsh et al. [30] have demonstrated such heterostructure consisting of styrene and phenylacetylene (PA). Unlike styrene nanostructures, no doubled nanostructures are observed for PA. However, Fig. 10.17 reveals an instance of a heteromolecular doubled 1D

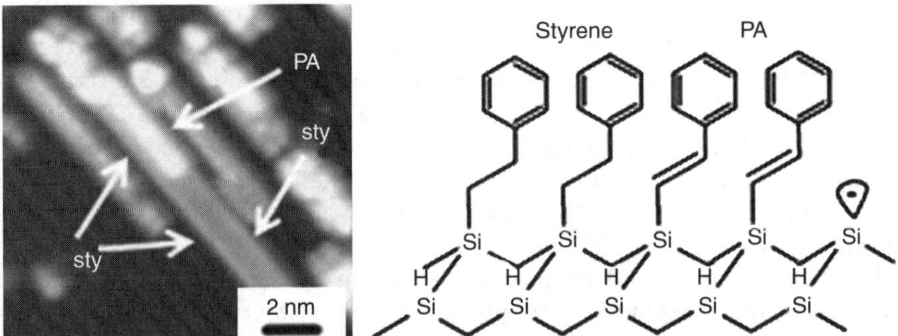

FIGURE 10.17 STM image showing the styrene–phenylacetylene heterostructures (left panel), and schematic model for a heterostructure (right panel) on Si(100)-(2 × 1)–H surface. Reproduced from Ref. 30 by permission of the American Chemical Society.

nanostructure. In this case, one side of the doubled nanostructure (the left side) is composed entirely of styrene, while the other side (the right side) consists of a block of styrene and a block of PA. Since only one binding site is available per silicon dimer for PA in the heteromolecular doubled nanostructure, each PA molecule must be bound to the surface via only one silicon atom in this case. It should be noted that the STM image of the PA portion of the doubled chain nanostructure appears identical to the other isolated PA nanostructures. Consequently, it can be concluded that the PA molecules in single 1D nanostructures are also bound to only one silicon atom on the surface, as shown in Fig. 10.17 (right panel) [30].

10.3.3 UPS of 1D Nanostructures on the Surface

The prospect of utilization of 1D nanostructure as a basic component of a molecular device essentially depends on the extent of interaction between the π-orbital and substrate electronic states, the π–π interaction between neighboring molecules in a nanostructure, and the effects of substituents on these interactions [40–42]. To understand these phenomena, it is essential to explore the valence states of such 1D nanostructures on the surface.

Valence electronic states of benzophenone, acetophenone, and acetone have been studied using UPS at room temperature [29]. Figure 10.18a(iii) shows the UPS spectrum of 1D structures of adsorbed benzophenone on the Si(100)-(2×1)–H surface at room temperature. The spectra of bare and H-terminated Si(100)-(2×1) are included as Fig. 10.18a(i) and (ii). The typical STM images of the surfaces corresponding to the spectra (ii) and (iii) are shown as Fig. 18b and c, respectively. All spectra are recorded with photon energy of 21.2 eV. The bare Si (100)-(2×1) surface is characterized by the surface state (SS) peak at ~0.8 eV below the Fermi level, as observed in the spectrum (i) [43]. For the H-terminated Si(100)-(2×1) surface (Fig. 10.18a(ii)), the peak at 0.8 eV disappeared and some weak features at higher binding energies (~1.9, 3.5, 5.6, and 7.6 eV) are observed, which is in agreement with the previous reports [44]. No peaks related to the DB sites appeared in the UPS spectrum because of very low concentration (roughly estimated to be ~0.4% of the surface silicon atoms) of the DB sites on the surface.

Although the concentration of the adsorbed molecules was estimated to be only ~6%, the UPS spectrum of these 1D nanostructures on H-terminated Si(100)-(2×1)–H surface clearly shows the intense peaks at 3.8, 6.8, 9.0, and 11.2 eV below the Fermi level [29]. Comparison of the UPS spectrum of benzophenone nanostructures on the surface (Fig. 10.18a(iii)) with that of the H-terminated surface (Fig. 10.18a(ii)) clearly indicates that these peaks arise from the nanostructure of the adsorbed benzophenone.

The UPS spectra of 1D nanostructures formed by acetophenone and acetone molecules help to explore the origin of the peaks observed for benzophenone nanostructures [29]. The comparison of the UPS spectra of acetone nanostructures with that of benzophenone (or acetophenone) nanostructures clearly indicates that the major peaks at 3.8, 6.8, 9.0, and 11.2 eV in the benzophenone spectrum arise due to the presence of the phenyl rings of the adsorbed molecules. The relative positions

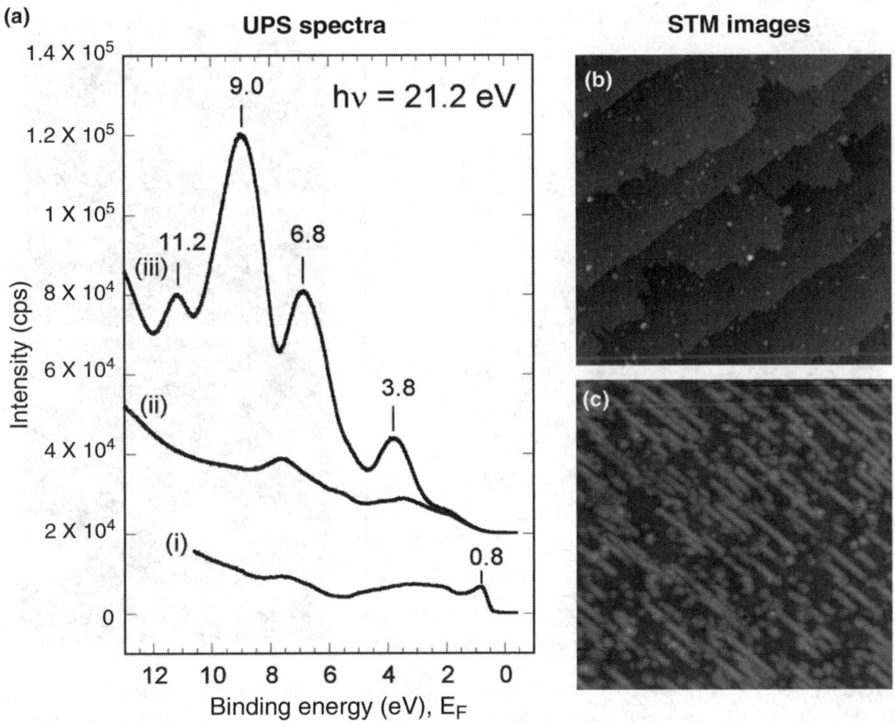

FIGURE 10.18 (a) UPS spectra of (i) bare Si(100)-(2×1), (ii) H-terminated Si(100)-(2×1), and (iii) one-dimensional nanostructures of adsorbed benzophenone on the Si(100)-(2×1)–H surface. For clarity, the base line of the spectra (ii) and (iii) was shifted upward by the same scale. The typical STM images of the surfaces corresponding to (ii) and (iii) are shown as (b) and (c), respectively. $V_{sample} = -2.4$ V (a) and –2.7 V (b). $I_{tunnel} = 0.2$ nA. Scan area = 80 × 80 nm^2. Reproduced from Ref. 29 by permission of the American Chemical Society.

of these peaks with respect to the higher binding energy (11.2 eV) peak seem very similar to that of solid benzophenone; that is, the energy of molecular orbitals related to phenyl rings does not change significantly after the adsorption of molecules on the Si(100)-(2×1)–H surface (Fig. 10.19). This suggests that the phenyl rings of the adsorbed molecule have little or no interaction with the surface silicon atoms. According to the assignments of the peaks observed for solid benzophenone and multilayer benzene on the silicon surface [44, 45], the peaks are qualitatively ascribed as 3.8 eV to Ph_{π}, 6.8 eV to $Ph_{\pi+\sigma}$, 9.0 eV to Ph_{σ}, and 11.2 eV to Ph_{σ} orbitals [45].

Although the UPS measurement does not give any direct evidence, one can predict a weak π–π interaction among the molecules of a nanostructure similar to (or a little higher than) the one that exists in benzophenone crystal. The relatively larger width of the lowest energy peaks and the absence of fine structure in Fig. 10.19a(iv) may relate to a higher π–π interaction between molecules in a nanostructure compared to

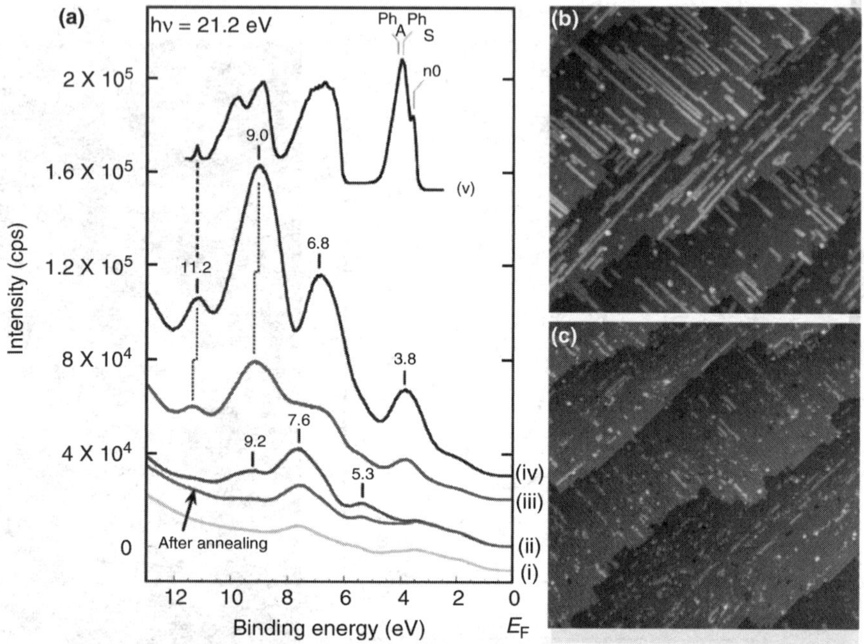

FIGURE 10.19 (a) UPS spectra of one-dimensional nanostructures of (i) H-terminated Si(100)-(2×1), (ii) adsorbed acetone (before and after annealing), (iii) acetophenone and (iv) benzophenone on the Si(100)-(2×1)–H surface at 300 K. The photoelectron spectrum of (v) solid benzophenone is also included for comparison. The spectrum of solid benzophenone was reproduced from Ref. 45 and aligned with the higher binding energy peak of benzophenone by the energy shift of ~1 eV. For clarity, the base line of the spectra was shifted. (b) and (c) show the STM images of the nanostructures of adsorbed acetone on the Si(100)-(2×1)–H surface and the same surface annealed to 600 K, respectively. Scan area = 80 × 80 nm^2. $V_{sample} = -2.5$ V; $I = 0.2$ nA. Reproduced from Ref. 29 by permission of the American Chemical Society.

the one that exists in benzophenone crystal. Indeed, a weak π–π interaction among the molecules separated by 3.8 Å has also been predicted by DFT calculation, where the dispersion of the π-band along the molecular assemblies was estimated to be ~0.4 eV [40].

10.4 CONCLUSIONS

Formation of organic nanostructures on semiconductor surfaces has particular interest from the perspective of molecular electronics. Unpaired dangling bond-initiated chain reaction has appeared as the most prominent and precise method for controlled fabrication of 1D organic nanostructure networks on the Si(100)-(2×1)–H surface. A variety of molecules with >C=C<, >C=O, –HC=O, and –C≡C– groups undergo DB-initiated chain reaction on the Si(100)-(2×1)–H surface and form 1D nanostructures on the surface. The successful growth of nanostructures is limited by the relative rate of forward and reverse chain reaction. The position, length, and composition, as well as the alignment of these nanostructures, are precisely controllable. Formation of different types of nanostructures, such as straight homogeneous, perpendicularly connected, and linearly connected heterogeneous nanostructures, has been demonstrated. The valence electronic states measurement suggests that the electronic states of phenyl π-orbitals stacked parallel to each other remain largely decoupled from the substrate electronic states, and the π–π interaction within the nanostructures formed by benzophenone is relatively weak. These studies regarding the controlled fabrication of 1D organic nanostructure networks on the H-terminated Si(100) surface are expected to have substantial impact on promoting further studies focusing on the functional nanoscale device.

ACKNOWLEDGMENT

This work is supported in part by the Grant-in-Aid for Scientific Research on Priority Areas "Electron transport through a linked molecule in nano-scale" (No. 17069006) from the Ministry of Education, Culture, Sports, Science and Technology (MEXT), Japan and the Program to Disseminate Tenure-Track System of the MEXT granted to Gunma University.

REFERENCES

1. Aviram, A.; Ratner, M. A. *Chem. Phys. Lett.* **1974**, *29*, 277.
2. Joachim, C.; Gimzewski, J. K.; Aviram, A. *Nature* **2000**, *408*, 541.
3. Nitzan, A.; Ratner, M. A. *Nature* **2003**, *300*, 1384.
4. (a) Ratner, M. *Nature* **2005**, *435*, 575. (b) Guisinger, N. P.; Yoder, N. L.; Hersam, M. C. *Proc. Natl. Acad. Sci. USA* **2005**, *102*, 8838.
5. Piva, P. G.; DiLabio, G. A.; Pitters, J. L.; Zikovsky, J.; Rezeq, M.; Dogel, S.; Hofer, W. A.; Wolkow, R. A. *Nature* **2005**, *435*, 658.
6. Eigler, D. M.; Schweizer, E. K. *Nature* **1990**, *344*, 524.
7. Stroscio, J. A.; Eigler, D. M. *Science* **1991**, *254*, 1319.
8. Lehn, J.-M. *Angew. Chem., Int. Ed. Engl.* **1988**, *27*, 89.
9. Lehn, J.-M. *Angew. Chem., Int. Ed. Engl.* **1990**, *29*, 1304.
10. Wolkow, R. A. *Annu. Rev. Phys. Chem.* **1999**, *50*, 413.
11. (a) Bent, S. F. *Surf. Sci.* **2002**, *500*, 879. (b) Hamers, R. J. *Nature* **2001**, *412*, 489.
12. Shen, T. C.; Wang, C.; Abeln, G. C.; Tucker, J. R.; Lyding, J. W.; Avouris, Ph.; Walkup, R. E. *Science* **1995**, *268*, 1590.
13. Boland, J. J. *J. Vac. Sci. Technol. A* **1992**, *10*, 2458.
14. Linford, M. R.; Fenter, P.; Eisenberger, P. M.; Chidsey, C. E. D. *J. Am. Chem. Soc.* **1995**, *117*, 3145.
15. Lopinski, G. P.; Wayner, D. D. M.; Wolkow, R. A. *Nature* **2000**, *406*, 48.

16. Kruse, P.; Johnson, E. R.; DiLabio, G. A.; Wolkow, R. A. *Nano Lett.* **2002**, *2*, 807.
17. Basu, R.; Guisinger, N. P.; Greene, M. E.; Hersam, M. C. *Appl. Phys. Lett.* **2004**, *85*, 2619.
18. DiLabio, G. A.; Piva, P. G.; Kruse, P.; Wolkow, R. A. *J. Am. Chem. Soc.* **2004**, *126*, 16048.
19. Kirczenow, G.; Piva, P. G.; Wolkow, R. A. *Phys. Rev. B* **2005**, *72*, 245306.
20. Hossain, M. Z.; Kato, H. S.; Kawai, M. *J. Am. Chem. Soc.* **2005**, *127*, 15030.
21. Hossain, M. Z.; Kato, H. S.; Kawai, M. *J. Phys. Chem. B* **2005**, *109*, 23129.
22. Basu, R.; Kinser, C. R.; Tovar, J. D.; Hersam, M. C. *Chem. Phys.* **2006**, *326*, 144.
23. Hossain, M. Z.; Kato, H. S.; Kawai, M. *J. Am. Chem. Soc.* **2007**, *129*, 12304.
24. Hossain, M. Z.; Kato, H. S.; Kawai, M. *J. Am. Chem. Soc.* **2007**, *129*, 3328.
25. Zikovsky, J.; Dogel, S. A.; Haider, A. B.; DiLabio, G. A.; Wolkow, R. A. *J. Phys. Chem. A* **2007**, *111*, 12257.
26. Dogel, S. A.; DiLabio, G. A.; Zikovsky, J.; Pitters, J. L.; Wolkow R. A. *J. Phys. Chem. C* **2007**, *111*, 11965.
27. Wang, Q. H.; Hersam, M. C. *J. Am. Chem. Soc.* **2008**, *130*, 12896.
28. Hossain, M. Z.; Kato, H. S.; Kawai, M. *J. Am. Chem. Soc.* **2008**, *130*, 11518.
29. Hossain, M. Z.; Kato, H. S.; Kawai, M. *J. Phys. Chem. C* **2009**, *113*, 10751.
30. Walsh, M. A.; Walter, S. R.; Bevan, K. H.; Geiger F. M.; Hersam, M. C. *J. Am. Chem. Soc.* **2010**, *132*, 3013.
31. Piva, P. G.; Wolkow, R. A.; Kirczenow, G. *Phys. Rev. Lett.* **2008**, *101*, 106801.
32. Cho, H. J.; Oh, H. D.; Kleinman, L. *Phys. Rev. B* **2002**, *65*, 081310.
33. Kang, J. K.; Musgrave, C. B. *J. Chem. Phys.* **2002**, *116*, 9907.
34. Takeuchi, N.; Kanai, Y.; Selloni, A. *J. Phys. Chem. C* **2010**, *114*, 3981.
35. Ferguson, G. A.; Than, C. T. L.; Raghavachari, K. *J. Phys. Chem. C* **2009**, *113*, 18817.
36. Choi, J. H.; Cho, J. H. *Phys. Rev. Lett.* **2009**, *102*, 166102.
37. Ferguson, G. A.; Than, C. T. L.; Raghavachari, K. *J. Phys. Chem. Lett.* **2010**, *1*, 679.
38. Pitters, J. L.; Dogel, I.; DiLabio, G. A.; Wolkow, R. A. *J. Phys. Chem. B* **2006**, *110*, 2159.
39. Tong, X.; DiLabio G. A.; Clarkin, O. J.; Wolkow, R. A. *Nano Lett.* **2004**, *4*, 357.
40. (a) Rochefort, A.; Boyer, P. *Appl. Phys. Lett.* **2006**, *89*, 092115. (b) Rochefort, A.; Boyer, P.; Nacer B. *Org. Electron.* **2007**, *8*, 1.
41. Rochefort, A.; Martel, R.; Avouris, Ph. *Nano Lett.* **2002**, *2*, 877.
42. Geng, W. T.; Oda, M.; Nara, J.; Kondo, H.; Ohno, T. *J. Phys. Chem. B* **2008**, *112*, 2795.
43. Gokhale, S.; Trichberger, P.; Menzel, D.; Widdra, W.; Dröge, H.; Steinrück, H.-P.; Birkenheuer, U.; Gutdeutsch, U.; Rösch, N. *J. Chem. Phys.* **1998**, *108*, 5554.
44. Ciraci, S.; Butz, R.; Oellig, E. M.; Wager, H. *Phys. Rev. B* **1984**, *30*, 711.
45. McAlduff, E. J.; Bunbury, D. L. *J. Electron Spectrosc. Relat. Phenom.* **1979**, *17*, 81.

CHAPTER 11

Formation of Organic Monolayers Through Wet Chemistry

DAMIEN AUREAU AND YVES J. CHABAL

11.1 INTRODUCTION, MOTIVATION, AND SCOPE OF CHAPTER

11.1.1 Background

Wet chemical cleaning has always been an intrinsic part of semiconductor processing [1, 2], but only relatively recently has wet chemistry been used to modify semiconductor surfaces for sensitive applications. Wet chemical functionalization is attractive because it can be performed in reasonably manageable environments and is scalable. Wet chemical cleaning, for instance, which constitutes ~30% of silicon processing in microelectronics, can be performed for entire cassettes of wafers in regular clean rooms. Clean chemicals have also been developed for such large-scale use. In general, wet chemical processing is simpler and more versatile than ultrahigh vacuum (UHV) methods, and can often lead to more perfect surfaces than what can be achieved in UHV [3–8].

Silicon has been central to the microelectronics industry for one main reason. Its oxide is stable and the interface between silicon and its oxide is electrically much better than for any other semiconductor surface, in particular germanium [9]. It is therefore not surprising that the early work on silicon surface functionalization has focused on attaching organic molecules to *silicon oxide*, using silanization for instance [10]. Typically, silanization is based on hydrolyzing molecules such as triethoxysilanes or trichlorosilanes with hydroxyl-terminated silicon oxide in the presence of water molecules. It has become apparent, however, that this process is strongly dependent on the density of initial surface hydroxyl groups, the amount of water molecules present in their vicinity, and temperature, leading to uncontrolled attachment [11]. These molecules also have a tendency to polymerize, making the formation of a single self-assembled monolayer (SAM) virtually impossible.

Functionalization of Semiconductor Surfaces, First Edition.
Edited by Franklin (Feng) Tao and Steven L. Bernasek.

An alternative process involves phosphonates that are not water sensitive and can therefore be attached on native oxide surfaces in ambient environments [12, 13]. However, this process is not yet well understood and does not readily lead to well-defined layers [13, 14].

The performance of the sensors in terms of reliability, efficiency, and sensitivity generally depends on the control of the immobilization chemistry of the probes [15]. Yet, most electronic applications such as biological and chemical sensors to date are using silane [16] or phosphonate [17] chemistry because of the electrical quality of the initial Si/SiO_2 interface, thereby suffering from poor SAM stability.

The work described in this chapter is based on the premise that *silicon oxide* is not a good starting point for controlled functionalization, primarily because of the poor control of and film quality resulting from silanization. Indeed, while the Si/SiO_2 interface is electrically excellent, the SiO_2/organic layer interface is unstable in aqueous environments needed for biological or other applications. The alternative to oxide would be a surface that is stable enough to prevent reoxidation under ambient conditions, yet that can react with organic molecules to form strong covalent bonds. In other words, a metastable atom or molecule must be attached to a clean surface, which is amenable to controlled further reaction [18]. For silicon, hydrogen and halogens have been considered. They can be attached via reasonably straightforward wet chemical procedures and can be further reacted in controllable fashion. Hydrogen-terminated silicon surfaces, for instance, can be produced very simply by HF etching, as described below in some detail.

Another substrate of interest is germanium because it exhibits noted advantages over silicon for microelectronics. Its hole carrier mobility is four times that in silicon, which is important for high-speed circuits and is of interest in CMOS technology where the p-channel component of Si has traditionally had less than ideal performance. Further, the 0.67 eV bandgap of Ge allows the absorption of infrared radiation, which makes Ge, or a SiGe alloy, a suitable rear absorber in a multijunction solar cell [19]. Methods yielding low defect density Ge surfaces have therefore been investigated. However, the high surface state density and the instability and water solubility of Ge oxide have proven to be significant drawbacks to the development of Ge-based technology [20]. Germanium surfaces can be hydrogen terminated in HF but they are much less stable than H-terminated silicon surfaces [21, 22], leading to chemically imperfect surfaces upon further processing. Halogenation of germanium surfaces appears to be more stable than H-terminated Ge [23, 24]. Sulfide passivation eliminates Ge oxides, but can lead to the formation of a GeS glass [25]. For completeness, however, comparison between wet chemical modification of germanium and silicon will be given in this chapter [9, 22, 26].

For semiconductors other than silicon and germanium, the situation is more complicated. Silicon carbide features a very stable native or thermally grown silicon oxide. However, removal of the oxide does not lead to hydrogenated or halogenated surfaces. For instance, HF etching leads to the formation of hydrophilic, hydroxyl-terminated surfaces [27] that cannot be readily reacted with organic molecules. In contrast to silicon and silicon carbide, III–V semiconductors have poor native oxides and cannot be easily hydrogen terminated in HF [28]. Instead, they have been passivated by group VI atoms (other than oxygen) or halogens. In particular, sulfur appears to stabilize GaAs and InP surfaces well [29], but the details of such passivation are not well understood. Therefore, organic functionalization of sulfur-passivated GaAs is not as well defined and controlled as needed for most devices.

Surface cleaning and termination constitute the first steps of surface functionalization. For silicon, the most common approach combines aggressive cleaning/oxidizing with peroxides and acids, followed by hydrogen termination obtained by HF etching. To better appreciate the issues associated with stabilizing semiconductor surfaces with hydrogen, it is therefore useful to consider the formation process and the stability of H-terminated *silicon* surfaces.

11.1.2 Formation of H-Terminated Silicon Surfaces

The unique properties of HF-etched Si surfaces have been recognized over a half-century but the mechanism leading to these hydrophobic surfaces has only been understood a little over two decades. It is now clear that hydrogen and not fluorine terminates the surface [30], and is responsible for hydrophobicity, chemical resistance, and low density of electronic states (i.e., high recombination velocity) [31]. The initial confusion and controversy about H versus F termination was understandable because a trace amount of fluorine is typically found on HF-etched surfaces and the Si–F bond is much stronger ($\sim$5 eV) than the Si–H bond ($\sim$3.5 eV). The key mechanism was first proposed by Ubara et al. [30, 32] on microcrystalline Si [33], and later confirmed by *ab initio* calculations [4]. In this mechanism summarized in Fig. 11.1, removal of the last oxygen atom in the SiO_2 overlayer does indeed lead to the formation of Si–F. However, although this bond is very strong, it is also very polar. The strong polarity polarizes the Si–Si back-bond of Si–Si–F, rendering it vulnerable to further attack by HF. The calculations of Trucks et al. [4] showed that the barrier for attaching H to the back Si of Si–Si–F and releasing the surface Si in the form of SiF_4 was only $\sim$1 eV, without including the facilitating effects of solvation. The resulting H-terminated surface is nonpolar and therefore quite resistant to further chemical attack by HF (i.e., there is no further etching in HF solution once the

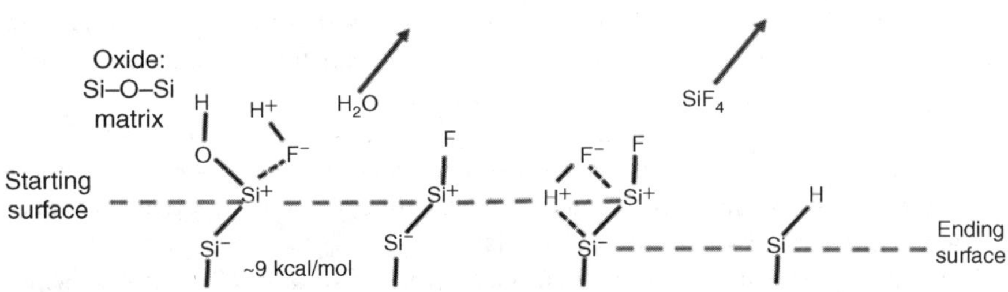

FIGURE 11.1 Mechanism leading to the formation of H-terminated silicon surface by HF etching. The last step of oxygen removal from SiO_2 involves HF attack of the Si–O bond, with removal of OH as H_2O and termination of the surface Si atom with fluorine. Further attack of the polarized $Si^{\delta-}$–$Si^{\delta+}$ leads to H termination [30].

surface is H-terminated) or by other chemical species as summarized later in this introductory section.

An important aspect of HF etching is the possibility to induce further etching (i.e., after the surface is H-terminated) *when heavily buffered HF solutions are used* (pH ~8) [3]. In such solutions (e.g., NH_4F), Si–Si bonds can be attacked slowly and selectively. As shown by several studies [5, 7, 8, 34, 35], the new etching mechanism involves the OH^- ions or related complexes with preferential removal of adstructures and attack of steps, leading to highly structured surfaces. In particular, atomically flat and perfectly monohydride-terminated Si(111) surfaces are obtained on well-oriented ⟨111⟩ Si surfaces, atomically straight and H-terminated steps are formed on vicinal ⟨111⟩ Si surfaces [8, 35], and well-defined flat surfaces (with double layer structures) [36] result from etching in NH_4F. The main mechanism proposed for the formation of atomically flat, monohydride-terminated surfaces is based on site-specific reactions [7], leading to a step flow etching. The exact chemical reactions have not been fully described, but it is clear that the *surface structure* plays a key role. The importance of surface structure has been emphasized here because it is central to surface chemistry and therefore surface chemical functionalization, as will be dramatically illustrated in section 11.6 of this chapter. It is also important when considering the stability of H-terminated surfaces as summarized below.

11.1.3 Stability of H-Terminated Silicon Surfaces

Stability and selective reactivity of H-terminated surfaces are essential components of the functionalization process. While stability of chemically passivated Si surfaces will be addressed in more detail in Chapter 12, we highlight here the essential findings for H-terminated Si surfaces.

It is well known that H-terminated Si surfaces, while mostly stable in clean solutions or gases, are degraded as soon as they become exposed to air. For instance, the surface recombination velocity increases by an order of magnitude when a Si sample is removed from the HF solution [31], and some initial oxidation is often observed after several hours in air. Yet, series of experiments performed in very well-controlled environments have revealed that H-terminated Si surfaces are completely stable in pure O_2, H_2O, and N_2 gases [1, 37]. Studies on flat and vicinal H-terminated Si(111) have shown that reactions with O_2, NH_3, and H_2O are initiated at steps, dependent on the step structure, and finally takes place on flat ⟨111⟩ terraces at temperatures above 300°C [38].

These findings indicate that, at room temperature, H-terminated Si surfaces are completely stable in clean gaseous environments even at atmospheric pressures. Degradation in air is therefore due to radicals, ozone, or other reactive species that then leave the surface susceptible to oxidation.

The stability of H-terminated Si surfaces in solution is more complex because solvation effects can lower the reaction barriers, the attempt frequencies are higher than those in vapor phase, and electrochemical effects are possible (hence the potential relevance of doping). For both gaseous and liquid environments, surface illumination can also modify the surface stability of H-terminated Si surfaces [39, 40].

11.1.4 Approach

This chapter is focused on the attachment of well-defined self-assembled monolayers. We therefore do not consider the large body of work in which molecules have been grafted on semiconductor surfaces when the quality of the layer is questionable or ill-defined, such as the case when attachment might depend on surface oxidation. This criterion is what excluded the consideration of silane grafting on silicon oxide, as mentioned above. Therefore, for all the SAMs and substrates considered here, a critical aspect of the wet chemical methods is the purity of chemicals and processing conditions. This is particularly important for advanced applications discussed in this chapter. For instance, there is often a competition between controlled modification and the tendency of surfaces (particularly silicon surfaces) to oxidize. Consequently, for most of the approaches described in this chapter based on anhydrous chemistry, the control of the environment is as important as the purity of chemicals themselves. In many cases described here, the use of Schlenk lines and glove boxes is therefore necessary to ensure that oxygen and moisture are minimized and deoxygenation of solutions by Ar bubbling is routinely performed.

Several excellent reviews of organic modifications of H-terminated Si surfaces have been published in the past decade [26, 41–46]. Many grafting mechanisms that are operational for silicon [26, 43, 44] or germanium [22, 26] surfaces of varied morphologies are similar, but there are important differences between flat surfaces [42, 45] and porous silicon [41] that need to be considered. The work described in this chapter specifically addresses functionalization of *atomically flat* Si(111) surfaces. Consequently, the focus is on the reaction of the Si–H bond itself rather than the attack of the Si–Si back-bonds during functionalization (which had to be invoked for porous Si functionalization) [41]. Although a subset of the mechanisms described in these excellent reviews is summarized, the role of structure (e.g., steric interactions) is emphasized here. In particular, the consequence of functionalizing an atomically flat surface will be illustrated in the last section of this chapter and shown to be responsible, for instance, for achieving F termination upon HF etching.

11.1.5 Outline

In this chapter, we first describe the main experimental methods that are particularly relevant to understanding the mechanisms of chemical functionalization. We critically review vibrational and photoelectron spectroscopy for chemical characterization, contact angle measurements and ellipsometry for determining hydrophobicity and index of refraction of the organic layers, photoluminescence and electrical measurements sensitive to electronic surface states, and atomic force microscopy (AFM) and low-energy electron diffraction (LEED) for monitoring substrate surface morphology and its stability during processing.

We then approach surface functionalization from a chemist point of view, bringing forth wet chemical methods to induce reactions such as catalyst-induced hydrosilylation, and summarizing how they have been applied to surfaces. Since the role of a catalyst is to activate a double (or triple) bond by temporarily tying up a dangling bond, similar reactions can be induced by the formation of a radical, either at the surface or on the alkene molecules. This can be optically or thermally activated by energetic photons or phonons that can directly break the surface Si–H bond or indirectly activate the reaction.

We first review the use and role of catalysts, and then the role of light and heat to activate the reactions of alkenes with H-terminated silicon surfaces, including internal surfaces of porous silicon. The use of Grignard or organolithium reagents is a natural extension of these ideas and is based on the nucleophilic character of the carbon involved in organometallic reagents able to react with the electrophilic silicon surface. A section is therefore devoted to Grignard functionalization of Si and Ge surfaces, including electrochemical or two-step methods via halogenated silicon surfaces. Although catalysts, photons, and Grignard molecules are effective to functionalize H-terminated surfaces, many of the relevant surface reactions can be thermally activated and then proceed through standard chemical transformations that are very specific to surfaces. Such surface reactions are much more dependent on steric interactions, local surface structures, and the chemical nature of the surface back-bonds than on thermodynamics driving solution reactions. The last part of this chapter is therefore devoted to thermally induced reactions on H-terminated silicon to form Si–O bonds through direct reactions with alcohols. This section shows how F-terminated silicon surfaces can be produced by HF etching, without however rekindling a 20-year-old controversy!

The chapter outline is graphically illustrated in Fig. 11.2.

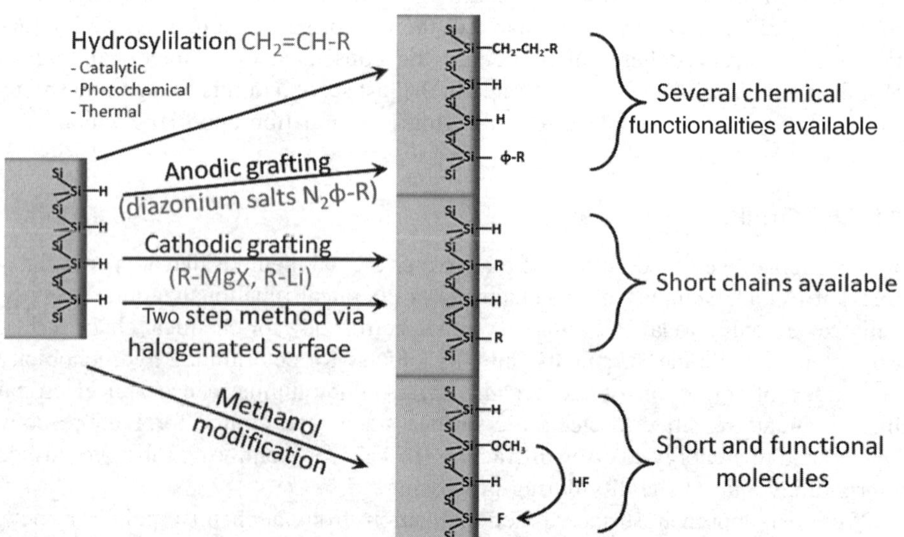

FIGURE 11.2 Schematic illustration and summary of the chemical modification pathways described in this chapter for monolayer functionalization of atomically flat Si–H (111).

11.2 TECHNIQUES CHARACTERIZING WET CHEMICALLY FUNCTIONALIZED SURFACES

As described in Chapter 2, reasonably well-established experimental techniques have been used for characterizing functionalized semiconductor surfaces. However, the information that can be derived from such techniques has not always been appreciated within the community, sometimes leading to overinterpretation. The goal of this section is therefore to critically discuss the main techniques typically used for understanding wet chemical surface functionalization, starting with spectroscopic techniques. In the end, it will be apparent that combining several techniques is important to gather the information necessary to guide the synthesis and to understand the self-assembly mechanisms.

11.2.1 X-Ray Photoelectron Spectroscopy

Detection of core levels of substrate atoms (Si, Ge, III–V), carbon atoms of the organic layer, head group atoms (N, O, S, etc.), and in some cases metal atoms can provide a quantitative measure of organic layer coverage and degree of chemical bonding. In particular, core level shifts can confirm the formation of interfacial Si–C bonds, the integrity of organic chain C–H_x bonds, and so on. In all reported work, XPS analysis is performed *ex situ*, that is, after transport of the sample in air from a glove box to the ultrahigh vacuum chamber. Issues associated with hydrocarbon contamination (adventitious carbon) on top of the organic layer and with contamination during evacuation of the load lock therefore need to be addressed since they can be serious in some cases (e.g., when the head group offers affinity to hydrocarbons, all surfaces contain adventitious carbon and oxygen traces). The main technical issues in the interpretation of the spectra revolve around the identification of impurities such as adventitious molecules containing carbon and oxygen, and the difficulty to detect and identify low-level oxidation of the substrate surface.

In the Si 2p region, it is also difficult to determine the surface stability just by studying the suboxides at the silicon surface. For instance, if a sample prepared in a glove box is quickly transferred to the vacuum chamber for XPS analysis, the surface may appear oxide free (i.e., stable), when in reality, it may not be thoroughly passivated and could oxidize further with time. In addition, the possible degradation of organic layers under intense X-ray irradiation by secondary electron emission must be checked by examining the time dependence of spectra.

The example of methyl-terminated silicon surfaces highlights the difficulty to get quantitative information just based on C 1s spectra, due to the contamination from the carbon atoms involved in the layer actually grafted on the surfaces. A low binding energy component ($\approx$284.3 eV) has been attributed to photoelectrons from carbon atoms directly bonded to the silicon substrate. Methyl-terminated surfaces, which can be prepared from CH_3MgX either by one-step anodic grafting or by two-step reaction via chlorination, exhibit varying amounts of adventitious carbons as determined by XPS (i.e., with C 1s peaks at energies higher than 285 eV) [47, 48]. For instance, Fig. 11.3 shows the C 1s region of a freshly prepared $SiCH_3$ surface

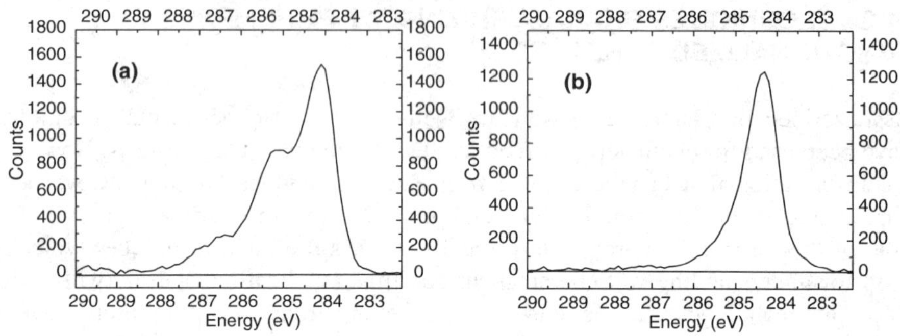

FIGURE 11.3 X-ray photoelectron spectra of the C 1s core level for methyl-terminated Si(111) surfaces prepared by anodic grafting of methylmagnesium according to the method presented in Section 11.4: (a) after rinsing in several solvents; (b) after annealing at 450°C in UHV for 5 min. Only carbon atoms linked to silicon (E_b < 285 eV) are observed after annealing.

(a) and after a 5 min annealing at 450°C (b), illustrating that physisorbed carbon species (contamination) can be removed by annealing [48, 49].

11.2.2 Infrared Absorption Spectroscopy

Several configurations are possible to perform IR spectroscopy. Much work has been done using multiple internal reflection (MIR) spectroscopy [50], where the organic layer is probed multiple times by the evanescent field. This configuration, which requires a special polishing of the edges, provides good sensitivity in a spectral range limited by the substrate absorption, and is particularly useful for polarization studies because all the components of the electric field are strong for typical internal angles (~45°) [51, 52]. It is a powerful method to perform *in situ* studies on semiconductor liquid interfaces [53, 54]. The main drawback for Si is that the spectral range is limited to frequencies above 900–1500 cm^{-1} depending on the sample length due to the substrate multiphonon absorption, which precludes the detection of Si–H bending vibrations (~620–650 cm^{-1}) and makes difficult the detection of Si–O–Si bonds (~950–1200 cm^{-1}). This geometry has therefore been successfully used mostly for characterizing Si–H, C–H_x, and C=O stretch modes [54].

A simple transmission geometry, typically performed at an incidence angle close to the Brewster angle [55], alleviates this problem and is ideal for *in situ* spectroscopy *during* measurements of the stability of organic layers in controlled gaseous environments as a function of temperature. However, good S/N is necessary when using transmission IR (TIR) spectroscopy because the IR beam interacts only *once* with the front and back of the sample [56]. Both MIR and TIR require double-side polished surfaces and take advantage of double-side functionalization.

An alternative to MIR that extends the spectral range and does not require double-side polished samples is an ATR technique whereby a Ge prism is brought in close contact with the surface under study (pressure needed) and an IR beam is internally

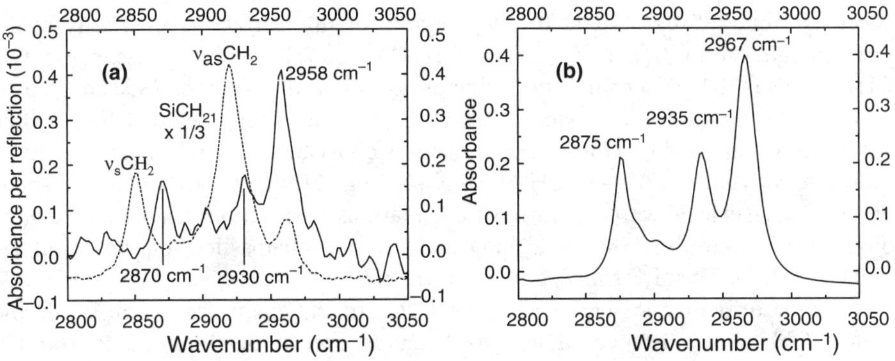

FIGURE 11.4 Infrared spectra in the ν_{CH} region of (a) Si–CH_2–CH_2–CH_3 (solid line) and Si–$C_{10}H_{21}$ (dashed line) surfaces obtained by MIR and (b) liquid Cl_3–Si–CH_2–CH_2–CH_3 obtained in transmission.

reflected within the Ge prism. The semiconductor surface in close proximity is then probed by the evanescent field. In this configuration, the incidence angle within the Ge must be larger than $\sin^{-1}(n_1/n_2)$, where n_1 is the refraction index of the semiconductor under study (e.g., $n_1 = 3.45$ for Si) and $n_2 = 4$ is the refraction index of Ge. In general, $\theta \sim 65°$, which is sufficient for $n_1 < 3.8$. This configuration provides a substantial enhancement for the components of vibrations perpendicular to the interface [51, 57]. Consequently, polarization studies are not well suited and some information is missing (or greatly suppressed) [27].

An intrinsic limitation of IR spectroscopy is the need for a reference sample or surface. In contrast to all other spectroscopic techniques, the raw spectrum is a measure of the transmitted intensity and not the absorbed radiation. The absorption spectrum needs to be extracted using a reference sample with a similar bulk substrate and a controlled surface composition. This process is a source of difficulties. It is also difficult to distinguish contributions of the head groups and potential physisorbed molecules (source of adventitious carbon) because these species are not well oriented, hindering the usefulness of studies based on polarized radiation.

Besides technical issues with the choice of geometries, IR spectroscopy faces issues associated with interpretation and analysis. For instance, the assignment of the CH_x region is not straightforward and can lead to errors, especially for short organic layers. In the case of nonfunctional alkyl chains, the CH_x region is usually fit with four distinct bands, the asymmetric and symmetric modes of CH_2 ($\nu_{as} \approx 2920\,cm^{-1}$ and $\nu_s \approx 2850\,cm^{-1}$) and CH_3 ($\nu_{as} \approx 2960\,cm^{-1}$ and $\nu_s \approx 2870\,cm^{-1}$). In reality, this spectral region involves Fermi resonances of the overtones of the deformation modes of both CH_2 and CH_3, which need to be taken into account. The Fermi resonance requires the addition of a fifth component at $\sim 2900\,cm^{-1}$. This situation is further complicated by the fact that the manifestation of these Fermi resonances depends on the chain length! Calculations must be done to take into account these various effects.

To illustrate this issue, we consider here the case of the propyl chains (three carbons) attached to the silicon, Si–CH_2–CH_2–CH_3. Figure 11.4a shows the infrared spectrum obtained by MIR in the ν_{CH} region for this surface (prepared from the propylmagnesium Grignard CH_3–CH_2–CH_2–MgX). While this chain contains two CH_2 groups in the alkyl chains, no IR bands appear around 2920 or 2850 cm^{-1}, as is typically observed for a longer chain ($SiC_{10}H_{21}$, as shown with a dashed-line spectrum in Fig. 11.4a). This latter spectrum is consistent with the spectra of liquid trichloropropylsilane, Cl_3–Si–CH_2–CH_2–CH_3 (Fig. 11.4b). In particular, the shape of the bands has been well explained by calculations [58]. The origin of this dramatic difference between the short and long hydrocarbon chains lies squarely with the reduced role of Fermi resonance in the former case.

Another difficulty arises because the CH_x stretch modes are often weak and undetectable. For instance, there has been no report of these modes on flat methoxylated surfaces until recently [59, 60], although they had previously been observed in porous silicon [61]. In general, the 1250 cm^{-1} deformation mode of the methyl group is easier to identify and has therefore been utilized to characterize SAMs on flat surfaces [59, 60, 62].

11.2.3 Secondary Ion Mass Spectrometry

There have been few studies of monolayers adsorbed on semiconductors using secondary ion mass spectrometry (SIMS), partly because alkylsilane monolayers generate a limited set of low mass fragment ions that are not structurally significant [63, 64]. This is because there is extensive cross-linking between the adsorbates, which leads to chain fragmentation during the SIMS process [63]. However, this is not the case for 1-alkene [65], 1-alkyne [65], 1-haloalkane [65], acid chloride [65, 66], and alkanethiol [67, 68] derived monolayers, in which a range of structurally significant fragments, and molecular and cluster ions are observed in the SIMS spectra.

SIMS is particularly useful to determine the monolayer adsorption sites that are difficult to identify with XPS. For example, XPS studies of Ga–S bonding in alkanethiolate SAMs adsorbed on GaAs(001) are complicated due to ambiguities in the spectra fitting in the relevant binding energy region and due to the weakness of the S 2p signal [69]. In contrast, SIMS does not suffer from such issues and is extremely surface sensitive. Thus, SIMS is an ideal technique to investigate adsorbate–substrate bonding. Using SIMS, McGuiness et al. [68] demonstrated that alkanethiols form direct S–(GaAs) attachments to bare GaAs(001) predominantly via As–S bonds with a small fraction of Ga–S bonds. SIMS has also been employed to characterize adsorbate–substrate bonding for monolayers prepared by scribing silicon under a series of 1-alkenes, 1-alkynes, and 1-haloalkanes [65, 66]. However, SIMS does not always provide an unambiguous determination of the monolayer–substrate binding, as illustrated by the recent work of Arafat et al. [70] on 1-alkenes and 1-alkynes on silicon nitride substrates.

Since SIMS is able to provide surface-specific molecular chemical information with high lateral resolution, it has been employed to monitor both qualitatively and quantitatively reactions in monolayers [64, 67, 70, 71]. SIMS is employed to follow not only the degradation reactions but also the functionalization of monolayers. For example, Arafat et al. [70] investigated the formation of amine-functionalized surfaces by reaction of acid-terminated monolayers using EDC/NHS chemistry. However, care must be taken in analyzing the resulting SIMS spectra and images because the chemical information obtained arises from the near-surface region (depth ~50 Å). Thus, SIMS may not provide information from the buried layers, such as in amine-functionalized monolayers.

11.2.4 Surface-Enhanced Raman Spectroscopy

Conventional Raman scattering is not sensitive enough to detect a single organic monolayer on atomically flat semiconductor surfaces. Typical modes associated with SAMs such as CH_x, Si–C, and Si–O are not very strong Raman modes. On the other hand, it is possible to obtain Raman spectra in the case of porous silicon [41] or on flat silicon when nanoparticles (gold or silver) are adsorbed on top of SAM layers [72]. An example is shown in Fig. 11.5 for ~15 nm diameter gold particles attached on amino-terminated monolayers on Si surfaces. In this spectrum, the modes associated with $\delta(NH_3^+)$ are clearly seen. Similar bands have been observed in amino acids linked to metal (gold, silver) nanoparticles [73].

11.2.5 Spectroscopic Ellipsometry

Spectroscopic ellipsometry (SE) is a powerful method for fast and routine characterization of SAMs on semiconductors, yielding both the thickness and index of refraction, as recently reviewed by Hinrichs et al. [74]. Once calibrated, it provides a sensitive measure of the quality of the SAM layer and of the interface quality for gross interface modifications (e.g., oxidation). The main drawback is that it is highly

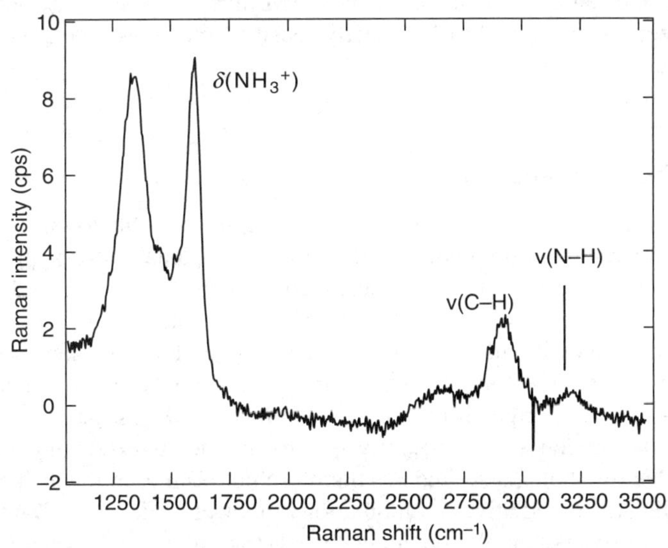

FIGURE 11.5 Raman spectra of amino-terminated silicon surface after deposition of gold nanoparticles.

model dependent and cannot therefore yield unambiguous parameters such as absolute thickness or precise dielectric function, particularly for inhomogeneous, imperfect layers. Consequently, SE is best used in conjunction with IR spectroscopy that provides more microscopic and chemical information.

11.2.6 X-Ray Reflectivity

Similarly to SE, X-ray reflectivity (XRR) provides an average measure of the entire surface, using standard Fresnel's equations. From these measurements, the height and surface roughness of the monolayer, the monolayer/silicon interface roughness, and the chain packing density can be extracted [75]. With XRR analysis, the entire surface of the sample is probed, while AFM and XPS look only at smaller parts of the surface. XRR is not readily able to measure the amount of oxidation that takes place, so other methods must be used to investigate this effect [76].

11.2.7 Contact Angle, Wettability

Contact angle (CA) measurements provide a convenient measure of the nature of SAM head groups. They are particularly useful to compare head groups with widely different hydrophilicities. The nature of functionalized surfaces ranges from highly hydrophilic surfaces (e.g., chemically prepared and hydrocarbon-free SiO_2 surfaces) that give CA values close to zero (complete spreading of the drop) to highly hydrophobic surfaces (CA $>110^{\circ}$) such as methyl-terminated long alkyl chain SAMs on oxide-free semiconductor surfaces.

The CA method provides an average value for SAM hydrophobicity, similarly to SE that provides an average value for the SAM index of refraction. Both methods therefore need to be coupled to chemically sensitive methods to deal with mixed or inhomogeneous SAMs.

11.2.8 Photoluminescence

The main advantage of photoluminescence is its high sensitivity to surface electronic states (i.e., recombination rates). It is therefore very valuable for applications that strongly depend on surface state densities, such as microelectronic-based sensors and photovoltaic devices. However, the nature of the defects cannot be extracted from such measurements without spectroscopy such as deep level transient spectroscopy (DLTS), inverse photoemission (IPES), and so on. Photoluminescence has been extensively used for porous Si because the PL level is very strong [41]. It is much weaker for flat silicon surfaces (partly due to nonradiative recombination of the generated electron–hole pairs) and has therefore not been used much. The nature of defects leading to nonradiative recombination is not well understood. Damage caused by the laser is also a potential problem that needs to be addressed. Nevertheless, there is detectable photoluminescence at 1130 nm on flat Si, which can be used as a sensitive probe of the electrical quality of the interface [77].

11.2.9 Electrical Measurements

Direct electrical tunneling measurements are probably the most sensitive technique for addressing the monolayer quality. Any defect will affect the transport from the substrate to the metal contact or vice versa. This method is robust and works well for reasonably good monolayers (i.e., with no shorts). The tunneling current is very sensitive to interface defects, although as for photoluminescence, the nature of these defects is not well known. They are thought to involve structural defects, dangling bonds, pinhole, oxidation, and other chemical impurities.

This technique is not widely used because it requires non-damaging and reliable metal deposition to establish contacts. Metal evaporation can indeed damage the organic layer. Currently, one of the most reliable methods is the use of Hg top contact [78, 79]. A recent review article addresses in detail the different issues associated with metal deposition for contact formation (monolayer properties, junction, semiconductor doping, etc.) in such systems [80].

11.2.10 Imaging Techniques

Atomic force microscopy is primarily used to measure the morphology of the surface after chemical modification to determine whether additional roughness is introduced in the process (due to oxidation, for instance) or whether the organic layer is homogeneous on a 0.1 nm scale. For short SAMs, the surface roughness remains, in general, close to that of the starting surface while for large molecules or long-chain molecules additional roughness is typically observed. For instance, indication of DNA hybridization [81] or changes of orientation of grafted azobenzene [82] have been observed with AFM.

For any tip-based technique, there is a possibility that the tip can interact with the SAM so that some organic molecules can bind to it. The nature of the head group must therefore be kept in mind when doing AFM since the affinity of the surface toward the tip depends on the head group. The ability of the head groups to interact with the tip can also be used to pattern surfaces [83]. As summarized in several review articles [43, 84], there are different ways to perform lithography of self-assembled monolayers using AFM tips, including local electrochemistry to remove part of the layers through anodization [85] or to graft molecules on small area through cathodic process [86].

Scanning tunneling microscopy (STM) is harder to perform on thick organic layers but has been used successfully on as-prepared H-terminated Si surfaces [45, 87], or after modification with short organic molecules such as $-CH_3$ [88] or $-C_6H_4R$ [45].

11.2.11 Electron and Atom Diffraction Methods

Low-energy electron diffraction is usually very sensitive to surface reconstruction, as demonstrated in thousands of surface science experiments under ultrahigh vacuum conditions. However, care must be taken in the interpretation of LEED patterns obtained for organic layers grafted to unreconstructed H-terminated Si surfaces. The observation of a sharp (1×1) pattern does not necessarily mean that a perfect surface

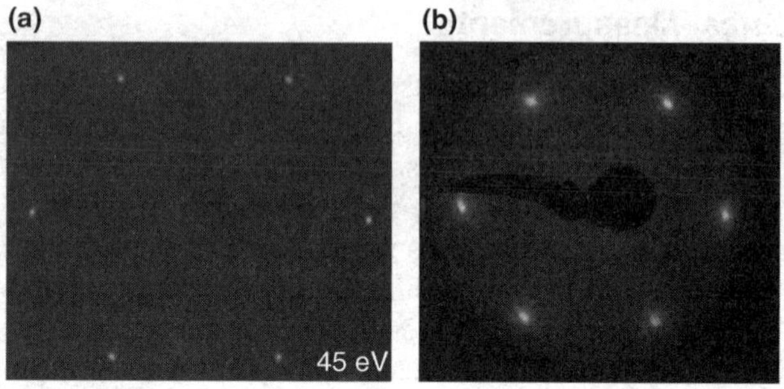

FIGURE 11.6 (a) LEED pattern presented in Ref. 37. (b) LEED pattern obtained at 37.7 eV on 1/3 $SiOCH_3$, 2/3 SiH.

termination has been established because the sensitivity to a thin adlayer composed primarily of carbon and hydrogen is poor if the substrate remains unreconstructed.

This is illustrated in Fig. 11.6a, which shows the LEED pattern observed for an unreconstructed Si(111)-(1×1) surface covered with a very homogeneous layer of methoxy molecules (each surrounded by six H atoms) arranged in a $\sqrt{(3)} \times \sqrt{(3)}$ pattern [60]. This pattern is identical to the (1×1) pattern obtained for the starting unreconstructed surface (completely monohydride terminated and atomically flat), and with the (1×1) pattern reported for a CH_3-covered, atomically flat Si(111) surface [48]. If hydrogen is removed from a H/Si(111)-(1×1) surface by beam methods, the top surface begins to reconstruct and the LEED pattern is affected as shown in Fig. 11.6b.

This example illustrates that, unless the top Si layer is reconstructed, it is not possible to conclude from a (1×1) LEED pattern that the organic overlayer is perfectly ordered and continuous.

A much more appropriate method to determine the homogeneity of an organic layer is He diffraction, as nicely illustrated by Scoles for self-assembled monolayers of alkanethiol on gold [89].

In conclusion for this methods section, it is clear that a combination of spectroscopy, imaging, electrical, and diffraction methods is needed to characterize the nature of functionalized semiconductor surfaces. In doing so, the information derived from each technique needs to be examined critically.

11.3 HYDROSILYLATION OF H-TERMINATED SURFACES

Since the main emphasis of this chapter is on the functionalization of H-terminated surfaces, in particular silicon surfaces, it is most appropriate to start with hydrosilylation, which is a well-known reaction in inorganic chemistry describing an addition of organic silicon hydride compounds to species with multiple bonds, particularly C–C, C–O, or C–N [90, 91].

The essential element of hydrosilylation, summarized in Equation 11.1, is the activation of the alkene, alkyne, or other unsaturated carbon compounds to permit reaction with the Si–H surface. The three main forms of activation on surfaces include the use of catalysts or Lewis acids, UV light, and temperature. The last two methods of activation involve a surface reaction and consequently are not derived from homogeneous hydrosilylation in solution that relies on a catalyst. The surface reaction can be written as

$$\mathrm{Si-H{+}H_2C=CH-R \xrightarrow{\text{catalyst},h\nu,\text{heat}};\ Si-CH_2-CH_2-R} \qquad (11.1)$$

For the attachment of alkyl chains to H-terminated silicon surfaces, the hydrosilylation mechanism provides an effective means of forming a strong Si–C bond between Si and alkyl chain molecules, starting with alkene or alkyne reagents. The resulting SAMs are both thermodynamically and kinetically stable due to the high bond strength and low polarity of the Si–C bond. The main requirements for this wet chemical process are that (i) the molecules remain liquid (hence some restriction on their length) and stable under the reaction conditions, and (ii) their functional groups be less reactive with the Si–H surface than with the targeted functions (alkene or alkyne termination). These requirements lead to two strategies: (1) protection of the functional groups during reaction (with subsequent deprotection to regain functionality), which often requires the synthesis of such protected species since they are not available commercially, and (2) further modification of the grafted molecules *without degradation of the initial Si/SAM interface* (i.e., mild conditions are necessary to avoid silicon oxidation or other unwanted interface reactions).

11.3.1 Catalyst-Aided Reactions

In homogeneous media (solution), catalytic hydrosilylation processes are commonly used. Such reactions can occur either by creation of a radical [90] or by polarization of the bond under consideration (predominant in most systems). The catalysts used may be transition metal complexes, nucleophiles (e.g., tertiary amine), or Lewis acids ($AlCl_3$, $EtAlCl_2$) [92–94]. The hydrosilylation process catalyzed by transition metal complexes was first proposed by Chalk and Harrod in 1965 as illustrated in Fig. 11.7 [92], and then expanded by using $EtAlCl_2$ (Fig. 11.8) and peroxides as shown in Fig. 11.9.

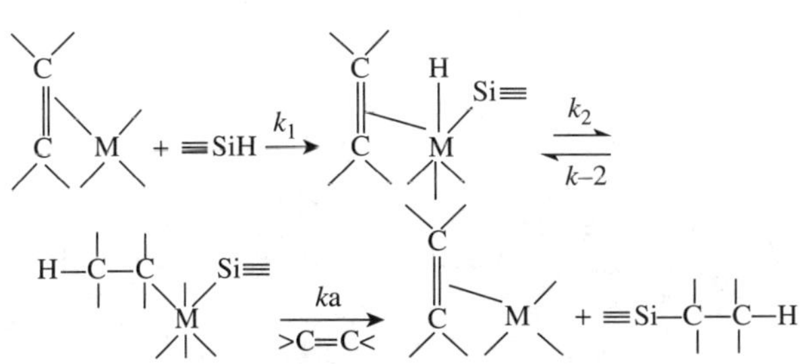

FIGURE 11.7 Mechanism proposed by Chalk and Harrod for the hydrosilylation of olefins catalyzed by group VIII metal complexes [92].

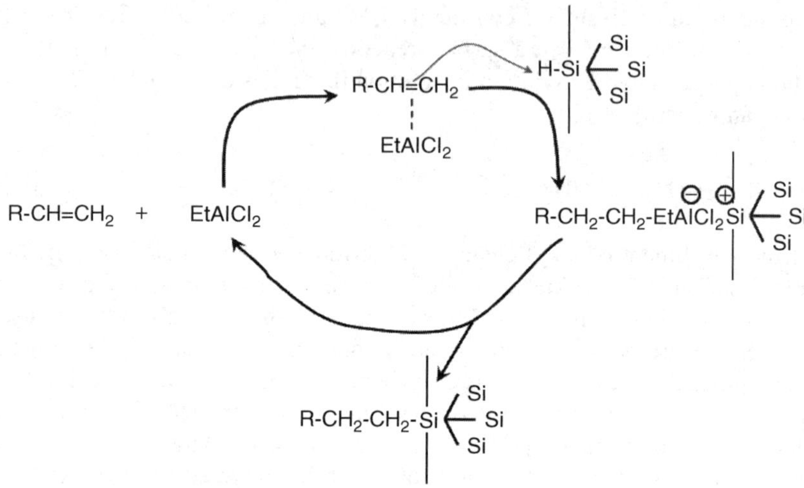

FIGURE 11.8 Schematic representation of the hydrosilylation mechanism using $EtAlCl_2$ to catalyze the organosilicon reaction.

Similar reactions have been performed on surfaces, initially on H-terminated porous silicon. Palladium-, rhodium-, or platinum-based catalysts have been used on surfaces but the process often leads to oxidation and metal reduction as thoroughly reviewed by Buriak [44] and Boukherroub and Szunerits [41].

In contrast, Lewis acid-catalyzed or -mediated hydrosilylation works very well to functionalize porous silicon. A catalyst of choice has been $EtAlCl_2$ because it is soluble in nonpolar solvents. It was first demonstrated by Boukherroub et al. and has been used extensively to functionalize porous and flat silicon surfaces [95, 96]. The

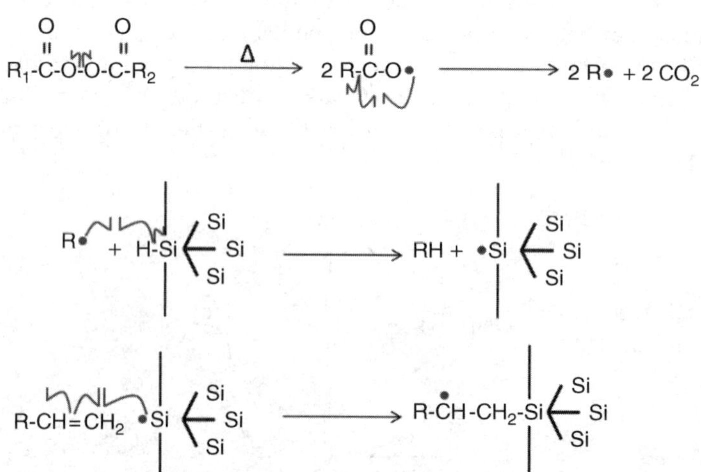

FIGURE 11.9 Schematic representation of the hydrosilylation mechanism using a peroxide agent to initiate a radical reaction.

mechanism is shown in Fig. 11.8, and follows the pathways determined by standard organosilicon chemistry [93].

The first example of functionalization of H-terminated Si surfaces by organic molecules was performed using radical initiators, such as diacyl peroxides [97]. Such catalysts are known to create radicals to initiate a chain reaction in polymers, for instance, and have been used early to perform hydrosilylation in organic chemistry [90]. When heated, the initiator undergoes homolytic cleavage to form acyloxy radicals that further decompose into carbon dioxide and an alkyl radical. The alkyl radical abstracts H• from a Si–H surface to produce a silicon radical on the surface. Such a coordinatively unsaturated "dangling bond" is known to react extremely rapidly with olefins that are in close proximity. The formation of a silicon–carbon bond (olefin attachment) is therefore the favorable next step. The resulting carbon-based radical is stabilized by abstracting a hydrogen atom either from the allylic position of an unreacted olefin or from a neighboring Si–H group. This method therefore often leads to a chain reaction on the surface. This mechanism and its associated reaction kinetics have been examined, confirmed, and quantified by first-principles calculations [98].

Hydride abstraction initiated hydrosilylation via an ionic pathway has also been proposed, using for instance triphenylcarbenium cations (Fig. 11.10) [99]. The cation is supposed to remove H^- from the surface, producing a positive charge on the silicon a surface leading to a nucleophilic attack by the electron-rich double bond. Such a process leads to Si–C formation and a β-silyl-substituted carbocation that can propagate the reaction by abstracting the hydrogen nearby.

In 2005, Arafat et al. used 4-(decanoate)-2,2,6,6-tetramethylpiperidinooxy (TEMPO) as a radical initiator to graft nonfunctional alkenes such as octadecene [100] or even unprotected carboxylic acids to Si(111) surfaces [101]. This approach is of particular interest because it makes it possible to attach directly organic layers with *functional* groups on silicon surfaces.

The hydrosilylation reaction can also be performed without any catalyst, if another activating medium (e.g., appropriate photons or thermal energy or electrochemistry activation) is used, as described in the following sections.

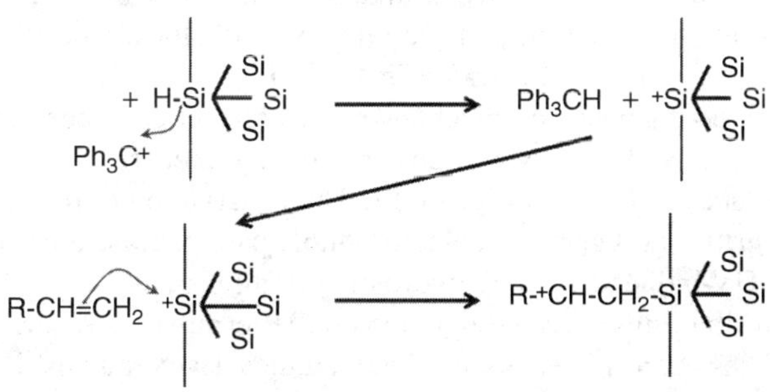

FIGURE 11.10 Schematic representation of the hydrosilylation mechanism using a carbocation to initiate an ionic reaction.

11.3.2 Photochemically Induced Reactions

The initial concept behind photoactivation of H-terminated surfaces was that photons with energy larger than the Si–H bond energy ($\sim$3.5 eV) could directly break this bond to form a silicon radical, thus activating the surface. Such a process would therefore require wavelengths shorter than 350 nm (3.55 eV). Clearly, highly anhydrous solutions (with the appropriate alkene) are required since silicon radicals strongly react with water to form silicon oxide. This simple picture motivated the pioneering work of Terry et al. [102]. His group used XPS to show that organic monolayers could be formed on silicon by illumination of freshly prepared Si–H surfaces in liquid pentene for 2 h with a mercury lamp ($\lambda = 185$ and 253.7 nm) [102]. They later grafted several other alkene and alkyne molecules using a variety of wavelengths and pointed out that silicon oxidation occurs in the presence of oxygen when wavelengths shorter than 350 nm are used [40]. Such observations support the idea that the first step could be an activation of the surface mediated by breaking Si–H bonds at short wavelengths. The reactive radical formed on the surface would then react, similarly to the radical obtained via the use of a radical, either with an oxidizing species leading to silicon oxidation or with a double bond leading to covalent grafting. The cross section for direct Si–H bond breakage is, however, very low for isolated Si–H (e.g., H-terminated surfaces in vacuum). The role of the solvent and the potential cooperative effects of the organic molecules are therefore critical for this mechanism.

Over the past decade, photochemical grafting of various alkene-based organic molecules has been performed using a wide range of photon energies [103, 104], including energies that are lower than the Si–H bond energy and even than the substrate bandgap. Since the photon energy used in these studies is significantly lower than the energy necessary to induce homolytic bond cleavage, it has been suggested that photoexcitation of substrate charges is likely responsible for the generation of active sites on the Si surface. The effectiveness of photons with energies below the Si–H bond energy (3.5 eV) indicates that mechanisms other than the simple mechanism outlined above may be operational [103]. In 2001, a mechanism involving the generation of excitons in the substrate was proposed by Stewart and Buriak to account for the observed wavelength dependence [105]. The main idea is the creation of electrons and holes in the substrate, which are then available to initiate the reaction and the nucleophilic attack from the electron-rich C=C. Both mechanisms are illustrated in Fig. 11.11.

This electron–hole generation mechanism was further corroborated by using two model surfaces with Si–H bonds grafted directly on either (i) an oxide-free Si or (ii) an oxidized surface as shown in Fig. 11.12. The observation of hydrosilylation *only on oxide-free surfaces* suggests that electron–hole pair excitons are important for the mechanism of UV-induced surface hydrosilylation [106].

Another contribution to surface reactions on H-terminated Si surfaces, involving chain reaction, has been suggested in several studies. Even if an excitonic process is at the origin of the grafting process, the propagation of functionalization may still involve radicals. The effect of a radical on the reaction of the neighboring Si–H site

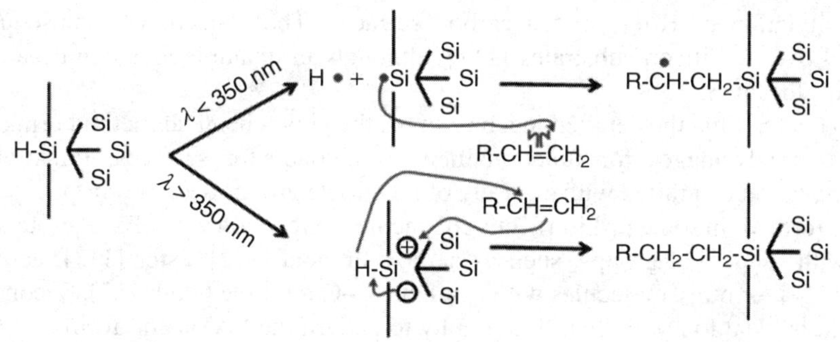

FIGURE 11.11 Schematic illustration of the two pathways for photochemically induced reactions for hydrosilylation of H-terminated Si surfaces: (top) radical-based mechanism; (bottom) exciton-based mechanism.

was first observed in ultrahigh vacuum by STM [107]. Confirmation of this mechanism in liquid environment is more difficult. Miramond and Vuillaume have shown that such surface reactions are influenced by the substrate doping although short wavelengths (257 nm) were still used in this work [108]. Grafting of highly n-doped silicon, for instance, led to the formation of more disordered monolayers that therefore exhibited larger leakage current densities ($>10^{-4}$ A/cm^2 compared to 10^{-7} A/cm^2 for others) when embedded in a silicon/monolayer/metal junction.

For completeness, we note that functionalization of other substrates, such as GaN, has been achieved by using this photochemical method [109].

Yet another fundamentally different mechanism has been suggested in the past few years by Hamers and coworkers. The observation of hydrosilylation reaction on H-terminated diamond or amorphous carbon using UV light with energies lower than the bandgap energy (e.g., 254 nm for reaction on diamond) [110] was used as evidence that the exciton model may not always be dominant. Instead, it was proposed that photoemission of electrons into the liquid was responsible for producing radicals on acceptor molecules that could then catalyze the

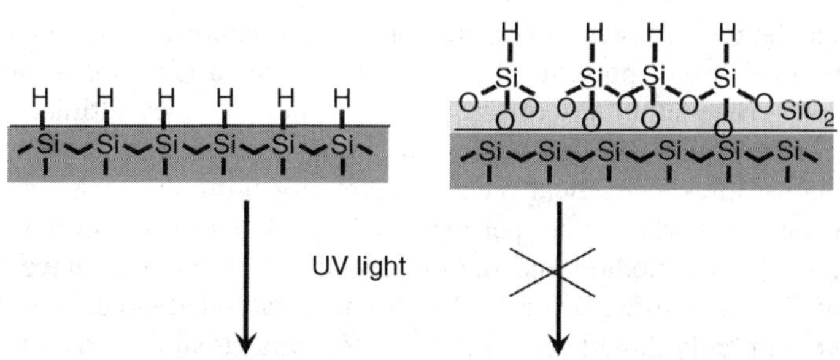

FIGURE 11.12 Schematic representation of the two model surfaces: (left) atomically flat, oxide-free H-terminated Si(111) surface; (right) oxidized Si functionalized with trichlorosilane, resulting in a H-terminated oxide surface.

hydrosilylation of H-terminated carbon surfaces. The concept was subsequently carried over to silicon substrates [111], although an unambiguous confirmation is difficult to obtain.

No matter what the detailed mechanism is, the photochemical method brings two important advantages for functionalizing semiconductor surfaces. First, alkene molecules can be grafted with a variety of functional groups because surface radicals can be formed (in some fashion), thus enhancing the reaction with alkene groups over that with functional groups, such as carboxylic acid [112], ester [113], activated ester [114], or other molecules with available C–C multiple bonds [82]. Second, the method is ideal to use optical lithography to pattern the SAMs and form molecular arrays [115, 116]. Recently, optical lithography has been applied, for instance, to process organic monolayers on oxide-free silicon substrates by using a photothermal laser at 514 nm [117].

11.3.3 Thermally Activated Reactions

Thermal activation of hydrosilylation reactions is a straightforward and interesting alternative to chemical or photochemical activation. This procedure typically involves immersing the H-terminated Si surfaces in an anhydrous solution of appropriate alkenes for extended periods of time (from hours to days) at moderate temperatures (from 150 to 200°C). The process can be well modeled by standard surface chemistry, whereby Si–H reacts with an alkene group through a transition state involving $Si \cdots C^{(1)}{-}H$ and $Si{-}H \cdots C^{(2)}H_2$, which results in the formation of $Si{-}C^{(1)}H_2{-}C^{(2)}H_3{-}\cdots$. Although this thermal process is sensitive to impurities because it is slow, there is an intrinsic advantage to the method that has not been fully appreciated. The higher solution temperatures typically used for such processing facilitate the removal of trace water, thus reducing the potential risk of surface *oxidation*. Therefore, simple methods and care in the execution of experiments make it possible to obtain surfaces with negligible interface oxide and high PL intensities (similar to H-terminated Si surfaces) [118]. Such surfaces are significantly more stable than H-terminated surfaces once produced because of the protective SAM layer, even if the majority of the surface (e.g., 60–70%) is still H-terminated.

Thermal grafting also provides an "annealing effect" on the substrate. Such annealing can help to suppress a degradation of the functionalization process due to the compensating effect present on p-type Si surfaces arising from the polishing process [119]. These subtle effects are important because they ultimately limit the performance of actual devices based on the hydrosilylation process. Thermal processing constitutes a practical means of circumventing these issues.

Chidsey and coworkers first performed thermal hydrosilylation in the early 1990s [97, 120]. The method and the quality of the monolayers have since been improved by Sieval by using several solvents and unsaturated molecules [121]. Such reactions were initially thought to involve surface radicals similar to those generated by catalysts or photochemical activation ($\lambda < 350$ nm) [44, 122]. It was also proposed that the first step of initiation was due to traces of water, oxygen, or other impurities [122, 123]. A similar mechanism has recently been suggested for

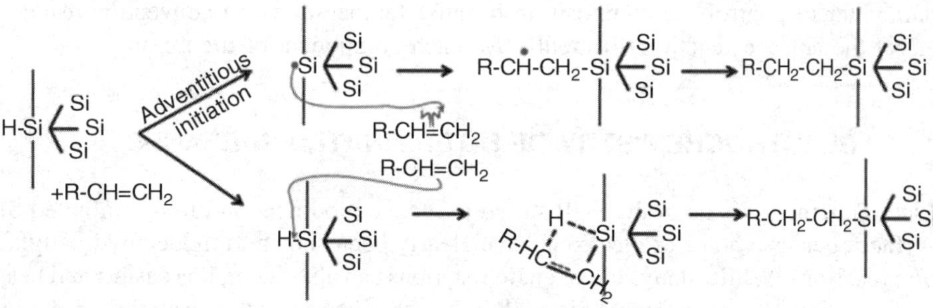

FIGURE 11.13 Possible reaction pathways for the thermal hydrosilylation: (top) radical-based mechanism; (bottom) concerted mechanism.

homogeneous hydrosilylation in solution, in which air could activate Si–H bonds instead of catalysts [124]. However, it is not clear how homolytic cleavage of a strong Si–H bond with bond energies on the order of 80–90 kcal/mol could occur at an appreciable rate at a temperature as low as 150°C. A more concerted mechanism was therefore proposed [41] and DFT calculations performed, as illustrated in Fig. 11.13 [125].

Recent studies based on thermal activation have highlighted the differences between the grafting of alkenes and alkynes on flat silicon surfaces. By using a mixed solution of alkene and alkyne or a molecule containing both an alkyne at one end and an alkene at the other, it was shown that triple bonds are preferentially attached over double bonds [126]. Furthermore, Scheres et al. have confirmed that an increased surface coverage is obtained with alkenyl chains on silicon, compared to what can be achieved with alkyl chains [127]. Interestingly, in addition to the classical thermal method [122], alkynes have also been grafted by very mild, room-temperature methods in the dark [128], producing better quality surfaces than alkenes [129].

The thermal grafting methods have been avoided partly for fear that thermal energy may activate the functional groups and induce them to react with SiH. Despite this potential problem, *high-quality* acid-terminated surfaces have been prepared by thermal grafting [130–132]. Carboxylic acid groups can then be modified and used for several applications such as metal deposition [130], biosensors [81, 132], electrical contacts [79, 133], or deposition of nanoparticles [131].

Practically, thermal grafting represents the most direct and easiest way to graft organic layers on surfaces. There is no need for special equipment or procedure such as UV light reactors, electrochemical cells, or addition of catalysts. The only requirement is hermetically closed glassware that can be purged with argon. There is no constraint on the geometry, which makes it possible to prepare samples of any size. Functionalization occurs on both sides and over large areas, which is not as easy for UV grafting at 254 nm, for instance, since quartz is required. A simple hot plate is sufficient to shake the solution when necessary, for example, in the case of mixed alkene molecules. To prepare mixed monolayers, temperature is an important parameter to control the composition because temperature and convection usually affect the surface reaction differently for each component of the mixture.

11.4 ELECTROCHEMISTRY OF H-TERMINATED SURFACES

Electrochemistry is naturally well suited to induce modification of H-terminated Si surfaces because holes or electrons have clearly been shown to induce hydrosilylation reactions. While many side or chain reactions can also occur, it is easiest and best to focus initially on the injection of holes or electrons into acceptor or donor molecules or groups that can then provide radicals as discussed below.

11.4.1 Cathodic Grafting

In cathodic grafting, an electron is created in the silicon substrate and can be captured by an acceptor species, such as the azo group (|N≡N$^+$–) of diazonium salts, as illustrated in Fig. 11.14. This process releases dinitrogen, thus producing an aryl radical [134]. Although diazonium salts are most commonly used, many other molecules can undergo the same process to become a radical. For instance, aryliodonium salts have been successfully grafted onto SiH [135]. Other halogenoalkanes of the form X–R, where X is a halogen, can also be reacted using this cathodic method because the halogen atom preferentially attracts the substrate electron to form X^- ions, thus releasing a radical R• [136].

Once formed, the radical can then abstract a hydrogen atom from the ≡SiH surface, producing ≡Si• that can easily recombine with many molecules, and in particular other radicals nearby. Since the functional group R can be tailored to a required functionality (e.g., R = Br, NO_2, COOH, CN, or C_nH_{2n+1} for $n = 1, 4, 12$), this method is useful to functionalize surfaces without formation of silicon oxide [137].

The main advantage of cathodic grafting is that it brings electrons to the surface, making it more resistant to nucleophilic attack by water and therefore to oxidation.

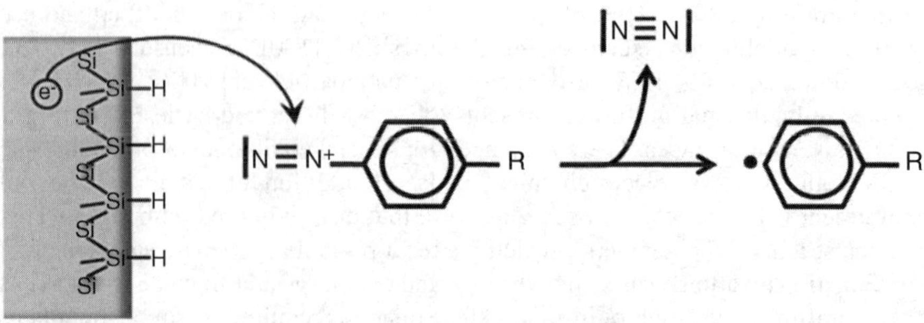

FIGURE 11.14 Radicalization of a diazonium salt by electron capture by the azo group: R can be Br, NO_2, COOH, CN, or C_nH_{2n+1} (n = 1, 4, 12).

A common problem, however, is the ease with which the salt radicals can polymerize. The challenge for such a procedure is therefore to control unwanted polymerization. This functionalization has also been applied to other surfaces such as GaAs or Pd [138].

11.4.2 Anodic Grafting

The anodic generation of an organic radical R• is shown schematically in Fig. 11.15. Once formed, the radical can react with the H-terminated surface, thus leading to hydrogen abstraction. A second radical may then be grafted on the remaining silicon dangling bond. This mechanism is similar to the exciton mechanism in the photochemical grafting described later in this section, whereby holes interact directly with the ≡SiH surface, leading to the formation of a radical ≡Si•.

Anodization of silicon surfaces by releasing a hole at the surface (or driving an electron deep in the substrate) unavoidably leads to formation of a layer of silicon oxide in the presence of water or, in the case of fluoride electrolytes, to the dissolution of silicon. If an organic modification is desired, it is therefore very important to operate under rigorously anhydrous conditions.

Whatever the dominant path may be, it is likely that the radical species R• and ≡Si• are both present, since there is an equilibrium formed usually written as

$$\equiv \mathrm{SiH} + \mathrm{R}\bullet \leftrightarrow \equiv \mathrm{Si}\bullet + \mathrm{RH}$$

Anodic grafting has been explored for alkyl, alkenyl, alkynyl, and aryl radicals using, for instance, Grignard electrolytes, and also for halogeno and cyano radicals

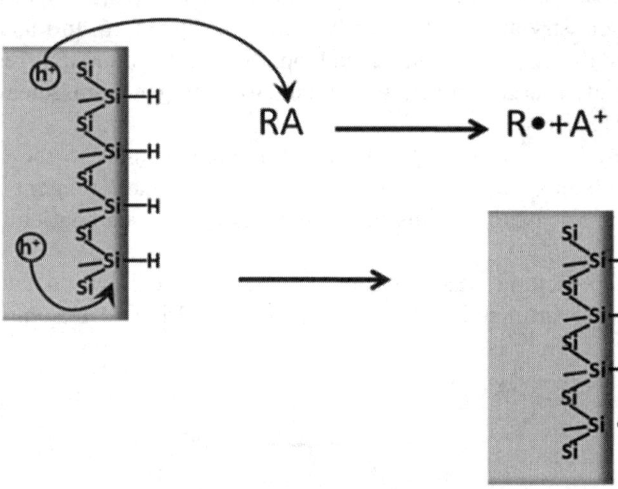

FIGURE 11.15 Possible pathways involving the radicalization of electron-donor species to initiate the anodic grafting of organic molecules: (top) ionization of molecule in solution; (bottom) ionization of surface Si atom.

using anhydrous acetonitrile electrolytes. Rapid grafting is generally obtained. However, the reaction is self-limiting only for the most inert radicals such as alkyl and ethynyl. For more reactive radicals, such as halogeno or cyano salts, side reactions may be observed, such as electropolymerization on the silicon surface, for example. Grignards are especially attractive precursors, since they ideally ensure anhydrous conditions by neutralizing trace amounts of water molecules and also constitute the electrolyte [139]. They are typically available with a variety of aliphatic and aromatic groups. There are, however, some restrictions on the type of possible substituents due to the high reactivity of the Grignard reagents with protonic molecules, that dissociate H^+ typically bound to nitrogen or oxygen [140].

For completeness, we note that Grignard grafting on H-terminated Si surfaces can also occur simply by immersion of a H/Si sample in an organomagnesium (RMgX) solution at room temperature on *porous silicon* [95, 141] or at 85°C during 16 h on flat silicon [95]. On atomically rough surfaces or in porous silicon, it appears that the mechanism involves the cleavage and reaction of Si–Si back-bonds without net consumption of SiH [41, 141]. On flat silicon, for which Si–Si bond breaking is energetically more costly due to strain, the modification associated with the consumption of Si–H remains surprising.

11.5 USE OF HALOGEN-TERMINATED SURFACES

A two-step procedure has been shown to facilitate this process. In the first step, the surface hydrogen is replaced by a halogen (typically chlorine). Once the surface is halogenated, the second step involves immersion in Grignard or organolithium solutions. The reaction that follows is typical of a classical Wurtz coupling described in inorganic chemistry and schematically shown in Fig. 11.16. In this reaction, the carbon linked to the magnesium has a nucleophilic character and thus reacts with the partially positively charged carbon atom linked to a halogen atom, releasing a longer alkane and a Mg halide.

A halogenated silicon atom behaves in a manner similar to the carbon atom in a halogenoalkane, namely, it reacts with the Grignard reagent to graft the organic ligand, releasing a transition metal halide as schematically shown in Fig. 11.17.

This two-step method was pioneered by Bansal and Lewis to attach organic monolayers on Cl-terminated silicon surfaces with Grignard reagents [142]. The

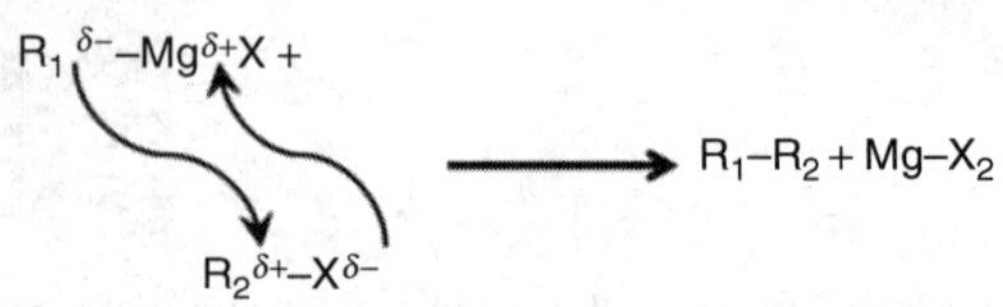

FIGURE 11.16 Mechanism of the Wurtz reaction between an organomagnesium R_1MgX and an organohalide R_2X [140].

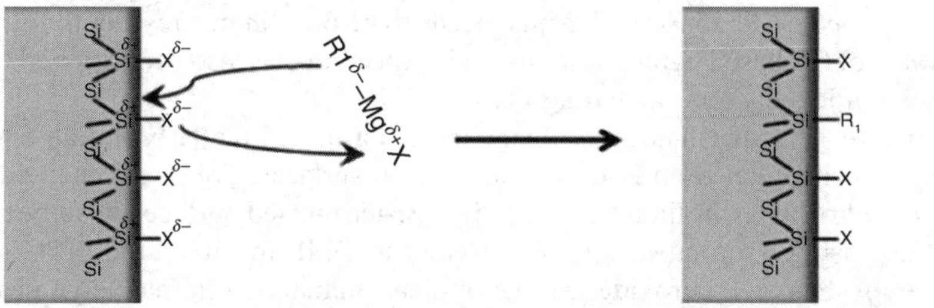

FIGURE 11.17 Possible mechanism of the reaction of an organomagnesium R_1MgX and a surface silicon halide Si–X.

method was improved to functionalize surfaces with low densities of electronic defects [143]. The exact mechanism involved in the two steps has been investigated in the past few years [62, 144]. Using this two-step method with electrochemistry, well-defined methyl-terminated surfaces have been obtained with high coverages [48].

In considering the quality of surfaces functionalized using the two-step process, it is important to understand the first step, or halogenation of H-terminated silicon surfaces. Chlorination, for instance, can be achieved in two ways on silicon surfaces:

- wet chemical process using PCl_5 (with benzoyl peroxide as a radical initiator in chlorobenzene or with UV);
- gas-phase process using Cl_2 gas at moderate temperatures.

Both methods can remove the hydrogen layer completely, as evidenced by the complete loss of Si–H stretch modes, and replace it with a chlorine layer without etching the silicon substrate, as determined by results on atomically flat H-terminated silicon surfaces [145]. In the case of gas-phase Cl_2, the reaction can be followed as a function of treatment time, confirming that chlorine substitutionally replaces hydrogen, leading to a newly observed blue shift of the remaining Si–H neighbors [146, 147].

The stability of chlorine-terminated silicon surfaces has been examined. Similarly to H-terminated surfaces, Cl-terminated Si surfaces are stable in water vapor. Despite early claims [148], this stability makes it impossible to form hydroxyl-terminated surfaces starting from Cl-terminated surfaces because water can only react with Si–Cl at temperatures ($\sim$300°C) at which Si–OH is not stable, leading to the incorporation of oxygen into the silicon lattice (i.e., oxidation) [145]. Once oxidized, the surface can of course support hydroxyl groups, hence the erroneous interpretation [148].

Chlorination of germanium surfaces is straightforward as well, thus making Ge amenable to Grignard chemistry. In addition to the methods developed for Si, chlorination of Ge surfaces can be achieved by immersion in hydrochloric acid [23]. Attachment of organic layers on chlorinated germanium was performed with similar procedures, such as ethylation [149] or alkane-linked Ge–S–C bonds obtained from alkanethiol [150]. However, in most cases, the stability of functionalized Ge surfaces has not been studied or demonstrated. Much needs to be done in this respect to fully assess the value of Ge-based hybrid systems. This procedure has also been applied to several III–V semiconductors, including Ga–P [151].

In closing, we note that halogens other than chlorine, particularly iodine and bromine, have been attached on H-terminated silicon surfaces. Following methods to brominate hydrosilanes in liquid phase [152], Br-terminated surfaces have been obtained [153], using *N*-bromosuccinimide (NBS) in DMF for 20 min at 60°C in the presence of benzoyl peroxide as the radical initiator. Alternatively, neat CCl_3Br can be used under thermal conditions (80°C for 30 min), with photolysis (300 nm UV for 20 min), or in the presence of benzoyl peroxide (60°C for 30 min) [153].

Iodination has been performed on porous silicon as reviewed by Boukherroub and Szunerits [41]. In this case, it appears that the reaction of the iodine gas involves the breaking of Si–Si bonds instead of Si–H bonds. In order to attack Si–H directly and to thus perfect a procedure that can be reliably used for modifying flat Si–H, iodination of silicon surfaces was carried out using iodoform (CHI_3) under visible light or thermal activation [154]. It was postulated that iodine radicals can be generated by light or heat activation, leading to a direct reaction with Si–H. Cai et al. used iodine dissolved in benzene, for instance, as a way to increase the Si–I coverage on flat surfaces [115].

Figure 11.18 summarizes the various methods developed for halogenating H-terminated silicon surfaces. Interestingly, it has not been possible to replace

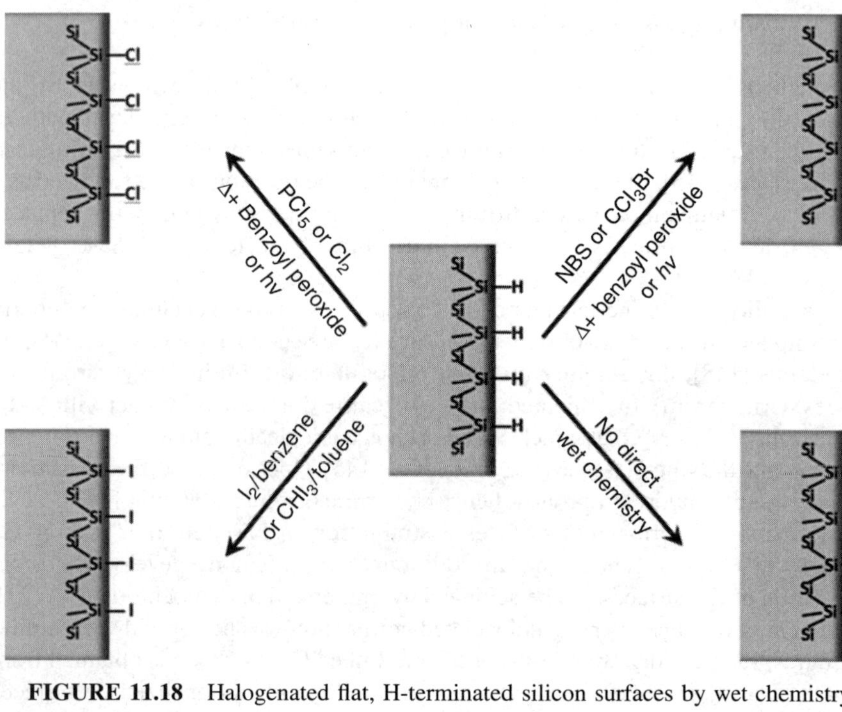

FIGURE 11.18 Halogenated flat, H-terminated silicon surfaces by wet chemistry.

hydrogen by fluorine, mostly because fluorinated liquid leads to hydrogen termination [30] and because gas-phase fluorine is very aggressive, leading to strong etching [155]. This difficulty in obtaining fluorinated silicon surfaces has motivated the work described in the next section, where a very recent method for partial fluorination of H-terminated silicon surfaces is presented. This method opens new possibilities for enhanced functionalization of silicon surfaces.

11.6 ALCOHOL REACTION WITH H-TERMINATED SI SURFACES

Chapter 12 is concerned with the stability of functionalized surfaces. Here, we focus on the reaction of alcohols (mostly methanol) with *atomically flat*, monohydride-terminated Si(111) surfaces. Indeed, the structure of the surface is shown to play an important role in both the degree of reaction and the properties of the functionalized surfaces.

It has been well established that immersion of H-terminated Si surfaces in alcohols leads to the grafting of alkoxy groups on Si surfaces, such as methoxy for methanol, forming Si–O–C linking bonds. The reaction can be thermally [156, 157] or photochemically induced [154, 158, 159], as described above for standard hydrosilylation, but has also been observed to take place at room temperature after long immersion times (~2–3 days), clearly ruling out the need for radical formation [160]. In fact, the preparation of monolayers linked via Si–O–C bonds to silicon surfaces has typically been achieved under mild conditions. As recently suggested by Shirahata using contact angle measurements [159], even long-chain alcohols can be grafted at 40°C, conditions under which alkene cannot be attached. The use of such mild conditions ensures that the functional groups of the grafted molecules remain intact [161, 162]. Alternatively, Zhu developed a two-step method to graft long-chain alcohols at 70°C, involving an intermediate Si–Cl surface [163], and extended this approach to attach polyethylene glycol species [164].

In most of these studies performed over the last decade, there is typically a small amount of interface oxide generated during the reaction as revealed by XPS or IR spectroscopic studies [154, 163, 165]. Furthermore, AFM investigations have also shown that pitting occurs on initially flat H-terminated Si(111) surfaces during the grafting of alcohol species in addition to some partial oxidation [157, 166]. Nonetheless, 1-decanol has been successfully grafted on Si(111) with little detectable silicon oxide at the interface [162, 167].

These two types of defects, oxidation and pitting, have undermined the potential of grafting alcohols on silicon surfaces. Pitting can be partially understood by invoking mechanisms proposed to address thermal grafting on porous silicon, such as Si–Si bond breaking during reaction with alcohols leading to subsequent etching of silicon [41, 168]. Oxidation, on the other hand, is most likely due to trace amounts of water in the alcohol solution, due to the high affinity of water to the OH group of alcohols (which is not the case for alkene molecules, for instance). To reduce the concentration of unwanted water, Boukherroub et al. suggested adding chlorotrimethylsilane to the reaction mixture to scavenge undesirable nucleophiles during wet chemical modification of Si(111)–H [157]. The use of highly anhydrous chemicals was also attempted, but until recently, some oxidation of the silicon could not be avoided during methanol reaction with H-terminated Si surfaces, even though the majority of the surface could clearly be perfectly methoxylated [160, 169].

The recent finding that immersion of H-terminated Si(111) surfaces in neat anhydrous methanol at 70°C inside a nitrogen-purged glove box leads to the formation of *oxide-free* methoxy-terminated surfaces [59, 60] confirms the role of trace amount of water in oxidizing the surface during alcohol reactions and highlights the importance of moderately elevated temperatures to reduce trace water contamination. Processing at this temperature also brings an unexpected result, namely, that only 1/3 of a monolayer can be reacted, the other 2/3 of the surface remaining hydrogen terminated. The resulting surface, schematically represented in Fig. 11.19, is composed of methoxy centers surrounded by six surface Si–H. The reason for this arrangement, as recently described and supported theoretically [60], is that once adsorbed, the methoxy group rotates around its axis, thus protecting the surrounding Si–H groups (Fig. 11.20). This effect is more pronounced at 70°C, and results in the formation of nanopatterned surfaces [60].

The fact that the surface remains *atomically flat* and *oxide free* after methoxylation provides a very special template for subsequent chemical reactions. For instance, immersion of such a methoxylated surface in HF produces a partially fluorine-terminated surface, that is, each methoxy group is replaced by fluorine, with no changes in the surrounding Si–H groups and no attack of Si–Si bonds (i.e., no pitting). The surface remains atomically flat, as can be seen for the reaction mechanism shown in Fig. 11.20. Once fluorinated, the surface can be hydroxylated by simple immersion in water. Again, only the Si–F bonds are attacked, not the surrounding Si–H groups or the Si–Si back-bonds. The process can be repeated multiple times, resulting in fluorine attachment by immersion in hydrofluoric acid and hydroxyl replacement of F by immersion in water, without affecting the ~2/3 monolayer H coverage.

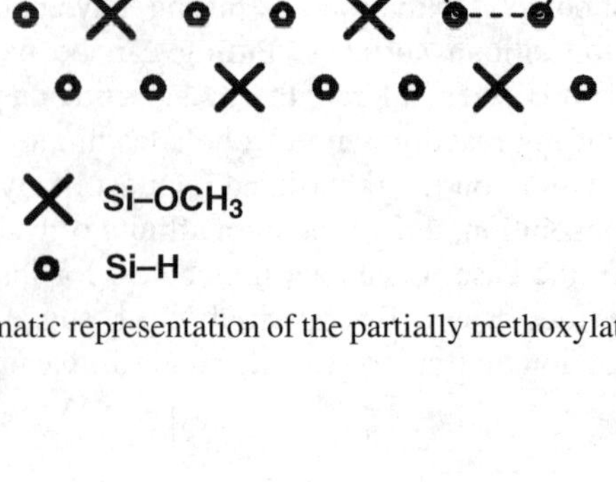

FIGURE 11.19 Schematic representation of the partially methoxylated Si(111) surface after reaction with methanol.

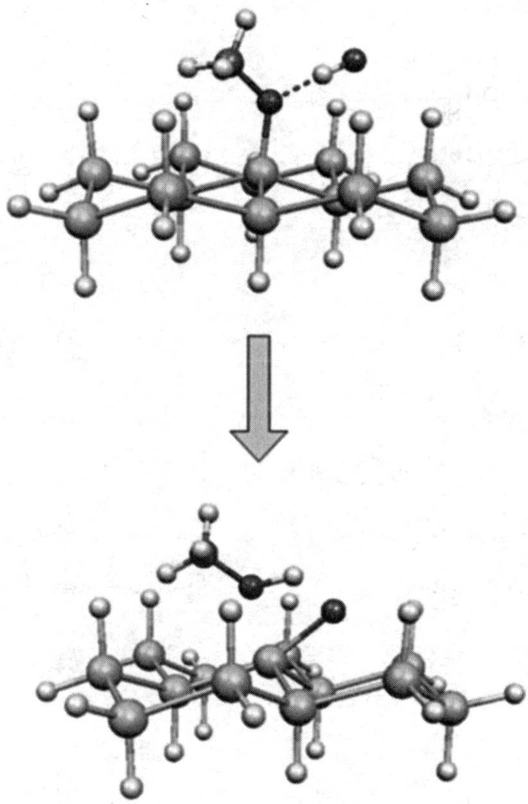

FIGURE 11.20 Theoretically calculated reaction of HF with methoxy-terminated silicon in an atomically flat surface with six Si–H species surrounding the methoxy.

The formation of a partially hydroxylated surface on an otherwise oxide-free surface opens up a host of possibilities, in particular for atomic layer deposition of high and low k dielectrics. For instance, HfO_2 and dimethyl-SiH have been deposited on oxide-free silicon [60]. In general, the presence of isolated OH groups on an otherwise hydrophobic (H-terminated) surface is of great value to attach molecules (e.g. phosphonates) that usually could only functionalize oxidized surfaces, which is of interest for silicon-based biosensors, for instance [170].

The fundamental mechanism for the above chemistry is determined by the structural and chemical perfection of the surface. Indeed, if there is any detectable oxide on the surface (characterized by interface atomic roughness), immersion in HF leads to H termination of the surface in the oxidized areas. This observation underscores the fact that the interface between crystalline silicon and amorphous SiO_2 is not atomically smooth. The mechanism for HF etching described at the beginning of this chapter is in fact only valid because the interface is rough. Figure 11.21 schematically illustrates this point. On the left, the fluorinated surface can be further attacked by HF to form a H-terminated surface. On the right, an atomically smooth and partially fluorinated surface cannot be further attacked, thus remaining fluorinated.

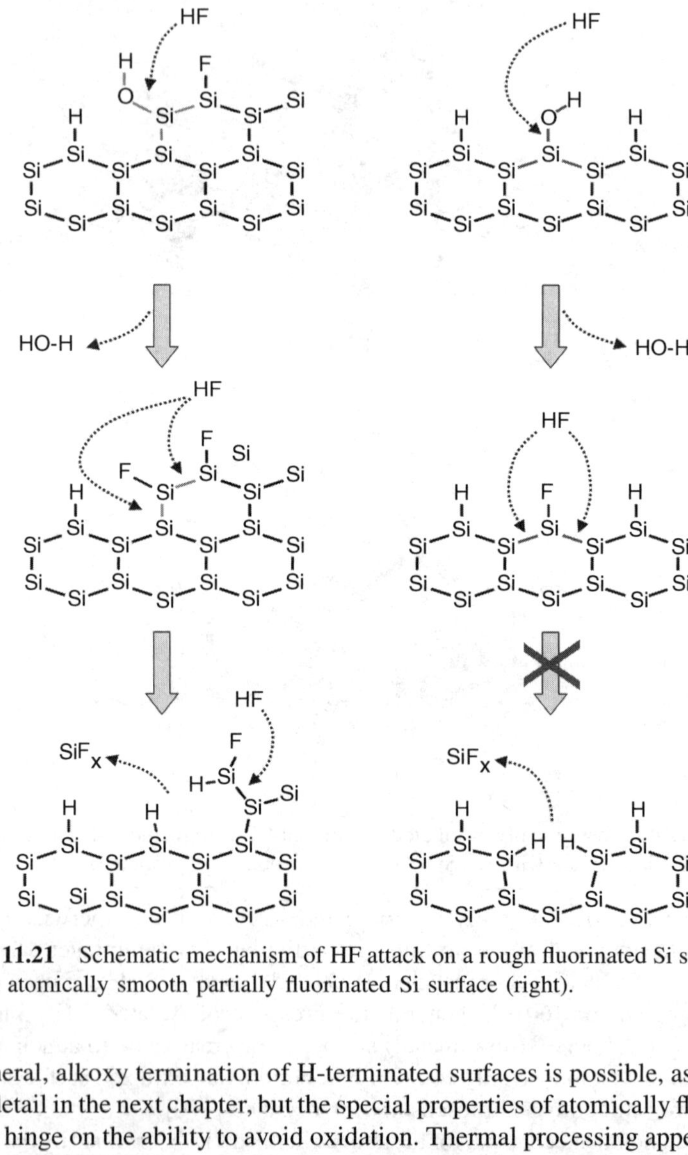

FIGURE 11.21 Schematic mechanism of HF attack on a rough fluorinated Si surface (left) and on an atomically smooth partially fluorinated Si surface (right).

In general, alkoxy termination of H-terminated surfaces is possible, as described in more detail in the next chapter, but the special properties of atomically flat surfaces critically hinge on the ability to avoid oxidation. Thermal processing appears to be a convenient method to prepare oxide-free surfaces, and is required to produce highly uniform nanopatterned surfaces.

Atomically flat and partially methoxy-terminated surfaces provide a model system to test a variety of physical phenomena such as vibrational and electronic dynamics, chemical reactions such as snap on chemistry, and chemical stability (e.g., increased stability of Si–H when surrounded by F). For instance, an unusual blue shift of the Si–H stretch mode in the vicinity of a halogen observed for chlorine [146] was recently explained theoretically and shown to lead to Si-H stabilization [147]. The ability to fluorinate a surface in a homogeneous nanopattern has made it possible to distinguish among several mechanisms, for instance [146, 147]. These surfaces are also a model system to test the interaction of electron, photon, atoms, and ions with a variety of atoms or molecules on *unreconstructed* surfaces. An example was given for LEED in the experimental section and it is anticipated that further work will be done using He scattering, scanning tunneling microscopy, grazing incidence X-ray scattering, and other methods.

11.7 OUTLOOK

The work summarized in this chapter underscores the importance of wet chemical methods for semiconductor surface functionalization. In some cases, the quality of functionalized surfaces greatly surpasses that of surfaces passivated in vacuum. While the mechanisms involved in wet chemical functionalization have been examined carefully, much remains to be done experimentally to improve the quality of surfaces, devise model systems, and provide precise enough information to distinguish among competing mechanisms. Theoretically, it is important to unravel all the various mechanisms including photochemical and thermal reactions to be able to control the quality of functionalized surfaces and to provide guidance for new systems.

Characterization is essential to the development of wet chemical functionalization. In general, a combination of surface spectroscopic, diffraction, and imaging techniques is critical to avoid misinterpretation potentially derived from a single method.

Among the various methods for wet chemical functionalization, thermally induced reactions provide interesting opportunities. The elevated temperature reduces the amount of water in anhydrous reactions, minimizing surface oxidation during hydrosilylation and alkylation reactions. It also improves nanopatterning of flat surfaces, and greatly influences the surface arrangement of bifunctional layers obtained from appropriate mixtures.

The formation of high-quality surfaces opens the door to fundamental studies of surface chemistry under a variety of environments (vacuum, vapor, solution). We anticipate that much work will be devoted to surface modification of starting model surfaces described in this chapter.

One area where the work summarized above will most impact is the fabrication of highly performing electrical devices that deal with organic or biological molecules, such as hybrid solar cells and biosensors (see Chapter 13). For these applications, control and quality of the starting surfaces are critical and can best be achieved by the wet chemical processes described in this chapter. The methods presented above constitute therefore the fundamental platform for the development of these technologies.

ACKNOWLEDGMENTS

This work was supported by the National Science Foundation (CHE-0911197) and partially by the Texas Higher Education Coordinating Board (NHAR Program). The authors are grateful to Professor Amy Walker and Dr. Oliver Seitz for critical input for this chapter.

REFERENCES

1. Chabal, Y. J.; Higashi, G. S.; Small, R. J. In *Silicon Wafer Cleaning Technology*. William Andrew Publishing, Norwich, NY, **2008**, p. 523.
2. Frank, M. M.; Chabal, Y. J. In *Into the Nano Era*. Springer, Berlin, **2009**, p. 113.
3. Higashi, G. S.; Chabal, Y. J.; Trucks, G. W.; Raghavachari, K. *Appl. Phys. Lett.* **1990**, *56*(7), 656.
4. Trucks, G. W.; Raghavachari, K.; Higashi, G. S.; Chabal, Y. J. *Phys. Rev. Lett.* **1990**, *65*(4), 504.
5. Jakob, P.; Chabal, Y. J. *J. Chem. Phys.* **1991**, *95*(4), 2897.
6. Jakob, P. et al. *Surf. Sci.* **1993**, *285*(3), 251.
7. Hines, M. A. *Int. Rev. Phys. Chem.* **2001**, *20*(4), 645.
8. Allongue, P. et al. *Electrochim. Acta* **2000**, *45*, 4591.
9. Rivillon Amy, S.; Chabal, Y. J. In *Advanced Gate Stacks for High-Mobility Semiconductors*, Vol. *27* Springer, Berlin, **2007**, p. 73.
10. (a) Haller, I. *J. Am. Chem. Soc.* **1978**, *100*(26), 8050. (b)Kurth, D. G.; Bein, T. *Langmuir* **1993**, *9*(11), **2965**.
11. Pasternack, R. M.; Rivillon Amy, S.; Chabal, Y. J. *Langmuir* **2008**, *24*(22), 12963.
12. (a) Bernasek, S. L.; Schwartz, J. *Langmuir* **1998**, *14*(6), 1367. (b)Mutin, P. H. et al. *Chem. Mater.* **2004**, *16*(26), 5670. (c) Ishida, T. et al. *Appl. Surf. Sci.* **2009**, *255*(21), 8824.
13. Nie, H.-Y. *Anal. Chem.* **2010**, *82*(8), 3371.
14. (a) Gouzman, I. et al. *Surf. Sci.* **2006**, *600*(4), 773. (b) Hauffman, T. et al. *Langmuir* **2008**, *24*(23), 13450. (c) Jayasundera, S. et al. *J. Phys. Chem. B* **2006**, *110*(37), 18121.
15. Lipshutz, R. J.; Fodor, S. P. A.; Gingeras, T. R.; Lockhart, D. J. *Nat. Genet.* **1999**, *21*(1), 20.
16. (a) Cloarec, J. P. et al. *Biosens. Bioelectron.* **2002**, *17*(5), 405. (b) Park, Y. D. et al. *Appl. Phys. Lett.* **2005**, *87*(24), 243509. (c) Park, J. Y. et al. *J. Chem. Phys.* **2009**, *130*(11), 114705. (d) Cho, J. H. et al. *Nat. Mater.* **2008**, *7*(11), 900.
17. Klauk, H.; Zschieschang, U.; Pflaum, J.; Halik, M. *Nature* **2007**, *445*(7129), 745.
18. Strother, T. et al. *J. Am. Chem. Soc.* **2000**, *122*(6), 1205.
19. (a) Guter, W. et al. *Appl. Phys. Lett.* **2009**, *94*(22), 223504. (b) King, R. R. et al. *Appl. Phys. Lett.* **2007**, *90*(18), 183516.
20. Montgomery, H. C.; Brown, W. L. *Phys. Rev.* **1956**, *103*(4), 865.
21. Rivillon, S.; Chabal, Y. J.; Amy, F.; Kahn, A. *Appl. Phys. Lett.* **2005**, *87*(25), 253101.
22. Loscutoff, P. W.; Bent, S. F. *Annu. Rev. Phys. Chem.* **2006**, *57*(1), 467.
23. Lu, Z. H. *Appl. Phys. Lett.* **1996**, *68*(4), 520.
24. Lu, Z. H.; Tyliszczak, T.; Hitchcock, A. P.; Dharma-wardana, M. W. C. *Surf. Sci.* **1999**, *442*(1), L948.
25. (a) Frank, M. M. et al. *Appl. Phys. Lett.* **2006**, *89*(11), 112905. (b) Anderson, G. W. et al. *Appl. Phys. Lett.* **1995**, *66*(9), 1123. (c) Houssa, M. et al. *Appl. Phys. Lett.* **2007**, *90*(22), 222105.
26. Buriak, J. M. *Chem. Rev.* **2002**, *102*(5), 1271.
27. Dhar, S. et al. *J. Am. Chem. Soc.* **2009**, *131*(46), 16808.

28. Pluchery, O.; Chabal, Y. J.; Opila, R. L. *J. Appl. Phys.* **2003**, *94*(4), 2707.
29. (a) Peisert, H.; Streubel, P.; Chassé, T.; Szargan, R. *Fresenius J. Anal. Chem.* **1997**, *358*(1), 201. (b) Eftekhari, G. *Vacuum* **2002**, *67*(1), 81.
30. Ubara, H.; Imura, T.; Hiraki, A. *Solid State Commun.* **1984**, *50*(7), 673.
31. Yablonovitch, E. et al. *Phys. Rev. Lett.* **1986**, *57*(2), 249.
32. Chabal, Y. J.; Higashi, G. S.; Raghavachari, K.; Burrows, V. A. *J. Vacuum Sci. Technol. A* **1989**, *7*(3), 2104.
33. Imura, T. et al. *Solid State Commun.* **1981**, *40*(2), 161.
34. Jakob, P.; Chabal, Y. L.; Raghavachari, K. *Chem. Phys. Lett.* **1991**, *187*(3), 325.
35. Jakob, P. et al. *Surf. Sci.* **1992**, *275*(3), 407.
36. Clark, I. T.; Aldinger, B. S.; Gupta, A.; Hines, M. A. *J. Phys. Chem. C* **2009**, *114*(1), 423.
37. (a) Hattori, T.; Nohira, H. In Chabal, Y. J.,Ed. *Fundamental Aspects of Silicon Oxidation.* Springer, Berlin, **2001**, p. 61. (b) Takagi, N.; Minami, N.; Furukawa, T.; Nishijima, M. *Surf. Sci.* **1993**, *297*(1), L43.
38. (a) Zhang, X. et al. *Appl. Phys. Lett.* **2001**, *79*(24), 4051. (b) Zhang, X. et al. *J. Vac. Sci. Technol. A* **2001**, *19*(4), 1725. (c) Dai, M. et al. *Nat. Mater.* **2009**, *8*(10), 825.
39. Ye, S. et al. *Surf. Sci.* **2001**, *476*(1–2), 121.
40. Cicero, R. L.; Linford, M. R.; Chidsey, C. E. D. *Langmuir* **2000**, *16*(13), 5688.
41. Boukherroub, R.; Szunerits, S. In Schmuki, P.; Virtanen, S., Eds. *Electrochemistry at the Nanoscale.* Springer, New York, **2009**, p. 183.
42. (a) Hamers, R. J. *Annu. Rev. Anal. Chem.* **2008**, *1*(1), 707. (b) Boukherroub, R. *Curr. Opin. Solid State Mater. Sci.* **2005**, *9*(1–2), 66.
43. Shirahata, N.; Hozumi, A.; Yonezawa, T. *Chem. Rec.* **2005**, *5*(3), 145.
44. Buriak, J. M. *Chem. Commun.* **1999**, *12*, 1051.
45. Wayner, D. D. M.; Wolkow, R. A. *J. Chem. Soc., Perkin Trans.* **2002**, *2*, 23.
46. Aswal, D. K. et al. *Anal. Chim. Acta* **2006**, *568*(1–2), 84.
47. (a) Terry, J. et al. *J. Appl. Phys.* **1999**, *85*, 213. (b) Fellah, S. et al. *Langmuir* **2002**, *18*(15), 5851.
48. Hunger, R. et al. *Phys. Rev. B* **2005**, *72*(4), 045317.
49. (a) Yang, F.Humboldt-Universität zu Berlin, Mathematisch-Naturwissenschaftliche Fakultät I, **2008**. (b) Aureau, D., *Ecole Polytechnique*, **2008**.
50. Harrick, N. J. *Internal Reflection Spectroscopy.* Interscience Publishers, **1967**.
51. Chabal, Y. J. *Surf. Sci. Rep.* **1988**, *8*(5–7), 211.
52. Chabal, Y. J. In Mirabella, F. M., Ed. *Internal Reflection Spectroscopy: Theory and Applications.* Marcel Dekker, **1993**, p. 791.
53. Queeney, K. T.; Fukidome, H.; Chaban, E. E.; Chabal, Y. J. *J. Phys. Chem. B* **2001**, *105*(18), 3903.
54. (a) Aureau, D.; Ozanam, F.; Allongue, P.; Chazalviel, J. N. *Langmuir* **2008**, *24*(17), 9440. (b) Moraillon, A.; Gouget-Laemmel, A. C.; Ozanam, F.; Chazalviel, J.-N. *J. Phys. Chem. C* **2008**, *112*(18), 7158.
55. Weldon, M. K.; Stefanov, B. B.; Raghavachari, K; Chabal, Y. J. *Phys. Rev. Lett.* **1997**, *79*(15), 2851.
56. Chabal, Y. J.; Hines, M. A.; Feijoo, D. *J. Vac. Sci. Technol. A* **1995**, *13*(3), 1719.

57. Milosevic, M.; Milosevic, V.; Berets, S. L. *Appl. Spectrosc.* **2007**, *61*(5), 530.
58. Durig, J. R.; Pan, C.; Guirgis, G. A. *Spectrochim. Acta A* **2003**, *59*(5), 979.
59. Michalak, D. J.; Amy, S. R.; Esteve, A.; Chabal, Y. J. *J. Phys. Chem. C* **2008**, *112*(31), 11907.
60. Michalak, D. J. et al. *Nat. Mater.* **2010**, *9*(3), 266.
61. Fidélis, A.; Ozanam, F.; Chazalviel, J. N. *Surf. Sci.* **2000**, *444*(1–3), L7.
62. Johansson, E.; Hurley, P. T.; Brunschwig, B. S.; Lewis, N. S. *J. Phys. Chem. C* **2009**, *113*(34), 15239.
63. Leggett, G.J. In Vickerman, J. C.; Briggs, D., Eds. *ToF-SIMS: Surface Analysis by Mass Spectrometry*. IM Publications/SurfaceSpectra Limited, Chichester/Manchester, **2002**, p. 573.
64. Roberson, S. V.; Fahey, A. J.; Sehgal, A.; Karim, A. *Appl. Surf. Sci.* **2002**, *200*(1–4), 150.
65. Lua, Y.-Y. et al. *Langmuir* **2002**, *18*(12), 4840.
66. Lua, Y.-Y. et al. *Langmuir* **2005** *21*(6), 2093.
67. (a) Zhou, C.; Walker, A. V. *Langmuir* **2007**, *23*, 8876. (b) Zhou, C.; Walker, A. V. *J. Phys. Chem. C* **2007**, *112*, 797.
68. McGuiness, C. L. et al. *J. Phys. Chem. C* **2007**, *111*, 4226.
69. McGuiness, C. L. et al. *ACS Nano* **2007**, *1*, 30.
70. Arafat, A. et al. *Langmuir* **2007**, *23*(11), 6233.
71. (a) Zhou, C. et al. *J. Phys. Chem. C* **2009**, *114*(12), 5400. (b) Zhou, C.; Trionfi, A.; Hsu, J. W. P.; Walker, A. V. *J. Phys. Chem. C* **2010**, *114*(20), 9362. (c) Zhou, C. et al. *Langmuir* **2009**, *26*(6), 4523.
72. Willets, K. A.; Van Duyne, R. P. *Annu. Rev. Phys. Chem.* **2007**, *58*(1), 267.
73. Suh, J. S.; Moskovits, M. *J. Am. Chem. Soc.* **1986**, *108*(16), 4711.
74. Hinrichs, K.; Gensch, M.; Esser, N. *Appl. Spectrosc.* **2005**, *59*(11), 272.
75. (a) Tidswell, I. M. et al. *Phys. Rev. B* **1990**, *41*(2), 1111. (b) Ishizaki, T. et al. *Langmuir* **2006**, *22*(24), 9962.
76. Lin, J. C. et al. *Eur. Phys. J.* **2009**, *167*(1), 33.
77. (a) Yang, F.; Hunger, R.; Rademann, K.; Rappich, J. *Physica Status Solidi c* **2010**, *7*(2), 161. (b) Aureau, D. et al. *J. Electroanal. Chem.*, **2010**, *646*(1–2), 33.
78. (a) Liu, Y.-J.; Waugh, D. M.; Yu, H.-Z. *Appl. Phys. Lett.* **2002**, *81*(26), 4967. (b) Seitz, O. et al. *J. Am. Chem. Soc.* **2007**, *129*(24), 7494. (c) Furuhashi, M. et al. *Jpn. J. Appl. Phys.* **2009**, *48*(5), 9.
79. Seitz, O. et al. *Langmuir* **2006**, *22*(16), 6915.
80. Yaffe, O. et al. *J. Phys. Chem. C* **2010**, *114*, 10270.
81. Touahir, L. et al. *Bioelectrochemistry* **2010**, *80*, 17.
82. Zhang, X. et al. *J. Phys. Chem. C* **2008**, *112*(22), 8288.
83. (a) Hoeppener, S. et al. *Adv. Mater.* **2002**, *14*(15), 1036. (b) Liu, S.; Maoz, R.; Schmid, G.; Sagiv, J. *Nano Lett.* **2002**, *2*(10), 1055.
84. (a) Kramer, S.; Fuierer, R. R.; Gorman, C. B. *Chem. Rev.* **2003**, *103*(11), 4367. (b) Rosa, L. G.; Liang, J. *J. Phys.:Condens. Matter* **2009**, *21*(48), 18.
85. (a) Ara, M.; Graaf, H.; Tada, H. *Appl. Phys. Lett.* **2002**, *80*(14), 2565. (b) Sugimura, H.; Nanjo, S.; Sano, H.; Murase, K. *J. Phys. Chem. C* **2009**, *113*(27), 11643.

86. Hurley, P. T.; Ribbe, A. E.; Buriak, J. M. *J. Am. Chem. Soc.* **2003**, *125*(37), 11334.
87. (a) Becker, R. S.; Higashi, G. S.; Chabal, Y. J.; Becker, A. J. *Phys. Rev. Lett.* **1990**, *65* (15), 1917. (b) Higashi, G. S.; Becker, R. S.; Chabal, Y. J.; Becker, A. J. *Appl. Phys. Lett.* **1991**, *58*(15), 1656. (c) Becker, R. S.; Becker, A. J.; Higashi, G. S.; Chabal, Y. J. *Scann. Microsc. Suppl.* **1993**, *7*, 269.
88. Yu, H. et al. *J. Phys. Chem. B* **2004**, *109*(2), 671.
89. (a) Scoles, G. *Atomic and Molecular Beam Methods*. Oxford University Press, New York, **1992**. (b) Chidsey, C. E. D.; Liu, G. Y.; Scoles, G.; Wang, J. *Langmuir 1990 6*(12), **1804**. (c) Bracco, G.; Scoles, G. *J. Chem. Phys.* **2003**, *119*(12), 6277.
90. Sommer, L. H.; Pietrusza, E. W.; Whitmore, F. C. *J. Am. Chem. Soc.* **1947**, *69*(1), 188.
91. (a) Marciniec, B. *Silicon Chem.* **2002**, *1*(3), 155. (b) Speier, J. L.; Webster, J. A.; Barnes, G. H. *J. Am. Chem. Soc.* **1957**, *79*(4), 974.
92. Chalk, A. J.; Harrod, J. F. *J. Am. Chem. Soc.* **1965**, *87*(1), 16.
93. Song, Y.-S.; Yoo, B. R.; Lee, G.-H.; Jung, I. N. *Organometallics* **1999**, *18*(16), 3109.
94. Asao, N.; Sudo, T.; Yamamoto, Y. *J. Org. Chem.* **1996**, *61*(22), 7654.
95. Boukherroub, R.; Morin, S.; Bensebaa, F.; Wayner, D. D. M. *Langmuir* **1999**, *15*(11), 3831.
96. Webb, L. J.; Lewis, N. S. *J. Phys. Chem. B* **2003**, *107*(23), 5404.
97. Linford, M. R.; Fenter, P.; Eisenberger, P. M.; Chidsey, C. E. D. *J. Am. Chem. Soc.* **1995**, *117*(11), 3145.
98. (a) Woods, M. et al. *J. Phys. Chem. B* **2005**, *109*(50), 24035. (b) de Smet, Louis C. P. M. et al. *J. Phys. Chem. B* **2005**, *109*(24), 12020.
99. Schmeltzer, J. M.; Porter, L. A.; Stewart, M. P.; Buriak, J. M. *Langmuir* **2002** *18*(8), 2971.
100. Arafat, S. N. et al. *Chem. Commun.* **2005**, *25*, 3198.
101. Perring, M. et al. *Langmuir* **2005**, *21*(23), 10537.
102. Terry, J. et al. *Appl. Phys. Lett.* **1997**, *71*(8), 1056.
103. (a) Sun, Q.-Y. et al. *Angew., Chem., Int. Ed.* **2004**, *43*(11), 1352. (b) Sun, Q. Y. et al. *J. Am. Chem. Soc.* **2005**, *127*(8), 2514. (c) de Smet, L. C. P. M. et al. *Appl. Surf. Sci.* **2005**, *252*(1), 24.
104. Sano, H. et al. *Jpn. J. Appl. Phys.* **2008**, *47*(7), 5659.
105. Stewart, M. P.; Buriak, J. M. *J. Am. Chem. Soc.* **2001**, *123*(32), 7821.
106. Langner, A. et al. *J. Am. Chem. Soc.* **2005**, *127*(37), 12798.
107. Eves, B. J.; Sun, Q.-Y.; Lopinski, G. P.; Zuilhof, H. *J. Am. Chem. Soc.* **2004**, *126*(44), 14318.
108. Miramond, C.; Vuillaume, D. *J. Appl. Phys.* **2004**, *96*(3), 1529.
109. Kim, H. et al. *Langmuir* **2006**, *22*(19), 8121.
110. Wang, X. et al. *J. Phys. Chem. C* **2010**, *114*(9), 4067.
111. Wang, X., Ruther, R. E.; Streifer, J. A.; Hamers, R.J. *J. Am. Chem. Soc.* **2010**, *132*(12), 4048.
112. Faucheux, A. et al. *Langmuir* **2006**, *22*(1), 153.
113. Boukherroub, R.; Wayner, D. D. M. *J. Am. Chem. Soc.* **1999**, *121*(49), 11513.
114. Yang, M. et al. *Langmuir* **2008**, *24*(15), 7931.
115. Cai, W. et al. *J. Phys. Chem. B* **2002**, *106*(10), 2656.
116. Mischki, T. K. et al. *Langmuir* **2006**, *22*(20), 8359.

117. Klingebiel, B. et al. *Langmuir* **2010**, *26*(9), 6826.
118. Aureau, D. et al. *J. Electroanal. Chem.* **2010**, *646*(1–2), 33.
119. Demierry, P.; Ballutaud, D.; Aucouturier, M.; Etcheberry, A. *J. Electrochem. Soc.* **1990**, *137*(9), 2966.
120. Linford, M. R.; Chidsey, C. E. D. *J. Am. Chem. Soc.* **1993**, *115*(26), 12631.
121. (a) Sieval, A. B. et al. *Langmuir* **1998** *14*(7), 1759. (b) Sieval, A. B.; Linke, R.; Zuilhof, H.; Sudhölter, E. J. R. *Adv. Mater.* **2000**, *12*(19), 1457.
122. Sieval, A. B. et al. *Langmuir* **2000**, *16*(26), 10359.
123. Mischki, T. K.; Lopinski, G. P.; Wayner, D. D. M. *Langmuir* **2009**, *25*(10), 5626.
124. Wang, J. et al. *J. Organomet. Chem.* **2008**, *693*(12), 2188.
125. Coletti, C. et al. *Langmuir* **2006**, *22*(24), 9949.
126. Ng, A. et al. *Langmuir* **2009**, *25*(24), 13934.
127. Scheres, L.; Giesbers, M.; Zuilhof, H. *Langmuir* **2010**, *26*(7), 4790.
128. Scheres, L.; Arafat, A.; Zuilhof, H. *Langmuir* **2007**, *23*(16), 8343.
129. Scheres, L.; Giesbers, M.; Zuilhof, H. *Langmuir* **2010**, *26*(13), 10924.
130. Seitz, O. et al. *J. Am. Chem. Soc.* **2009**, *131*(50), 18159.
131. Aureau, D. et al. *J. Phys. Chem. C* **2010**, *114*(33), 14180.
132. Dietrich, P. et al. *Appl. Phys. A* **2008**, *93*(2), 285.
133. Magid, I. et al. *J. Phys. Chem. C* **2008**, *112*(18), 7145.
134. de Villeneuve, C.; Henry Pinson, J.; Bernard, M. C.; Allongue, P. *J. Phys. Chem. B* **1997**, *101*(14), 2415.
135. Dirk, S. M. et al. *Langmuir* **2005**, *21*(24), 10899.
136. Gurtner, C.; Wun, A. W.; Sailor, M. J. *Angew. Chem., Int. Ed.* **1999**, *38*, 1966.
137. (a) Allongue, P. et al. *Electrochim. Acta* **1998**, *43*(19–20), 2791. (b) Hunger, R. et al. *J. Phys. Chem. B* **2006**, *110*(31), 15432.
138. Stewart, M. P. et al. *J. Am. Chem. Soc.* **2003**, *126*(1), 370.
139. Teyssot, A. et al. *Electrochim. Acta* **2002**, *47*(16), 2565.
140. Silverman, G. S.; Rakita, P. E. *Handbook of Grignard Reagents*. Marcel Dekker, New York, **1996**.
141. Kim, N. Y.; Laibinis, P. E. *J. Am. Chem. Soc.* **1998**, *120*(18), 4516.
142. Bansal, A. et al. *J. Am. Chem. Soc.* **1996**, *118*(30), 7225.
143. Bansal, A.; Lewis, N. S. *J. Phys. Chem. B* **1998**, *102*(7), 1067.
144. (a) Rivillon Amy, S. et al. *J. Phys. Chem. C* **2007**, *111*(35), 13053. (b) Solares, S. D. et al. *J. Am. Chem. Soc.* **2006**, *128*(12), 3850.
145. Rivillon, S. et al. (a) *Appl. Phys. Lett.* **2004**, *85*(13), 2583. (b) *Appl. Phys. Lett.* **2005**, *87*(17), 253101.
146. Ferguson, G. A.; Rivillon, S.; Chabal, Y.; Raghavachari, K. *J. Phys. Chem. C* **2009**, *113*(52), 21713.
147. Ferguson, G. A.; Aureau, D.; Chabal, Y. J.; Raghavachari, K. *J. Phys. Chem. C* **2010**, *114*(41),17644.
148. Damlencourt, J. F. et al. *J. Mater. Sci.* **2003**, *14*(5), 379.
149. Cullen, G. W.; Amick, J. A.; Gerlich, D. *J. Electrochem. Soc.* **1962**, *109*(2), 124.

150. Ardalan, P.; Musgrave, C. B.; Bent, S. F. *Langmuir* **2009** *25*(4), 2013.
151. Mukherjee, J.; Peczonczyk, S.; Maldonado, S. *Langmuir* **2010**, *26*(13), 10890.
152. Gansle, P. B. et al. *Microchem. J.* **1997**, *55*(2), 222.
153. He, J. et al. *Chem. Phys. Lett.* **1998**, *286*(5–6), 508.
154. Vadakkan Thomas, J.; Daniel, M. *Chem. Phys. Chem.* **2002**, *3*(11), 973.
155. Leonel, R. A. et al. *J. Micromech. Microeng.* **2007**, *17*(2), 384.
156. Cleland, G.; Horrocks, B. R.; Houlton, A. *J. Chem. Soc., Faraday Trans.* **1995**, *91*, 4001.
157. Boukherroub, R. et al. *Langmuir* **2000**, *16*(19), 7429.
158. Hacker, C. A.; Anderson, K. A.; Richter, L. J.; Richter, C. A. *Langmuir* **2004**, *21*(3), 882.
159. Shirahata, N.; Hozumi, A.; Sakka, Y. *J. Nanosci. Nanotechnol.* **2009**, *9*(1), 666.
160. Michalak, D. J. et al. *J. Phys. Chem. B* **2006**, *110*(41), 20426.
161. Kim, N. Y.; Laibinis, P. E. In *Inorganic Materials Synthesis*. American Chemical Society, Washington, DC, **1999**, p. 157.
162. Florent, T. et al. *Adv. Mater.* **2008**, *20*(20), 3931.
163. Zhu, X. Y. et al. *Langmuir* **2000**, *16*(17), 6766.
164. Zhu, X. Y. et al. *Langmuir* **2001**, *17*(25), 7798.
165. Saito, N. et al. *J. Vac. Sci. Technol. A* **2004**, *22*(4), 1425.
166. Dusciac, D. et al. *Surf. Sci.* **2007**, *601*(18), 3961.
167. Wallart, X.; Henry de Villeneuve, C.; Allongue, P. *J. Am. Chem. Soc.* **2005**, *127*(21), 7871.
168. Bateman, J. E.; Eagling, R. D.; Horrocks, B. R.; Houlton, A. *J. Phys. Chem. B* **2000**, *104*(23), 5557.
169. (a) Bateman, J. E.; Horrocks, B. R.; Houlton, A. *J. Chem. Soc., Faraday Trans.* **1997**, *93*(14), 2427. (b) Hory, M. A. et al. *Thin Solid Films* **1995**, *255*(1–2), 200. (c) Haber, J. A. et al. *J. Phys. Chem. B* **2000**, *104*(43), 9947. 170. Thissen, P. et al., *Angewandte Chemie* (under review).

CHAPTER 12

Chemical Stability of Organic Monolayers Formed in Solution

LESLIE E. O'LEARY, ERIK JOHANSSON, AND NATHAN S. LEWIS

12.1 REACTIVITY OF H-TERMINATED SILICON SURFACES

12.1.1 Background

The H–Si surface is important to the electronics and photovoltaics industries because Si–H is the starting point for many Si-based devices. In turn, the electronic and chemical properties of the H-terminated Si surface affect the properties of subsequent Si surfaces and interfaces [1–3]. Dangling or weak bonds present at the surface will affect minority-carrier ("excited-state") processes, and will thus affect devices such as field-effect transistors and photovoltaics. Furthermore, the topography of, as well as the presence of adsorbed chemical contaminants on, a H-terminated Si surface greatly affects the electronic properties of Si/SiO$_x$ interfaces formed from this initial H-terminated Si surface [4, 5]. Organic contaminants are difficult to remove from the Si surface, and different cleaning procedures have been shown to result in a variety of contaminant fingerprints. Hence, significant effort has been directed to understand the reactivity of the H–Si surface.

This section will explore the reactivity of the H-terminated Si surface with O_2, H_2O, alcohols, metals, amines, and thiols. The synthesis, as well as the physical and electronic characterization, of the H–Si surfaces will be reviewed briefly, but the reader is encouraged to consult Chapter 3 for a more detailed account of the preparation and characterization of H-terminated Si surfaces.

12.1.1.1 Synthesis of H-Terminated Si Surfaces The H-terminated Si surface can be formed by a variety of procedures, including vacuum techniques, electrochemical etching, and chemical etching [6–14]. H-terminated surfaces prepared by vacuum cleavage of Si, followed by exposure to hydrogen, have

Functionalization of Semiconductor Surfaces, First Edition.
Edited by Franklin (Feng) Tao and Steven L. Bernasek.

been studied using high-resolution electron energy loss spectroscopy (HREELS) and other ultrahigh vacuum techniques [10, 11]. Low-energy electron diffraction (LEED) studies have indicated that the surface order does not change upon exposure of a freshly cleaved surface to hydrogen.

Although such techniques enable investigation of the properties of very pure H-terminated surfaces, vacuum cleavage of bulk crystals is not suitable for large-scale processing. A significant amount of effort has thus been devoted to elucidating the behavior of solution-based routes to H-terminated Si surfaces. The composition of the Si surface following HF etching has been studied using multiple internal reflection Fourier transform infrared (MIR-FTIR) spectroscopy. The presence of IR absorption bands ascribable to Si–H, Si–H_2, and Si–H_3 moieties indicated that aqueous HF etching of Si(111) yielded microscopically rough surfaces, but linewidth analysis suggested that the surfaces were locally ordered [12–14].

The ability to create atomically flat Si(111) surfaces using a wet chemical approach (Fig. 12.1) was discovered in 1989. The method built on the observation that HF etching of Si(111) and Si(100) induced microscopic roughness to the surfaces [12, 15]. Following these studies, the effects of pH were examined to elucidate the effects of the HF(aq) concentration on Si surface quality [15, 16].

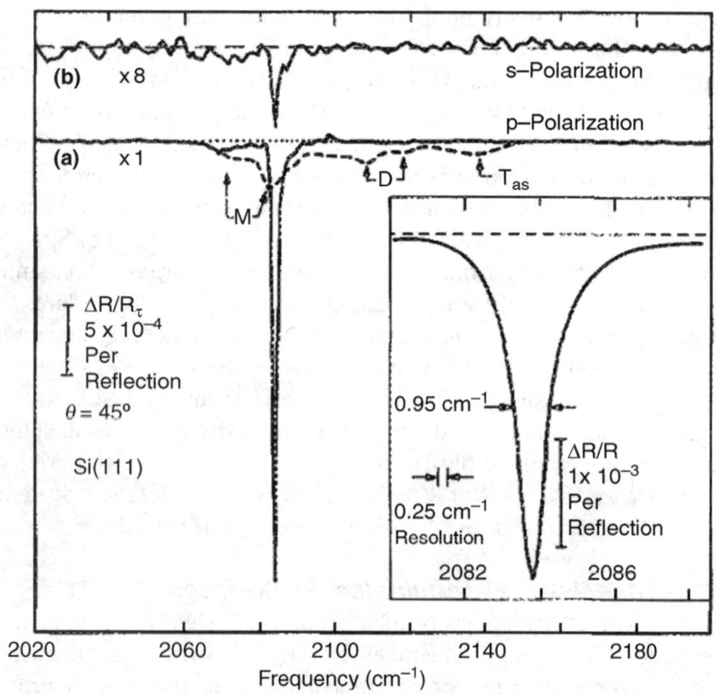

FIGURE 12.1 Internal reflection IR spectra of fluoride-etched Si(111) surfaces etched in acidic, dilute HF solution (100:1 H_2O:HF) (dashed curve) and buffered HF (pH 9–10) (solid curve) probed under p-polarized light (a) and s-polarized light (b) [16]. Inset shows high-resolution spectrum of Si(111) etched in buffered HF. Reproduced from Ref. 16 by permission of the American Institute of Physics, © 1990.

Etching in buffered HF, in which the pH was adjusted using either NH_4OH(aq) or HCl(aq), produced very different results than when the Si was etched in HF(aq) alone. Specifically, when the pH was greater than 4, a dramatic change occurred in the Si–H region of the FTIR spectra. A single, dominant peak, attributable to the Si–H stretching vibration, was produced with no peaks ascribable to the di- and trihydride species that were observed on surfaces etched in acidic HF(aq) solutions.

In alkaline solution, the rate-determining step of the fluoride etch on Si is OH^- attack. Simulations have shown that the relative reaction rates of OH^- attack on features commonly encountered on a Si(111) surface, including kinks, dihydride, and monohydride steps, can differ by as much as eight orders of magnitude. The anisotropy of the basic fluoride etch is significantly decreased by the introduction of trace metal, oxygen, or other impurities into the etching solution [17, 18]. Introduction of O_2 into an NH_4F(aq) etching solution greatly reduces the fidelity of the resulting Si(111) surface, producing Si–H_2 and Si–H_3 surface sites, as well as adding to the incorporation of O atoms into the Si–Si back-bonds. Scanning tunneling microscopy (STM) images clearly show the effect of the presence of oxygen in the NH_4F(aq) etching solution. Without oxygen, the surface is dominated by large terraces of Si(111), whereas rough surfaces are produced by O_2-containing NH_4F(aq) etching solutions [17, 18].

To further investigate the quality of the Si(111) surface, vacuum STM images and LEED patterns were obtained [15]. LEED patterns suggested that both HF(aq) and NH_4F(aq) etching resulted in high-quality (1×1) surfaces with defect densities of 0.5%. Vacuum STM data, however, showed clear differences in the surface morphology produced by HF(aq) versus NH_4F(aq) etchants, consistent with the presence of Si monohydride, dihydride, and trihydride peaks for HF(aq)-etched Si(111) surfaces, and consistent with only a sharp Si–H stretch in the IR spectrum of the NH_4F(aq)-etched surfaces. For NH_4F(aq)-etched surfaces, flat terraces extending thousands of Ångströms were observed with less than 1% surface contamination. The level of surface contamination seemed to not be intrinsic to the NH_4F(aq) etching procedure. X-ray photoelectron (XP) spectra further confirmed that neither oxygen nor carbon constitute significant contaminants on surfaces prepared through the NH_4F(aq) etch [17, 19].

The techniques commonly used to characterize Si interfaces electronically were developed to study the Si/SiO_2 or Si/metal interface [22]. The formation of Si/SiO_2 structures and Si/metal junctions, however, rarely preserves the H-terminated Si surface. Consequently, almost all techniques used to study the H–Si surface are contactless [2, 23], or form only soft contacts [24–26]. Using a contactless device, the H-terminated Si has been shown to be very well passivated electronically. Measurements taken in HF(aq) or in other strong acids yielded minority-carrier lifetimes >10 ms, suggesting that defect densities were as low as 1 per 40 million surface atoms [2]. The possibility that measurements in various solvents yielded long minority-carrier lifetimes due to band bending was investigated extensively, and it was found that one must proceed cautiously when correlating minority-carrier lifetimes to surface defect densities [27–30]. The highly ideal, flat, H-terminated Si(111) surface is attractive from an experimental point of view as it is a very well-defined system.

12.1.2 Reactivity of H–Si

The chemistry of the initial, H-terminated, Si surface greatly affects the properties of subsequent surfaces and interfaces [1–3]. The reactivity of unreconstructed Si(111) and Si(100) H-terminated Si surfaces has been thoroughly investigated. The (100) surface is of interest because the interatomic spacing of the Si atoms is well matched to the lattice constant of SiO_2. This behavior has led to Si(100) being the predominant orientation of Si wafers used in the electronics industry [31]. However, the (111) surface is an attractive model system for scientific investigation because of the ability of the NH_4F(aq) etch to form well-defined, high-quality H–Si(111) surfaces. In the following sections, the reactivity of H-terminated Si surfaces will be discussed, with emphasis on systems in which the reactivity affects the electronic properties of the resulting surfaces.

12.1.2.1 Aqueous Acidic Media The reactivity of H–Si with aqueous acidic media has been extensively evaluated [32–35]. Perhaps one of the most well-known examples of H–Si reacting with an aqueous acidic solution is the "SC-2 clean," in which H–Si is reacted with a $H_2O:H_2O_2:HCl$ mixture to ultimately yield a Si surface that is free of organic or inorganic contaminants. The chemistry of H–Si in aqueous acidic media is highly dependent on both the acid and the presence of any additional reactants or reagents. Depending on the solution composition, H–Si surfaces are protected in some acidic media, oxidize in other media, and dissolve in yet others. In contrast to alkaline HF etching described above, reactivity of Si in acidic media is highly isotropic.

In the absence of fluoride ions or oxidizing reagents, H–Si is stable in concentrated acids. This behavior is expected, because the high $[H^+]$ keeps the surface fully H-terminated. H–Si surfaces immersed in strong acids exhibit excellent electronic properties [2]. The lifetime of photogenerated minority carriers was measured using a contactless method, wherein the carrier density was measured using an inductively coupled RF bridge. The surface recombination velocity (SRV), which is related to the electronic defect density on the surface, was very low for such surfaces. Low SRV values were observed in a number of strong acids, and the SRV was a function of acid molarity. Furthermore, changing the acid molarity induced reversible changes to the surface recombination velocity. From the dependence on $[H^+]$, it was suggested that H–Si surfaces contain a finite number of strained, or otherwise weak, bonds, that act as recombination centers. The recombination centers resembled Lewis bases in an acid–base equilibrium with the surrounding solution, and the observed effect of acid molarity on SRV was ascribed to neutralization (H termination) of these defect sites [2].

In contact with an acidic aqueous solution that contains oxidants, the near-surface layer of such H-terminated Si(111) surfaces is rapidly converted to 1–2 nm of SiO_x [34]. The SiO_x passivates the surfaces to further oxidation, because HF-free solutions that contain oxidizers do not etch SiO_x. As part of the Si cleaning procedure developed at RCA laboratories, oxidation of H–Si in HCl(aq)/H_2O_2(aq) has been widely utilized [35, 36]. The RCA cleaning procedure removes a variety of organic contaminants and leaves low levels of metal contaminants.

The incorporation of fluoride ions into acidic media causes etching of Si and SiO_x. SiO_x etches very quickly in HF(aq); however, H–Si surfaces are more stable, etching at rates <1 Å/min. The etch rate increases dramatically when oxidizers are added to the HF(aq) solution [35], and numerous systems composed of H_2O, HF, and an oxidizer have been employed to rapidly etch Si [37]. Specifically, the HF(aq)/ HNO_3(aq) system etch rate is mm/min [36].

12.1.2.2 Aqueous Basic Media The H-terminated Si surface is unstable in alkaline media. The etching of Si in alkaline media has been studied extensively [32]. The alkaline etching of Si is anisotropic and thus it proceeds very differently from etching in acidic, fluoride-containing, media [33]. The etch rates of single-crystal Si have been investigated as a function of crystallographic orientation (Table 12.1) [38]. The use of a large hemispherical Si crystal allowed for monitoring of the etch rates at higher index planes. The orientation-dependent etch rates were functions of $[OH^-]$ and temperature. The alkaline etching of Si also depends on both $[H_2O]$ and $[OH^-]$. In organic alkaline solutions, the etch rate goes to zero as $[H_2O]$ approaches zero. In all cases, the etch rate of the (111)-oriented crystal face was considerably lower than the etch rates of the other faces. A reaction scheme for the dissolution of (100) oriented Si is presented in Fig. 12.2.

The replacement of H atoms with $-OH^-$ groups polarizes the Si back-bonds such that H_2O can attack. The rate-limiting step determined whether a surface was left either H-terminated or OH-terminated. The mechanism of Fig. 12.2 explains why the alkaline etches are anisotropic, because the (111) surface is much more stable with only one bond left to be OH-terminated, while three are involved in back-bonding. With only one –OH group, the back-bonds are less activated, and H_2O is less likely to attack and break the back-bonds.

TABLE 12.1 Orientation-Dependent Etch Rates (Relative to (110)) in mm/min as a Function of [KOH] at 70°C [38]

Crystallographic Orientation	KOH Concentration 30%	40%	50%
(100)	0.797 (0.548)	0.599 (0.463)	0.539 (0.619)
(110)	1.455 (1.000)	1.294 (1.000)	0.870 (1.000)
(210)	1.561 (1.072)	1.233 (0.953)	0.959 (1.103)
(211)	1.319 (0.906)	0.950 (0.734)	0.621 (0.714)
(221)	0.714 (0.491)	0.544 (0.420)	0.322 (0.371)
(310)	1.456 (1.000)	1.088 (0.841)	0.757 (0.871)
(311)	1.436 (0.987)	1.067 (0.824)	0.746 (0.858)
(320)	1.543 (1.060)	1.287 (0.995)	1.013 (1.165)
(331)	1.160 (0.797)	0.800 (0.619)	0.489 (0.563)
(530)	1.556 (1.069)	1.280 (0.989)	1.033 (1.188)
(540)	1.512 (1.039)	1.287 (0.994)	0.914 (1.051)
(111)	0.005 (0.004)	0.009 (0.007)	0.009 (0.010)

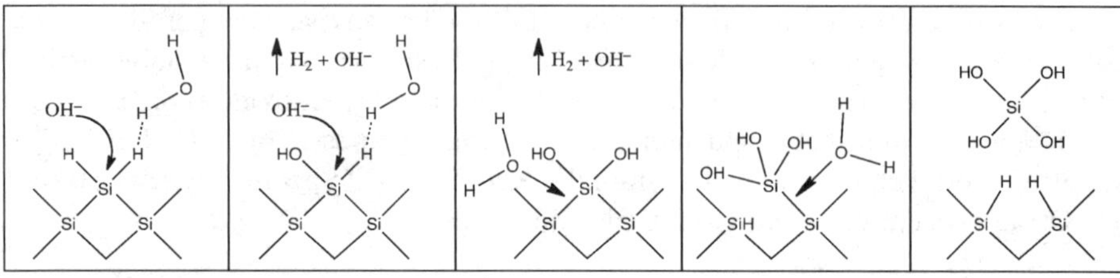

FIGURE 12.2 Proposed reaction scheme for the chemical dissolution of Si in basic media. Redrawn from Ref. 32.

In alkaline aqueous solutions that contain an oxidizing agent, hydrogen-terminated Si surfaces rapidly convert to Si covered with a thin overlayer of SiO_x. The conversion of H–Si to SiO_x has been studied extensively for RCA-1, $NH_4OH/H_2O/H_2O_2$ [35–37]. The amount of oxidizer present in the alkaline aqueous solution determines whether a SiO_x overlayer will form and thus protect the Si from further alkaline etching. If only small amounts of oxidizer are present, the surface will be continuously etched by the hydroxide ions.

12.1.2.3 Oxygen-Containing Environments

Electronics must be robust in air, so the stability of HF-etched Si surfaces in oxygen-containing environments has been investigated quite thoroughly. The mechanism of the initial oxidation of H–Si surfaces is not fully understood, but many important aspects have been observed and described. The growth rate of SiO_x on Si(100) was measured to be 0.2 nm/decade (hours), with a limiting oxide thickness of <1.4 nm as determined using XPS [39]. Conflicting results exist as to the effect of microscopic roughness on the rates of initial oxidation, but recent data suggest that the oxidation rates depend on the surface roughness [40, 41].

Oxidation of H-terminated Si in the presence of O_2 has also been investigated for the special case of oxidation in a pure oxygen, low-pressure (1–15 mTorr) environment at elevated temperatures (530–590 K). The reaction rate constants between O_2 and various Si surface moieties, such as monohydrides and dihydrides present on various crystal phases, differed by as much as 20-fold. The monohydride-terminated surfaces were substantially more stable than the other surfaces. Si(111) surfaces oxidized significantly more slowly than Si(100) surfaces, and dihydride steps on Si (111) surfaces oxidized more rapidly than monohydride steps. Flat Si(111) surfaces were the most stable. It was suggested that neither O_2 nor H_2O alone could explain the rapid rates of native oxide formation commonly observed, and perhaps radicals or other trace contaminants are responsible for the observed rate of oxide formation [41]. Curiously, the initial oxidation of H–Si surfaces proceeds without loss of atop Si–H bonds. Oxygen inserts into the Si back-bonds without the formation of Si–OH species [41, 42].

12.1.2.4 Alcohols

Alcohols are commonly used solvents for chemical reactions and cleaning procedures; thus, their interaction with H-terminated Si is of great interest. The H–Si surface is stable in a number of solvents including acetonitrile [43], toluene and mesitylene [43], and diethyl ether [44]; however, H–Si surfaces react with various alcohols [28, 43, 45, 46]. Early investigations of reactions between H–Si and alcohols utilized porous Si, which provided for improved signal to noise in spectroscopic studies and allowed further investigation through the luminescence properties of porous Si. Methanol was attached to porous Si through electrochemical processes involving partial dissolution of the Si substrate [47], though simply exposing the H–Si to methanol has been seen to be sufficient for CH_3O– attachment. Various studies have reported that attachment of alcohols to H–Si proceeds with partial oxidation of the underlying substrate [48, 49], while others have successfully formed Si–O–C bonds without oxidation of the substrate [50, 51]. The reactivity between alcohols and H–Si was used to form ferrocene-modified Si electrodes. The ferrocene was covalently linked to H–Si(111) simply by immersion in a ferrocene derivative that contained a terminal –OH group. Immersion in other alcohols also yielded modified surfaces, although the bonding and surface quality were not thoroughly explored [43]. The Si–O–C bond is prone to hydrolytic cleavage, regenerating the original alcohol and leaving the surface Si–OH terminated; thus, this approach is not generally used to effect subsequent modification of Si surfaces.

Protic and deuterated MeOH were used in a detailed study of Si–O–C formation on Si(111) [28, 46]. Partially methoxy-terminated Si surfaces were synthesized, with the remaining atop sites H-terminated and no observable incorporation of O atoms into the Si back-bonds. Prolonged exposure to methanol at room temperature led to some subsurface oxidation, while shorter exposures to methanol at 65°C resulted in no detectable (less than 3%) subsurface oxidation. Approximately one-third of the Si atop sites were methoxylated, in agreement with steric considerations (Fig. 12.3). A likely reaction mechanism for the formation of methoxy termination of Si(111) was suggested based on polarized FTIR spectra (Fig. 12.4); however, it was observed that upon exposure to CD_3OD, a Si–D species was present on the surface. The experimental data available did not allow for determination of a likely reaction mechanism to account for the origin of the Si–D species.

The electronic effects of alcohol termination of Si surfaces have also been investigated. The methoxylation of Si improved the anodic stability [52]. Exposure to methanol vapors effected both reversible and irreversible changes to H-terminated Si. Results from both capacitance–voltage and contact potential measurements indicated that surface methoxylation shifted the band-edge positions such that the electron affinity was decreased. The effects of methoxylation on the SRV were investigated as a function of the sample history, allowing for the separation of effects due to band bending and passivation of the Si surface. Low effective SRV was observed due to band bending, and it was demonstrated that care must be taken when using methanolic systems for characterization of Si surface properties [27].

12.1.2.5 Metals Highly rectifying metal contacts on Si can be used as solar cells and photodiodes. In addition, an understanding of the Si/metal interface is crucial to understanding the formation of ohmic contacts to Si [53]. The ability to form Si/metal contacts without disrupting the H termination of Si is also of great

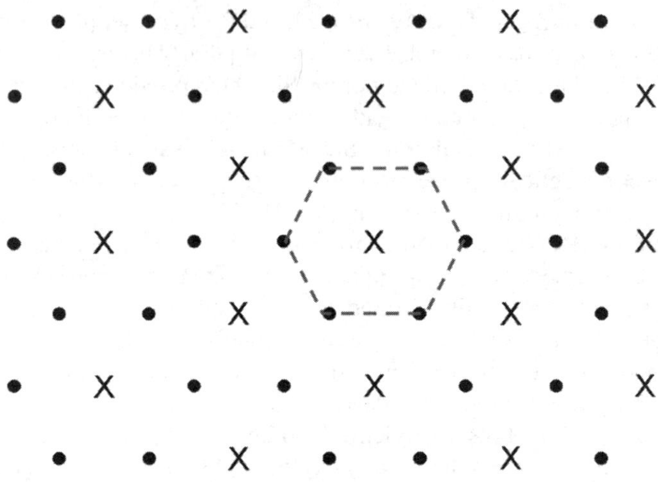

FIGURE 12.3 Si(111) with a saturation coverage of 66% Si–H (dots) and 33% Si–OCH_3 (crosses). Si–OCH_3 groups block all nearest neighbor atoms (denoted by hexagon) by the 360° thermal rotation of the methyl group about the Si–O bond. Adapted from Ref. 28.

interest because such methods allow for nondestructive characterization of the interface, dopant densities, and surface quality across Si wafers. Unfortunately, the electronic properties of Si/metal junctions are, with a few exceptions, highly nonideal and are dominated by the junction properties rather than by the properties of the metal and the bulk Si. Ideally, the barrier height, Φ_B, of a Si/metal junction should be linearly dependent on the work function, Φ_m, of the contacting metal. However, for Si, the barrier height is only weakly dependent on Φ_m [22, 54–56]. This behavior has been ascribed to surface states, likely metal silicides, at the Si/metal interface.

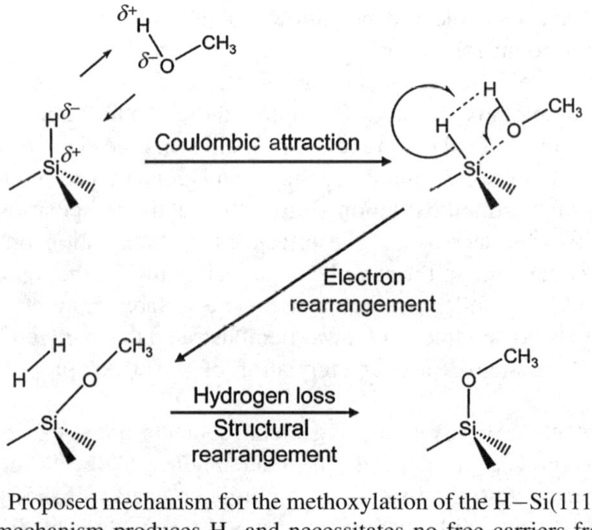

FIGURE 12.4 Proposed mechanism for the methoxylation of the H–Si(111) surface in neat methanol. The mechanism produces H_2 and necessitates no free carriers from the bulk Si. Reproduced from Ref. 46 by permission of the American Chemical Society, © 2006.

The barrier height can be correlated with the metal work function, though the dependence is weak (Fig. 12.5).

An important exception is the formation of Si/Hg junctions [26, 57–59]. Si/Hg silicides do not form under ambient conditions, as has been shown via XPS [59]. Hence, the Si/Hg junction has proven to be an important tool for soft-contact, nondestructive, electrical characterization of Si surfaces. Impedance spectroscopy and current density versus voltage (*J*–*V*) data have shown that the Si/Hg junction is near ideal, in contrast to other metal/Si junction formation schemes.

12.2 REACTIVITY OF HALOGEN-TERMINATED SILICON SURFACES

12.2.1 Background

Halogen-terminated silicon, that is, Cl–Si, Br–Si, and I–Si, has been studied extensively because halogen etching and termination was, initially, integral to understanding the reconstructions of Si(111) and Si(100) surfaces [60, 61]. This work was all performed under UHV conditions, with UHV-cleaned Si and molecular gases. Later it was discovered that Si surfaces in contact with Br_2/ROH or I_2/ROH solutions displayed very low surface recombination velocities [27, 62–64]. While neither the surface-bound species nor the mechanism for the surface lifetime

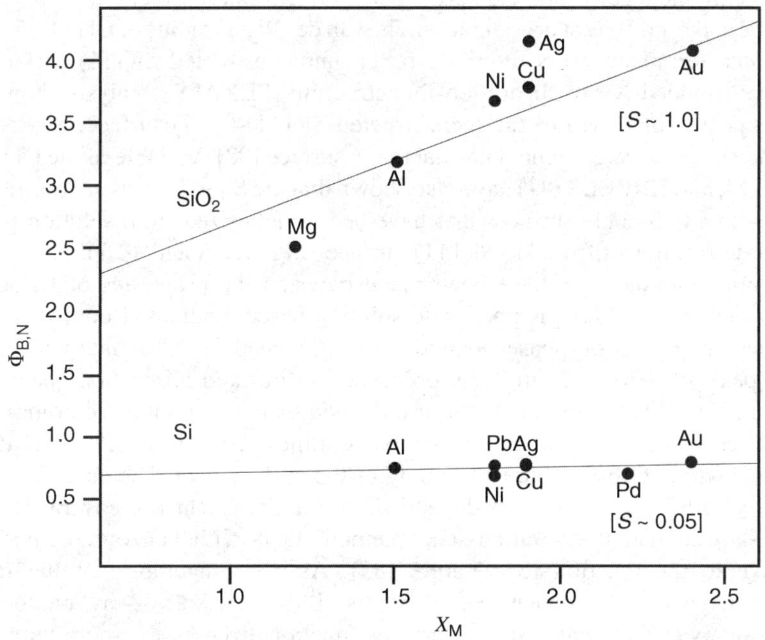

FIGURE 12.5 Barrier height at Si/metal and SiO_2/metal junctions as a function of the metal electronegativity. The slope, *S*, is inversely related to the extent of Fermi level pinning, where an ideal junction would have $S = 1.0$. Adapted from Ref. 56 by permission of the American Physical Society, © 1969.

enhancement was known, this behavior sparked much interest, and motivated research in the field. Finally, more recent interest in halogenated silicon surfaces has been spurred because although the halogen-terminated surfaces are more reactive than H-terminated surfaces, they can nevertheless be handled briefly under ambient conditions [65]. Accordingly, the halogenated Si surfaces serve as convenient and scalable intermediates for the solution-state syntheses of bound organic monolayers [66–70], and as a reactive surface for the deposition of inorganic films via atomic layer deposition (ALD) [71–75].

This section will focus primarily on the reactivity of halogenated surfaces in environments, for example, air, water, and organic solvents, that are typically encountered when the surfaces are used as intermediates for further functionalization. Many methods can be used to produce each halogenated surface, so a brief discussion of the differences in the resulting halogenated surfaces, and their effect on the subsequent surface reactivity, will also be presented.

12.2.1.1 Synthesis of Cl-Terminated Surfaces Several methods allow for the synthesis of Cl-terminated Si surfaces, including gas-phase reactions with diatomic molecular halogens under UV irradiation [69, 76, 77], elevated temperatures [78–80], and room temperature [77, 81], in addition to solution-state methods like PCl_5 with a radical initiator [67, 77, 82–86]. These methods have been shown to result in nominally similar Si–Cl surface bonding; however, the coverage, homogeneity, and etch pit density vary for the different functionalization methods and for different crystal faces. The van der Waals radius of Cl is 1.75 Å, so Cl atoms can terminate every atop site on an unreconstructed Si(111) surface [87]. Surface extended X-ray absorption fine structure (SEXAFS) analysis showed that after exposure of Cl_2(g) to the reconstructed Si(111)-(7×7) surface, Si–Cl bonds were present and were oriented normal to the surface [88]. Variable-angle FTIR [77], STM [81], and HREELS [81] have also shown that the Si–Cl bonds are normal to the surface on Cl–Si(111) surfaces that have been synthesized from solution-phase or gas-phase reactions of the H–Si(111) surface (Figs. 12.6 and 12.7).

Detailed comparisons have been made between the properties of halogenated Si surfaces prepared by gas-phase and solution-based methods. Room-temperature gas-phase and solution-phase methods both produced Si–Cl bonds with a vibrational peak FWHM of $5\,cm^{-1}$, as observed via infrared absorption spectroscopy (IRAS) [77]. This narrow peak width indicated extended regions of homogeneous Si–Cl termination. At short Cl_2 exposure times, partial Si–Cl coverage was observed without observable roughening of the surface [87]. The chemical shift of the Si–Cl stretching mode was dependent on the Si–Cl surface coverage. A STM study showed that the solution-state method, PCl_5 with benzoyl peroxide, was sensitive to the reaction temperature [83]. As the temperature of the reaction increased, so did the density of etch pits. Based on soft X-ray photoelectron spectroscopy (SXPS) data, Si–Cl surfaces functionalized by the room-temperature gas-phase reaction gave comparable Si–Cl coverage to surfaces produced by a solution-state reaction pathway. While the solution-based method resulted in a higher amount of oxygen, 1.4 ± 0.6 ML as observed via XPS, no SiO_x was detectable

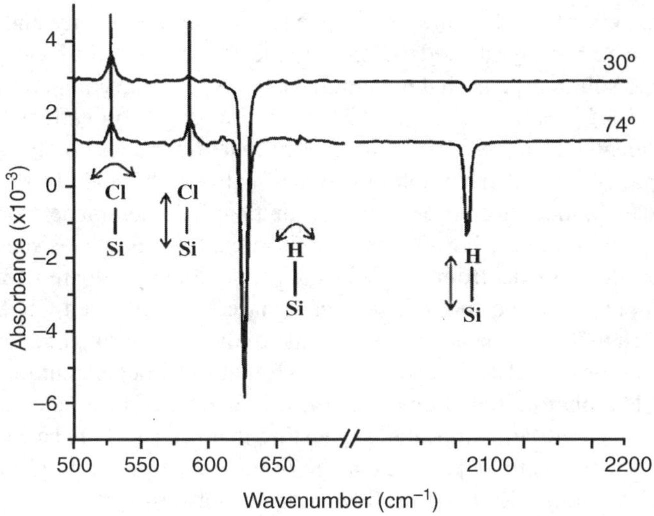

FIGURE 12.6 Transmission FTIR of the Cl–Si(111) surface (referenced to the H–Si(111) surface) produced using a wet chemical technique [77]. Si–Cl and Si–H (subtracted) bend modes are observed at both 30° and 74° off normal. The Si–Cl and Si–H stretch modes are polarization dependent and are observed only at 74° off normal, indicating that these modes are perpendicular to the surface. Reproduced from Ref. 77 by permission of the American Institute of Physics, © 2005.

in the SXPS spectra. The presence of oxygen was attributed to solvent workup steps that were not necessary for the gas-phase reaction. IRAS is more sensitive for oxide detection, and based on integration of the 1080 cm^{-1} peak, gas-phase chlorination gave ~3% of a ML of Si–O–Si, whereas the solution-phase chlorination produced <15% Si–O–Si coverage. In other IRAS studies, very low surface oxidation, 1.2%, or 0.15 ± 0.01 Å of subsurface oxide, was observed to form during the solution-phase chlorination reaction [89]. The observed differences in oxidation could thus be due to brief exposure of the surface to ambient air as opposed to the inherent properties of the reaction sequence. Based on vibrational shifts, the oxidation was presumed to be

FIGURE 12.7 Constant-current STM images showing full coverage halogenated Si(111) surfaces with little roughening. (a) Cl–Si(111), 50 × 50 nm^2, −1 V sample bias, 40 pA; (b) Cl–Si(111), 6.5 × 6.5 nm^2, −0.5 V, 50 pA; (c) Br–Si(111), 6.5 × 6.5 nm^2, 0.4 V, 400 pA. Reproduced from Ref. 81.

in isolated patches [77]. Despite the difference in etch pit density and initial SiO_x coverage, an SXPS study showed that the surfaces prepared by the room-temperature gas-phase and solution-phase halogenation routes oxidized at comparable rates [84]. The initial oxide formation during the Cl termination step did not affect the electronic trap state density at the resulting Si−C terminated surfaces. In fact, CH_3CH_2−Si(111) surfaces formed from solution-functionalized Cl−Si(111) showed only a slightly higher oxidation after >3 days in air than surfaces made from gas-phase Si−Cl precursor surfaces; however, the SRV after 24 h in air was measured to be lower for surfaces made from the solution-phase route compared to that of the CH_3CH_2−Si(111) surface formed from gas-phase functionalized Cl−Si(111).

Differences in Si−Cl surface coverage, and in the surface roughness that resulted from various types of Cl_2(g) reactions, were found to be dependent on the reaction conditions. The photoinitiated chlorination of silicon led to broadening and a red shift of the Si−Cl vibrational peaks, suggesting that such Si(111) surfaces were not fully covered with Cl atoms [77]. A change in the temperature of chlorination of Si (100) with UV, from 350 to 423 K, led to an increase in surface roughness, from 1.70 ± 0.54 to 7.9 ± 7.0 Å [90]. The initial H-terminated surface had a roughness of 1.61 ± 0.14 Å. Significant etching by Cl_2 occurred only above 350 K and 10 Torr.

The temperature-programmed desorption (TPD) analysis of Cl−Si(111) synthesized via a solution-based benzoyl peroxide-initiated PCl_5 route closely matched that of high-coverage, gas-phase functionalized Cl−Si(111), although some differences were observed in the high-temperature desorption [67, 82, 91]. TPD of high-coverage Si surfaces showed desorption of $SiCl_x$, where $x = 2$ and 3, beginning at 670 K, with some $x = 4$ and 2 desorbing at 850 K [67] or 950 K [91]. At low Cl coverage, $SiCl_2$ was the main product, and desorbed at 950 K [91]. A recombinative desorption mechanism was proposed, in which $Si{-}Cl + Cl \rightarrow SiCl_2$ and $SiCl_2 + Cl \rightarrow SiCl_3$. The Cl-terminated Si surfaces were demonstrated to be stable for >12 h in inert atmosphere [80, 92].

12.2.1.2 Synthesis of Br-Terminated Surfaces Several methods, including Br_2(g) or Br_2(soln) and solution-based bromination reagents such as N-bromosuccinamide (NBS), have been used to obtain Br-terminated silicon surfaces [66]. It has been suggested that most of these functionalization methods proceed by the same mechanism. For example, it is thought that the NBS reaction generates *in situ* small amounts of Br_2, which is presumed to be the actual brominating agent. Solution-based bromination has been achieved using NBS in DMF, with a radical initiator present, upon heating to 60°C for 20 min. Bromochloroform can also be used under thermolysis (80°C for 30 min), photolysis (300 nm UV for 30 min), or radical initiation (benzoyl peroxide at 60°C for 30 min) to produce Si−Br terminated Si surfaces [66]. All of these methods produced the same Si−Br shift in XPS and showed C and O present on the surface, attributable to adventitious C and O sources.

12.2.1.3 Synthesis of I-Terminated Surfaces Unlike Cl- and Br-terminated surfaces, iodine was not observed to terminate every atop site on Si(111), so such surfaces have always been found to be chemically inhomogeneous. Solution-state iodination of silicon surfaces has been accomplished through several methods. Si(001) surfaces functionalized by submersion in I_2/benzene gave ~33% coverage of Si atop sites [68]. Surfaces functionalized in this manner showed surface photovoltage effects suggesting the presence of midgap surface states. SEXAFS data of surfaces resulting from the reaction of I_2 with the reconstructed Si(111)-(7×7) surface showed that I binds in the same fashion as Cl–Si(111), forming Si–I bonds oriented normal to the Si surface [88].

Exposure of vacuum-cleaned Si(100)-(2×1) surfaces to CH_3I resulted in mixed CH_3 and I termination [93–95]. Approximately one CH_3I adsorbed per two Si atop atoms. The surface was characterized through a variety of techniques, including kinetic uptake measurements [93], TPD [93], Auger electron spectroscopy (AES) [93], HREELS [94], and IRAS [95].

Differences in the reactivity of Cl-, Br-, and I-terminated silicon in air or toward solvent and other reactants have not been thoroughly documented. The different factors that affect etching by molecular diatomic halogens, however, have been well investigated. While measurements of the activation energy for desorption are not statistically different between Si–Br and Si–Cl surfaces, Br potentially has a lower activation energy for desorption from Si than Cl does [96]. The activation energy for desorption of Cl was observed to be 2.3 ± 0.3 eV, while that of Br was 1.8 ± 0.4 eV. In addition, during bromination the higher adsorption probability and back-bond breaking probability should be considered. The adsorption probability was 1.7 times that of Cl, and the back-bond breaking probability was 2.5 times that of Cl.

Finally, caution is advised in interpretation of some data for halogenated surfaces. Br–Si bonds were observed to break during Rutherford backscattering (RBS) and AES experiments [97]. In addition, while monochlorinated sites were stable under UV illumination, dichlorinated sites were not [90].

12.2.2 Reactivity of Halogenated Silicon Surfaces

12.2.2.1 Halogen Etching When clean, reconstructed Si surfaces or H-terminated Si surfaces are exposed to diatomic halogens, etching can occur, leaving a halogen-terminated surface [60, 61, 72, 77, 92]. Alternatively, exchange of hydrogen for halogen termination on H-terminated Si surfaces occurs [61, 80, 92]. Both of these mechanisms result in a halogen-terminated Si surface; however, the choice of reaction conditions dictates whether the initial surface structure will be preserved. For example, Si is etched upon exposure to HCl in vacuum at high temperatures, but at lower temperatures, HCl exposure results in Si–Cl termination [79].

Solution-based Cl termination of Si via radical-initiated PCl_5 (~0.7 M PCl_5 in chlorobenzene with a catalytic amount of benzoyl peroxide) results, by STM, in an increased density of etch pits relative to functionalization at room temperature with Cl_2(g) [83]. The density of etch pits was dependent on the reaction temperature. IRAS was used to characterize surfaces functionalized via the solution method, and the Si–Cl stretch (586 cm^{-1}) and bend (626 cm^{-1}) were observed [89].

12.2.2.2 Aqueous Media Exposure of Cl-terminated Si to 2 Torr of pure water vapor for 1 h, did not lead to surface degradation. HREELS showed some peak broadening, but upon heating to 100–200°C, the original spectrum was restored [81]. After vapor exposure and heating, no reaction, or adsorbed species, were observed via HREELS or STM. A more in-depth study focused on the effect of temperature during water vapor exposure [98]. IRAS was used to monitor the reaction of D_2O at Cl–Si(111) and Cl–Si(100) surfaces. At $T < 300$°C, no reaction was observed. At >300°C, Si–Cl modes diminished and a broad feature, indicative of Si–O–Si, grew in at 1050 cm^{-1}. The Si(100) surface was less stable, with oxidation apparent at 250°C. With increasing temperature, the oxide modes observed on Cl–Si(111) split into the transverse optical (TO) and longitudinal optical (LO) modes of an extended SiO_2 structure at 1060 and 1240 cm^{-1} (Fig. 12.8). A 1 min exposure to D_2O at 450°C resulted in an overlayer having ~1/3 of the thickness of a native oxide, that is, 5.5 Å of oxide. No O–D stretch was observed at any temperature tested, for a 1 min D_2O exposure. After D_2O expose and an additional 15 min heating period, additional oxidation and the appearance of a peak at 2760 cm^{-1} were observed. It was presumed that additional reaction with physisorbed D_2O effected the increase in overlayer thickness to 8.6 Å of oxide and 0.5 ML of Si–OD. Reaction of Si–Cl with H_2O vapor was not a viable method to form oxide-free, OH-terminated Si, because the temperature required to produce Si–OH also caused oxidation of the Si surface.

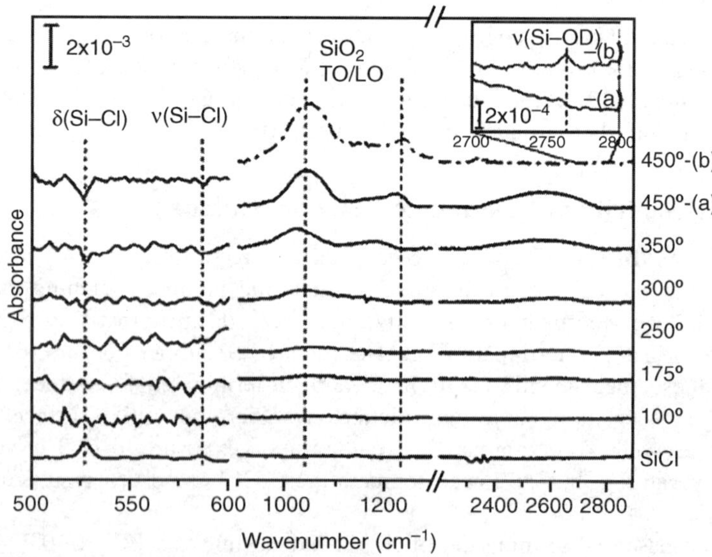

FIGURE 12.8 Transmission FTIR spectra of pure D_2O exposure to Cl–Si(111) surfaces at increasing temperatures. Below 300°C, no reaction is observed; however, at greater than 300°C, Si–O–Si stretching modes at 1050 cm^{-1} appear. O–D stretch modes are not observed at any temperature after a 1 min exposure without a 15 min anneal at $T = 450$°C. Reproduced from Ref. 98 by permission of the American Institute of Physics, © 2005.

Immersion of Cl–Si in water that contained pyridine, or even a 5 s dip in pure water, resulted in irreversible oxidation of the surface. After immersion, hydroxide termination, with a peak at 825 cm^{-1}, with O insertion into the Si–Si back-bonds, with a peak at 1120 cm^{-1}, was observed via HREELS [81] (Fig. 12.9) and via IRAS [80]. The Si–Cl vibrational peak intensity also diminished. Surface hydroxylation is of interest as a starting surface for ALD; however, this method does not lead to a hydroxylated surface that is free of back-bonded O. The inversion layer, created by the electron-withdrawing chlorine at the Si surface [92], decreased when the surface was immersed in H_2O [81], as evidenced by a narrowing of the elastic peak observed by HREELS. AM1 Si cluster models showed that Si–Cl bonding was preserved, with O insertion into back-bonds, and concomitant H_2 production, as the thermodynamically favored reaction pathway (Fig. 12.10). Cl-terminated Si has been used as an intermediate in the deposition of SiO_2 by ALD. The two-step growth process uses H_2O and $SiCl_4$ [71, 99]:

(a) $\text{Si–OH}^* + \text{SiCl}_4 \rightarrow \text{SiO–SiCl}_3^* + \text{HCl}$

(b) $\text{Si–Cl}^* + \text{H}_2\text{O} \rightarrow \text{Si–OH}^* + \text{HCl}$

where * denotes a surface species. The growth was slow and required large amounts of H_2O and high temperatures, likely because of the low reactivity of Si–Cl with water vapor. Upon addition of pyridine, however, the temperature required for growth decreased from >600 to 300 K, and the saturation exposure of H_2O was reduced from $\sim 10^9$ to $\sim 10^4$ L [71]. It was suggested that pyridine complexes with the H of the surface Si–OH moieties, allowing for nucleophilic attack of the $SiCl_4$.

Br-terminated Si is not stable in fluoride-containing aqueous solutions. The fluoride etch acts by attacking the polarized Si back-bonds, and completely removes Br from Br-terminated silicon surfaces [97]. It is very likely that an analogous process would be observed for Cl- and I-terminated surfaces.

12.2.2.3 Oxygen-Containing Environments The Cl-terminated Si surfaces are relatively stable in air, compared to the air stability of a vacuum-cleaned Si surface. Transport in air for up to several minutes does not lead to observable surface contamination [83]. The conductivity of Si–Cl surfaces decayed rapidly upon continued exposure to air, with decay time constants of <15 min or >1 h, depending on the humidity [92]. The Si : Cl ratio of such surfaces calculated by XPS decreased by 50% after 20 min in air [100]. Upon exposure to air for extended periods of time (>15 min), oxidation was observed at Cl-terminated Si surfaces via IRAS and HREELS [80, 81]. After 1 h in air, HREELS spectra showed clearly visible peaks at $\sim$1110 and 850 cm^{-1}, which indicated back-bonded O in a relatively advanced state of oxidation. This behavior indicated that the oxidation occurred in patches. Unlike the oxidation resulting from immersion in water, air oxidation led to very few hydroxylated Si surface sites. Instead, O insertion into back-bonds was observed, with H-terminated Si peaks at 2250 cm^{-1} (Fig. 12.9). STM images showed that the oxide grew initially at step edges, but upon exposure to air for an extended

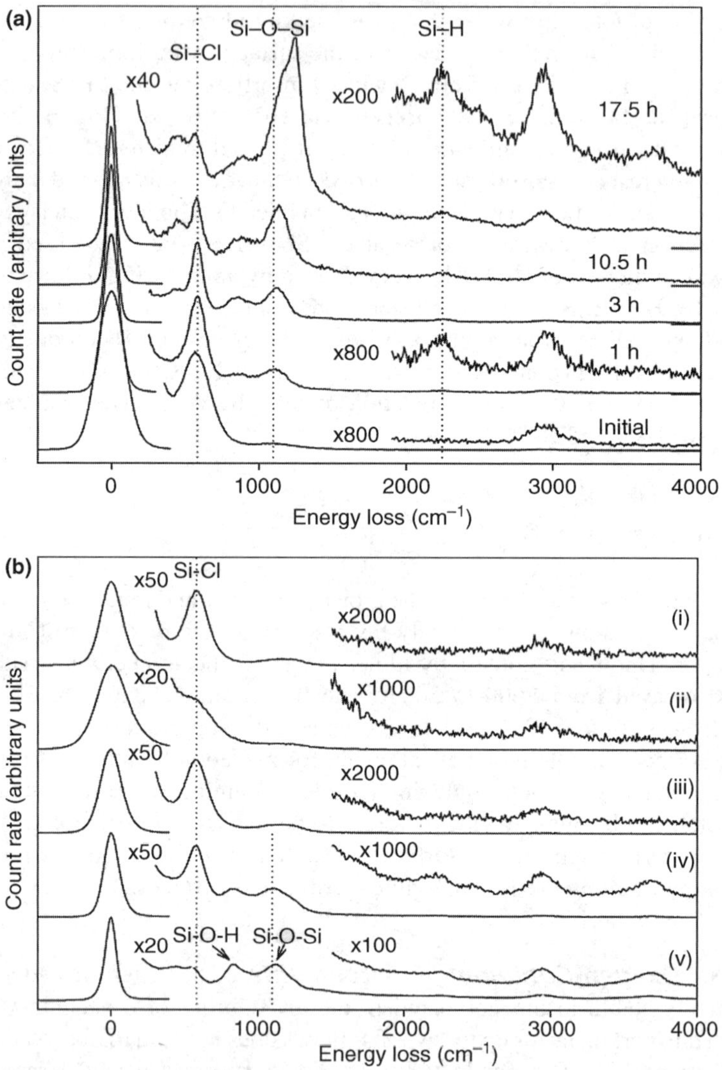

FIGURE 12.9 (a) Oxidation of Cl–Si(111), observed using HREELS, by exposure to air for various times. (b) HREELS spectra of Cl–Si(111) (i), after exposure to 7×10^9 L water vapor (ii), after water vapor exposure and heating to ~200°C (iii), 5 s dip in deoxygenated H_2O (iv), and water dip for 1 min (v). Figures reproduced from Ref. 81.

period of time, the oxide grew in patches that extended from the step edges. After >17 h in air, the surface resembled a "native oxide," although some Si–Cl sites persisted. After >24 h, Cl was no longer detected by XPS [79]. The contact angle also changed, from 60° to 40°, during oxidation in air. Humidity plays a role in oxidation in air, in that the surface oxidation rate increases as humidity increases. Consistently, the time to degrade the surface conductivity decay increased from 15 min to over an

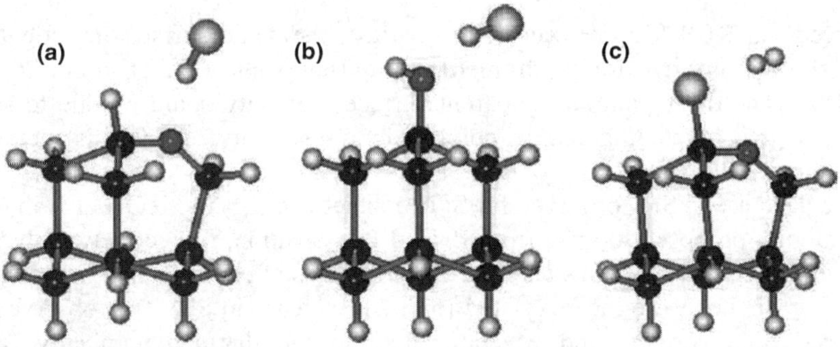

FIGURE 12.10 Possible reaction products from the reaction of the Cl–Si(111) surface with water. (a) Oxygen insertion in the Si–Si back-bond with hydride termination of the Si atop site and evolution of HCl, (b) hydroxylation of the Si atop site with release of HCl, and (c) oxygen insertion in the Si–Si back-bond with H_2 evolution and retention of the atop Si–Cl. Reproduced from Ref. 81.

hour with changes in humidity [92]. The factors contributing to oxidation of Si–Cl surfaces in air are complex, and are analogous to those involved in the oxidation of Si–H in air [80]. Water plays a role, but as mentioned earlier, pure water vapor does not cause oxidation. Few studies offer a direct comparison of the stability of different crystal faces of halogen-terminated Si; however, in a comparison of the stability in air of Cl–Si(111) and Cl–Si(100), the (111) surface was more stable [79, 80]. The difference in stability was proposed to arise from the chemical inhomogeneity, or only partial chlorination of the Cl–Si(100) [80].

Br-terminated Si(111) has also been reported to be only mildly moisture sensitive and stable in air for up to several minutes [66].

12.2.2.4 Alcohols Subsequent to the observation of very low SRV values for Si surfaces in contact with I_2/MeOH, the reactivity of halogenated surfaces with alcohol was thoroughly investigated. In addition, the surface chemistry of Si in MeOH is interesting because MeOH is a common rinse solvent for solution-phase Si surface functionalization chemistries.

When Cl-terminated Si was immersed in a 0.1 M alcohol solution and heated to 60–70°C for >14 h, alcohols were observed to attach to the Si surface [69]. This method of surface functionalization has been demonstrated on Si(111) and Si (100). On Si(111), the addition of pyridine catalyzed the reaction, although the effect on rate was minimal. XPS analysis of Si–OR surfaces functionalized via this method indicated the presence of some suboxides. After 15 h at 60°C in either octanol or octadecanol, well-ordered organic films were produced on the resultant Si surfaces [79]. The contact angle was observed to increase from 70° on the Cl-terminated surface to >102° for the Si–OR surfaces. An increase in thickness, measured using ellipsometry, corroborated the increase in contact angle after reaction with alcohols. Sonication of Cl-terminated Si in MeOH for 10 min resulted in oxide patches of $\leq$0.2 Å or 1.6% surface area, and Si–O–C vibrational modes, as detected by IRAS. The consumption of 28% of the Si–Cl sites was observed via RBS [89]. However, in contrast, the same surface investigated by XPS showed no reaction with methanol during sonication at room temperature [82]. The discrepancy in apparent surface reactivity could be due to solvent purity, degree of etching during chlorination, sensitivity of analysis method, or some other factor.

The decrease in SRV observed for Si in contact with Br_2 in MeOH or I_2 in MeOH was initially proposed to arise from Br and I passivation, respectively. Subsequent work indicated that some Si sites were Br-terminated after immersion in Br_2/MeOH, and other Si sites were methoxylated [63]. An increase in SRV was observed after extended time periods, and was attributed to the thermodynamically favored oxidation of Si–Br sites by dissolved O_2 or residual H_2O.

An XPS comparison of I–Si(100) surfaces functionalized from benzene and from methanolic I_2 solutions showed decreased C and higher I coverage on the surfaces that had been functionalized in benzene [68]. Immersion of Si in the I_2/benzene solution gave ~33% of a monolayer while I_2/MeOH gave only 18%. The decrease in SRV can be attributed to repulsion of carriers from the surface, as a result of band bending induced by the large barrier height contact between the Si and the I_2/MeOH solution (Table 12.2) [27, 101]. At longer immersion times, the SRV increased [63], consistent with the thermodynamically favored oxidation of Si–I sites. As the surface oxidized, electronic defect states could pin the Fermi level and decrease the band bending, thus increasing the effective SRV. Upon immersion of H–Si(111) in I_2/MeOH, both I- and CH_3O-terminated sites were observed [68, 102, 103]; however, isotope labeling studies showed that the CH_3O– termination occurred only during immersion in I_2/MeOH, not during a subsequent rinse [102, 104]. This suggested that alkoxylation was not occurring through Si–I sites, but via reaction of CH_3OH with H-terminated sites, with I_2 acting as an oxidant.

12.2.2.5 Other Solvents Alkylated surfaces formed via the two-step halogenation alkylation reaction were studied using IRAS and RBS [89]. TO and LO (1034 and 1100 cm^{-1}) SiO_2 oxide modes were observed after submersion of Cl–Si(111) surfaces in an alkyl Grignard solution at 80°C, and oxidation was thought to occur in patches. Ring opening of the THF, through a mechanism similar to autoxidation with O_2 and light, was thought to play a role in the observation of Si–O–C modes (Fig. 12.11). When Cl-terminated surfaces were immersed in anhydrous, inhibitor-free THF at 80°C for 3 h, <25% of the Si–Cl was consumed, and 5 Å (20% of a monolayer) of Si–OR was observed. In another study, Cl–Si(111) was immersed in 1.0 M CH_3CH_2MgCl in THF-d_8 at 70°C, and no C–D was observed via grazing angle ATR [105]. The cause of the discrepancy is unknown, but could be due to impurities. Reemergence of H-terminated sites was observed after alkylation reactions on Cl-terminated silicon [89, 105, 106]. Isotope labeling studies have shown that the hydride originates from the β-hydrogen on CH_3CH_2–Si(111) [105]; however, in the absence of a β-hydrogen, H termination is still observed. It has been suggested, although not demonstrated, that this could originate from the solvent [106].

TABLE 12.2 Lifetimes of Si Samples [101]

	τ(μs)	S (cm/s)
$NH_4F_{(aq)}$-etched Si(111)		
(1) CH_3OH–0.05 M I_2	600 ± 300	16 ± 8[a]
(2) N_2	12 ± 1	810 ± 70
(3) CH_3OH–0.05 M I_2	700 ± 200	14 ± 4[a]
(1) CH_3OH–0.05 M Fc–0.05 M Fc^+	700 ± 200	14 ± 4[a]
(2) N_2	140 ± 40	70 ± 20
(3) CH_3OH–0.05 M Fc–0.05 M Fc^+	600 ± 20	16 ± 1[a]
(4) THF–0.05 M $Me_{10}Fc$–0.01 M $Me_{10}Fc^+$	101 ± 8	98 ± 8
(1) THF–0.05 M I_2	270 ± 90	40 ± 10
(2) N_2	3.8 ± 0.2	2600 ± 100
(3) THF–0.05 M I_2	300 ± 100	30 ± 10
(1) THF–0.05 M Fc–0.005 M Fc^+	180 ± 90	50 ± 30
(2) N_2	2.9 ± 0.3	3400 ± 300
(3) THF–0.05 M Fc–0.005 M Fc^+	220 ± 160	40 ± 30
(4) THF–0.05M $Me_{10}Fc$–0.01 M $Me_{10}Fc^+$	3.0 ± 0.2	3300 ± 200
(1) THF–0.05 M $Me_{10}Fc$–0.01 M $Me_{10}Fc^+$	4 ± 1	2400 ± 600
(2) N_2	3.4 ± 0.1	2870 ± 90
(3) THF–0.05 M $Me_{10}Fc$–0.01 M $Me_{10}Fc^+$	3.0 ± 0.4	3300 ± 400
(4) THF–0.05M Fc–0.005 M Fc^+	250 ± 60	40 ± 10
Air–oxidized Si(111)		
(1) CH_3OH–0.05M I_2	500 ± 200	19 ± 8[a]
(2) N_2	12 ± 2	800 ± 100
(3) CH_3OH–0.05 M I_2	500 ± 200	18 ± 8[a]
(4) THF–0.05 M $Me_{10}Fc$–0.01 M $Me_{10}Fc^+$	23 ± 5	420 ± 90
(1) CH_3OH–0.05 M Fc–0.05 M Fc^+	300 ± 100	30 ± 10[a]
(2) N_2	9 ± 2	1100 ± 200
(3) CH_3OH–0.05M Fc–0.05M Fc^+	220 ± 40	44 ± 8
(4) THF–0.05 M $Me_{10}Fc$–0.01 M $Me_{10}Fc^+$	8 ± 1	1200 ± 200

[a] Because the lifetimes measured for these surfaces were not statistically different from lower bounds of the bulk lifetime deduced from measurements of the same wafer in conc. $H_2SO_{4(aq)}$, the computed values represent upper bounds on the actual surface recombination velocity.

As with H-terminated silicon, any passage of anodic current on Cl–Si(111) would likely oxidize the Si; however, this expectation has not been borne out by experiment. An irreversible reduction of Cl–Si(111) in THF has been observed at −2.5 V versus Ag/Ag^+ [106]. Upon successive sweeps, the intensity of this reduction decreased, indicating that the reducible species was in finite supply at the electrode.

I-terminated surfaces are more susceptible to nucleophilic attack than is the H-terminated silicon surface [68]. Reactivity of I–Si(111) has been observed with terminal olefins in the presence of light. This behavior is in contrast to that of Cl-terminated silicon, which does not react with ethylene, even under UV irradiation [100]. Using this approach, the 2,2,2-trifluoroethyl-undec-10-enoate was attached to I-terminated Si

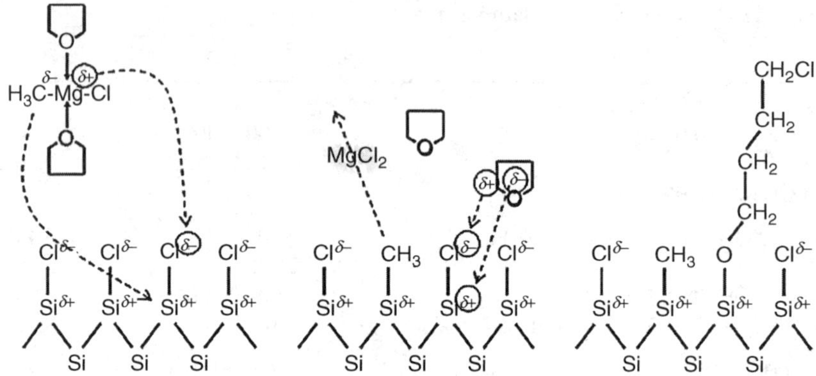

FIGURE 12.11 Proposed reaction mechanism of Grignard reagent addition and THF ring opening at the Si–Cl surface. Reproduced from Ref. 89 by permission of the American Chemical Society, © 2007.

surfaces. This functionalization was reported to be faster, and to occur under irradiation with longer wavelength light, than hydrosilylation of H-terminated silicon. Photopatterning was also demonstrated. Doping highly affected this reaction, suggesting that photoexcited carriers created Si^+ sites for attack by a nucleophile.

Amine functionalization has been observed on clean Si(111)-(7×7) surfaces in UHV [107]. In addition, amines were observed to react with Cl–Si(100) and porous Cl-terminated surfaces in vacuum at 450 K and in solution with porous Cl-terminated Si [70]. The surface species was thought to be a bridged amine, Si–N(R)–Si, with loss of HCl in the reaction, because no N–H was observed via IR. The amine-functionalized surface was thermally stable up to 750 K in UHV [70]. In another gas-phase example, exposure of Cl–Si(100) to NH_3(g), at 348 K and below 100 Torr, produced Si-bound primary amines [90]. Assignment of the surface species was based on the XPS shift of the amine-derived N. Upon heating to 500–1100 K, a transition to secondary and, finally, to tertiary amines was observed (Fig. 12.12). This behavior could explain the observation of secondary amines on surfaces that were functionalized at a higher temperature. A saturating coverage of 0.33 ML of Si–NH_2 was observed; this coverage was attained within 5 min and, below 100 Torr, was independent of pressure. The reactivity of Cl–Si(100) with NH_3 was markedly different from that of H–Si(100), in that no reaction was observed at H-terminated silicon. The higher reactivity of Cl–Si was attributed to a lower activation energy. At higher pressures, oxidation by H_2O contaminants dominated the surface reactivity. Although the bond energy of Si–Cl is very high (3.7 eV versus 2.5 eV of Cl–Cl and 2.3 eV of Si–Si), the Si–Cl bond is very polarized. Incoming ammonia could form a dative bond to Si, facilitating the reaction. First-principles calculations of ammonia adsorption on Cl–Si(001) have shown that reaction with release of HCl is highly thermodynamically unfavorable [108]. It was suggested that a surface bound Si–NH_3^+ with bound Cl^- remaining on the surface is more likely, but experimentally no Si–$NH_3^+Cl^-$ was observed [90].

As would be expected in small molecule chemistry, the presence of an organic base, such as pyridine, promotes nucleophilic attack at Cl-terminated Si (Fig. 12.13).

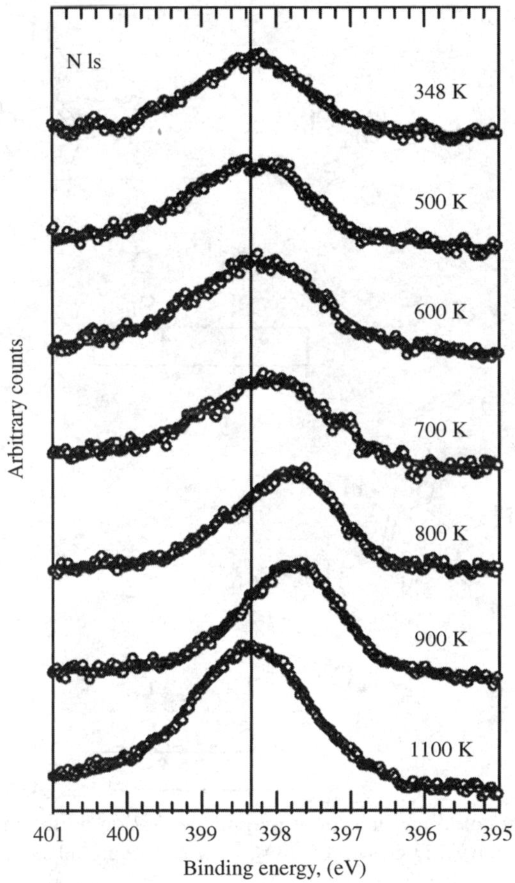

FIGURE 12.12 High-resolution XP spectra of the N 1s peak after exposure of a Cl−Si(100) surface to NH_3 (10 Torr NH_3 in N_2 at 100 Torr for 10 min) at 348 K. Transition from primary to secondary to tertiary amines is observed with annealing at the temperatures indicated. Reproduced from Ref. 90.

Pyridine accelerates the reaction between Cl-terminated silicon and water [81], including water vapor [71], as well as reactions with alcohols [69, 109]. Pyridine has also been used to catalyze dense OH-terminated polyethylene glycol (PEG) brush formation on Cl-terminated silicon [109].

12.2.2.6 Metals Because halogen-terminated Si surfaces are, in general, a reactive intermediate for surface functionalization, X−Si/metal heterojunctions are rare. The Cl−Si(111) and Cl−Si(100) surfaces have been used as starting surfaces for the ALD formation of Al- and Hf-based dielectrics [98]. The resulting surfaces were relatively unreactive with water; however, Si−M bonds were readily formed. After exposure to trimethylaluminum (TMA), Al−CH_3 and Si−CH_3 modes were readily visible via IRAS.

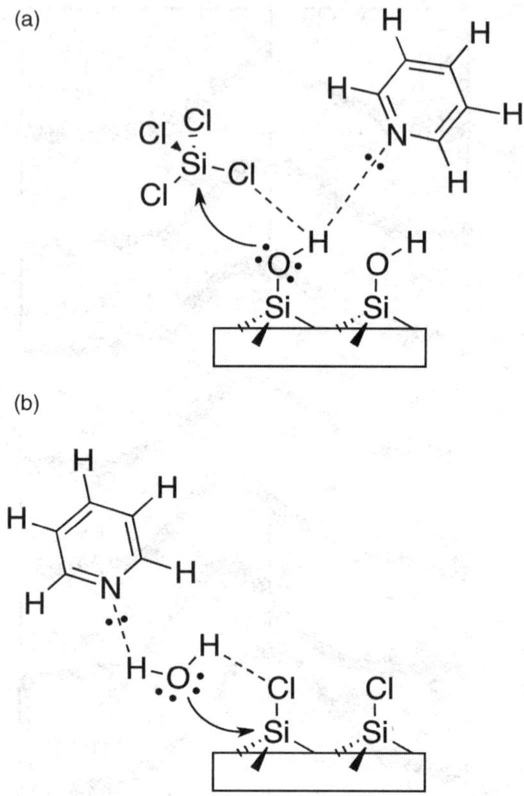

FIGURE 12.13 Proposed mechanism of pyridine activation during the $SiCl_4$ (a) and H_2O (b) half-reactions of the ALD growth of SiO_2. Redrawn figure adapted from Ref. 71.

12.3 CARBON-TERMINATED SILICON SURFACES

12.3.1 Introduction

The chemical mechanisms of Si–C bond formation have been discussed in the previous chapter. Hence, this chapter focuses on the reactivity between Si interfaces and common reactants, including water under neutral, acidic, and alkaline conditions, ambient atmosphere, metals, and a variety of common organic solvents, including alcohols. The physical and electronic properties of C-terminated Si surfaces are described below, as a starting point to elucidate the effects of reactivity of the Si–C surface under commonly occurring conditions.

The reactivity of Si–H, Si–Br, and Si–Cl surfaces depends on the surface roughness and on the crystal orientation. In addition, for C-terminated Si surfaces, incomplete termination (i.e., a mix of Si–C and Si–H terminated atop sites) and physical barriers imposed by large molecules also affect the surface reactivity. The fully methyl-terminated Si(111)-(1×1) surface (CH_3–Si(111)) provides a unique opportunity to obtain detailed insight into the inherent chemical reactivity and electrical properties of the Si–C bond. A comparison between the atop Si atom spacing of 3.84 Å [53] and the van der Waals radius of the methyl group (~2.5 Å) suggests that $-CH_3$ is the only saturated hydrocarbon that can terminate every atop site on the unreconstructed Si(111) surface [85, 110, 111]. The conversion of atomically flat H–Si(111) to CH_3–Si(111), via the Cl–Si(111) surface, produces an atomically flat, CH_3–Si(111) surface with nominally unity coverage (Fig. 12.14) [67, 85, 112]. The CH_3–Si(111) surface can also be synthesized through the anodic addition of CH_3MgI to H–Si(111) [110]. Because the smallest physical barrier to interacting with the Si–C bond is presented by the CH_3–Si(111) surface, its similarities to, and differences from, other C-terminated surfaces will be highlighted.

Isolation of the stability introduced by a physical barrier to reactants is more difficult. The deconvolution of the effects of a physical barrier and chemical stability have been attempted [113]. It is important to understand effects of a physical barrier on stability, as well as the consequences for common applications.

12.3.2 Structural and Electronic Characterization of Carbon-Terminated Silicon

The structural characterization of C-terminated Si surfaces will be briefly reviewed, as it serves to highlight the properties and differences among Si–C terminated surfaces.

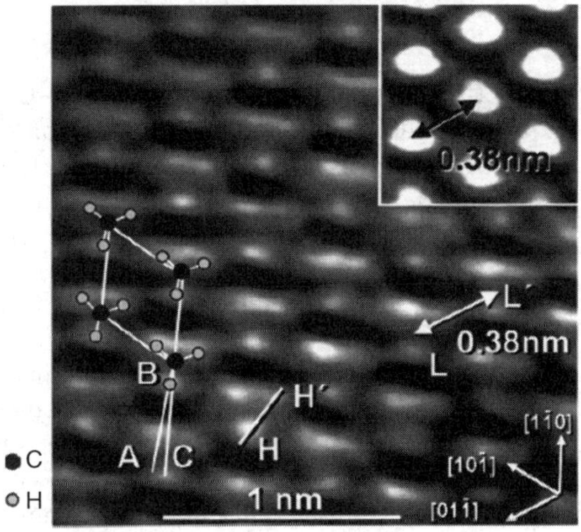

FIGURE 12.14 STM image of CH_3–Si(111) surface (4.7 K, sample bias $V_s = -2.5$ V, and constant current of 0.050 nA). The center to center distance of the $-CH_3$ groups (L to L′) is 0.38 nm. The low-index planes, denoted in the bottom right, were determined by X-ray crystallography. The surface unit cell is outlined by the parallelogram, with $-CH_3$ groups superimposed at each vertex. The angle between A and C is $7 \pm 3°$. The inset (top right) is the same surface imaged at 77 K. Figure reproduced from Ref. 112 by permission of the American Chemical Society, © 2005.

12.3.2.1 Structural Characterization of CH_3–Si(111) Two methods have been used to synthesize CH_3–Si(111) surfaces [67, 110]. One is through anodic addition of CH_3MgI to H-terminated Si surfaces [110]. The other, more widely utilized approach, is through surface halogenation followed by alkylation. In the later approach, a highly reactive Si–Cl or Si–Br surface is synthesized and then exposed to alkyl Grignard or organolithium reagents [67]. The resulting CH_3–Si(111) surface has been thoroughly characterized. The chlorination/alkylation procedure results in an unreconstructed Si(111) surface, as evidenced by LEED [114]. The absence of Si–H, Si–O, and Si–Cl stretching modes in the FTIR spectrum of CH_3–Si(111), along with the appearance of a polarization-dependent –CH_3 umbrella mode, suggested that Si–C bonds terminated every atop Si atom on the unreconstructed Si(111) surface [115]. The polarization dependence of the methyl umbrella vibrational modes showed that the –CH_3 group was attached normal to the Si(111) surface plane. The Si–C bond has been directly observed through photoelectron spectroscopy (PES) [85, 116], HREELS [82], and FTIR spectroscopy [115]. Surface-sensitive XPS and ultraviolet photoelectron spectroscopy (UPS), in conjunction with use of the substrate-overlayer model, indicated a –CH_3 group coverage of one monolayer [85, 116–118]. Low-temperature STM images confirmed the high degree of order, and unity coverage, of the CH_3–Si(111) surfaces over large-area terraces [112]. Thus, the CH_3–Si(111) surface is an ideal surface to understand the effects of kinetic stabilization of the Si–C bond without the confounding effects of the physical barrier imposed by a long-chain alkane or by additional surface functionality.

12.3.2.2 Structural Characterization of Other Si–C Functionalized Surfaces A large body of literature describes the structural characterization of Si–C surfaces functionalized with groups other than methyl. The reactivities of all such surfaces, aside from CH_3–Si(111), are affected by a combination of Si–C site reactivity, non-Si–C functionalized site reactivity, the physical barrier imposed by large molecules, and the terminating chain functionality.

Difference in Coverage The fraction of saturated C-terminated Si atop sites is nominally unity only for the CH_3–Si(111) surface. For all other Si(111) surfaces functionalized with alkanes or alkenes, the Si–C coverage is less than unity because of steric factors. For surfaces functionalized using the two-step reaction or by hydrosilylation, the non-Si–C terminated sites are most commonly terminated by Si–H bonds [105, 106, 119]. While the CH_3–Si(111) surface coverage was measured to be 100% termination of Si atop sites, the C_2H_5–Si(111) surface had a maximum of approximately 80% coverage [86, 120]. As expected, the coverage decreased as the steric bulk of the alkyl group increased [106]. The maximum coverage of surfaces functionalized via hydrosilylation, which commonly employs large molecules, is approximately 50% [111, 119]. Clearly, the CH_3–Si(111) surface is an attractive choice for the isolation and study of the Si–C bond.

Difference in Coverage Pattern The hydrosilylation reaction proceeds through a chain reaction on the surface, while for the two-step mechanism each site functionalization reaction is only influenced sterically by closely adjacent sites. The hydrosilylation surface chain reaction therefore leads to patches of functionalization at low reaction conversion (Fig. 12.15) [121]. In contrast, the two-step mechanism is thought to give between random and self-avoiding coverage with bulky functional groups, because of the irreversibility, mechanistic independence, and steric considerations of each reaction event. This difference in mechanism, along with the tendency to proceed to higher coverage, implies that very few adjacent sites are not Si–C terminated on Si surfaces functionalized using the two-step halogenation/alkylation procedure [86, 89]. A decreased rate of surface oxidation was observed on partially alkylated surfaces relative to that of Si–H surfaces [122]. It was suggested that if lateral SiO_x propagation proceeds through adjacent Si–H sites, the presence of even incomplete Si–C termination would retard the oxide growth. Although clear examples are lacking on the effect of adjacent site

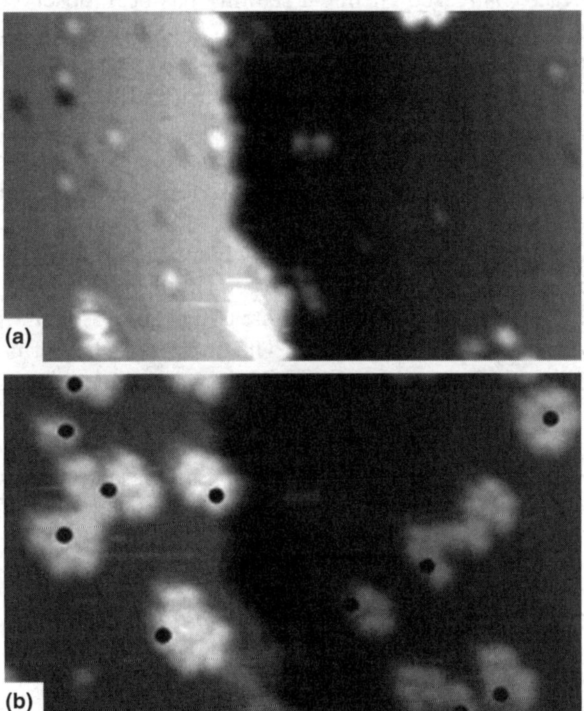

FIGURE 12.15 Occupied state STM images (215 Å × 215 Å, −2.1 V, 44 pA) of a H–Si (111) surface with isolated dangling bonds created by desorption activated with the STM tip (a) before styrene addition and (b) after exposure to 12 L of styrene. The black dots in (b) mark the initial dangling bonds showing these sites initiate hydrosilylation of styrene. Maximum height range is ~5 Å (a) and ~10 Å (b). Figure reproduced from Ref. 121 by permission of the American Chemical Society, © 2002.

identity on surface stability, the degree and identity of non-C-terminated Si atop sites are likely to affect the reactivity of such surfaces.

Difference in Monolayer Density and Thickness Despite a possible lower coverage of functionalized Si surface sites, an increase in the size of the surface-bound species can also produce a physical barrier on the surface. An increase in the monolayer thickness and density, but not in the total coverage of Si–C bonds on Si atop sites, has been obtained by chemical functionalization of ester-terminated alkane monolayers on Si [113]. This highly protected overlayer increased the stability of silicon toward oxidation under ambient conditions. The monolayer protected the underlying substrate not via kinetic stability of the Si–C bond, but primarily via physical separation between reactant and the silicon surface sites.

Difference in Crystal Face and Surface Roughness As the density of dangling bonds decreased, either because of surface roughness or due to the structure of a specific crystal face, increased chemical stability has been observed for Si–H and Si–X terminated surfaces [98, 123]. CH_3–Si(111) surfaces are better passivated toward oxidation in air than are CH_3–Si(100) surfaces [124, 125]. The unreconstructed Si(100) surface has a higher density of dangling bonds than does the Si(111)-(1×1) surface, $1.36\times10^{15}\,cm^{-2}$ compared to $7.83\times10^{14}\,cm^{-2}$ on Si (111). Hence, on the Si(100) surface, the Si atop sites are more available for attack by oxidants, because sterically there is not enough space for –CH_3 termination of every dangling bond.

Hydrosilylation proceeds to give nominally identical monolayers on Si(111) and Si(100) surfaces [126, 127]. TPD results showed similar thermal stability on Si(100) compared to that on Si(111); however, increased disorder of the monolayer was observed via contact angle measurements. The observed behavior is consistent with the increased disorder of the initial H–Si(100) surface, which contained SiH, SiH_2, and SiH_3 moieties. Although a rigorous comparison of the stability of hydrosilylation-functionalized Si(100) and Si(111) surfaces has not been conducted, a large difference between the two crystal faces may not be present, because the dominant stabilization mechanism for hydrosilylation-functionalized surfaces is a physical barrier, which should be similar on both hydrosilylated crystal faces.

12.3.2.3 Electronic Characterization of Alkylated Silicon

Surface Recombination Velocity C-terminated Si surfaces, specifically those synthesized via the two-step halogenation/alkylation reaction [67, 82], can be prepared with very low electronic defect. The CH_3–Si(111) surface exhibited stable SRV values of <40 cm/s for >30 days, in ambient air, despite the growth of some SiO_x [67, 122, 128]. A combination of SXPS and UPS measurements indicated that the low SRV was due to low electronic defect densities rather than due to band bending [114]. Even surfaces functionalized with bulky groups that are capable of only sparse coverage of the Si surface showed SRV values <100 cm/s

[106]. These data thus showed that the Si–C bond efficiently passivated the Si surface, and that the electronic properties of these functionalized surfaces were stable in air over long periods of time, unlike the behavior of the Si–H terminated surfaces.

Surface Dipole The presence of an ~400 mV dipole (Fig. 12.16) at the CH_3–Si (111) surface has been observed via synchrotron photoemission spectroscopic measurements in UHV. The dipole has been attributed to the electronegativity difference between the C and H of the –CH_3 groups [114]. In addition, a comparison of the barrier height measurements made at n-Si–H/Hg and n-Si–CH_3/Hg contacts shows an ~0.55 eV dipole [59, 129]. In the electrodeposition of Cd and Pb on H–Si versus CH_3–Si, the dipole was manifested as the presence or absence of an oxidative stripping peak [130].

Electronic Resistance An insulating layer between two conducting phases presents a barrier to electron transfer. Experimentally, such a barrier will appear as a series resistance, as has been observed for dense organic monolayers on Si and on other electrode surfaces [132–135]. If the dominant mechanism of electron transfer through a barrier is tunneling, the observed electron-transfer rate constant, k, should depend on the thickness of the barrier, as given by

$$k = k_0 \exp[-\beta(d - d_0)], \tag{12.1}$$

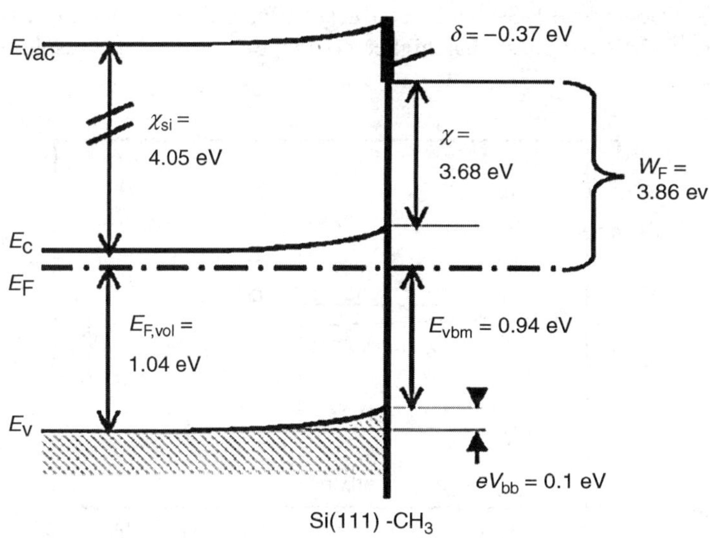

FIGURE 12.16 Energy-band diagram of the CH_3–Si(111) surface. The electron affinity, χ, of CH_3–Si(111) was measured as ~3.7 eV via photoemission spectroscopic measurements, indicating the presence of a surface dipole, δ, of ~400 mV. The Si is near flat-band conditions, $eV_{bb} = 0.1 \pm 0.1$ eV, at the CH_3–Si(111)/vacuum interface. Figure reproduced from Ref. 114 by permission of the journal article authors.

where k_0 is the rate constant in the absence of a tunneling barrier, b is the material-dependent electron tunneling constant ($\sim 1.0\ \text{Å}^{-1}$ for saturated alkanes) [136–138], and $(d - d_0)$ is the thickness of the barrier. Interestingly, this relationship predicts that electron-transfer rates across Si-bound monolayers should also be highly dependent on the identity of the molecules [59]. A chain length-dependent series resistance was observed for the photoelectrochemical systems Si(111)–C_xH_{2x+1}/$Me_2Fc^{0/+}$ (CH_3OH) and Si(111)–C_xH_{2x+1}/$K_3Fe(CN)_6$/$K_4Fe(CN)_4$(aq), where $x = 1 - 6$ [131, 139]. The expected exponential dependence of the electron-transfer rate on chain length was not observed (Fig. 12.17), so it was suggested that solution redox couples could penetrate to some extent into the alkyl monolayers [140].

Low interfacial resistance is important for surface electrochemical stability, as well as for device efficiency. The branching ratio of Faradaic current associated with hole transfer to Fe^{2+}, and that associated with oxidation of Si, was altered by introducing a resistive interfacial element in the form of the alkylated overlayer. Ultimately, rapid anodic oxidation of the Si substrate was observed in the presence of very resistive monolayers [139].

12.3.3 Reactivity of C-Terminated Silicon Surfaces

Functionalization of Si with the two-step halogenation/alkylation reaction, or through anodic deposition, can produce surfaces with short alkyl chains ($\leq C_4$). These two mechanisms, in addition to hydrosilylation and cathodic deposition of diazonium compounds, can also produce surfaces that have thicker organic monolayers. Electrochemical [139] and electroless [106, 122, 125] stability experiments with shorter chains suggest that higher coverage improves the resulting surface

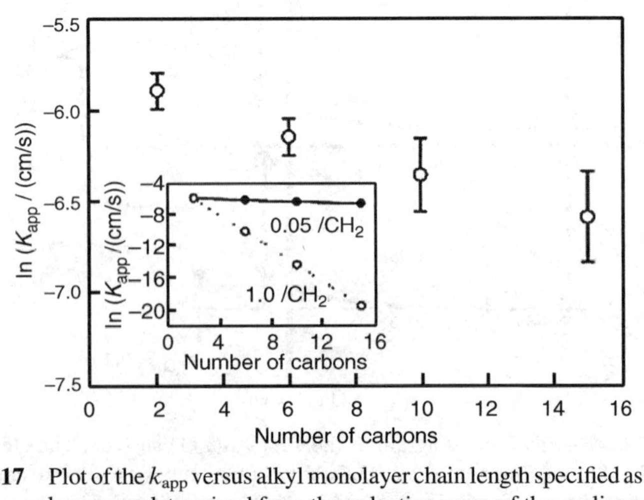

FIGURE 12.17 Plot of the k_{app} versus alkyl monolayer chain length specified as the number of carbons, n. k_{app} values were determined from the reduction wave of the cyclic voltammograms based on a convolution integral. The inset shows the measured dependence on chain length, $\gamma = 0.05\ CH_2^{-1}$, in comparison to the prediction assuming $\beta = 1.0\ CH_2^{-1}$. Taken from Ref. 140.

stability. However, electroless experiments on longer alkyl chains ($\geq C_6$) showed that surfaces with a thicker organic monolayer are more stable [113]. These results seem contradictory; however, a higher degree of disorder is observed in short-chain monolayers, such that small oxidizing agents, like water, could more easily penetrate such barriers. In this regime of oxidizing agents close to the surface, the total coverage of Si atop sites that have a kinetically stable Si–C bond is very important. On the other hand, with thicker organic monolayers, the long-chain alkanes are more ordered, and are better at physically blocking oxidizing agents from the surface. These surfaces also all have nominally the same total coverage, so it is expected that the stability of such systems would increase with the overlayer thickness. A difference was also observed between stability in electrochemical and electroless experiments. In experiments where Faradaic current is passed, the additional resistance that is imposed by an organic monolayer can affect the stability of the resulting surface.

12.3.3.1 Thermal Stability of Alkylated Silicon The thermal decomposition mechanism, and thus the thermal stability of alkylated Si surfaces, depends on the length of the alkyl chain. CH_3–Si(111) surfaces are stable up to at least 440°C, but CH_3CH_2–Si(111) surfaces decompose at <300°C under UHV conditions [120, 141]. The thermal stability of C_xH_{2x+1}–Si(111) surfaces decreases with increasing chain length, but plateaus at $\geq C_6$ [141]. Upon heating of surfaces functionalized with $\geq C_2$ to over 300°C in UHV, Si–H moieties were observed. The suggested mechanisms of desorption included β-hydride elimination, to leave Si–H, and a minority reaction that results in C–C bond cleavage, to leave CH_3–Si [120, 142, 143]. These reactions were found to occur at 250–300°C, while the subsequent desorption of the remaining CH_3–Si occurred at higher temperatures. These findings highlight the unique thermal stability of the CH_3–Si(111) surface. Under ambient pressure, heating without surface oxidation has proven difficult, and even in nominally reducing environments, Si in higher oxidation states is observed [141].

12.3.3.2 Stability in Aqueous Conditions

Neutral Aqueous Conditions The Si–C terminated surfaces are stable in neutral, O_2-free, H_2O. Initially, this behavior was tested by immersion of surfaces functionalized by hydrosilylation into boiling H_2O for 1 h. Integration of the C–H stretching region using IRAS indicated that the monolayer remained intact. However, the ~30% of a monolayer thought to be tethered through a Si–O–C bond was hydrolyzed [144]. This observation is consistent with contact angle measurements, XPS, AFM, and ellipsometry data that have indicated less than 24 h stability when silanes bonded to oxidized Si surfaces are immersed in water [145, 146]. In contrast, immersion of Si surfaces thought to have no Si–O–C bonds into boiling water for 1 h and then room-temperature water for 3 days produced no effect on the observed surface IR vibrational mode intensities [147]. Furthermore, in the case of $C_{18}H_{37}$–Si(111), immersion into boiling water for 7 h resulted in negligible SiO_x formation [119]. When comparing $C_{10}H_{21}-Si(111)$ to $EtOOCC_{10}H_{20}-Si(111)$ to a thicker and more dense, branched $(C_5H_{11})_2C(OH)C_9H_{18}-Si(111)$ surface, HREELS, contact potential, and differential capacitance measurements all indicated that more dense monolayers led to less SiO_x formation and to improved stability [113].

When passing anodic current across the Si/H_2O junction, a high concentration of electron-deficient sites (holes) is formed. This process, in the presence of good nucleophiles (H_2O, OH^-), leads to rapid SiO_x formation. Notably, under such conditions, Si–H terminated surfaces are extremely unstable, but Si–C terminated surfaces are remarkably stable [139]. Specifically CH_3–n-Si(111) and C_2H_5–n-Si(111) can sustain photoanodic current densities of 1 mA/cm^2 for over 60 min in aqueous solutions with negligible change in open-circuit voltage (V_{OC}) or fill factor, indicating negligible formation of surface states or SiO_x. For longer chains, $\geq C_4$, the photoelectrode performance decayed within the same 60 min time period, although the decay occurred much less rapidly than that of Si–H terminated surfaces. This behavior is consistent with the hypothesis that the additional resistance imposed by the organic monolayer affected the branching ratio of productive to deleterious Faradaic processes (Fig. 12.18).

The Si–C termination not only passivated the Si surface toward oxidation, but also stabilized the band-edge positions as the pH was changed. A pH dependence of band-edge positions is a classic phenomenon observed for metal oxides in H_2O, and has also been reported for Si. It was suggested that the Si oxide protonation/deprotonation equilibrium shifts the Si band-edge positions. With pH, however, such shifts are minimal for Si–C terminated surfaces, and were shown to be correlated with the fraction of atop Si sites terminated by C. The reduction of $MV^{2+/\cdot+}$ (methyl viologen), a pH-independent and water-soluble redox couple, at methyl-terminated n-Si(111) surfaces was in fact shown to be essentially independent of pH [148].

Acidic Aqueous Conditions Hydrogen-terminated Si is stable in concentrated acids, in the absence of oxidants [2]. C-terminated Si surfaces also exhibit good stability in aqueous acidic conditions. HREELS data have shown that surfaces functionalized with long alkyl chains via hydrosilylation remained intact after 7 h in 0.1 M H_2SO_4 [113]. In fact, dilute solutions of acids, such as HCl(aq) and trifluoroacetic acid (TFA) in THF, are routinely used to clean Si surfaces after functionalization, or to carry out secondary functionalization reactions [110, 126, 149]. Organic monolayers have even been shown to resist exposure to very concentrated and/or corrosive acids, like pH 2 H_2SO_4 for 2.5 h with no detectable oxide growth via XPS [82] or boiling 2.5 M H_2SO_4 in 10% water in dioxane [119]. In addition, the photoanodic performance of such Si surfaces was not affected by operation at pH 1.5 [148].

Very concentrated acids have been observed to disrupt the Si–C monolayer. Immersion of $CH_3-Si(111)$ surfaces in 36 M H_2SO_4 gave very low SRV values while the crystal was submerged; however, upon a second exposure to ambient air, the carrier lifetime decayed rapidly, suggesting that the monolayer had been damaged [128].

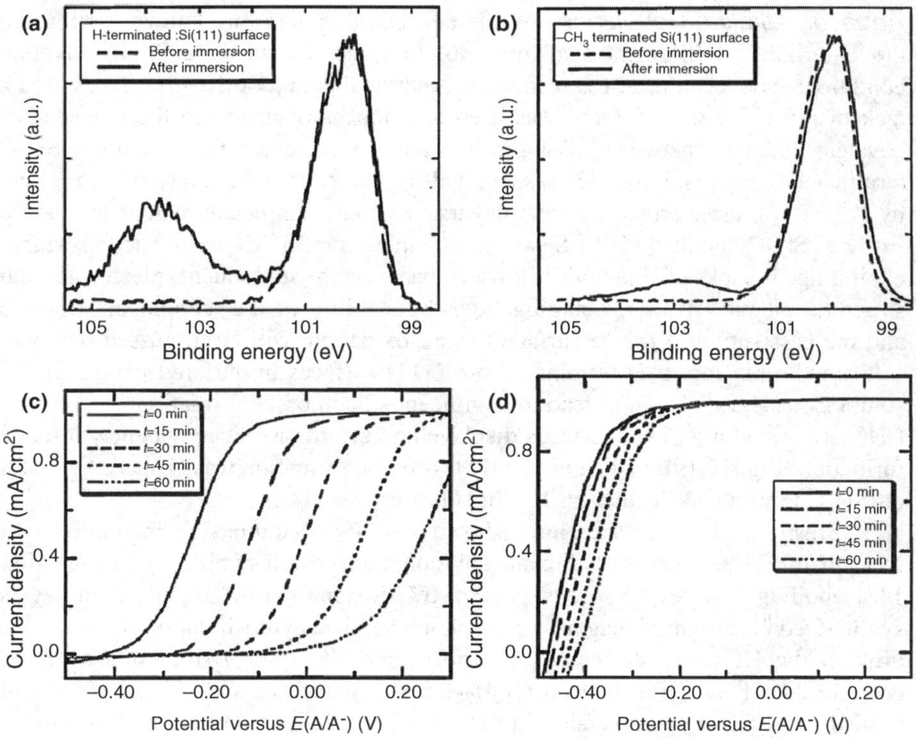

FIGURE 12.18 High-resolution XP spectra of the Si 2p region of H−Si(111) (a) and CH_3−Si(111) (b) surfaces before and after immersion in 0.35 M $K_4Fe(CN)_6$–0.05 M $K_3Fe(CN)_6$(aq) solution for 10 min. The CH_3−Si(111) surface is better passivated toward SiO_x growth. Time-dependent *J–E* behavior of H−Si(111) (c) and CH_3–Si(111) (d) surfaces in contact with $Fe(CN)^{3-/4-}$(aq) solution. *J–E* data were collected at a scan rate of 50 mV/s in a three-electrode configuration versus a Pt wire reference electrode poised at the Nernstian potential of the redox couple, $E(A/A^-)$. Figures reproduced from Ref. 139 by permission of the American Chemical Society, © 1998.

Thorough investigations of the stability of Si−C monolayers in dilute, oxidizing acidic conditions are still warranted, however.

Fluoride-containing aqueous acidic solutions are known to etch Si slowly, and have been shown to have little effect on oxide-free, Si−C terminated surfaces. The presence of silicon oxide species enabled partial removal of the Si−C monolayer, and fluoride-containing solutions were therefore used to quantify the oxidation of the silicon substrate [82, 119, 134, 147]. Differences in monolayer characteristics have been studied by observing the effects of HF(aq) solutions on the Si substrate. C_{10}−Si(111) functionalized surfaces synthesized via the chlorination/alkylation route or by photochemical hydrosilylation were more robust to immersion in NH_4F following 1 h in boiling water than those made via Lewis acid-catalyzed hydrosilylation [147]. The increased stability was attributed to a higher surface coverage, giving more dense organic monolayers in the more stable systems.

Aqueous Alkaline Conditions While concentrated acids are known to preserve the H-terminated Si surface, alkaline solutions are known to etch Si. Alkaline conditions have been used for secondary functionalization of Si–C surfaces, and a lack of reactivity toward OH^- was used as a means of deducing the presence of covalent linkages between alkyl groups and Si surfaces. For example, Si–C terminated surfaces have been subjected to aqueous 10% NH_4OH, to effect hydrolysis of ester-protected monolayers, without significant disruption of the surface Si–C bond [150]. Several examples of Si–C terminated surfaces exhibiting a lack of reactivity toward base, with subsequent electronic and structural characterization detailing both the stability of the monolayer structure and the passivation of the Si surface toward oxidation, will be discussed below.

Cathodic biasing of H-terminated n-Si(111) surfaces in buffered pH 11 MV^{2+} solutions has been shown to lead to significant silicon oxide formation. In contrast, CH_3-terminated n-Si(111) surfaces displayed no electrochemical evidence of oxide formation (Fig. 12.19), and significantly less oxide formation was observed by XPS on Si–C terminated Si than on H–Si(111) surfaces [148].

Hydrosilylated Si surfaces have shown stability to immersion in boiling 1 M NH_4OH in a 10% water/90% dioxane solution, but showed significant damage after 1 h in boiling 1 M NaOH in 10% water/90% dioxane [119]. Organic monolayers synthesized via Grignard reagents or via photochemical hydrosilylation showed a 6% drop in the IR C–H stretching absorption intensity after 10 min in 2 M KOH (Fig. 12.20) [147]. XPS data on $C_{12}H_{25}$–Si(111) surfaces showed the growth of 0.5 ML of SiO_x after immersion in pH 13 KOH solutions, but showed almost no change in Si–C coverage, suggesting oxidation, but not etching, of the silicon [82].

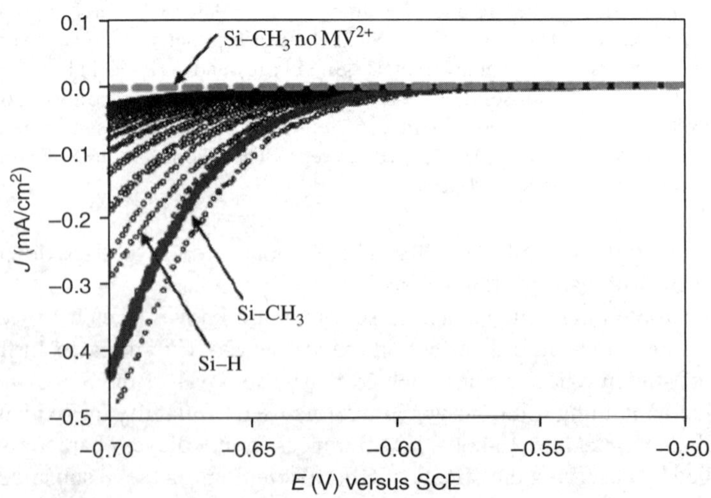

FIGURE 12.19 *J–E* behavior of H–Si(111) (open circles) and CH_3–Si(111) (filled circles) upon repeated cycling in the dark at 10 mV/s in contact with a 10 mM MV^{2+}(aq) solution buffered at pH 11.0. The dark *J–E* behavior of the CH_3–Si(111) surface in contact with aqueous solution with no MV^{2+} present is shown for comparison (dashed line). Reproduced from Ref. 148 by permission of the American Chemical Society, © 2006.

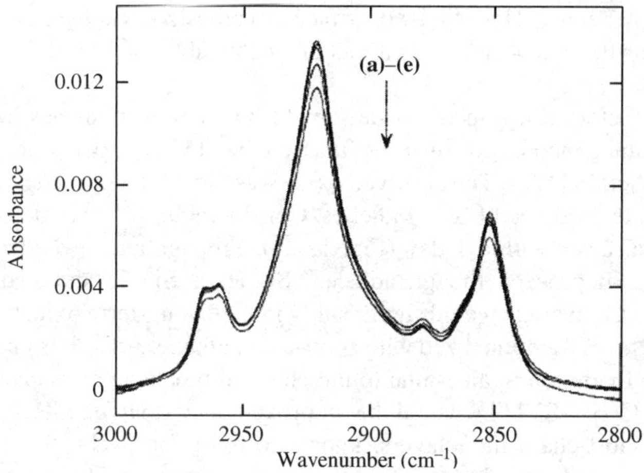

FIGURE 12.20 FTIR spectra of the alkyl C–H stretching region of $C_{10}H_{21}$–Si(111) prepared via Grignard reaction with a H–Si(111) surface. (a) As prepared followed by a trichloroethane rinse, and through sequential treatment of (b) sonication in CH_2Cl_2 for 1 min and boiling in CH_3Cl for 1 h; (c) boiling in H_2O for 1 h; (d) immersion in 40% NH_4F(aq) for 10 min followed by 3 days in H_2O; (e) immersion in 2 M KOH for 10 min. The integrated area progressively decreases through each treatment. Figure reproduced from Ref. 128 by permission of the American Institute of Physics, © 2000.

An increased monolayer thickness resulted in an increase in monolayer stability in the presence of NaOH(aq). Although the monolayer remained intact after immersion in NaOH, the underlying Si was still subject to oxidation, although much less so than unprotected Si [134]. The passivating properties of Si–C termination also extend to porous silicon [151]. In summary, C termination of Si atop sites impedes Si oxidation and dissolution in aqueous alkaline solutions, due to both the kinetic stability of the Si–C bond and the physical barrier presented by the monolayer.

12.3.3.3 Stability of Si–C Terminated Surfaces in Air C termination of Si surfaces significantly improves passivation toward SiO_x formation in ambient air compared to the behavior of H-terminated Si surfaces. For surfaces functionalized via a two-step chlorination/alkylation procedure, the rate of SiO_x formation correlated with increased steric bulk [122]. For example, after 12 h in air, CH_3–Si(111) surfaces had formed 0.08 ML of SiO_x, C_2H_5–Si(111) had formed 0.27 ML of SiO_x, and $C_6H_5CH_2$–Si(111) had formed 0.52 ML SiO_x. This behavior was attributed to differences in the Si–C bond coverage on the various functionalized Si surfaces. Si–C termination has been shown to decrease the rate of oxidation on a time scale of weeks [122, 125]. The carrier lifetime at CH_3–Si(111) remained high, 300 ms, even after 1 month of exposure to air, despite the growth of some oxide [128]. A similar result was observed with surfaces functionalized by the reduction of aryl diazonium salts, with surfaces showing growth of 0.5 ML of SiO_x within 10 days, but no additional oxidation for greater than 6 months [134, 152].

C_6H_{13}–Si(111) and C_6H_{13}–Si(100) surfaces, synthesized via hydrosilylation, were exposed to air for over 4 months without any detectable growth of silicon oxide via XPS [153].

Multiple factors can impede oxide growth. Oxidation in air has been shown to proceed inhomogeneously at surfaces functionalized via the two-step halogenation/alkylation reaction [122]. The observed oxide was mostly composed of Si^{+} and Si^{3+}, with negligible presence of Si^{4+} moieties. Considerable Si^{4+} was detected at Si–H and Si–Cl surfaces within 1 day (Fig. 12.21). Propagation of oxidation has been hypothesized to proceed though adjacent Si–H or Si–Cl sites, so that even a moderate Si–C coverage would significantly impede long-term oxidation to the Si^{4+} state. For surfaces functionalized with short-chain alkanes, $<C_6$, it is unlikely that the resistance to oxidation in air is due to the physical blockage of oxidants or else the stability of C_2H_5–Si(111) would be improved over that of CH_3–Si(111). The disorder of short-chain monolayers, known to exist for C_2H_5–Si(111) [86], thus prevents the monolayers from acting as a significant physical barrier to reactants.

12.3.3.4 Stability of Si–C Terminated Surfaces in Alcohols H-terminated Si surfaces yield excellent photoelectrochemical performance in methanolic electrochemical cells, but H–Si termination is not stable in methanol [28]. Current density versus potential measurements showed that alkylated n-Si(111) electrodes in contact with $CH_3OH–Me_2Fc^{+/0}$ yielded very similar J–V properties to hydride-terminated n-Si(111) electrodes. Differential capacitance versus potential data yielded very similar barrier heights for the two systems. No significant surface recombination was observed; however, the alkyl monolayers did impose additional series resistance to charge transfer at the Si/liquid junction [131]. C-terminated Si proved to be more stable than H-terminated Si upon addition of 20% (v/v) H_2O in MeOH. Within 90 min of photoanodic activity, a highly resistive layer had formed on hydride-terminated n-Si(111) surfaces, but alkylated n-Si(111) continued to perform well until the experiment was arbitrarily ended 22–25 h later (Fig. 12.22). Si–H and Si–Cl sites at surfaces with less than complete Si–C termination were susceptible to the formation of alkoxy surface species [89, 153, 154].

12.3.3.5 Stability in Other Common Solvents Little information is available on the inherent Si–C bond in common organic solvents, and many solvents and reactants have been used successfully together with Si–C functionalized surfaces [126, 149]. In early work, a decrease in C–H stretching intensity was observed after boiling alkylated Si surfaces in $CHCl_3$ for 2 h, but this behavior is now thought to have been due to cleavage of Si–O–C bonds [119, 144]. Subsequently, boiling or sonication in organic solvents, such as $CHCl_3$, has shown >90% of the monolayer retention, as determined via XPS and IR [82, 147]. The functionalized Si surfaces were less stable at high temperatures, and the contact angles decreased after boiling $C_{18}H_{37}$–Si(111) in toluene at 111°C for 4 days [153]. Instability of the Si–C functionalized surfaces in organic solvents has sometimes been attributed to the presence of water. Acetonitrile was more harmful to Si–C

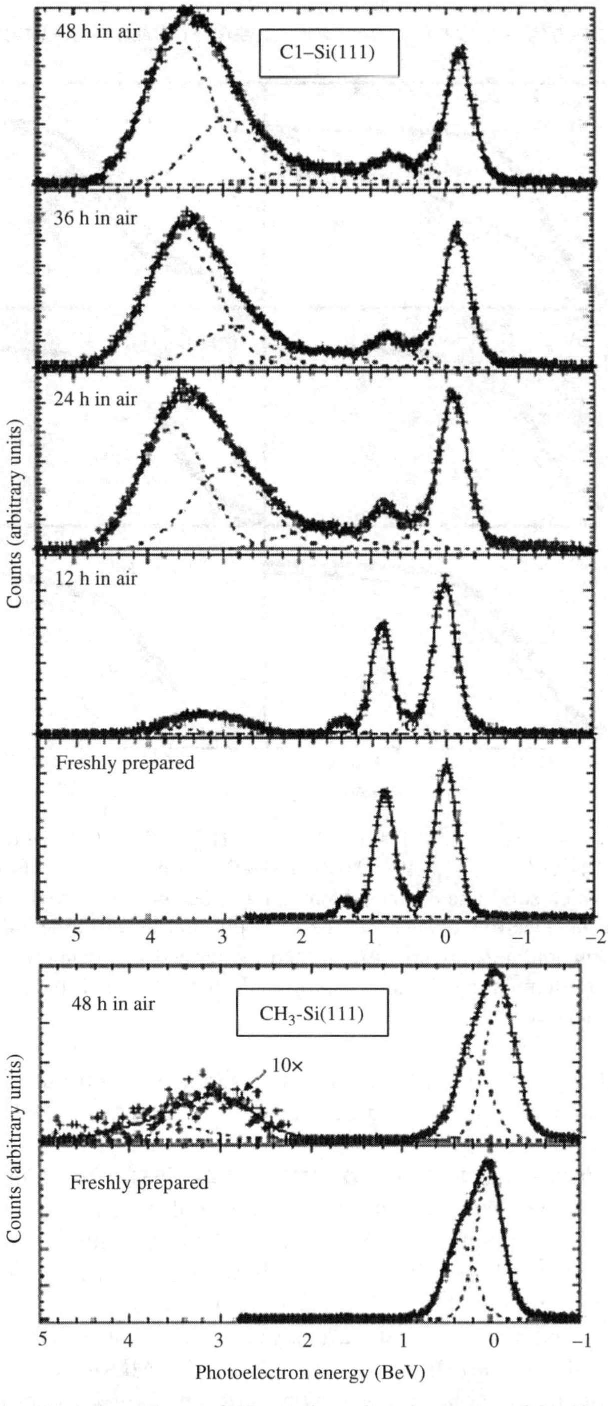

FIGURE 12.21 Time-dependent soft X-ray photoelectron spectra of Cl–Si(111) and CH_3–Si (111) surfaces with exposure to air. Spectra are shown after background removal, spin-orbit stripping, and peak fitting. Cross marks are raw data and dashed lines are fits. Note the appearance of oxidized Si signals after shorter exposures to lab air in the Cl–Si(111) spectra. In addition, higher oxides are observed at Cl–Si(111) surfaces than CH_3–Si(111) surfaces after exposure to air. Figures reproduced from Ref. 122 by permission of the American Chemical Society, © 2006.

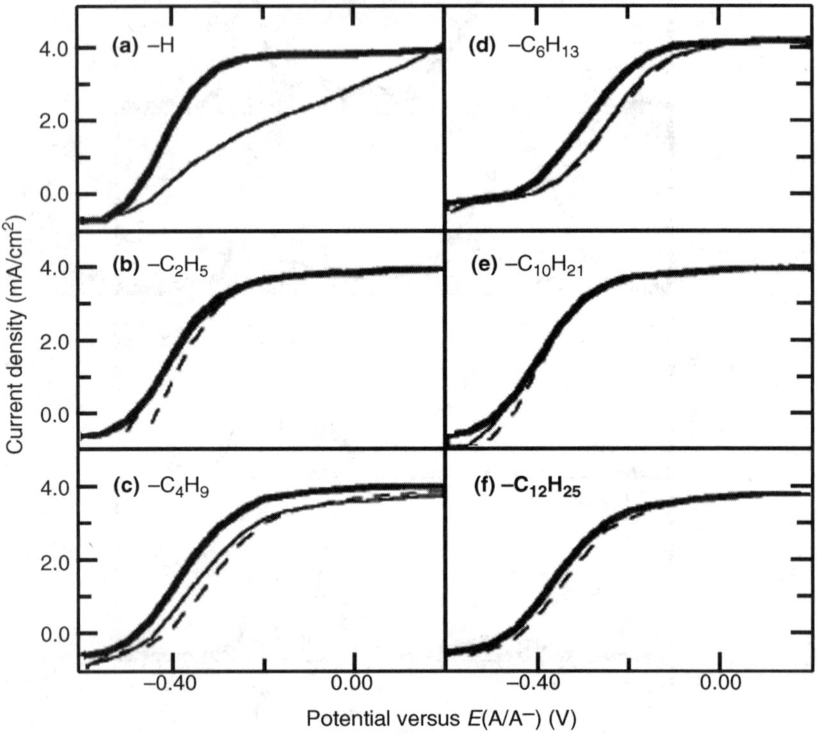

FIGURE 12.22 *J–E* behavior at H–Si(111) and C_xH_{2x+1}–Si(111) functionalized surfaces in contact with CH_3OH–1.0 M $LiClO_4$–0.050 M Me_2Fc–~1 mM $Me_2Fc^+PF_6$ containing 20% (v/v) H_2O. The thick solid line denotes the initial *J–E* behavior observed immediately after adding water to the electrolyte solution. The thin solid line denotes the *J–E* behavior after 90 min of illumination at $J_{SC} = 4.0$ mA/cm^2, and the dashed line denotes the *J–E* behavior after 22–24 h of illumination. Figures reproduced from Ref. 131 by permission of the American Chemical Society, © 1998.

functionalized surfaces than DMF. This observation was attributed to the greater difficulty in removing trace water from acetonitrile than from DMF [113].

12.3.3.6 Silicon–Organic Monolayer–Metal Systems Metal silicides are produced at most metal/silicon contacts, and this holds true for most metal/CH_3–Si junctions as well. Evaporation of metals such as Au or Cu onto CH_3–Si(111) led to silicide formation and Fermi level pinning; however, the –CH_3 group affected the formation of the silicides [155]. For Au evaporation onto a bare Si(111)-(7×7) surface, $SiAu_3$ formation was observed at any amount of evaporated Au. However, on CH_3–Si(111), silicide formation occurred only at ≥0.5 ML of Au. Partial disruption of the CH_3–Si(111) monolayer was observed despite protection of the Si surface. Si/Au barrier heights remained unchanged from those observed on the pinned Si (111)-(7×7) surface. Very similar observations were made for Si/Cu junctions (Fig. 12.23); Cu silicide was observed on CH_3–Si(111) surfaces after ≥2 ML of

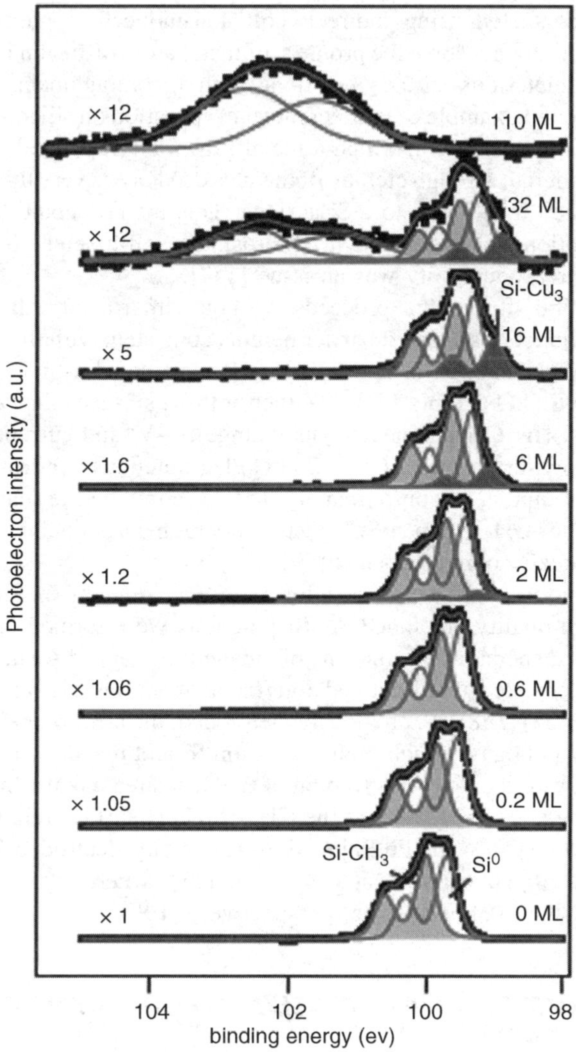

FIGURE 12.23 XP spectra ($E_{ex} = 150$ eV) of Cu deposition onto the CH_3–Si(111) surface. Intensities have been scaled as indicated. Contributions from the bulk Si^0, CH_3-bound surface components (Si–CH_3), and copper silicide ($SiCu_3$) are noted. Taken from Ref. 155.

Cu was deposited. Again, silicide formation was observed after any deposition of Cu on the Si(111)-(7×7) surface. Evaporated Cu contacts on the CH_3–Si(111) surface also gave barrier heights that were identical to those of evaporated Cu on Si(111)-(7×7). One difference between Au and Cu evaporation was that the $SiCu_3$ remained buried at the Cu/Si interface, rather than floating atop the deposited metal film, as in the case of Au deposition.

There have been attempts to produce "softer," less destructive evaporated contacts on semiconductor surfaces. GaAs, which is also commonly limited by Fermi level pinning, was contacted using indirect, collision-induced, cooled electron beam evaporation. This did not solve the problem of metal atom diffusion into the material, but did limit deleterious surface reactions with incoming metal atoms or clusters [156]. A second example of indirect metal evaporation on silicon was the lift-off, float-on (LOFO) technique, in which a metal film was evaporated on a glass slide. The film was undercut through etching of the glass, to leave a very thin, floating metal film that was then transferred to a Si surface. This method could be used without apparent disruption of an octadecyltrichlorosilane self-assembled monolayer, although contact reproducibility was an issue [157].

Hg/Si junction formation proceeds without interfacial silicide formation. Hg/H–Si(111) junctions showed barrier heights consistent with the electron affinity of Si and the work function of Hg: 0.3 eV at H-terminated n-Si(111) and 0.8 eV at H-terminated p-Si(111) (Table 12.3). Neither of these surfaces showed any evidence of reaction with Hg. Capacitance versus voltage (*C–V*) and current density versus voltage (*J–V*) measurements of CH_3–Si(111)/Hg junctions confirmed the presence of an interfacial dipole, of approximately 0.55 eV compared to that of H–Si(111), attributable to the –CH_3 group. n-Si/Hg gave very high barrier heights of 0.91 eV, and p-Si gave practically ohmic contacts [59].

CH_3 termination of Si surfaces allows further freedom in Si/metal junction formation. High-quality, unpinned Au/Si junctions were formed on CH_3–Si(111) surfaces though dispersion and sintering of butanethiol-capped Au nanoparticles. For H–Si, this scheme led to ill-defined interfacial reactions giving highly resistive barriers (Fig. 12.24) [158]. As previously mentioned, the electrodeposition of Cd and Pb on CH_3–Si(111) gave much higher solution/Si and metal/Si barrier heights, as evidenced by lowered oxidative stripping of the deposited metals. In addition, much fewer nucleation sites are present at the CH_3–Si(111) surface relative to the H–Si (111) surface. Both Cd/Si and Pb/Si junctions formed by electrodeposition gave high barrier heights, of $\Phi_B \sim 0.8$ eV, at CH_3–n-Si(111), whereas at H–n-Si(111) the junctions gave $\Phi_B \leq 0.4$ and 0.59 eV, respectively [130].

12.4 APPLICATIONS AND STRATEGIES FOR FUNCTIONALIZED SILICON SURFACES

The previous three sections have explored the stability and reactivity of H-, halogen-, and C-terminated silicon surfaces. Compared to UHV-cleaned surfaces, these functionalized Si surfaces all exhibit remarkable stability. However, Si-based devices made today are expected to last for 10–20 years of operation under ambient conditions. The only surface that yields close to that sort of performance is the C-terminated Si surface. In this final section on Si, we discuss strategies to utilize organic monolayers on Si to further enhance the surface stability during device fabrication and operation.

A very old technique for surface stabilization is the facilitation of charge transfer to turn over productive electrochemical events. This is the final of three surface stabilization techniques presented in this chapter: kinetic stability of the Si–C bond,

TABLE 12.3 Measured Junction Properties at CH_3–Si(111) in Contact with Hg [59]

Surface Functionalization	Number of Tested Junctions	$\ln J_0$ $(A/cm^2)^a$	$q\Phi_{b,\,j\text{-}v}(eV)^a$	n^a	$V_{bi}(V)^b$	$q\Phi_{b,\text{Mott-schottky}}$ $(eV)^b$	$\Delta q\Phi_b(eV)^c$
n-Si, Si–H	7	-21 ± 3^d	0.17 ± 0.02 $(0.3 \pm 0.1)^{d,e}$	2.6 ± 0.8^d			
n-Si, SiOx	10	-13 ± 2	0.73 ± 0.06	1.2 ± 0.1	0.39 ± 0.07	0.65 ± 0.07	0.08 ± 0.09
n-Si, Si–CH_3	17	-17.1 ± 0.5	0.85 ± 0.02	1.6 ± 0.2	0.65 ± 0.02	0.91 ± 0.02	-0.06 ± 0.03
n-Si, Si–C_2H_5	10	-16 ± 3	0.82 ± 0.07	2.1 ± 0.8	$0\ 63 \pm 0.08$	0.88 ± 0.07	-0.07 ± 0.10
n-Si, Si–*iso*-C_3H_7	20	-14 ± 5	0.8 ± 0.1	1.6 ± 0.3	0.64 ± 0.19	0.93 ± 0.20	-0.2 ± 0.2
n-Si, Si–C=C–CH3	9	-18 ± 1	0.87 ± 0.01	1.8 ± 0.4	0.57 ± 0.04	$0\ 84 \pm 0.01$	0.01 ± 0.03
p-Si, Si–H	5	-15.8 ± 0.8	0.82 ± 0.02	1.16 ± 0.03	0.6 ± 02	0.8 ± 02	-0.01 ± 0.08
p-Si, SiOx	10	-7 ± 2	0.56 ± 0.04	1.2 ± 0.1	0.5 ± 02	0.7 ± 02	-0.14 ± 0.2
p-Si, Si–CH3	20	$\geq -7^d$	$<0.15^d$				

[a] Determined from the forward bias portion of *J-V* data, assuming thermionic emission-diffusion.
[b] Measured from the reverse bias portion of C^{-2} vs. *V* data.
[c] $q\Phi_{b,j\text{-}v} - q\Phi_{\text{Mott-schottky}}$.
[d] Measured at $T = 85$ K.
[e] The value in parentheses is after correction.

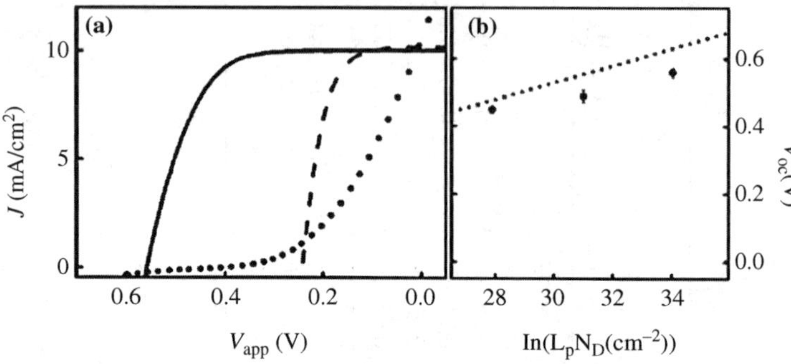

FIGURE 12.24 (a) *J–E* behavior for H–Si(111) (dotted line) and CH_3–Si(111) (solid line) in contact with sintered Au nanoparticles and H–Si(111) (dashed line) in contact with evaporated Au (b) Measured V_{oc} values versus $\ln(L_pN_D)$ for n-Si/Au devices made with CH_3-terminated n-Si (111) and sintered Au NP films. The dotted line represents the bulk-recombination-limited V_{oc}. Figures reproduced from Ref. 158 by permission of the American Chemical Society, © 2008.

physical barrier to reactants, and facilitated charge transfer to effectively compete with surface oxidation. This section will deal with four methods to facilitate charge transfer, which in some cases also utilize the physical barrier method to protect against oxidation. The methods to be discussed include covalent attachment of redox-active species, deposition of conductive or redox-active polymers, deposition of metals, and deposition of other nonmetallic inorganic films.

12.4.1 Tethered Redox Centers

Electrochemical oxidation of silicon occurs when deleterious oxidation events occur competitively with the productive charge-transfer event. Facilitation of productive charge transfer can prevent passivation of the Si. This behavior has been demonstrated by the tethering of a redox-active center to n-type silicon. A ferrocene-derivatized silane was attached to n-Si(111), and the photoanodic performance was evaluated [159–161]. It was shown that charge transfer to $Ru(NH_3)_6^{2+/3+}$, $Fe(CN)_6^{4-/3-}$, and I^-/I_3^- could be facilitated by the bound ferrocene. Current–voltage cycling of the bound ferrocene produced nearly identical data for 1000 cycles, whereas bare photoanodes could not produce even one reproducible voltage sweep. Furthermore, functionalized electrodes were stable, to within 10% of the original photocurrent, for >5 h of continuous illumination with a turnover number of $>10^{-5}$. The photocurrent at bare electrodes decayed by >90% within 5 min of operation under anodic conditions.

In addition to bound individual redox-active species, stability enhancements have been demonstrated with polymeric methyl viologen-based derivatives [162]. When illuminated in solution with a redox reagent, $Fe(CN)_6^{4-}$, the poly(methyl viologen)-coated n-Si electrodes exhibit enhanced stability. Coatings with a non-redox-active, but charged, backbone, derived from $(CH_3O)_3Si(CH_2)_3N^+(CH_3)_3$, also exhibited enhanced stability presumably because electrostatic binding interactions drew solution redox ions close to the electrode. In this fashion, direct charge transfer to the redox species was not inhibited, but a physical barrier blocked molecules that could undergo deleterious oxidation reactions with the underlying n-Si. In all cases, the photocurrent stability enhancement was a balance between the physical barrier and charge-transport limitations (Fig. 12.25).

12.4.2 Conductive Polymer Coatings

The previous section discussed the use of long-chain alkanes to form a physical barrier to chemical oxidants. Although this approach could be extrapolated to long chains, such long chains introduce a detrimental effect on the photoelectrode efficiency, in the form of a series resistance, because thicker barriers must be conductive in order to pass charge across the interface.

The deposition of polypyrrole (PPy) on small bandgap semiconductors such as n-GaAs, n-CdTe, n-CdSe, and n-CdS has resulted in improved stability under both cyclic voltammetric and continuous illumination photoanodic conditions [163, 164]. Specifically, polypyrrole-coated GaAs has been shown to be stable for >100 h in methanol (Fig. 12.26), whereas bare GaAs is stable under the same conditions for only a few minutes. Peeling of the polymer was a problem for these systems, and eventual oxidation was attributed to deleterious reactions at sites of peeling. The peeling was worse in water; PPy-coated photoanodes were stable only for 10–20 min in water. Within that 10–20 min, an efficiency of 10.5% was achieved for the aqueous oxidation of $Fe(CN)^{3-/4-}$ under 170 mW/cm^2 illumination.

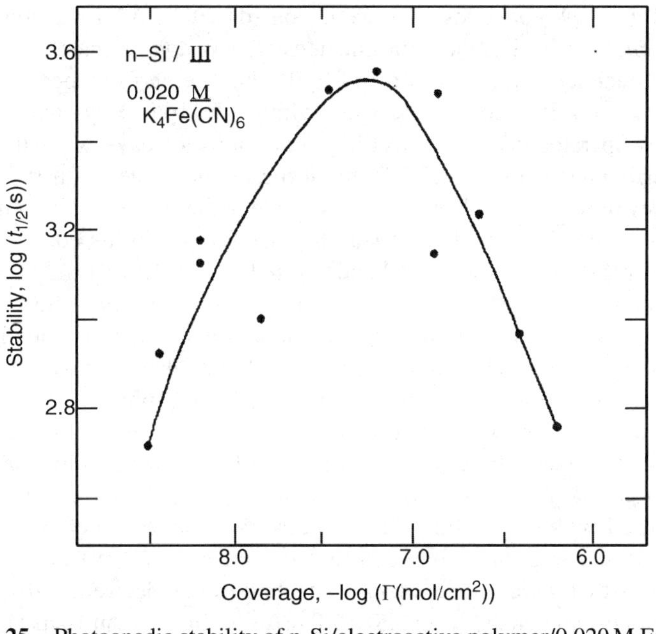

FIGURE 12.25 Photoanodic stability of n-Si/electroactive polymer/0.020 M $Fe(CN)_6^{4-}$ (aq) junctions at constant light intensity (1.0 mW/cm^2 at 632.8 nm) as a function of polymer coverage in mol/cm^2 of electroactive monomeric units. Figure reproduced from Ref. 162 by permission of the American Chemical Society, © 1984.

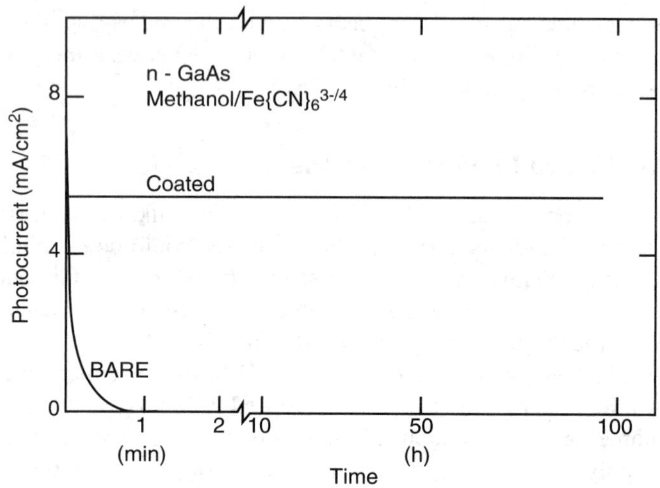

FIGURE 12.26 J_{SC} versus time characteristics showing the photoanodic stability of bare and polypyrrole-coated n-GaAs electrodes in contact with methanolic 0.1 M Et_4NBF_4/0.2 M $Fe(CN)_6^{3-/4-}$ solution. Figure taken from Ref. 164.

In contact with various redox couples, PPy was also demonstrated to increase the stability of n-Si photoanodes that were coated with 5 Å of vacuum-evaporated Pt [165, 166]. The initial photocurrent density was approximately 7% higher for n-Si/Pt photoanodes compared to n-Si/PPy/Pt, but the performance of the surfaces that did not have a PPy coating decayed within 4 h [165]. PPy-coated n-Si/Pt was continuously operated in an aqueous I^-/I_3^- cell for over 6 days, when the experiment was arbitrarily ended (Fig. 12.27). Without the Pt, electrodes with PPy were stable for only a few minutes, and bare n-Si was stable for less than 1 min in this medium. The effects of metal deposition on stability are further discussed below. The V_{OC} observed for n-Si/Pt/PPy was 0.425 V, and varied by less than 100 mV for a change in redox potential of 1.25 eV. In contact with an acidic aqueous solution of $FeSO_4$/$FeNH_4(SO_4)_2$, the n-Si/PPy with no Pt was stable for 120 h, with a decline of ~30% in photovoltage [167]. The stabilization effect has also been observed for PPy electrodeposited on amorphous, hydrogenated silicon [168].

Organic monolayers have been used to enhance the adherence of the PPy films to semiconductor surfaces. Electropolymerization of pyrrole was initiated at a pyrrole-functionalized silane self-assembled monolayer [169]. Films deposited onto this surface showed enhanced stability in contact with an aqueous $FeCl_3$/$FeCl_2$/1 M HCl solution, and showed stable photocurrent densities of 7.8 mA/cm^2 for >25 h. In contrast, films that were not covalently attached to the electrode surface showed a drop in photocurrent density from 6.5 to 1.8 mA/cm^2 in less than 18 h. The increase in stability was attributed to a drastic decrease in the peeling of the films (Fig. 12.28). This idea has been applied more recently to oxide-free semiconductor surfaces in which H-terminated Si surfaces were functionalized with an alkyllithium chain

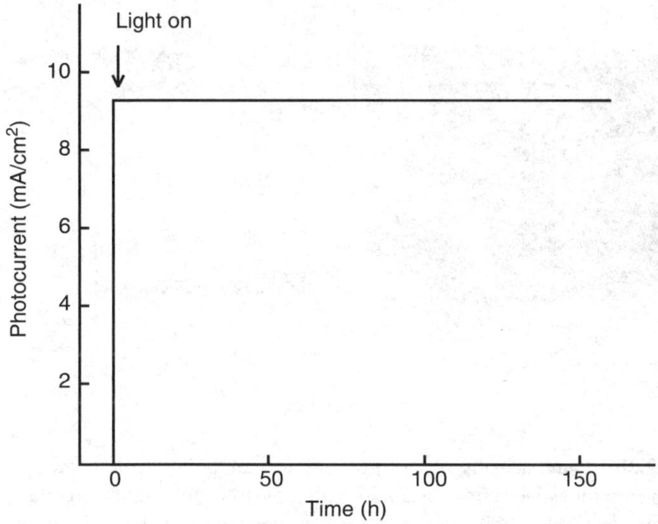

FIGURE 12.27 J_{SC} versus time characteristics showing the photoanodic stability for n-Si/Pt/polypyrrole electrodes in contact with 0.1 M KI, 0.01 M I_2, and 0.1 M KCl(aq), and illuminated at 75 mW/cm^2 by a tungsten halogen lamp. Polypyrrole was 250 monolayers thick. Taken from Ref. 165.

bearing a pyrrole. Polymer films electrodeposited at the pyrrole-functionalized surface were smoother than polymer films electrodeposited at a Si–H surface, as observed using AFM, and the diode quality factor dropped from 2.3 (at noncovalently linked Si/PPy junctions) to 1.5 (Fig. 12.29). The decrease in ideality factor is consistent with an increase in sites of contact between the Si and PPy [170]. Polythiophene derivatives have been electropolymerized at surfaces that were functionalized with a di(2-thienyl)carbinoyl obtained from the reaction of thienyl lithium with an acid-terminated organic monolayer that was formed via hydrosilylation [170]. Surface states that were suggested to dominate interfacial energetics were observed by capacitance versus voltage methods. The maximum V_{OC} observed for the n-Si/polythiophene system was 390 mV. It was suggested that the low V_{OC} was attributable to film inhomogeneities, under which the current preferentially flowed through regions of low barrier height. Alternatively, adhesion and stability were improved by first depositing a Au film [171].

In addition to the demonstration of enhanced photoanodic stability, conductive polymers can make highly rectifying and ohmic, non-Fermi-level-pinned contacts to both n- and p-Si (Fig. 12.30) [173, 174]. Ring-opening metathesis polymerization (ROMP) was used to make well-defined Si/polyacetylene (PA) junctions. PA can be doped with either K metal or I_2 to give n- and p-type PA, respectively. The exchange current density, J_0, was 6×10^{-7} and 2.4×10^{-8} A/cm^2 at p-Si/$(CHK_{0.06})_x$ and n-Si/$(CHI_{0.17})_x$, respectively, whereas at most metals J_0 is typically on the order of 10^{-3} A/cm^2. Higher rectification was thus observed at Si/PA junctions than at metal/Si junctions. n-Si/poly$(CH_3)_3$Si-COT contacts exhibited V_{OC} of up to 0.64 V [174].

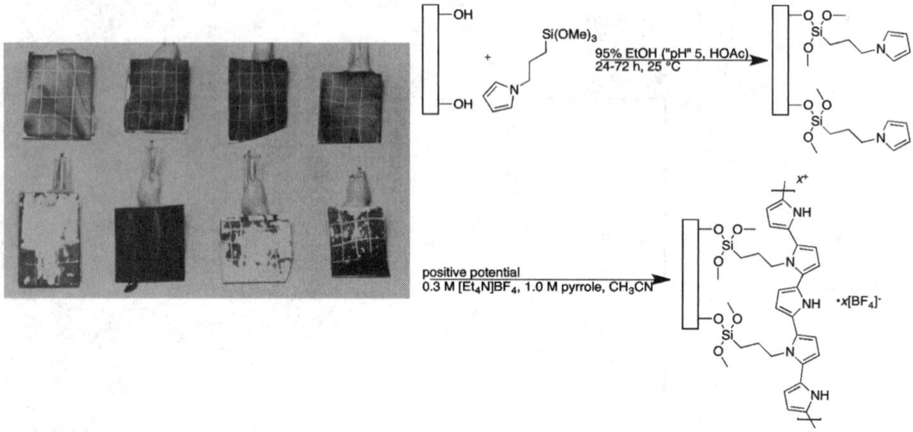

FIGURE 12.28 Comparisons of the peel-off test of polypyrrole-coated n-Si electrodes. The top four electrodes were first modified with a pyrrole-functionalized silane (modification scheme shown at right), while polypyrrole was deposited on bare Si for the bottom four electrodes. Figures reproduced from Ref. 169 by permission of the American Chemical Society, © 1982.

12.4.3 Metal Films

12.4.3.1 Stability Enhancement Improvement in the stability of Si and other small bandgap semiconductors has been achieved by facilitation of charge transfer through the deposition of metal overlayers onto the electrode surface [175–177]. Accordingly, anodic and photoanodic stability of n-GaP, n-Si, p-GaP, and p-Si has been achieved by use of evaporated Au, Pd, or Ag films [175, 176]. The devices exhibited very low V_{OC} and no dependence of V_{OC} on the redox potential of the solution, although this could be avoided by depositing thinner metal layers without loss of the added stabilization [175, 177].

A combination of metallic and nonmetallic overlayers, such as Pt on a thin SiO_2, inhibits further electronic passivation [177, 178]. At 10 Å deposition, the junction electrical properties were still determined by the contacting solution; however, at 20 or 40 Å of film thickness, the interfacial energetics were dominated by the Si/Pt junction.

12.4.3.2 Deposition on Organic Monolayers As discussed earlier, evaporation of metal, even on Si–C terminated silicon, leads to silicide formation and Fermi level pinning at Si/metal junctions [155]. Evaporated metal penetrates $-CH_3$ terminated monolayers on Si [179–182]. In some cases, increasing the physical barrier to metal penetration, by increasing the monolayer chain length, has no effect [182]. Soft contacting techniques such as sintered Au nanoparticles [158], electrodeposition of Cd and Pb [130], LOFO [183], and polymer-assisted lift-off (PALO) [184] show some promise; however, recently, utilization of monolayer functionality for improved Si/metal junctions has been explored.

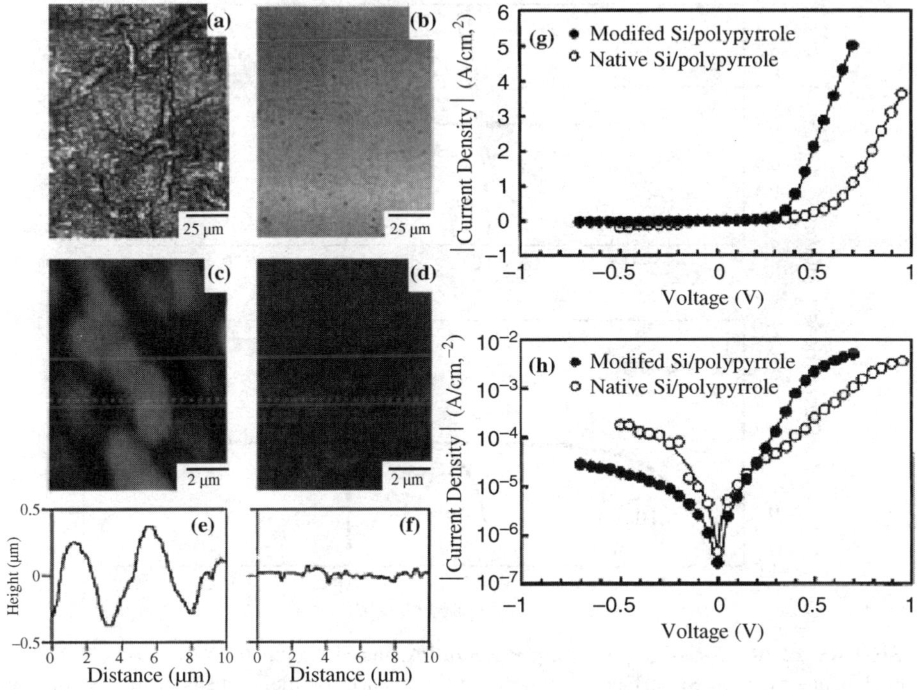

FIGURE 12.29 Optical micrographs (a, b) and AFM images (c, d) for polypyrrole films electrodeposited on H–Si(111) (a, c) and 5-(*N*-pyrrolyl)pentyl–Si(111) (b, d). Cross-sectional images (e, f) are taken from the dotted lines denoted in the AFM images (c, d). *J*–*E* (g) and log (*J*)–*E* (h) behavior of polypyrrole/n-Si(111) junctions formed at H–Si(111) (open symbols) and 5-(*N*-pyrrolyl)pentyl–Si(111) (filled symbols) surfaces. Reproduced from Ref. 172 by permission of the American Institute of Physics, © 1999.

The morphology of evaporated films was observed to change with changes in the functionality of the monolayer [185, 186]. Dithiodiacid and dithiodisilane reagents have been used to functionalize oxidized Al and Si surfaces, respectively. Evaporated Au films on treated surfaces were observed to be continuous at lower thicknesses than on bare surfaces, as determined via film resistivity measurements and TEM images. This behavior was attributed to the interaction of incoming Au atoms with the terminating thiol groups [185]. E-beam evaporated Au films on oxidized Si surfaces treated with (3-mercaptopropyl)trimethoxysilane exhibited a surface roughness factor of ~2 Å over 1 mm^2 area [186].

The interaction of terminal functionality with incoming metal atoms affords an opportunity to prevent metal penetration to the Si surface during the deposition step. ToF-SIMS has been a major tool in determination of metal penetration through monolayers versus accumulation at the monolayer–vacuum interface. ToF-SIMS data on Au evaporation onto $–CH_3$ terminated monolayers on Au showed no change in the positive ion spectrum, even after deposition of 50 EL (equivalent layers or

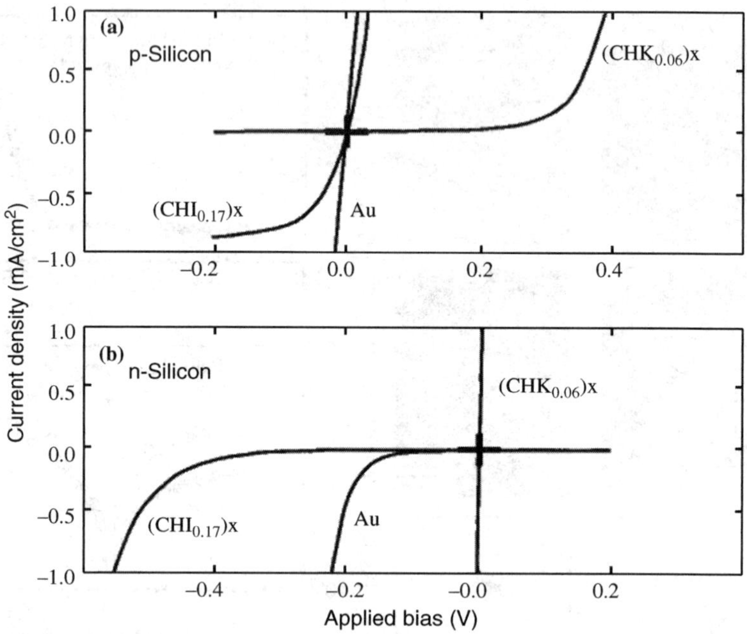

FIGURE 12.30 *J–E* behavior at Si/p-PA, Si/n-PA, and Si/Au junctions formed by ROMP of cyclooctatetraene at Si surfaces. p-type PA forms nearly ohmic contacts with p-Si (a) and highly rectifying contacts with n-Si (b). Likewise, n-type PA forms highly rectifying contacts with p-Si (a) and ohmic contacts with n-Si (b). Evaporated Au forms ohmic contacts with p-Si (a) and weakly rectifying contacts with n-Si (b). Reproduced from Ref. 173.

metal atoms/SAM molecule) of Au atoms (Fig. 12.31) [179]. In addition, very little change was observed in the negative ion spectra, and no new ion peaks were observed. This behavior suggests that the monolayers "float" on top of the deposited Au.

An interaction between evaporation-deposited Cu atoms and a monolayer organic functional end group (OFG), –C(O)OH, was observed by XPS [187]. High-resolution XP spectra of the O 1s region showed peaks at 532.8 and 531.9 eV, representing the C(O)$\underline{\text{O}}$H and C($\underline{\text{O}}$)OH peaks, respectively. After deposition of 1 nm of Cu, the region showed peaks at 531.6 and 530.8 eV, representing the C($\underline{\text{O}}$)OCu and C(O)$\underline{\text{O}}$Cu, respectively.

Interactions between metal atoms and OFGs are specific. As the OFG identity is changed, the ratio of vacuum-evaporated Au monolayer penetration and deposition at the monolayer–vacuum interface also changes [179]. ToF-SIMS, AFM, and CPAFM have been used to analyze the extent to which monolayer penetration occurred in each system of interest. AFM images revealed that in the case of metal penetration, the structure of the Au(111) substrate was still clearly visible. If, however, the deposition occurred above the monolayer, mushroom-like Au particles

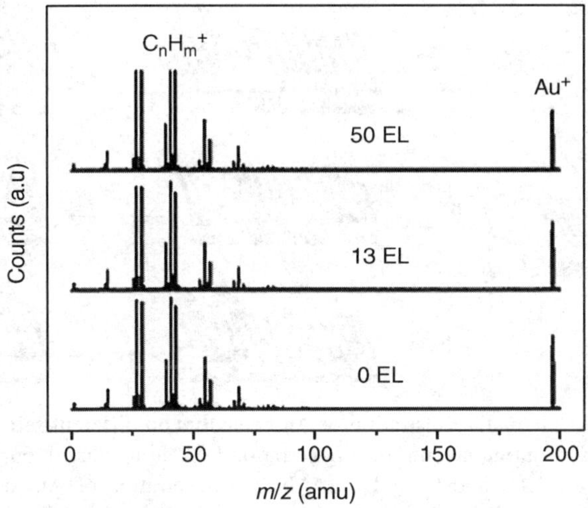

FIGURE 12.31 Positive ion ToF-SIMS spectra of CH_3-terminated films with Au dose for 0, 13, and 50 equivalent layers of Au. The absence of any change in the spectra suggests the Au is penetrating the CH_3-terminated monolayers. Reproduced from Ref. 179 by permission of the American Chemical Society, © 2006.

were observable, or a more disordered film was seen. CPAFM was used to reveal shorts, indicative of Au filaments spanning the monolayer with some deposition at the monolayer/vacuum interface. Less Au penetrated to the substrate when the OFG was changed from $-CH_3$ to $-C(O)OCH_3$ to C(O)OH to $C(O)O^-K^+$ (Fig. 12.32). This behavior is consistent with the expectations for the different physical barriers produced by the interactions between the metals and the different OFGs.

Grazing incidence FTIR (GI-FTIR) has been used to observe the interactions between metal atoms and OFGs. Upon deposition of Au and Al onto thiol-terminated SAMs, the monolayer peaks decreased in intensity, but neither a peak broadening nor a peak shift was observed, as would be expected upon disruption of the monolayer [188]. Deposition on the aryl-terminated monolayers resulted in a peak broadening and a peak shift, but without much change in peak intensity. This behavior indicated penetration of metal atoms through the monolayer. Consistently, deposition of Ti destroyed all of the monolayers tested.

Al was observed to react with –OH terminated monolayers to completion, rather than penetrating to the monolayer/substrate interface [189]. A dielectric Al layer was formed at low Al coverages. XP spectra at various deposition thicknesses showed a shift of the O binding energy, from 533.4 eV to a broadened peak at 532.6 eV, suggesting formation of an Al—O species. In addition, at low deposition thicknesses, the Al peak was very broad, at 75.2 binding eV (BeV). At high Al thickness, a metallic peak was observed to grow in at 72.9 BeV, with retention of the nonmetallic Al—O species. IR spectra before and after deposition also showed changes in the C—O stretching frequency upon conversion from C—OH to C—OAl, and a peak assignable to Al—O species, at 850 cm^{-1}, appeared. By comparison, deposition on

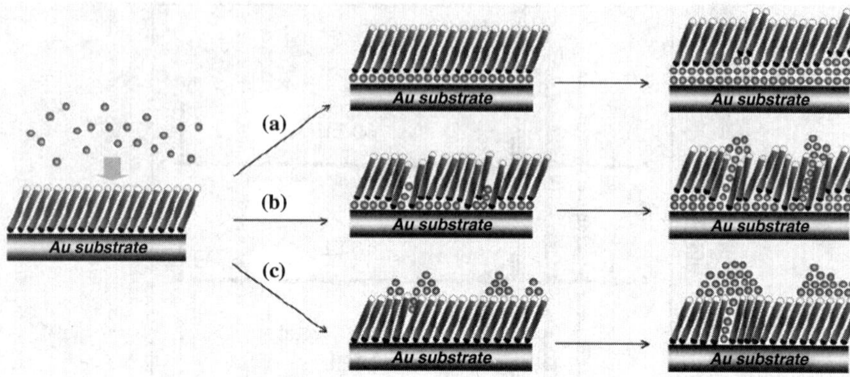

FIGURE 12.32 Schematic illustration of Au behaviors on different self-assembled monolayers (SAMs). (a) Continuous, uniform penetration leading to smooth buried underlayers is seen for Au on the CH_3 and CO_2CH_3 SAMs; (b) penetration of Au to form conducting filaments with connected overlayer clusters, seen for Au on the CO_2H SAM; (c) high interaction of incoming Au with monolayer OFGs, as seen for Au on the K-modified CO_2CH_3 and CO_2H films, leads to nucleation of overlayer clusters with initial scattered penetration into the self-assembled monolayer and, at higher coverages, penetration leading to formation of scattered conducting filaments connected to overlayer clusters in mushroom-like overlayer deposits and contiguous overlayer islands. Figure reproduced from Ref. 179 by permission of the American Chemical Society, © 2006.

–OCH_3 terminated monolayers showed penetration of the monolayer via observation of Al–S via XPS and ToF-SIMS. Interaction of different noble metals with the –OCH_3 terminated monolayer was explored using ToF-SIMS, infrared reflection spectroscopy, and XPS [180]. Al penetrated the monolayer until the monolayer substrate interface was 1:1 Al/Au. In contrast, Cu and Ag continuously partitioned deposition through, and on top of, the monolayer. Au only deposited through the monolayer, and the monolayer was observed to float on top.

The *J–V* characteristics provide an additional indicator of the extent of monolayer penetration [179, 182]. For example, Au has been evaporated onto –CH_3 and –SH terminated monolayers formed by hydrosilylation of Si [179]. *J–V* and capacitance–voltage measurements on the Au/monolayer/Si junctions showed similar effective barrier heights and built-in potentials regardless of the nature of the OFG. Nanoscale electrical measurements via ballistic electron emission microscopy (BEEM) closely matched the macroscopic measurements, and threshold voltages were approximately the same across all of the samples, including evaporated films on H-terminated silicon. The current decreased by over 80% for functionalized samples, with a marked difference between –CH_3 and –SH terminated monolayers because of the difference in metal penetration (Fig. 12.33). TEM showed much higher Au penetration for the –CH_3 OFGs; however, Au penetration of the monolayer was observed for all surface treatments (Fig. 12.33).

Terminal functionality of monolayers has been utilized to facilitate other metal deposition techniques, such as transfer printing (TP) and nanotransfer printing (nTP) [190–192]. These techniques utilize the interaction between a metal, for example Au, and the terminal end of an organic monolayer; 1,8-octanedithiol on GaAs served as an early example to demonstrate transfer of an evaporated metal film [190]. By introduction of functionality to the organic monolayer, direct evaporation is avoided. Instead, the metal films are evaporated onto a treated PDMS stamp that has poor adhesion to the metal, and the films are then transferred to a surface terminated with strongly interacting functional groups, to make the desired metal/organic monolayer/semiconductor structures. The use of TP to contact Si with Au yields a pure metal–insulator–semiconductor structure [193]. The tunneling through the monolayer was shown to be purely temperature independent, in contrast to the behavior of Au evaporated on –SH monolayers. This method also readily forms nanoscale metal patterns with the master pattern being formed in the PDMS stamp, thus avoiding repetitive, costly lithography steps. SAMs of a thiol-terminated silane, $(CH_3)_3Si(CH_2)_3SH$, were formed on oxidized Si for the transfer of Au films to Si surfaces (Fig. 12.34) [191]. nTP can make Cu contacts as well, with one additional processing step [192].

The stamp approach to contact patterning can also be used to pattern the initial monolayer [194]. In this method, the stamp catalyzes hydrosilylation at the points of contact. Gold nanoparticles then deposited preferentially on the patterned monolayers. Pd nanoparticle–OFG interactions have been used to seed electroless Ni deposition; however, the resulting junction properties very closely resembled those of evaporated Ni [195].

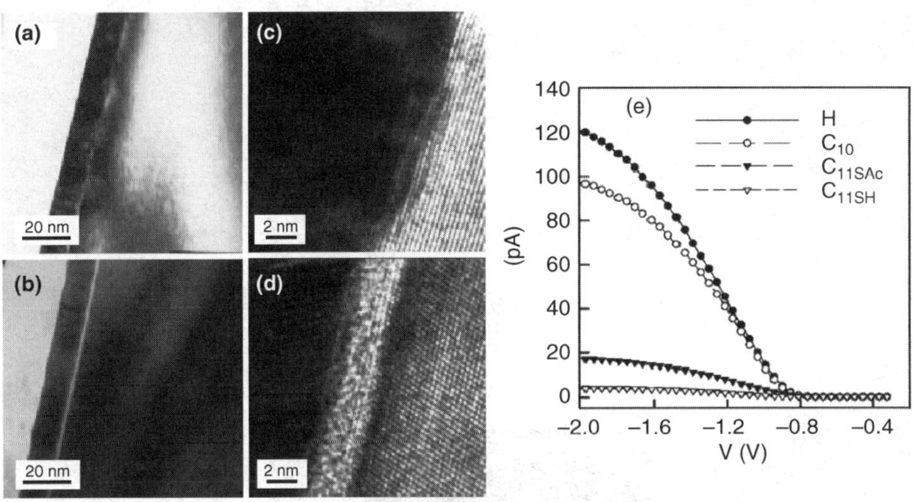

FIGURE 12.33 Cross-sectional TEM images of evaporated Au/organic monolayer/Si(111) interface. (a, c) Au/$C_{10}H_{13}$–Si(111) at low (bright field) and high magnification (phase contrast images); (b, d) Au/$HSC_{11}H_{22}$–Si(111) at low and high magnification. (e) Corresponding averaged BEEM spectra for (a), (b), and Au/H–Si(111) and Au/$CH_3C(O)SC_{11}H_{22}$–Si(111) junctions. Figures reproduced from Ref. 182 by permission of the American Chemical Society, © 2008.

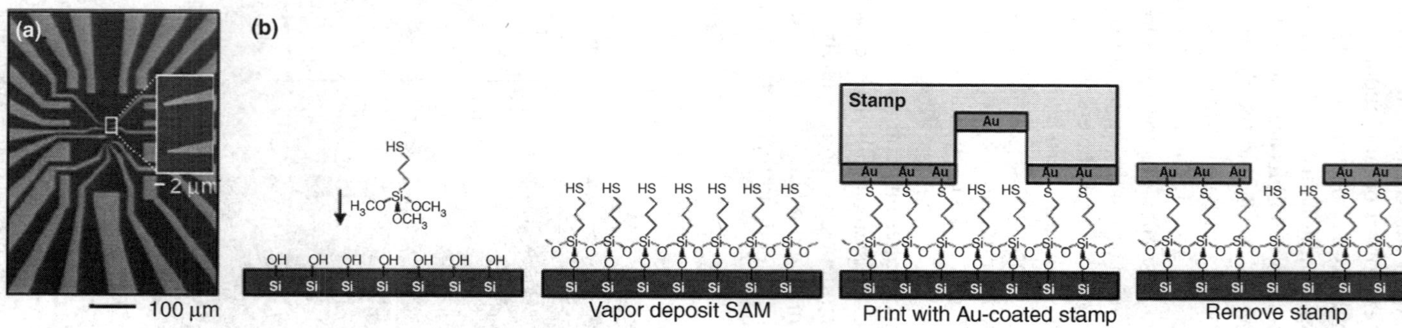

FIGURE 12.34 (a) Optical micrograph of Au pattern formed by nanoscale transfer printing. (b) Scheme for creating nanoscale transfer printed Au pads on Si surfaces. Figure reproduced from Ref. 191 by permission of the American Chemical Society, © 2002.

Physical vapor deposition of metal on patterned monolayers consisting of more than one OFG has utilized the differential extent of monolayer penetration to pattern contacts [181]. In this method, the metal atoms deposit at the monolayer–substrate interface over a portion of the monolayer, and on top of the monolayer over the rest of the surface.

12.4.4 Semiconducting and Nonmetallic Coatings

12.4.4.1 Stability Enhancement Nonmetallic coatings have been used frequently to stabilize Si under anodic or photoanodic conditions. ITO [196, 197], other doped SnO_2 films [198], boron phosphide [199], thallium(III) oxide [200], TiO_2 [201], Fe_2O_3 [202], and even silicides [203] have been used. These techniques have yielded high-efficiency devices; for example, the n-Si/thallium(III) oxide/ferricyanide electrochemical cell achieved 11.0% efficiency under natural light. An early problem was deposition of thick overlayers. For example, deposition of TiO_2 onto n-Si showed only the photoelectrochemical properties of the TiO_2 [201]. Thus, methods such as ALD that can deposit thinner semiconducting films are desirable.

12.4.4.2 Deposition on Si by ALD In addition to thin protective layers, ALD on Si is of interest for transistors. The fundamental limitations on the use of SiO_2 as a dielectric in transistors cause problems for the future scale of electronics [73]. The thickness required, low dielectric constant, k, and high leakage currents make SiO_2 not well suited for low-power applications or for the next stages of miniaturization of electronics. As low interfacial trap state densities are crucial, it would be difficult to create a more ideal interface than that of a well-prepared Si/SiO_2 interface, which has a typical midgap interface state density of 2×10^{10} states/cm^2. Most interfaces fabricated using ALD have between one and two orders of magnitude higher interfacial trap state density. In addition, these interfaces exhibit substantial flat-band voltage shifts, in excess of 300 mV. For high-quality devices, no silicide or metal oxide phases should be at or near the channel interface, because silicide and most M–Si bonding will lead to poor leakage currents and low electron channel mobilities.

ALD is thus an attractive method to circumvent these difficulties by controlling the interfacial chemistry. Unfortunately, due to a lack of reactivity, ALD requires extreme conditions to yield film growth on Si–H [204]. ALD generally utilizes very high reactivity, but self-limiting, reactions for controlled material preparation. The Si–H surface simply does not have the inherent required reactivity. Long initial exposure times, of 3600 s, with H_2O, or an initial exposure to a metal precursor, $Al(CH_3)_3$ for example, are thus necessary to initiate uniform growth on Si [204]. Si oxidation was observed to occur during this time, due to impurity M–OR species in the metal feed [205]. After metal binding, oxidation by H_2O was catalyzed. Oxidation during growth was more pronounced on the Si(100) face. One TMA exposure catalyzed the growth of 1.2 Å SiO_x.

Just as the evaporation of metal onto end-functionalized organic monolayers affected the extent of penetration to the substrate and the final metal film morphology, a similar effect occurs in ALD [206, 207]. ALD of $Ti[N(CH_3)_2]_4$ at SAMs of alkyltrichlorosilanes with –OH, $-NH_2$, and $-CH_3$ OFGs was studied as a step toward understanding the deposition of titanium nitride on organic monolayers [206]. The reaction proceeded at every temperature tested. Angle-resolved XPS was used to observe whether Ti deposition occurred at the SiO_2/SAM or SAM/vacuum interface (Fig. 12.35). Deposition on $-CH_3$ terminated monolayers penetrated to the SAM/SiO_2 interface, and partitioned to both interfaces at –OH functionalized SAMs. The $-NH_2$ terminated SAM showed reactivity solely at the SAM/vacuum interface. In addition, deposition rates at $-CH_3$ terminated SAMs were severely attenuated, and islanding was observed using AFM [111]. Growth at –OH and $-NH_2$ surfaces was less attenuated, but growth was not uniform until thicker depositions were performed.

In situ and *ex situ* IR and XPS have been used to observe metal interactions with OFGs, during ALD of metallic Cu using $[Cu(sBu\text{-}amd)]_2$ as a precursor [208]. IR data showed that a carboxylic acid-terminated monolayer interacts with Cu atoms in a bidentate fashion without the formation of a SiO_x layer (Fig. 12.36). This work

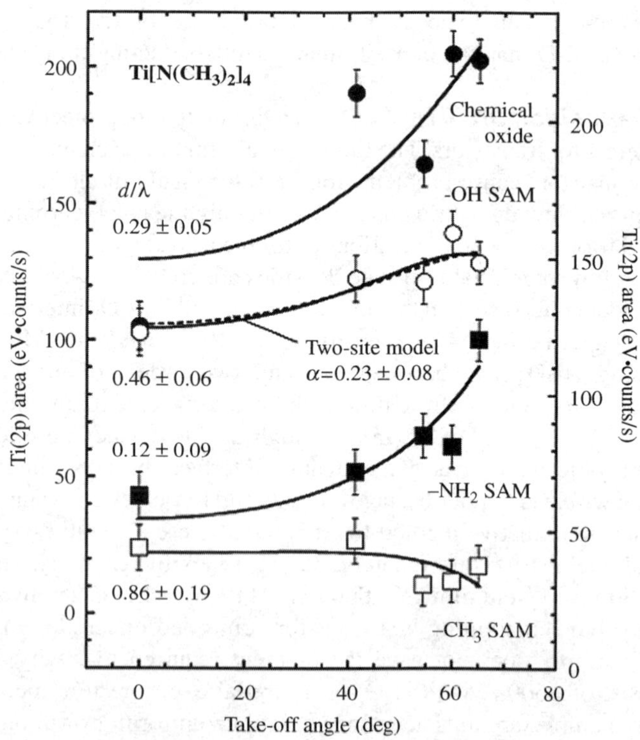

FIGURE 12.35 Angle-dependent integrated Ti 2p intensity for the adsorption of ALD precursor $Ti[N(CH_3)_2]_4$ on Si chemical oxide (filled circles, left axis), –OH terminated SAM (open circles, left axis), $-NH_2$ terminated SAM (filled squares, right axis), and $-CH_3$ terminated SAM (open squares, right axis). The curves represent a fit to a model that assumes that the Ti is uniformly distributed at a depth d from the surface, and the inelastic mean free path of the Ti 2p photoelectrons is λ. The value d/λ is given for each case. Reproduced from Ref. 206 by permission of the American Chemical Society, © 2005.

serves as a demonstration that high-quality metallic contacts can be made to Si using ALD. Unlike deposition on H-terminated Si surfaces, Al_2O_3 can be deposited at –COOH terminated monolayers on Si, without the formation of oxide [209]. TIR observation of Al_2O_3 growth showed formation of a high-quality film. The absence of any shifts or changes in the C–H region of the IR showed no disruption of the monolayer. Stringent conditions in monolayer preparation, and interaction of ALD reagents with the –COOH rather than penetration to the SAM/Si interface, thus prevented formation of SiO_x species. Characteristic SiO_x vibrational modes only appeared upon heating to >400°C.

12.5 CONCLUSIONS

Si–C termination renders Si surfaces much more stable to a larger range of working conditions than is observed for H- or halogen-terminated silicon surfaces. The increased stability stems from the kinetic stability of Si–C bonds and from the physical barrier imposed by a well-ordered organic film. Increased stability does not necessarily have to produce a loss of beneficial electronic properties. Beyond the stability imparted by the Si–C group, functional organic monolayers have great utility in the fabrication of advanced device structures. Classic problems with Si, including oxidation in air and Fermi level pinning in contact with most metals, can be overcome by the clever use of organic monolayers. Global device issues such as

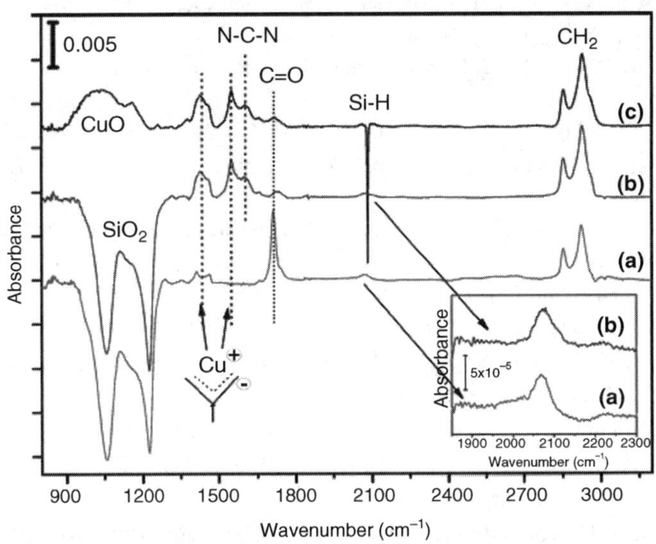

FIGURE 12.36 Absorption spectra of ALD deposition of Cu metal at a COOH-terminated SAM on Si. (a) COOH-terminated SAM referenced to the Si native oxide; (b, c) COOH-terminated SAM with 20 Cu ALD cycles referenced to the native oxide and H-terminated surfaces, respectively. Figure reproduced from Ref. 208 by permission of the American Chemical Society, © 2009.

high-cost patterning can be avoided by the use of cheaply patterned monolayers. A thorough understanding of the surface chemistry of Si thus enables the formation of well-defined, electronically preserved interfaces and, ultimately, more efficient devices in a variety of optoelectronic applications.

REFERENCES

1. Yablonovitch, E.; Gmitter, T. J. *Solid State Electron.* **1992**, *35*, 261–267.
2. Yablonovitch, E.; Allara, D. L.; Chang, C. C.; Gmitter, T.; Bright, T. B. *Phys. Rev. Lett.* **1986**, *57*, 249–253.
3. Heller, A. *Solar Energy* **1982**, *29*, 153–162.
4. Grunthaner, F. J.; Grunthaner, P. J. *Mater. Sci. Rep.* **1986**, *1*, 65–160.
5. Grundner, M.; Jacob, H. *Appl. Phys. A* **1986**, *39*, 73–82.
6. Rappich, J.; Lewerenz, H. J. *J. Electrochem. Soc.* **1995**, *142*, 1233–1237.
7. Dittrich, T.; Rauscher, S.; Bitzer, T.; Aggour, M.; Flietner, H.; Lewerenz, H. J. *J. Electrochem. Soc.* **1995**, *142*, 2411–2413.
8. Rappich, J.; Jungblut, H.; Aggour, M.; Lewerenz, H. J. *J. Electrochem. Soc.* **1994**, *141*, L99–L102.
9. Bitzer, T.; Gruyters, M.; Lewerenz, H. J.; Jacobi, K. *Appl. Phys. Lett.* **1993**, *63*, 397–399.
10. Ibach, H.; Rowe, J. E. *Surf. Sci.* **1974**, *43*, 481–492.
11. Mayer, T. M.; Lampe, F. W. *J. Phys. Chem.* **1974**, *78*, 2429–2433.
12. Burrows, V. A.; Chabal, Y. J.; Higashi, G. S.; Raghavachari, K.; Christman, S. B. *Appl. Phys. Lett.* **1988**, *53*, 998–1000.
13. Chabal, Y. J.; Higashi, G. S.; Raghavachari, K.; Burrows, V. A. *J. Vac. Sci. Technol. A* **1989**, *7*, 2104–2109.
14. Clark, I. T.; Aldinger, B. S.; Gupta, A.; Hines, M. A. *J. Phys. Chem. C* **2010**, *114*, 423–428.
15. Higashi, G. S.; Becker, R. S.; Chabal, Y. J.; Becker, A. J. *Appl. Phys. Lett.* **1991**, *58*, 1656–1658.
16. Higashi, G. S.; Chabal, Y. J.; Trucks, G. W.; Raghavachari, K. *Appl. Phys. Lett.* **1990**, *56*, 656–658.
17. Hines, M. A. *Annu. Rev. Phys. Chem.* **2003**, *54*, 29–56.
18. Wade, C. P.; Chidsey, C. E. D. *Appl. Phys. Lett.* **1997**, *71*, 1679–1681.
19. Cicero, R. L.; Linford, M. R.; Chidsey, C. E. D. *Langmuir* **2000**, *16*, 5688–5695.
20. Jakob, P.; Chabal, Y. J.; Raghavachari, K. *Chem. Phys. Lett.* **1991**, *187*, 325–333.
21. Jakob, P.; Dumas, P.; Chabal, Y. J. *Appl. Phys. Lett.* **1991**, *59*, 2968–2970.
22. Cowley, A. M.; Sze, S. M. *J. Appl. Phys.* **1965**, *36*, 3212–3220.
23. Linnros, J. *J. Appl. Phys.* **1998**, *84*, 275–283.
24. Angermann, H.; Kliefoth, K.; Flietner, H. *Appl. Surf. Sci.* **1996**, *104*, 107–112.
25. Wittmer, M.; Freeouf, J. L. *Phys. Lett. A* **1993**, *173*, 190–194.
26. Wittmer, M.; Freeouf, J. L. *Phys. Rev. Lett.* **1992**, *69*, 2701–2704.
27. Gstrein, F.; Michalak, D. J.; Royea, W. J.; Lewis, N. S. *J. Phys. Chem. B* **2002**, *106*, 2950–2961.

28. Michalak, D. J.; Amy, S. R.; Esteve, A.; Chabal, Y. J. *J. Phys. Chem. C* **2008**, *112*, 11907–11919.

29. Michalak, D. J.; Gstrein, F.; Lewis, N. S. *J. Phys. Chem. C* **2008**, *112*, 5911–5921.

30. Michalak, D. J.; Gstrein, F.; Lewis, N. S. *J. Phys. Chem. C* **2007**, *111*, 16516–16532.

31. Ciampi, S.; Harper, J. B.; Gooding, J. J. *Chem. Soc. Rev.* **2010**, *39*, 2158–2183.

32. Lehmann, V. *Electrochemistry of Silicon*. Wiley-VCH, Weinheim, 2002.

33. Tuck, B. *J. Mater. Sci.* **1975**, *10*, 321–339.

34. Henderson, R. C. *J. Electrochem. Soc.* **1972**, *119*, 772.

35. Kern, W.; Puotinen, D. A. *RCA Rev.* **1970**, *31*, 187–206.

36. Kern, W. *J. Electrochem. Soc.* **1990**, *137*, 1887–1892.

37. Kern, W. *RCA Rev.* **1978**, *39*, 278–308.

38. Sato, K.; Shikida, M.; Matsushima, Y.; Yamashiro, T.; Asaumi, K.; Iriye, Y.; Yamamoto, M. *Sens. Actuators A* **1998**, *64*, 87–93.

39. Mende, G.; Finster, J.; Flamm, D.; Schulze, D. *Surf. Sci.* **1983**, *128*, 169–175.

40. Niwano, M.; Kageyama, J.; Kurita, K.; Kinashi, K.; Takahashi, I.; Miyamoto, N. *J. Appl. Phys.* **1994**, *76*, 2157–2163.

41. Zhang, X.; Garfunkel, E.; Chabal, Y. J.; Christman, S. B.; Chaban, E. E. *Appl. Phys. Lett.* **2001**, *79*, 4051–4053.

42. Yasaka, T.; Takakura, M.; Sawara, K.; Uenaga, S.; Yasutake, H.; Miyazaki, S.; Hirose, M. *IEICE Trans. Electron.* **1992**, *E75C*, 764–769.

43. Cleland, G.; Horrocks, B. R.; Houlton, A. *J. Chem. Soc., Faraday Trans.* **1995**, *91*, 4001–4003.

44. Yu, H.-Z.; Morin, S.; Wayner, D. D. M.; Allongue, P.; Henry de Villeneuve, C. *J. Phys. Chem. B* **2000**, *104*, 11157–11161.

45. Boukherroub, R.; Morin, S.; Sharpe, P.; Wayner, D. D. M.; Allongue, P. *Langmuir* **2000**, *16*, 7429–7434.

46. Michalak, D. J.; Rivillon, S.; Chabal, Y. J.; Esteve, A.; Lewis, N. S. *J. Phys. Chem. B* **2006**, *110*, 20426–20434.

47. Warntjes, M.; Vieillard, C.; Ozanam, F.; Chazalviel, J. N. *J. Electrochem. Soc.* **1995**, *142*, 4138–4142.

48. Bateman, J. E.; Eagling, R. D.; Horrocks, B. R.; Houlton, A. *J. Phys. Chem. B* **2000**, *104*, 5557–5565.

49. Hory, M. A.; Herino, R.; Ligeon, M.; Muller, F.; Gaspard, F.; Mihalcescu, I.; Vial, J. C. *Thin Solid Films* **1995**, *255*, 200–203.

50. Hacker, C. A.; Anderson, K. A.; Richter, L. J.; Richter, C. A. *Langmuir* **2004**, *21*, 882–889.

51. Bateman, J. E.; Horrocks, B. R.; Houlton, A. *J. Chem. Soc., Faraday Trans.* **1997**, *93*, 2427–2431.

52. Chazalviel, J. N. *J. Electroanal. Chem.* **1987**, *233*, 37–48.

53. Sze, S. M. *The Physics of Semiconductor Devices*, 2nd ed. Wiley, *New York*, 1981.

54. Kampen, T. U.; Monch, W. *Surf. Sci.* **1995**, *331*, 490–495.

55. Kampen, T. U.; Schmitsdorf, R. F.; Monch, W. *Appl. Phys. A* **1995**, *60*, 391–394.

56. Kurtin, S.; McGill, T. C.; Mead, C. A. *Phys. Rev. Lett.* **1969**, *22*, 1433–1436.

57. Severin, P. J.; Poodt, G. J. *J. Electrochem. Soc.* **1972**, *119*, 1384.
58. Schaffer, P. S.; Lally, T. R. *Solid State Technol.* **1983**, *26*, 229–233.
59. Maldonado, S.; Plass, K. E.; Knapp, D.; Lewis, N. S. *J. Phys. Chem. C* **2007**, *111*, 17690–17699.
60. Villarrubia, J. S.; Boland, J. J. *Phys. Rev. Lett.* **1989**, *63*, 306–309.
61. Aldao, C. M.; Weaver, J. H. *Prog. Surf. Sci.* **2001**, *68*, 189–230.
62. Horanyi, T. S.; Pavelka, T.; Tutto, P. *Appl. Surf. Sci.* **1993**, *63*, 306–311.
63. M'Saad, H.; Michel, J.; Lappe, J. J.; Kimerling, L. C. *J. Electron. Mater.* **1994**, *23*, 487–491.
64. Stephens, A. W.; Green, M. A. *Solar Energy Mater. Solar Cells* **1997**, *45*, 255–265.
65. Hamers, R. J. *Surf. Sci.* **2006**, *600*, 3361–3362.
66. He, J.; Patitsas, S. N.; Preston, K. F.; Wolkow, R. A.; Wayner, D. D. M. *Chem. Phys. Lett.* **1998**, *286*, 508–514.
67. Bansal, A.; Li, X. L.; Lauermann, I.; Lewis, N. S.; Yi, S. I.; Weinberg, W. H. *J. Am. Chem. Soc.* **1996**, *118*, 7225–7226.
68. Cai, W.; Lin, Z.; Strother, T.; Smith, L. M.; Hamers, R. J. *J. Phys. Chem. B* **2002**, *106*, 2656–2664.
69. Zhu, X.-Y.; Boiadjiev, V.; Mulder, J. A.; Hsung, R. P.; Major, R. C. *Langmuir* **2000**, *16*, 6766–6772.
70. Bergerson, W. F.; Mulder, J. A.; Hsung, R. P.; Zhu, X.-Y. *J. Am. Chem. Soc.* **1999**, *121*, 454–455.
71. Klaus, J. W.; Sneh, O.; George, S. M. *Science* **1997**, *278*, 1934–1937.
72. Buriak, J. M. *Chem. Rev.* **2002**, *102*, 1271.
73. Wilk, G. D.; Wallace, R. M.; Anthony, J. M. *J. Appl. Phys.* **2001**, *89*, 5243.
74. Guha, S.; Cartier, E.; Gribelyuk, M. A.; Bojarczuk, N. A.; Copel, M. C. *Appl. Phys. Lett.* **2000**, *77*, 2710.
75. Edge, L. F.; Schlom, D. G.; Brewer, R. T.; Chabal, Y. J.; Williams, J. R.; Chambers, S. A.; Hinkle, C.; Lucovsky, G.; Yang, Y.; Stemmer, S.; Copel, M. C.; Hollander, B.; Schubert, J. *Appl. Phys. Lett.* **2004**, *84*, 4629–4631.
76. Terry, J.; Mo, R.; Wigren, C.; Cao, R.; Mount, G.; Pianetta, P.; Linford, M. R.; Chidsey, C. E. D. *Nucl. Instrum. Methods Phys. Res. B* **1997**, *133*, 94.
77. Rivillon, S.; Chabal, Y. J.; Webb, L. J.; Michalak, D. J.; Lewis, N. S.; Halls, M. D.; Raghavachari, K. *J. Vac. Sci. Technol. A* **2005**, *23*, 1100–1106.
78. Florio, J. V.; Robertson, W. D. *Surface Science* **1969**, *18*, 398–427.
79. Li, Z.; Kamins, T. I.; Li, X.; Williams, R. S. *Surf. Sci.* **2004**, *554*, L81–L86.
80. Rivillon, S.; Amy, F.; Chabal, Y. J.; Frank, M. M. *Appl. Phys. Lett.* **2004**, *85*, 2583–2585.
81. Eves, B. J.; Lopinski, G. P. *Surf. Sci.* **2005**, *579*, L89–L96.
82. Bansal, A.; Li, X. L.; Yi, S. I.; Weinberg, W. H.; Lewis, N. S. *J. Phys. Chem. B* **2001**, *105*, 10266–10277.
83. Cao, P.; Yu, H.; Heath, J. R. *J. Phys. Chem. B* **2006**, *110*, 23615–23618.
84. Webb, L. J.; Lewis, N. S. *J. Phys. Chem. B* **2003**, *107*, 5404–5412.
85. Webb, L. J.; Nemanick, E. J.; Biteen, J. S.; Knapp, D. W.; Michalak, D. J.; Traub, M. C.; Chan, A. S. Y.; Brunschwig, B. S.; Lewis, N. S. *J. Phys. Chem. B* **2005**, *109*, 3930–3937.

86. Yu, H. B.; Webb, L. J.; Solares, S. D.; Cao, P. G.; Goddard, W. A.; Heath, J. R.; Lewis, N. S. *J. Phys. Chem. B* **2006**, *110*, 23898–23903.

87. Ferguson, G. A.; Rivillon, S.; Chabal, Y. J.; Raghavachari, K. *J. Phys. Chem. C* **2009**, *113*, 21713–21720.

88. Citrin, P. H.; Rowe, J. E. *Surf. Sci.* **1983**, *132*, 205–211.

89. Amy, S. R.; Michalak, D. J.; Chabal, Y. J.; Wielunski, L.; Hurley, P. T.; Lewis, N. S. *J. Phys. Chem. C* **2007**, *111*, 13053–13061.

90. Finstad, C. C.; Thorsness, A. G.; Muscat, A. J. *Surf. Sci.* **2006**, *600*, 3363–3374.

91. Gupta, P.; Coon, P. A.; Koehler, B. G.; George, S. M. *Surf. Sci.* **1991**, 249.

92. Lopinski, G. P.; Eves, B. J.; Hul'ko, O.; Mark, C.; Patitsas, S. N.; Boukherroub, R.; Ward, T. R. *Phys. Rev. B* **2005**, *71*, 125308.

93. Gulteben, H.; Lucas, S. R.; Cheng, C. C.; Choyke, W. J.; T., Y. J. J. *Surf. Sci.* **1991**, *257*, 146–156.

94. Colaianni, M. L.; Chen, P. J.; Gutleben, H.; T.,Y. J. J. *Chem. Phys. Lett.* **1992**, *191*, 561–568.

95. Kong, M. J.; Lee, S. S.; Lyubovitsky, J.; Bent, S. F. *Chem. Phys. Lett.* **1996**, *263*, 1–7.

96. Tanaka, M.; Yamakawa, E.; Shirao, T.; Shudo, K. *Phys. Rev. B* **2003**, *68*, 165411.

97. Cowan, P. L.; Golovchenko, J. A.; Robbins, M. F. *Phys. Rev. Lett.* **1980**, *44*, 1680–1683.

98. Rivillon, S.; Brewer, R. T.; Chabal, Y. J. *Appl. Phys. Lett.* **2005**, *87*, 173118.

99. Klaus, J. W.; Ott, A. W.; Johnson, J. M.; George, S. M. *Appl. Phys. Lett.* **1997**, *70*, 1092–1094.

100. Nishiyama, K.; Tanaka, Y.; Harada, H.; Yamada, T.; Niwa, D.; Inoue, T.; Homma, T.; Osaka, T.; Taniguchi, I. *Surf. Sci.* **2006**, *600*, 1965–1972.

101. Royea, W. J.; Michalak, D. J.; Lewis, N. S. *Appl. Phys. Lett.* **2000**, *77*, 2566–2568.

102. Haber, J. A.; Lauermann, I.; Michalak, D.; Vaid, T. P.; Lewis, N. S. *J. Phys. Chem. B* **2000**, *104*, 9947–9950.

103. Reddy, A. J.; Chan, J. V.; Burr, T. A.; Mo, R.; Wade, C. P.; Chidsey, C. E. D.; Kimerling, L. C. *Physica B* **1999**, *274*, 468–472.

104. Haber, J. A.; Lewis, N. S. *J. Phys. Chem. B* **2002**, *106*, 3639–3656.

105. Johansson, E.; Hurley, P. T.; Brunschwig, B. S.; Lewis, N. S. *J. Phys. Chem. C* **2009**, *113*, 15239–15245.

106. Nemanick, E. J.; Hurley, P. T.; Brunschwig, B. S.; Lewis, N. S. *J. Phys. Chem. B* **2006**, *110*, 14800–14808.

107. Wolkow, R. A.; Avouris, P. *Phys. Rev. Lett.* **1988**, *60*, 1049–1052.

108. Lange, B.; Schmidt, W. G. *Surf. Sci.* **2008**, *602*, 1207–1211.

109. Zhu, X.-Y.; Jun, Y.; Staarup, D. R.; Major, R. C.; Danielson, S.; Boiadjiev, V.; Gladfelter, W. L.; Bunker, B. C.; Guo, A. *Langmuir* **2001**, *17*, 7798–7803.

110. Fidelis, A.; Ozanam, F.; Chazalviel, J. N. *Surf. Sci.* **2000**, *444*, L7–L10.

111. Sieval, A. B.; van den Hout, B.; Zuilhof, H.; Sudholter, E. J. R. *Langmuir* **2001**, *17*, 2172–2181.

112. Yu, H. B.; Webb, L. J.; Ries, R. S.; Solares, S. D.; Goddard, W. A.; Heath, J. R.; Lewis, N. S. *J. Phys. Chem. B* **2005**, *109*, 671–674.

113. Bin, X. M.; Mischki, T. K.; Fan, C. Y.; Lopinski, G. P.; Wayner, D. D. M. *J. Phys. Chem. C* **2007**, *111*, 13547–13553.

114. Hunger, R.; Fritsche, R.; Jaeckel, B.; Jaegermann, W.; Webb, L. J.; Lewis, N. S. *Phys. Rev. B* **2005**, *72*, 045317.

115. Webb, L. J.; Rivillon, S.; Michalak, D. J.; Chabal, Y. J.; Lewis, N. S. *J. Phys. Chem. B* **2006**, *110*, 7349–7356.

116. Terry, J.; Linford, M. R.; Wigren, C.; Cao, R. Y.; Pianetta, P.; Chidsey, C. E. D. *J. Appl. Phys.* **1999**, *85*, 213–221.

117. Greenwood, N. N.; Earnshaw, A. *Chemistry of the Elements*, 2nd ed. Reed Educational and Professional Publishing Ltd., Oxford, UK, 1997.

118. Himpsel, F. J.; McFeely, F. R.; Talebibrahimi, A.; Yarmoff, J. A.; Hollinger, G. *Phys. Rev. B* **1988**, *38*, 6084–6096.

119. Linford, M. R.; Fenter, P.; Eisenberger, P. M.; Chidsey, C. E. D. *J. Am. Chem. Soc.* **1995**, *117*, 3145–3155.

120. Jaeckel, B.; Hunger, R.; Webb, L. J.; Jaegermann, W.; Lewis, N. S. *J. Phys. Chem. C* **2007**, *111*, 18204–18213.

121. Cicero, R. L.; Chidsey, C. E. D.; Lopinski, G. P.; Wayner, D. D. M.; Wolkow, R. A. *Langmuir* **2002**, *18*, 305–307.

122. Webb, L. J.; Michalak, D. J.; Biteen, J. S.; Brunschwig, B. S.; Chan, A. S. Y.; Knapp, D. W.; Meyer, H. M.; Nemanick, E. J.; Traub, M. C.; Lewis, N. S. *J. Phys. Chem. B* **2006**, *110*, 23450–23459.

123. Gosalvez, M. A.; Sato, K.; Foster, A. S.; Nieminen, R. M.; Tanaka, H. *J. Micromech. Microeng.* **2007**, *17*, S1–S26.

124. Nemanick, E. J.; Hurley, P. T.; Webb, L. J.; Knapp, D. W.; Michalak, D. J.; Brunschwig, B. S.; Lewis, N. S. *J. Phys. Chem. B* **2006**, *110*, 14770–14778.

125. O'Leary, L. E.; Johansson, E.; Brunschwig, B. S.; Lewis, N. S. *J. Phys. Chem. B* **2010**, *114*, 14298–14302.

126. Sieval, A. B.; Demirel, A. L.; Nissink, J. W. M.; Linford, M. R.; van der Maas, J. H.; de Jeu, W. H.; Zuilhof, H.; Sudholter, E. J. R. *Langmuir* **1998**, *14*, 1759–1768.

127. Sun, Q. Y.; de Smet, L.; van Lagen, B.; Wright, A.; Zuilhof, H.; Sudholter, E. J. R. *Angew. Chem., Int. Ed.* **2004**, *43*, 1352–1355.

128. Royea, W. J.; Juang, A.; Lewis, N. S. *Appl. Phys. Lett.* **2000**, *77*, 1988–1990.

129. Yaffe, O.; Scheres, L.; Puniredd, S. R.; Stein, N.; Biller, A.; Lavan, R. H.; Shapisman, H.; Zuilhof, H.; Haick, H.; Cahen, D.; Vilan, A. *Nano Lett.* **2009**, *9*, 2390–2394.

130. Maldonado, S.; Lewis, N. S. *J. Electrochem. Soc.* **2009**, *156*, H123–H128.

131. Bansal, A.; Lewis, N. S. *J. Phys. Chem. B* **1998**, *102*, 1067–1070.

132. Sondag-Huethorst, J. A. M.; Fokkink, L. G. J. *Langmuir* **1992**, *8*, 2560–2566.

133. Sondag-Huethorst, J. A. M.; Fokkink, L. G. J. *J. Electroanal. Chem.* **1994**, *367*, 49–57.

134. Allongue, P.; de Villeneuve, C. H.; Pinson, J.; Ozanam, F.; Chazalviel, J. N.; Wallart, X. *Electrochim. Acta* **1998**, *43*, 2791–2798.

135. Osaka, T.; Matsunaga, M.; Kudo, S.; Niwa, D.; Shacham-Diamand, Y.; Jaegermann, W.; Hunger, R. *J. Electrochem. Soc.* **2007**, *154*, H919–H926.

136. Becka, A. M.; Miller, C. J. *J. Phys. Chem.* **1992**, *96*, 2657–2668.

137. Chidsey, C. E. D. *Science* **1991**, *251*, 919–922.

138. Sachs, S. B.; Dudek, S. P.; Hsung, R. P.; Sita, L. R.; Smalley, J. F.; Newton, M. D.; Feldberg, S. W.; Chidsey, C. E. D. *J. Am. Chem. Soc.* **1997**, *119*, 10563–10564.

139. Bansal, A.; Lewis, N. S. *J. Phys. Chem. B* **1998**, *102*, 4058–4060.

140. Yu, H. Z.; Boukherroub, R.; Morin, S.; Wayner, D. D. M. *Electrochem. Commun.* **2000**, *2*, 562–566.

141. Faucheux, A.; Yang, F.; Allongue, P.; de Villeneuve, C. H.; Ozanam, F.; Chazalviel, J. N. *Appl. Phys. Lett.* **2006**, *88*, 193123.

142. Faucheux, A.; Gouget-Laemmel, A. C.; Allongue, P.; de Villeneuve, C. H.; Ozanam, F.; Chazalviel, J. N. *Langmuir* **2007**, *23*, 1326–1332.

143. Sung, M. M.; Kluth, G. J.; Yauw, O. W.; Maboudian, R. *Langmuir* **1997**, *13*, 6164–6168.

144. Linford, M. R.; Chidsey, C. E. D. *J. Am. Chem. Soc.* **1993**, *115*, 12631–12632.

145. Vandenberg, E. T.; Bertilsson, L.; Liedberg, B.; Uvdal, K.; Erlandsson, R.; Elwing, H.; Lundstrom, I. *J. Colloid Interface Sci.* **1991**, *147*, 103–118.

146. Smith, E. A.; Chen, W. *Langmuir* **2008**, *24*, 12405–12409.

147. Boukherroub, R.; Morin, S.; Bensebaa, F.; Wayner, D. D. M. *Langmuir* **1999**, *15*, 3831–3835.

148. Hamann, T. W.; Lewis, N. S. *J. Phys. Chem. B* **2006**, *110*, 22291–22294.

149. Boukherroub, R.; Wayner, D. D. M. *J. Am. Chem. Soc.* **1999**, *121*, 11513–11515.

150. Bocking, T.; Salomon, A.; Cahen, D.; Gooding, J. J. *Langmuir* **2007**, *23*, 3236–3241.

151. Buriak, J. M.; Allen, M. J. *J. Am. Chem. Soc.* **1998**, *120*, 1339–1340.

152. Allongue, P.; de Villeneuve, C. H.; Pinson, J. *Electrochim. Acta* **2000**, *45*, 3241–3248.

153. Sieval, A. B.; Linke, R.; Zuilhof, H.; Sudholter, E. J. R. *Adv. Mater.* **2000**, *12*, 1457–1460.

154. Langner, A.; Panarello, A.; Rivillon, S.; Vassylyev, O.; Khinast, J. G.; Chabal, Y. J. *J. Am. Chem. Soc.* **2005**, *127*, 12798–12799.

155. Hunger, R.; Fritsche, R.; Jaeckel, B.; Webb, L. J.; Jaegermann, W.; Lewis, N. S. *Surf. Sci.* **2007**, *601*, 2896–2907.

156. Haick, H.; Niitsoo, O.; Ghabboun, J.; Cahen, D. *J. Phys. Chem. C* **2007**, *111*, 2318–2329.

157. Moons, E.; Bruening, M.; Shanzer, A.; Beier, J.; Cahen, D. *Synth. Met.* **1996**, *76*, 245–248.

158. Maldonado, S.; Knapp, D.; Lewis, N. S. *J. Am. Chem. Soc.* **2008**, *130*, 3300.

159. Bocarsly, A. B.; Walton, E. G.; Wrighton, M. S. *J. Am. Chem. Soc.* **1980**, *102*, 3390–3398.

160. Bolts, J. M.; Bocarsly, A. B.; Palazzotto, M. C.; Walton, E. G.; Lewis, N. S.; Wrighton, M. S. *J. Am. Chem. Soc.* **1979**, *101*, 1378–1385.

161. Bocarsly, A. B.; Walton, E. G.; Bradley, M. G.; Wrighton, M. S. *J. Electroanal. Chem.* **1979**, *100*, 283–306.

162. Rosenblum, M. D.; Lewis, N. S. *J. Phys. Chem.* **1984**, *88*, 3103–3107.

163. Noufi, R.; Tench, D.; Warren, L. F. *J. Electrochem. Soc.* **1980**, *127*, 2310–2311.

164. Noufi, R.; Tench, D.; Warren, L. F. *J. Electrochem. Soc.* **1981**, *128*, 2596–2599.

165. Skotheim, T.; Lundstrom, I.; Prejza, J. *J. Electrochem. Soc.* **1981**, *128*, 1625–1626.

166. Skotheim, T.; Petersson, L.-G.; Inganas, O.; Lundstrom, I. *J. Electrochem. Soc.* **1982**, *129*, 1737–1741.

167. Noufi, R.; Frank, A. J.; Nozik, A. J. *J. Am. Chem. Soc.* **1981**, *103*, 1849–1850.

168. Skotheim, T.; Lundstrom, I.; Delahoy, A. E.; Kampas, F. J.; Vanier, P. E. *Appl. Phys. Lett.* **1982**, *40*, 281–284.

169. Simon, R. A.; Ricco, A. J.; Wrighton, M. S. *J. Am. Chem. Soc.* **1982**, *104*, 2031–2034.

170. Fabre, B.; Lopinski, G. P.; Wayner, D. D. M. *J. Phys. Chem. B* **2003**, *107*, 14326–14335.

171. Fan, F.-R. F.; Wheeler, B. L.; Bard, A. J.; Noufi, R. *J. Electrochem. Soc.* **1981**, *128*, 2042–2025.

172. Vermeir, I. E.; Kim, N. Y.; Laibinis, P. E. *Appl. Phys. Lett.* **1999**, *74*, 3860–3862.

173. Sailor, M. J.; Klavetter, F. L.; Grubbs, R. H.; Lewis, N. S. *Nature* **1990**, *346*, 155–157.

174. Sailor, M. J.; Ginsburg, E. J.; Gorman, C. B.; Kumar, A.; Grubbs, R. H.; Lewis, N. S. *Science* **1990**, *249*, 1146–1149.

175. Nakato, Y.; Ohnishi, T.; Tsubomura, H. *Chem. Lett.* **1975**, *19*, 883–886.

176. Menezes, S.; Heller, A.; Miller, B. *J. Electrochem. Soc.* **1980**, *127*, 1268–1273.

177. Howe, A. T.; Hawkins II, R. T.; Fleisch, T. H. *J. Electrochem. Soc.* **1986**, *133*, 1369–1375.

178. Contractor, A. Q.; Bockris, J. O. M. *Electrochim. Acta* **1984**, *29*, 1427–1434.

179. Zhu, Z.; Daniel, T. A.; Maitani, M.; Cabarcos, O. M.; Allara, D. L.; Winograd, N. *J. Am. Chem. Soc.* **2006**, *128*, 13710–13719.

180. Walker, A. V.; Tighe, T. B.; Cabarcos, O. M.; Reinard, M. D.; Haynie, B. C.; Uppili, S.; Winograd, N.; Allara, D. L. *J. Am. Chem. Soc.* **2004**, *126*, 3954–3963.

181. Zhou, C.; Nagy, G.; Walker, A. V. *J. Am. Chem. Soc.* **2005**, *127*, 12160–12161.

182. Kuikka, M. A.; Li, W.; Kavanagh, K. L.; Yu, H.-Z. *J. Phys. Chem. C* **2008**, *112*, 9081–9088.

183. Vilan, A.; Shanzer, A.; Cahen, D. *Nature* **2000**, *404*, 166–168.

184. Shimizu, K. T.; Fabbri, J. D.; Jelincic, J. J.; Melosh, N. A. *Adv. Mater.* **2006**, *18*, 1499–1504.

185. Allara, D. L.; Hebard, A. F.; Padden, F. J.; Nuzzo, R. G.; Falcone, D. R. *J. Vac. Sci. Technol. A* **1983**, *1*, 376–382.

186. Mahapatro, A. K.; Scott, A.; Manning, A.; Janes, D. B. *Appl. Phys. Lett.* **2006**, *88*, 151917–151919.

187. Czanderna, A. W.; King, D. E.; Spaulding, D. *J. Vac. Sci. Technol. A* **1991**, *9*, 2607–2613.

188. de Boer, B.; Frank, M. M.; Chabal, Y. J.; Jiang, W.; Garfunkel, E.; Bao, Z. *Langmuir* **2004**, *20*, 1539–1542.

189. Fisher, G. L.; Walker, A. V.; Hooper, A.; Tighe, T. B.; Bahnck, K. B.; Skriba, H. T.; Reinard, M. D.; Haynie, B. C.; Opila, R. L.; Winograd, N.; Allara, D. L. *J. Am. Chem. Soc.* **2002**, *124*, 5528–5541.

190. Loo, Y.-L.; Lang, D. V.; Rogers, J. A.; Hsu, J. W. P. *Nano Lett.* **2003**, *3*, 913–917.

191. Loo, Y.-L.; Willett, R. L.; Baldwin, K. W.; Rogers, J. A. *J. Am. Chem. Soc.* **2002**, *124*, 7654–7655.

192. Felmet, K.; Loo, Y.-L.; Sun, Y. *Appl. Phys. Lett.* **2004**, *85*, 3316–3318.

193. Guerin, D.; Merckling, C.; Lenfant, S.; Wallart, X.; Pleutin, S.; Vuillaume, D. *J. Phys. Chem. C* **2007**, *111*, 7947–7956.

194. Mizuno, H.; Buriak, J. M. *J. Am. Chem. Soc.* **2008**, *130*, 17656–17657.

195. Lynch, J. E.; Pehrsson, P. E.; Leonard, D. N.; Calvert, J. M. *J. Electrochem. Soc.* **1997**, *144*, 1698–1703.

196. Hode, G.; Dubow, T. J.; Rajeshwar, K. *J. Am. Chem. Soc.* **1983**, *105*, 324–330.

197. Thompson, L.; Dubow, J.; Rajeshwar, K. *J. Electrochem. Soc.* **1982**, *129*, 1934–1935.

198. Decker, F.; Fracastoro-Decker, M.; Badawy, W.; Doblhofer, K.; Gerischer, H. *J. Electrochem. Soc.* **1983**, *130*, 2173–2179.

199. Ginley, D. S.; Baughman, R. J.; Butler, M. A. *J. Electrochem. Soc.* **1983**, *130*, 1999–2002.

200. Switzer, J. A. *J. Electrochem. Soc.* **1986**, *133*, 722–728.

201. Kohl, P. A.; Frank, S. N.; Bard, A. J. *J. Electrochem. Soc.* **1977**, *124*, 225–229.

202. Morisaki, H.; Ono, H.; Dohkoshi, H.; Yazawa, K. *Jpn. J. Appl. Phys.* **1980**, *19*, L148–L150.

203. Fan, F.-R. F.; Keil, R. G.; Bard, A. J. *J. Am. Chem. Soc.* **1983**, *105*, 220–224.

204. Frank, M. M.; Chabal, Y. J.; Green, M. L.; Delabie, A.; Brijs, B.; Wilk, G. D.; Ho, M.-Y.; Da Rosa, E. B. O.; Baumvol I. J. R.; Stedile, F. C. *Appl. Phys. Lett.* **2003**, *83*, 740–742.

205. Frank, M. M.; Chabal, Y. J.; Wilk, G. D. *Appl. Phys. Lett.* **2003**, *82*, 4758–4760.

206. Killampalli, A. S.; Ma, P. F.; Engstrom, J. R. *J. Am. Chem. Soc.* **2005**, *127*, 6300–6310.

207. Dube, A.; Sharma, M.; Ma, P. F.; Ercius, P. A.; Muller, D. A.; Engstrom, J. R. *J. Phys. Chem. C* **2007**, *111*, 11045–11058.

208. Seitz, O.; Dai, M.; Aguirre-Tostado, F. S.; Wallace, R. M.; Chabal, Y. J. *J. Am. Chem. Soc.* **2009**, *131*, 18159–18167.

209. Li, M.; Dai, M.; Chabal, Y. J. *Langmuir* **2009**, *25*, 1911–1914.

CHAPTER 13

Immobilization of Biomolecules at Semiconductor Interfaces

ROBERT J. HAMERS

13.1 INTRODUCTION

Many emerging applications in biotechnology and biological/environmental sensing require surfaces that are highly selective and stable. The integration of biomolecules such as DNA and antibodies onto semiconductor surfaces provides several potential benefits. For example, semiconductors such as silicon are extremely pure and homogeneous, and can be used as highly reproducible substrates. Semiconductors such as silicon can be conveniently processed, etched, and manipulated using the tools developed for microchip fabrication, and translated toward biochips. In principle, it should be possible to take advantage of the unique properties that semiconducting materials can offer. Of these, one of the most important is the ability to amplify small changes in voltage or charge density and thereby convert biological information into measurable electrical signals in a highly parallel and low-cost manner. As one example, an inexpensive electronic voltmeter can read potential to four significant digits and costs less than $10. A current-generation microprocessor has over 2 billion transistors, each of which is a sensitive amplifier of current or voltage. If biological systems could be truly integrated with semiconductor-based materials, it could provide new opportunities for highly parallel biological detection. In the near term, bioelectronic devices are of interest for applications such as real-time monitoring of pathogens [1].

Self-assembled monolayers on gold have been widely studied because gold-thiol chemistry provides a convenient pathway toward well-ordered molecular layers [2–4]. Yet, many emerging applications require semiconducting substrates because of their unique electronic, optical, or mechanical properties. The integration of organic and biological molecules with covalent materials such as silicon and diamond presents new opportunities for making ultrastable highly selective surfaces.

Functionalization of Semiconductor Surfaces, First Edition.
Edited by Franklin (Feng) Tao and Steven L. Bernasek.

While silicon is the foundation of the microelectronics industry, diamond is an extraordinarily stable material even in water and is also a transparent, wide bandgap semiconductor. In each case, the challenge is to start with highly stable materials and make them "smart" by developing and implementing new surface modification strategies.

Here we will review recent and ongoing research on the formation of biomolecular interfaces to semiconducting materials.

13.2 MOLECULAR AND BIOMOLECULAR INTERFACES TO SEMICONDUCTORS

Silicon has long been a substrate of high interest because of the possibility of combining the extensive infrastructure of microelectronics for patterning and device fabrication with the study of biological systems [5–8]. The ability to convert biological information directly into electrical signals without the need for fluorescence or optical detection schemes is extremely attractive and remains a driving force behind much of the research in biological surface chemistry of semiconductors.

Of all the semiconductors, silicon is by far the most heavily studied because of its widespread use as the foundation for the vast majority of today's microelectronic devices such as microprocessors and memory chips. In addition, silicon has attracted attention as a substrate because the availability of relatively low-cost crystalline material with extremely high purity can provide a reliable starting point. Furthermore, etching and deposition procedures provide the ability to construct MEMS (microelectromechanical system) structures such as flow channels [9]. The ability to grow single-crystalline silicon nanowires has also led to a number of nanowire-based biosensing systems [10].

One of the biggest challenges associated with interfacing biological systems with semiconductor materials arises because while microelectronic device processing is typically done under dry environments and in tightly controlled atmospheric conditions, biological studies are performed under aqueous (and typically salty) environments. This presents two problems. First, controlling the surface chemistry to yield organic-to-semiconductor interfaces that will be stable under these conditions. Second, the electrical properties of semiconductors are often quite sensitive to the presence of electronic states lying within the bandgap, such that successful implementation of direct biological-to-electrical signal transduction requires careful attention to both the chemical and the electrical properties at the interface [11, 12].

13.2.1 Functionalization Strategies

In general, one of the biggest challenges in making biological interfaces to semiconductors is that most semiconductor materials are not thermodynamically stable in the presence of water. For example, silicon reacts with water to form SiO_2, which in turn slowly hydrolyzes in H_2O to form silicic acid ($Si(OH)_4$) and other water-soluble silicon species [13–17]. Similarly, the interfacial bonds used in linking (bio)molecular systems to bulk semiconductors, such as Si–O–C [18] or Ti–O–C linkages, are often subject to slow hydrolysis in aqueous environments. Consequently, the design of stable biomolecular architectures requires careful attention to the stability of the interface.

There are four main approaches that are used to produce molecular layers for biological applications of semiconductors:

1. Use of the most stable materials and chemical bonds possible (e.g., diamond).
2. Multidentate bonding to the surface (e.g., phosphonate and/or siloxane layers).
3. Lateral cross-linking between surface-bound molecules to form a two-dimensional surface net [19].
4. Use of molecules bearing hydrophobic regions that act as barrier layers to keep reactive species away from the interface [19].

These four approaches are used to varying extents in all successful surface functionalization strategies. Almost all successful methods have some degree of lateral cross-linking and use hydrophobic barrier layers. Yet these different strategies differ greatly in the level of control, reproducibility, versatility, and stability that they provide. Figure 13.1 summarizes the most commonly used approaches to functionalizing semiconductor surfaces. They include the use of organosilane compounds (Fig. 13.1a), phosphonic acid derivatives (Fig. 13.1b), grafting of terminal alkenes to hydrogen-terminated surfaces (Fig. 13.1c), grafting of terminal alkenes to hydroxylated surfaces of metal oxides (Fig. 13.1d), and use of aryldiazonium salts (Fig. 13.1e).

13.2.2 Silane Derivatives

Since silicon oxidizes readily and is widely used as the gate material in microelectronic devices, one approach to making biomolecular interfaces on silicon has been to intentionally form a surface oxide layer and to construct biomolecular interfaces to the oxide. This is the simplest approach and has been the most widely used. The oxide that forms spontaneously upon contact with air and/or water, typically referred to as the "native" oxide, has a thickness on the order of 1 nm [20, 21] but has poor electrical quality due to the presence of large number of unsaturated "dangling bonds" at the interface and in the oxide. Hence, more often studies of this type have formed thicker, more stoichiometric SiO_2 layers using thermal processing to oxidize the silicon [22]. While very early studies of silicon-based biosensors used non-functionalized SiO_2 (measuring instead changes in ion activity or pH) [5], in order to produce selective interfaces it is necessary to link more complex molecules such as DNA or proteins directly to the surface to provide surface-linked biomolecular recognition elements.

One common approach to forming biomolecular interfaces has been the use of silane derivatives, typically alkoxysilanes or alkylchlorosilanes, two of which are depicted in Fig. 13.1a [23–32]. These molecules react with hydroxyl groups on

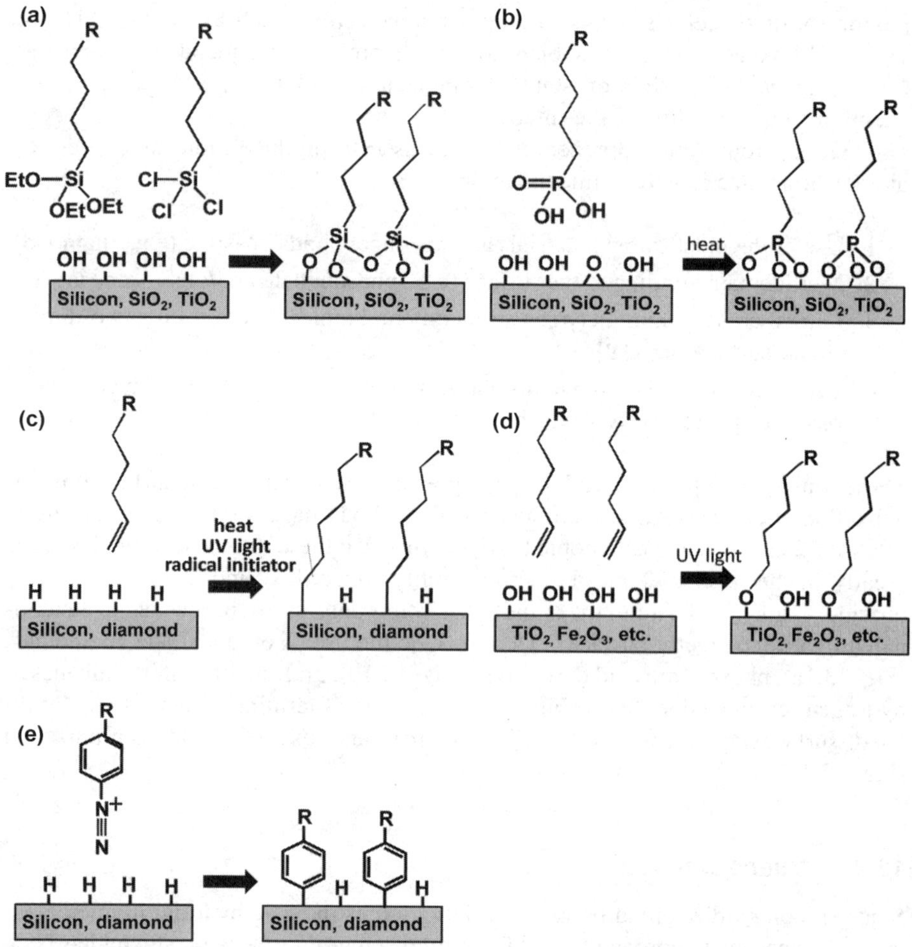

FIGURE 13.1 Schematic illustration of different schemes for making organic interfaces to semiconductors for biological application. (a) Use of alkoxysilanes and chlorosilanes, (b) phosphonic acid derivatives, (c) alkene grafting to H-terminated surfaces, (d) alkene grafting to hydroxide-covered surfaces of metal oxides, and (e) grafting via aryldiazonium compounds.

surfaces to form Si–O–X linkages, where X corresponds to the substrate material. Oxidized silicon surfaces expose substantial number of silanol groups (Si–OH) that are reactive toward the silanes [26]. The silane layers are typically deposited from solutions in methanol or other polar solvent. The deposition reactions are complicated by the fact that small amounts of water are necessary for the reaction to occur; some water is always present on the starting surfaces as a physisorbed layer, but water is also present (often unintentionally) in the solvent [33]. Thus, hydrolysis occurs in the bulk and also at the surface. Hydrolysis in the bulk produces polymeric materials that deposit on the surface in an inhomogeneous manner, forming clumps of deposited materials [34]. Reaction with the surface –OH bonds produces surface-bound monolayers with the extent of cross-linking versus surface binding varying depending on the number of hydroxyl groups on the surface and on the precise nature of the alkoxysilane chosen.

One difficulty with silane chemistry is that reaction with the underlying surface relies on the presence of surface –OH groups, while trace amounts of water in the liquid-phase reaction medium also cause hydrolysis and agglomeration of silane polymers in the liquid phase that then deposit onto the surface [34]. Controlling the density of hydroxyl groups on surfaces is challenging because the kinetics of the associated hydrolysis reactions are slow [35]. One strategy that is commonly used is to heat the surfaces in a strongly oxidizing bath such as "piranha" solution. (*Note*: Piranha solutions are extremely dangerous and should only be used with appropriate safety precautions.) The concentration of surface –OH groups has also been increased by exposure to bromine before oxidation, presumably by exposing a more reactive, polar bond [36]. With care, very high quality monolayers can be produced.

Once formed, the silane layers face a second challenge based on the fact that the Si–O–C and Si–O–Si linkages are thermodynamically unstable in the presence of water [13–17, 36, 37]. Thus, having a very hydrophobic barrier that keeps water away from the interface is crucial. As depicted in Fig. 13.2, water and other reactive species that penetrate the hydrocarbon layer can attack the underlying interface bonds, leaving the molecular functionalization layer to desorb as silicic acid derivative. Once initiated at even a single point, additional surface bonds are directly exposed to the aqueous layer and readily react, essentially "unzipping" the molecular layer from the surface. Hydrolysis of Si–O–C bonds is catalyzed under acidic conditions [38] and Si–O–Si hydrolysis is catalyzed under basic conditions. While most biological studies are performed near neutral pH, the chemical procedures used in subsequent steps to link biomolecules to the surface are often strongly acidic or basic. One important feature of silane layers is the extensive cross-linking; this strongly improves the layer stability by increasing the total number of bonds that must be hydrolyzed before molecules will completely hydrolyze and leave the surface. While a single molecule only needs hydrolysis of three bonds to be removed

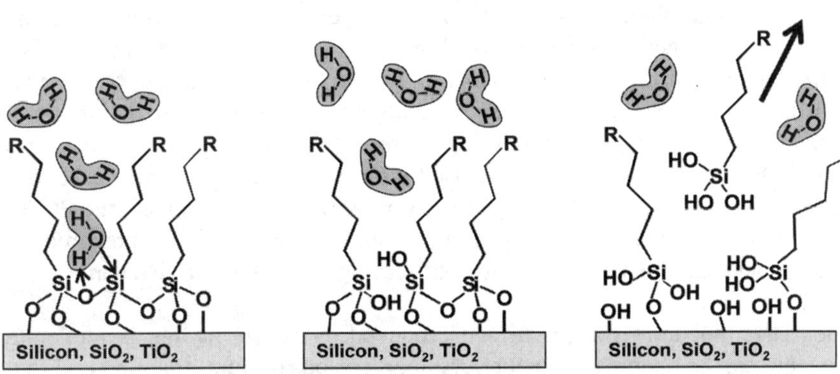

FIGURE 13.2 Hydrolytic instability of siloxane-based monolayers due to penetration of water and other reactive species to the organic–inorganic interface.

from the surface, a cross-linked pair requires four, and higher degrees of cross-linking yield even greater stability. One common derivative, aminopropyl triethoxysilane (APTES), is widely used as a way of forming surfaces terminated with primarily amine (–NH_2) groups. Glass surfaces functionalized in this manner are available commercially from major manufacturers for biochip applications.

13.2.3 Phosphonic Acids

Phosphonic acids and their salts can bind strongly to many surfaces [39–45]. White it is widely recognized that the alkylsilane chemistry requires –OH groups to proceed, evidence suggests that the phosphonic acid chemistry on silicon can occur without surface –OH groups [43]. Since metal oxides such as silicon and TiO_2 often have bridging oxygens (like that shown in Fig. 13.1b), the ability to graft in the absence of water or hydroxyls may provide a more reproducible method of attachment, avoiding the solution-phase hydrolysis that causes clumping of organosilanes. XPS studies have indicated that the binding is tridentate [45]. This approach has been successfully used to graft cell-binding proteins to titanium surfaces [40] and as a starting point for linking proteins to TiO_2 via nonspecific interactions with the ammonium dodecyl phosphate [46].

13.2.4 Alkene Grafting

Because of the irreproducibility and hydrolytic instability of the organosilicon chemistry, substantial efforts have been made to eliminate the need for a surface oxide altogether, by using alternative reactions to form direct silicon–carbon linkages. The goal of a direct Si–C linkage was motivated by the fact that hydrolysis reactions are often driven by strongly polar bonds; for example, Si–O–Si (silica, silicones) and C–O–C (ether) bonds are more resistant to hydrolysis than Si–O–C bonds. Similarly, chemical compounds containing direct Si–C bonds are much more resistant to hydrolysis than those with Si–O–C linkages [47]. These factors all suggest that bonding organic layers directly to the silicon surface without intervening oxygen atoms may yield improved stability, as well as possibly providing a more electronically conductive interface. Working directly on silicon also allows one to take full advantage of the fact that wet chemical etching using NH_4F forms extraordinarily flat Si(111) surfaces with all exposed silicon "dangling bonds" terminated with H atoms [48], providing an almost perfectly homogeneous starting surface. While the Si(001) surface is used almost exclusively in microelectronics processing, the Si(111) surface is generally preferred for biological studies. This difference arises primarily because Si(111) surfaces can be easily prepared with atomic flatness due to the highly anisotropic nature of the NH_4F chemical etch [48, 49].

The "H-terminated" silicon surface produced by NH_4F etching is moderately reactive but can be handled in air for very short periods of time, enabling one to use benchtop chemistry for subsequent chemical functionalization. Most reactions on H-terminated silicon have used terminal alkenes. The reaction of alkenes with

H-terminated silicon is not spontaneous, but must be activated. Linford et al. showed that terminal alkenes can be thermally activated to link directly to the silicon surface [50, 51]. Additional studies have shown that the reaction can be activated using radical initiators to remove a surface H atom (thereby creating a reactive Si "dangling bond") [50, 51], or by using energetic photons, typically at 254 nm [52–56].

Figure 13.3 outlines the procedure typically used for photochemical grafting of alkenes onto H-terminated Si surfaces [52–56]. Studies using X-ray photoelectron spectroscopy (XPS) and Fourier transform infrared spectroscopy (FTIR) show that in most cases the grafting reactions are self-limiting, stopping after a coverage of approximately one monolayer, although multilayer formation is possible in some cases [57, 58]. While initial studies of alkene grafting focused primarily on simple alkenes, the range of opportunities expands greatly using bifunctional molecules [52–54, 59–64]. Sieval et al. showed these grafting methods could be extended to enable multistep surface chemical synthesis by using ω-functionalized 1-alkenes; however, because many functional groups of interest react directly with silicon (including both amines and carboxylic acids), it is necessary to use protecting groups [61]. The use of protecting groups is complicated because care must also be exerted in the choice of deprotection conditions to ensure that the harsh chemicals do not etch the underlying silicon or the Si–C bonds. Additionally, since the protecting groups are sometimes larger than the size of the alkyl chain, the protecting groups can inhibit good packing of the molecules on the surface, resulting in a molecular layer that has poor stability in aqueous media. Figure 13.3b illustrates some of the molecules that have been grafted successfully [53, 54, 65, 66]. They include pure alkenes and molecules bearing additional functional and/or protecting groups. For example, TFAAD and tBoc contain protected amine groups and thereby provide routes to amine-terminated surfaces. The molecule with the dmt (dimethoxytrityl) group can be deprotected to yield alcohol-terminated surfaces [65], while UAME (undecylenic acid methyl ester) is an excellent route to carboxylic acid-modified surfaces, and oligomers of ethylene glycol form protein-resistant surfaces [66, 67]. This combination of molecules provides many different pathways for linking DNA, proteins, and other biomolecules to the surface.

While most studies of alkene grafting have been performed on group IV semiconductors, alkenes will also photochemically graft to metal oxides such as TiO_2 [58, 68, 69].

13.3 DNA-MODIFIED SEMICONDUCTOR SURFACES

13.3.1 DNA-Modified Silicon

One attractive feature of DNA chemistry is that automatic DNA synthesizers make it simple to produce DNA oligonucleotides terminated with reactive terminal groups. While termination with amine groups is common, one difficulty in this approach is that primary amines are also reactive with atmospheric CO_2. Consequently, many

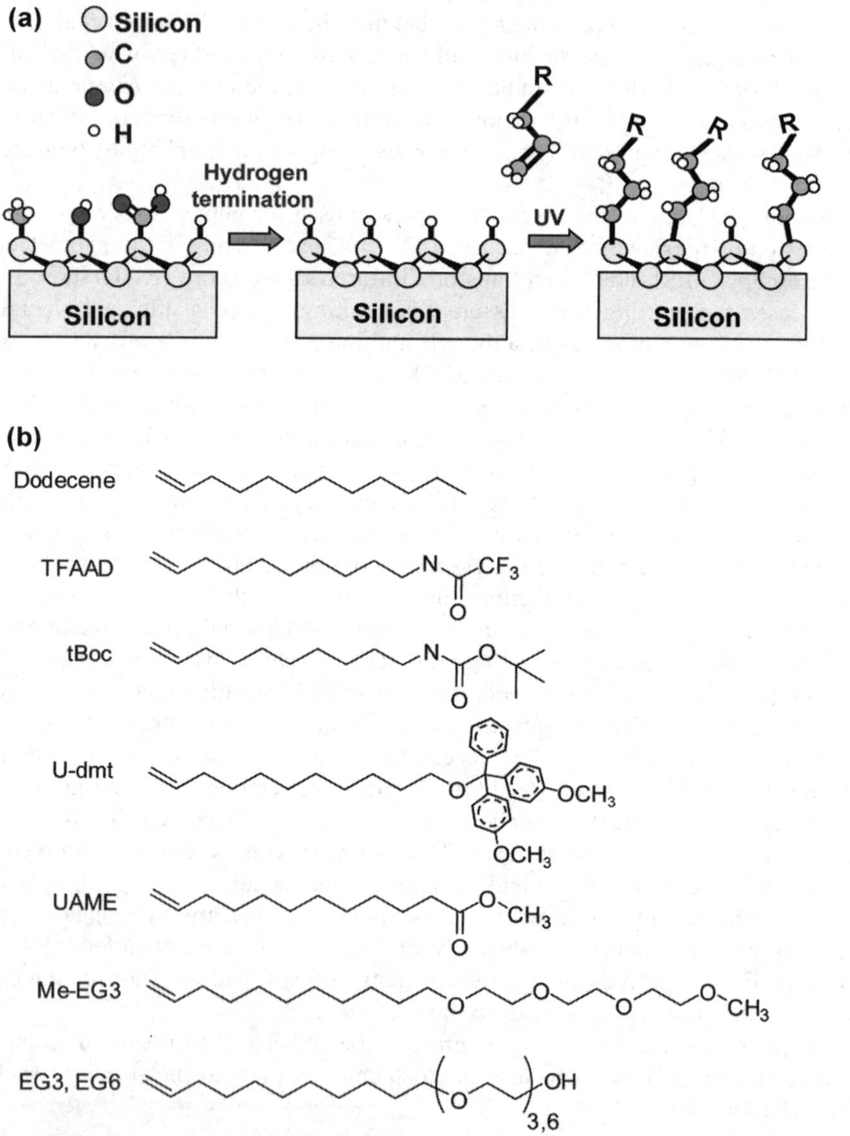

FIGURE 13.3 (a) Alkene grafting to H-terminated surfaces of silicon. (b) Summary of some of the bifunctional alkenes that have been grafted to silicon surfaces.

studies have instead used DNA terminated with thiol (–SH) groups. This approach has the additional advantage that the surface linkage can be facilitated using heterobifunctional cross-linkers such as sulfosuccinimidyl 4-(*N*-maleimidomethyl)cyclohexane 1-carboxylate (Sulfo-SMCC), which has one functional group (*N*-hydroxysuccinimide) that reacts with primary amines to form amide linkages and a second group (maleimide)

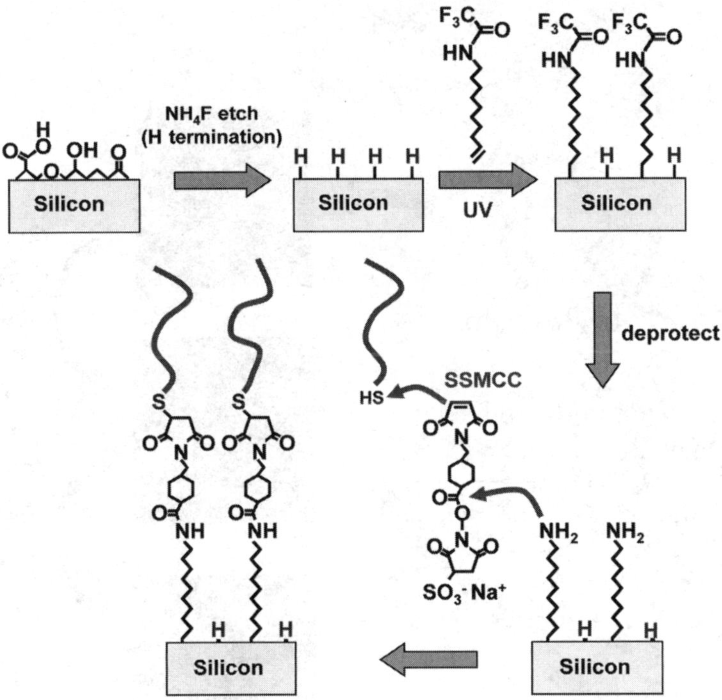

FIGURE 13.4 Overall summary of pathway to DNA-modified silicon surface.

that will link to sulfhydryl (thiol, –SH) groups. Figure 13.4 depicts this overall pathway on silicon. We note that because the linkers are often larger than the alkyl chains of the underlying functionalization layer (and the DNA molecules larger yet), after SSMCC there are likely unreacted but inaccessible amine groups remaining, and after linking of DNA there are likely unreacted maleimide groups as well. Because each successive step in this multistep pathway involves molecules of increasing size, there are often advantages in using initial functionalization layers in which molecules having chemically unreactive terminal groups (such as EG3 or EG6 depicted in Fig. 13.3b, which resist protein binding) are mixed with molecules having reactive end groups (such as TFAAD) that will serve as covalent attachment points. Making such mixed layers can reduce the undesired effects of unreacted but inaccessible functional groups and can aid in making the surfaces resistant to nonspecific binding while retaining their ability to react with specific biomolecular targets of interest [66].

Making mixed layers can also enhance the reactivity of the remaining surface attachment points due to steric blocking that can occur in dense molecular layers. For example, Böcking et al. prepared layers by thermal hydrosilylation of 10-undecenoic acid and then linked peptides to the surface by first using 1-ethyl-3-[3-dimethylaminopropyl]carbodiimide (EDC) followed by *N*-hydroxysuccinimide (NHS) to produce a NHS-terminated surface, which was then reactive toward a simple

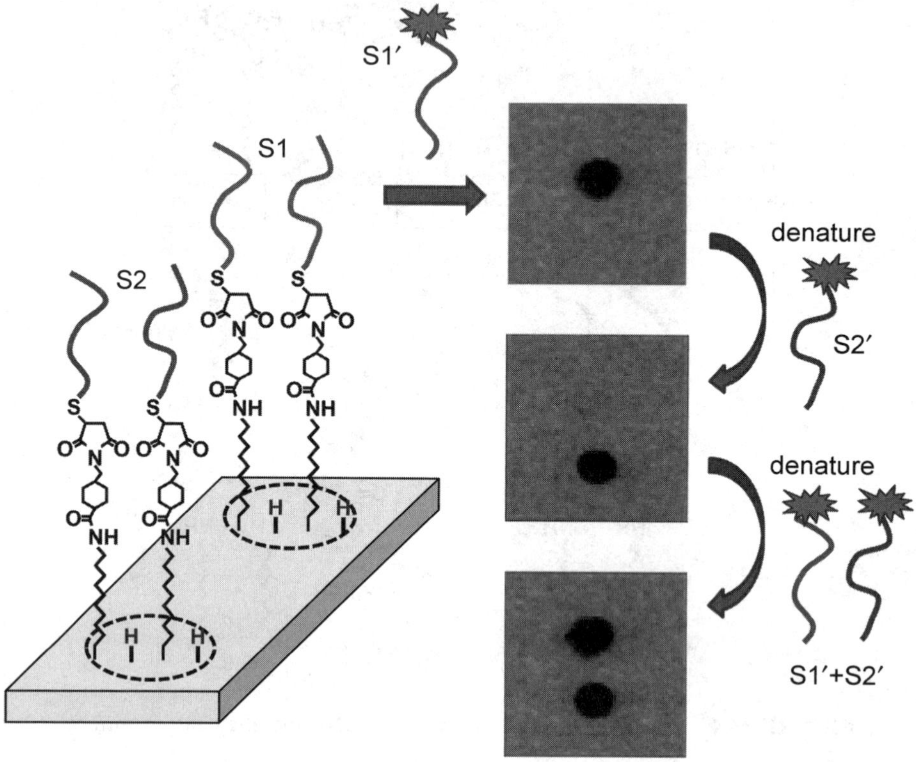

FIGURE 13.5 Testing selectivity and reversibility of hybridization of DNA oligonucleotides covalently grafted onto the Si(001) surface. Figure adapted from Ref. 53.

tripeptide. They found that at high coverages the exposed carboxylic acid groups were difficult to activate. This behavior might arise from strong hydrogen bonding between adjacent molecules. Yet, if the hydrosilylation reaction was terminated early, the samples showed significant levels of oxidation that led to mechanical failure of the devices due to dissolution of the underlying nanostructured Si [70].

Figure 13.5 depicts data showing the interaction of a DNA-modified Si surface with complementary and noncomplementary DNA targets in solution. In this case, the surface was linked to two different sequences, S1 and S2, in two spots on the surface. The samples were then covered with a fluorescently labeled complement, S1′, and the fluorescence measured. The images in Fig. 13.5 are negatives, so that the dark region corresponds to high fluorescence intensity. Figure 13.5 shows that sequence S1′ binds to the surface region functionalized with S1 but not to the surface reaction functionalized with sequence S2, which is a 4-base mismatch (out of 16) to S1. To test the reversibility of binding, the sample was then denatured and exposed to a fluorescently labeled complement to S2 (labeled as S2′), and in this case the other surface region shows high fluorescence intensity (dark in the image). Finally, the sample was denatured and exposed to a mixture of both targets. These and other studies show that surfaces of semiconductors such as silicon can be functionalized with DNA in a manner that gives excellent selectivity, even able to detect a single-base mismatch [71]. With sufficiently densely packed layers, repetitive hybridization–denaturation cycles can yield losses of approximately 1% per cycle. This value is sufficiently low that silicon can be used as a high-quality substrate for interfacing DNA and other biomolecules to surfaces.

The selectively DNA-modified surfaces can be evaluated by comparing the bonding of surface-tethered oligonucleotides with molecules in solution having complementary and noncomplementary DNA sequences. One of the most important goals is to be able to distinguish between perfectly complementary sequences and those having mismatches of a single base. These so-called single nucleotide polymorphisms (SNPs) are believed to be important in factors underlying the genetic basis of disease [72, 73]. Discriminating between perfect matches and single-base mismatch has been achieved at silicon surfaces using fluorescence [71] and electrochemical [74] methods.

One of the primary motivations for interfacing biomolecules to semiconductors such as silicon is to take advantage of the unique electronic properties of semiconductors to make high-sensitivity devices such as transistors, which are essentially charge amplifiers. Devices of this type have been made from silicon for many years, but have not found widespread applications in large part because they suffer from chemical and electrical instability due to the factors discussed above. In addition, while studies on silicon demonstrate the benefits of using photochemical grafting to prepare complex interfaces the utility of silicon in aqueous environments is limited by the fact that penetration of water to the interface hydrolyzes Si–C and Si–O–C bonds, eventually removing the molecular layers from the substrate. This process can be slowed by using hydrophobic alkyl chains, but ultimately the molecular layers formed on silicon are only kinetically stable. To make truly ultrastable molecular layers, a different approach is necessary, starting with materials having greater intrinsic stability.

13.3.2 DNA-Modified Diamond

Carbon-based materials are of great interest in part because the properties can be controlled by varying the hybridization [75]. Diamond is of particular interest because in addition to having outstanding mechanical and optical properties, it is a semiconductor, having a bandgap of 5.5 eV. Diamond's semiconducting properties enable the creation of hybrid bioelectronic devices such as biologically actuated field-effect transistors, or "bio-FETs" [76]. While diamond in bulk form remains prohibitively expensive for most practical applications, thin diamond coatings <1 μm in thickness can be easily and inexpensively deposited on a wide range of materials including metals, silicon, and transparent quartz; these diamond films yield the chemical stability of diamond in a practical and scalable manner. Amorphous carbon and "diamond-like" carbons with mixed sp^2–sp^3 hybridization can also be deposited as thin-film coating materials that facilitate integration with a wider range of other materials [75, 77, 78].

Carbon surfaces are typically cleaned using atomic hydrogen from a hydrogen plasma, yielding "H-terminated" surfaces [79]. H-terminated diamond is very reproducible and stable even in air. However, this high stability also makes H-terminated diamond surfaces difficult to functionalize. One approach to diamond functionalization is the use of aryldiazonium salts, as depicted in Fig. 13.1e [80–83]. Nitrobenzenediazonium salts can be reduced to make amine-terminated diamond surfaces, enabling coupling chemistry like that depicted above for silicon to be applied to diamond [82, 83].

A more versatile approach for modifying diamond has been to use photochemical grafting of organic alkenes. The procedure for carrying out the photochemical grafting is similar to that on silicon described earlier (see below), the only notable difference being the initial cleaning process to make the diamond surface chemically uniform, with all exposed surface bonding linked to hydrogen atoms. This H-terminated diamond is typically made by exposing the diamond sample to a hydrogen plasma [79]. The H-terminated diamond is covered with a thin film of suitable reactant and then illuminated with ultraviolet light, typically at 254 nm for ~12 h [84, 85]. Using the TFAAD molecule shown in Fig. 13.3b, DNA-modified diamond samples were prepared using chemistry similar to that shown for silicon.

The outstanding properties of diamond as a robust substrate for functional surfaces are apparent from the greatly increased stability of DNA on diamond compared with other substrates, as shown in Fig. 13.6 [85]. In this comparison, several different substrates were modified with organic molecules terminating in a primary amine group. These amine groups were then linked to thio-modified DNA oligonucleotides using the heterobifunctional linker SSMCC. The stability and selectivity of the resulting DNA-modified surfaces were then evaluated by exposing the DNA-modified diamond surfaces to repeated cycles of hybridization and denaturation using fluorescently labeled targets with complementary and noncomplementary sequences [85]. In addition to the DNA-modified diamond, the graph shows data for surfaces produced from commercial amine-modified glass slides (Corning GAPS II), amine-terminated self-assembled monolayers on gold, and surfaces prepared by photochemically grafting molecular layers to H-terminated Si(111) and to H-terminated glassy carbon. All of these examples started with amine-terminated molecular layers and then used identical DNA chemistry. Consequently, they have identical molecular structures except for the two atoms (labeled "A" and "B" in Fig. 13.6a) that form the molecule–substrate interface.

Figure 13.6b shows that the two carbon substrates—diamond and glassy carbon—clearly exhibit superior stability. These results particularly highlight the importance of the substrate-to-molecule bond in controlling the overall stability of biomolecular interfaces, as the extraordinary stability of the C–C bond at the interface provides unprecedented chemical stability to the biomolecular layers.

13.3.3 DNA on Metal Oxides

Recent studies showed that alkenes will graft to surfaces of TiO_2 and other metal oxides [58, 69], presumably linking via surface –OH groups. Using the TFAAD molecule depicted in Fig. 13.3b, DNA-modified TiO_2 surfaces were prepared using

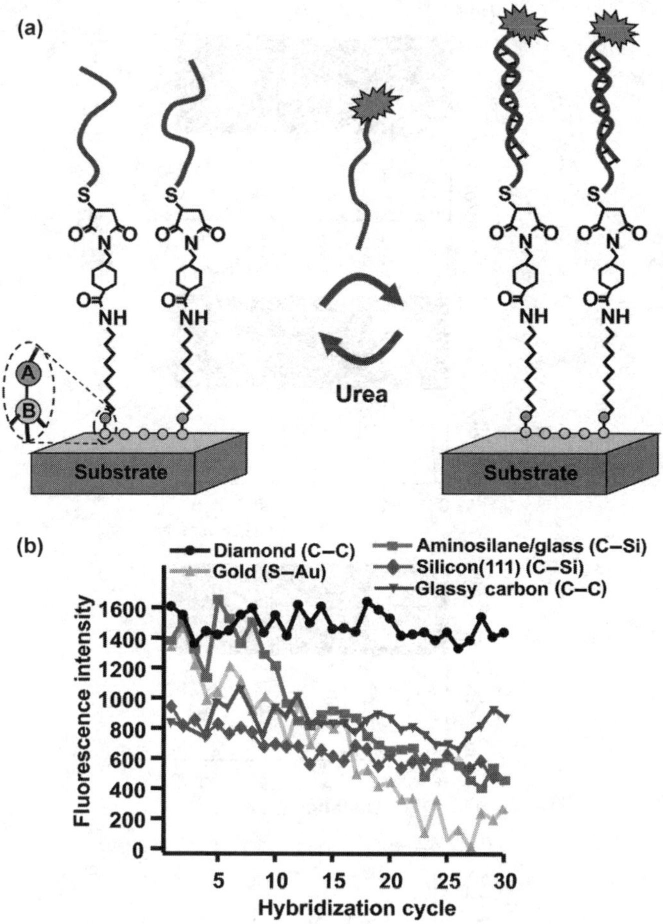

FIGURE 13.6 Comparison of stability in repeated cycles of hybridization and denaturation of several different materials. In each case, the sample was functionalized to produce an amine-terminated surface, followed by identical methods of covalent grafting of thio-modified DNA to the surface. The samples were subjected to repeated cycles of hybridization with a fluorescently labeled complementary sequence, followed by denaturation in 8.3 M urea. The different absolute signal intensities at the start of the experiment are a consequence of the different optical properties of the materials and have no important consequence. Note the excellent stability of diamond and glass carbon in comparison with the other materials. Figure adapted from Ref. 85.

TFAAD, deprotecting, and using SSMCC as a linker [69]. Figure 13.7 shows some of the results obtained with two different oligonucleotides. Tests of the stability in repeated cycles of hybridization and denaturation showed that the surfaces have good stability, although not as stable as DNA on diamond. In this particular case, the Ti–O–C linkages at the surface are likely susceptible to hydrolysis, so that the molecular layer stability hinges on having some partial cross-linking of TFAAD on

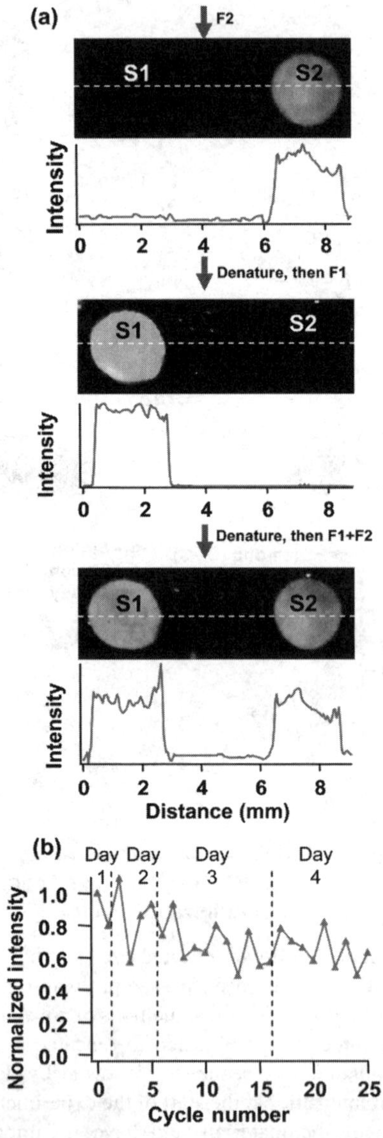

FIGURE 13.7 DNA hybridization on DNA-modified TiO_2 surface prepared by photochemical grafting of TFAAD to TiO_2. Two surface regions were modified with two different oligonucleotides (S1 and S2), each having a 16-base hybridization region and different from one another by 4 bases. Fluorescence images and intensity profiles are shown after exposure to the fluorescently labeled complements F2 (fluorescent complement to S2) and F1 (fluorescence complement to S1). Figure adapted from Ref. 69.

the surface. Nevertheless, TiO_2 is an attractive material for some applications because Ti metal can be easily deposited onto a wide range of materials (including polymers), potentially facilitating the integration of DNA and other biomolecules with polymers and other materials.

13.4 PROTEINS AT SURFACES

13.4.1 Protein-Resistant Surfaces

One of the biggest challenges in working with biological systems is that many proteins undergo nonspecific adsorption on surfaces. This nonspecific adsorption is often accompanied by denaturation of the proteins and leads to a variety of undesired consequences, including fouling of biosensors and generation of an inflammatory response. The use of such short-chain ethylene glycol (EG) oligomers to reduce nonspecific binding has been well established in studies of self-assembled monolayers on gold [86–90] and can also be applied to semiconductors such as silicon and diamond [66, 67, 91]. As depicted in Fig. 13.8a, alkenes with terminal EG groups effectively repel nonspecific binding, even when as few as three EG groups are present (as in Fig. 13.8a). Longer oligomers such as EG6 are even more effective. By making mixed monolayers incorporating EG oligomers and molecules having reactive functional groups (such as TFAAD), the overall selectivity of the surface toward the target of interest can be maximized [67].

Figure 13.8b depicts data for nanocrystalline diamond films modified with four different surface treatments, interacting with four different proteins; these data show that even short oligomers such as EG6 are highly effective at reducing nonspecific binding of proteins [66, 67]. Grafting of the EG6 oligomers reduces the nonspecific binding of avidin on nanocrystalline diamond to less than 0.02 monolayer [67], a level similar to that observed with self-assembled monolayers on gold [87].

XPS studies can also be useful in identifying protein adsorption in a quantitative manner. For example, Fig. 13.8c shows XPS data for an EG6-functionalized diamond surface and a H-terminated diamond surface, both shown before and after exposure to fibrinogen in solution. An XPS spectrum of a drip-dried film of fibrinogen is also shown. The presence of the fibrinogen can be inferred from the presence of an N 1s peak in the XPS spectrum of the H-terminated diamond after immersion in fibrinogen. However, there is no detectable N 1s or S 2p for the EG6-modified sample. Based on the signal-to-noise ratio of the experiments, from these data we infer that the EG6 monolayer reduces the nonspecific binding of fibrinogen by >97%. Both XPS and fluorescence methods show that diamond surfaces modified with EG oligomers have superior stability to those made on silicon or gold. Figure 13.9 shows fluorescence measurements testing the ability of EG-modified surfaces of silicon, gold, and diamond to resist the nonspecific binding of avidin over longer periods of time. While at short times silicon is the best (likely because of its atomically flat structure), over a period of 10 days in water both the silicon and gold surfaces degrade, while the EG-modified diamond surface retains its ability to resist

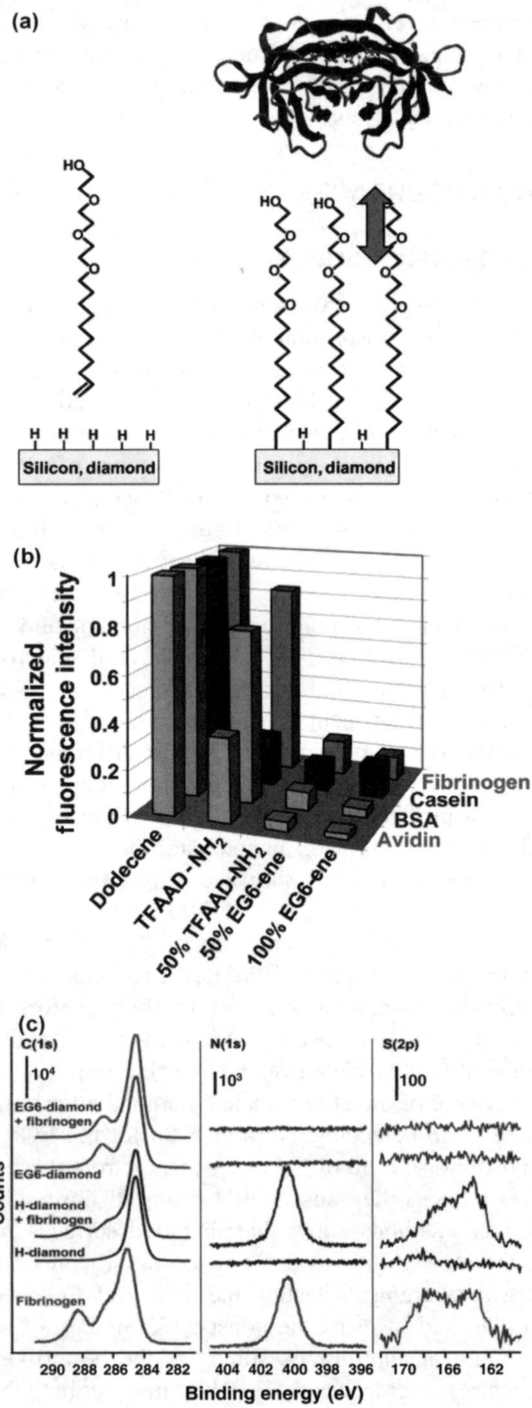
(a)
HO
O
H H H H H
Silicon, diamond
Silicon, diamond
(b)
Normalized fluorescence intensity
1
0.8
0.6
0.4
0.2
0
Fibrinogen
Casein
BSA
Avidin
Dodecene
TFAAD-NH2
50% TFAAD-NH2 50% EG6-ene
100% EG6-ene
(c)
C(1s)
N(1s)
S(2p)
10^4
10^3
100
EG6-diamond + fibrinogen
EG6-diamond
H-diamond + fibrinogen
H-diamond
Fibrinogen
Counts
290 288 286 284 282
404 402 400 398 396
170 166 162
Binding energy (eV)

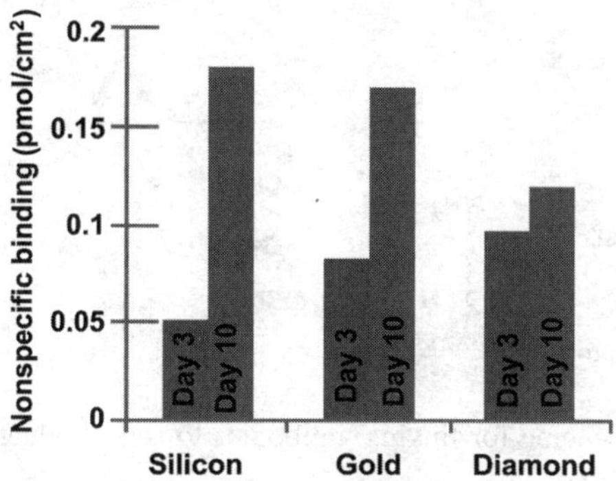

FIGURE 13.9 Time-dependent changes in the ability of EG layers on silicon, gold, and diamond to resist nonspecific binding of avidin. Note that while functionalized silicon is best at short times, the stability of diamond allows the molecular functionalization layers to persist for long periods of time and therefore retains its effectiveness for longer periods of time. Figure adapted from Ref. 66.

nonspecific binding. Corresponding XPS measurements showed no significant change in the surface chemistry of EG-modified diamond [66]. In contrast, EG-modified silicon and gold samples both exhibited significant increases in nonspecific binding and evidence for partial loss of the EG layers [66, 67, 87, 91].

13.4.2 Protein-Selective Surfaces

Just as DNA can be used to link to its complementary sequence, proteins are often able to recognize specific molecules. The ability to functionalize semiconductor surfaces with proteins, particularly antibodies, is of interest for many applications in sensing, medical diagnostics, and the rapidly growing field of proteomics.

Antibodies contain two major domains: the F_{ab} region is primarily responsible for biomolecular recognition, while the F_c region is rich in carboxylic acid groups. Unfortunately, existing chemistries are better suited to linking to the antibody's amine groups than to the carboxylic acid groups. Existing biochips often use Protein G to selectively immobilize antibodies via the F_c region, but this does not provide the type of strong, covalent attachment desired. Consequently, linking through the amine groups in the F_{ab} region is a reasonable, albeit less than ideal compromise. Hartl et al. covalently linked proteins to diamond using succinic anhydride to convert the amine-

FIGURE 13.8 Nonspecific binding of proteins at semiconductor surfaces. (a) Use of alkenes terminated with ethylene glycol groups to make protein-resistant surfaces on silicon and diamond. (b) Comparison of different mixed layers to resist nonspecific binding of avidin. Figure adapted from Ref. 66. (c) XPS measurements of H-terminated and EG6-terminated diamond, before and after exposure to fibrinogen. Note that immersion into fibrinogen leads to large N 1s and S 2p peaks on the H-terminated sample, but no detectable change on the EG6-modified sample. This establishes that the EG layer is very effective at resisting nonspecific binding of fibrinogen. Figure adapted from Ref. 91.

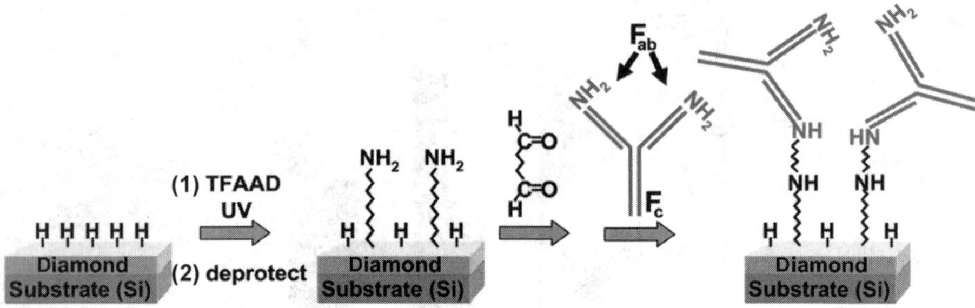

FIGURE 13.10 One scheme for linking antibodies to semiconductor surfaces via the F_{ab} region.

terminated surface to a carboxylic acid-modified surface, and then using EDC and NHS to link to the amine groups of two different proteins: green fluorescent protein (GFP) and the catalase enzyme [92]. Their work demonstrated that it is possible to make highly robust protein interfaces to diamond that retained their biological activity. Figure 13.10 illustrates another pathway toward antibody-modified surfaces, using glutaraldehyde to couple between amine-terminated surfaces and the amine groups of the antibody. This approach has been used to make antibody-selective surfaces on diamond and has also been used to make cell-selective surfaces [91]. By linking *E. coli* antibodies to a diamond surface, it is possible to selectively capture *E. coli* cells. This may provide an important opportunity for detection of biological pathogens via continuous monitoring.

The ultimate measure of the selectivity of biomolecular interface is the ratio of specific binding (binding of a particular target molecule in solution) to nonspecific binding. By making mixed monolayers in which molecules bearing a reactive group of interest (such as a protected amine for subsequent modification) are distributed among EG oligomers, the overall ratio of specific to nonspecific binding (S/NS ratio) can be significantly improved. Experimental measurements using the biotin–avidin system showed that the monolayers formed using $\sim 10\%$ of a protected amine in 90% of ethylene glycol in the parent liquid yielded the best overall selectivity; these surfaces were able to selectively bind avidin even in undiluted chicken serum [67]. These experiments demonstrate that multiple properties can be optimized using mixed monolayers.

13.5 COVALENT BIOMOLECULAR INTERFACES FOR DIRECT ELECTRICAL BIOSENSING

13.5.1 Detection Methods on Planar Surfaces

One of the most intriguing applications of biomolecular surface chemistry on semiconductors is the potential development of label-free schemes to convert biological information directly into electrical signals [6, 7, 11, 22, 76, 93–100].

The basis for semiconductor-based sensing schemes is that the conductivity of semiconductors can be very sensitive to the presence of small charges at or near the semiconductor surfaces. Under appropriate conditions (ones that, unfortunately, are extremely difficult to achieve in aqueous media), some transistors are even able to detect the charge of a single electron [101]. In biological applications, the essential idea is that biomolecules, ions, and other charged species located near the surface of the semiconductor create electric fields that penetrate a short distance (typically 10–1000 nm) into the semiconductor bulk. These penetrating electric fields repel and attract charge carriers consisting of negatively charged electrons in the conduction band and positively charged "holes" (electron vacancies) in the valence band, thereby changing the conductivity of the semiconductor in the region closest to the solid–liquid interface. These changes in conductivity can be measured perpendicular to the interface (between the bulk semiconductor and an electrode in the adjacent liquids phase) or parallel to the interface by using two electrical contacts on the semiconductor surface, as depicted in Fig. 13.11.

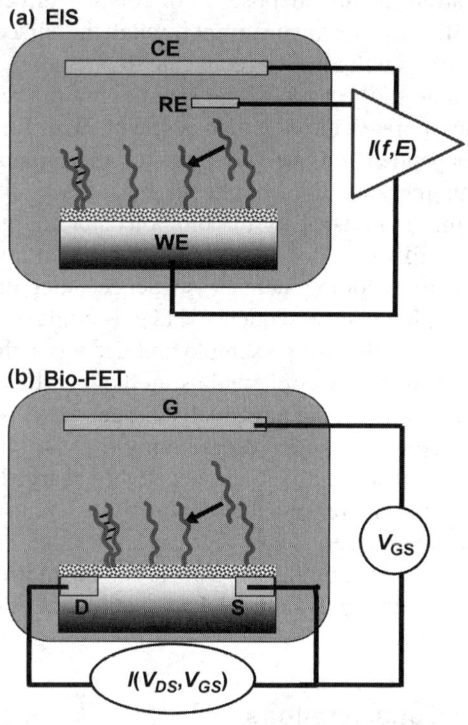

FIGURE 13.11 Two methods for achieving electrical detection of biological binding events at functionalized semiconductor surfaces. (a) Electrochemical impedance spectroscopy, measuring the current as a function of frequency (*f*) and the electrochemical potential (*E*) in a three-electrode potentiostat geometry. (b) Biologically sensitive field-effect transistor (bio-FET), measuring the drain–source current as a function of the drain–source voltage and the gate–source voltage.

In the former technique, the electrical impedance perpendicular to the interface is measured using a technique known as electrochemical impedance spectroscopy (EIS). EIS measures the amplitude and phase shift of the current flow induced by a sinusoidally modulated voltage applied using the sample as the working electrode in a three-electrode electrochemical cell, as depicted in Fig. 13.11a [7, 8, 11]. The EIS data are frequency-dependent complex numbers that are often plotted as figures in the complex plane and sometimes fit to electrical circuit models to extract parameters such as the interfacial capacitance and solution resistance.

Measurements parallel to the surface typically involve a configuration like a field-effect transistor (FET), as in Fig. 13.11b. Two electrodes (referred to as the "source" and "drain" electrodes) are placed on the sample and used to measure the conductivity parallel to the interface; a third electrode (typically an in-solution reference electrode) can be used to adjust the initial state, or "operating point" of the transistor for maximum sensitivity [22, 93, 94, 96, 102, 103]. The current between source and drain is then measured as a function of the drain–source voltage and the gate–source voltage.

The FET configuration requires deposition of contacts directly onto the front of the sample and can suffer from leakage current through the adjacent liquid, but can be more sensitive. The EIS configuration conveniently requires only one electrical contact that can be made to the back of the sample, but the resistance of the bulk substrate becomes incorporated into the measurement. Both EIS and FET measurements are strongly dependent on the proximity of the biomolecule to the semiconductor, since both measure the change in impedance in the semiconductor space-charge region that is induced by the biomolecular charge.

Figure 13.12 shows EIS data for DNA grafted onto n-type (a) and p-type (b) silicon after exposure to a complementary sequence, after denaturing, and after exposure to a noncomplementary sequence [11]. Binding of DNA significantly increases the impedance on the n-type sample and decreases the impedance on the p-type sample. This can be easily explained using the fact that on n-type substrates the charge carriers are electrons; the negatively charged DNA repels these electrons from the near-surface region, leaving it more highly resistive. In contrast, on p-type substrates the charge carriers are holes; the negatively charged DNA attracts these holes to the surface and increases the conductivity. With positively charged molecules, the situation reverses. Thus, from a knowledge of the doping of the semiconductor the charge of the binding molecule can be inferred. Indeed, measurements using positively charged human IgG antibodies showed just such a reversal [99].

13.5.2 Sensitivity Considerations

A critical factor controlling the sensitivity of bioelectronic sensing systems is that the distribution of charges in an ionic medium leads to a characteristic length scale, the Debye length, which represents the distance over which a locally varying electrostatic field (due to a charged biomolecule or surface) drops to $1/e$ of its initial value.

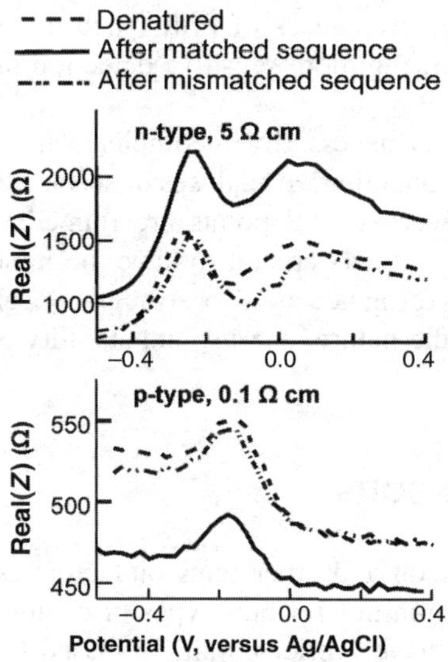

FIGURE 13.12 Electrochemical impedance spectroscopy measurements of DNA-modified silicon surfaces after exposure to complementary (matched) sequence and noncomplementary (mismatched) sequence. Data are shown on both n-type and p-type silicon samples. Binding of the negatively charged DNA molecule to its surface-bound complement increases the impedance on n-type samples and decreases the impedance on p-type samples. Figure adapted from Ref. 11.

The Debye length is given by

$$\kappa^{-1} = \left(\frac{\varepsilon_0 \varepsilon k T}{2q^2 I}\right)^{1/2}$$

where κ^{-1} is the Debye length, ε_0 is the permittivity of free space, ε is the permittivity of the aqueous medium, and q is the electron charge. $I = \sum_i c_i z_i^2$ is the ionic strength, with c_i and z_i representing the concentration and charge of each species present. At physiological salt concentrations, the Debye length is very short, approximately 0.8 nm. Consequently, in order to have high sensitivity it is crucial for the biological species to be linked very close to the semiconductor surface [6, 104]. While in principle the sensitivity can be enhanced by using lower ionic strength solutions (and thereby extending the Debye length), most biomolecular interactions in nature take place in high ionic strength environments and working in low ionic strength solutions can alter these interactions. However, a compromise can be reached, using salt concentration on the order of 10 mM, yielding a Debye length of $\sim$3 nm [6, 104]. While the requirements to achieve high sensitivity in bio-to-electronic transduction are daunting, clever approaches to achieving the transduction continue to emerge. For example, Fritz et al. detected DNA at nanomolar concentrations by using two sensors in parallel and measuring the difference between them [74]. In this case, many causes of drift (such as temperature changes) are effectively nulled out, allowing more sensitive detection of the DNA hybridization events [74].

The above discussion is necessarily incomplete, and the interested reader is referred to more thorough quantitative analyses of surface charge distributions [6, 7, 74, 93, 104, 105]. However, several points are immediately clear: First, that to achieve maximum sensitivity it is crucial to have the molecular layers as short as possible. Second, the performance is also strongly dependent on the quality of surface and in particular the nature, energy, and stability of electronic states at the surface.

13.6 NANOWIRE SENSORS

The ability to prepare silicon and other semiconductors as one-dimensional nanowires brings with it opportunities for new types of chemical biological sensors. A number of recent studies have fabricated silicon-based biosensors using Si nanowires [10, 106, 107]. Li et al. [107] made label-free DNA sensors using planar Si nanowires that were oxidized to produce a ~ 3 nm thick SiO_2 coating, followed by the use of silane chemistry. Cui et al. argued that it should be possible to achieve single-molecule detection [10]. For example, Li et al. linked short DNA oligonucleotides (12-mers) to silicon nanowires using silane chemistry [107]. By measuring the nanowire conductance and the surface photovoltage effect, they succeeded in obtaining a sensitivity of 25 pM. Moreover, they were able to easily distinguish between perfectly complementary sequences and single-base mismatches. One of the key potential advantages of nanowire sensors is that the nanowires are smaller than the thickness of the semiconductor space-charge region. Consequently, nanowire sensors might be expected to have a larger fractional change in conductivity compared with larger planar sensors. However, because of the small size of the nanowires the currents involved are very small, and great care must be taken to ensure that the measured signals arise from binding to the nanowire surface rather than to the contacts or other exposed surfaces. Additionally, while high sensitivity can be achieved, the small size of the sensor also makes the effective capture volume very small. Yet, this remains an intense area of continued activity.

13.7 SUMMARY

Covalently grafted molecular layers on semiconductors provide the opportunity to create highly stable, functional interfaces. The overall utility of these layers is often connected to their long-term stability, driving a desire for increasingly robust interfaces. Silicon, long the foundation of microelectronics, is very useful for short-term studies and is readily available, inexpensive, and can be easily processed using standard techniques of microelectronics processing. However, new applications such as environmental sensing, homeland security, and detection of water-borne pathogens continue to lead to the development of more robust materials able to withstand use for long periods of time. Thin-film diamond is one of the attractive candidates because of its outstanding mechanical properties and chemical stability. The successful integration of biological molecules with diamond, silicon, and semiconductor materials in a manner that takes advantage of the unique electronic properties of these materials is difficult, but holds great promise for new and emerging applications in chemical and biological sensing and biomedical diagnostics.

ACKNOWLEDGMENTS

The work described here was supported in part by the National Science Foundation Grants CHE-0314618, CHE-0613010, CHE-0911543, DMR-0210806, and DMR-0706559. The authors acknowledge fruitful collaborations and interactions with Dr. James Butler and Professor Lloyd M. Smith throughout this work.

REFERENCES

1. Castner, D. G.; Ratner, B. D. *Surf. Sci.* **2002**, *500*(1–3), 28.
2. Nuzzo, R. G.; Dubois, L. H.; Allara, D. L. *J. Am. Chem. Soc.* **1990**, *112*, 558.
3. Nuzzo, R. G.; Allara, D. L. *J. Am. Chem. Soc.* **1983**, *105*(13), 4481.
4. Laibinis, P. D.; Whitesides, G. M.; Allara, D. L.; Tao, Y.-T.; Parikh, A. N.; Nuzzo, R. G. *J. Am. Chem. Soc.* **1991**, *113*, 7152.
5. Bergveld, P. *IEEE Trans. Biomed. Eng.* **1970**, *17*, 70.
6. Bergveld, P. *Biosens. Bioelectron.* **1991**, *6*(1), 55.
7. Souteyrand, E.; Cloarec, J. P.; Martin, J. R.; Wilson, C.; Lawrence, I.; Mikkelsen, S.; Lawrence, M. F. *J. Phys. Chem. B* **1997**, *101*(15), 2980.
8. Souteyrand, E.; Martin, J. R.; Martelet, C. *Sens. Actuators B* **1994**, *20*, 63.
9. Beebe, D. J.; Mensing, G. A.; Walker, G. M. *Annu. Rev. Biomed. Eng.* **2002**, *4*, 261.
10. Cui, Y.; Wei, Q. Q.; Park, H. K.; Lieber, C. M. *Science* **2001**, *293*(5533), 1289.
11. Cai, W.; Peck, J. R.; van der Weide, D. W.; Hamers, R. J. *Biosens. Bioelectron.* **2004**, *19*(9), 1013.
12. Kar, S.; Miramond, C.; Vuillaume, D. *Appl. Phys. Lett.* **2001**, *78*(9), 1288.
13. Casey, W. H.; Lasaga, A. C.; Gibbs, G. V. *Geochim. Cosmochim. Acta* **1990**, *54*(12), 3369.
14. Sjoberg, S. *J. Non-Cryst. Solids* **1996**, *196*, 51.
15. Chazalviel, J. N. *J. Phys. IV* **1994**, *4(C1)*, 117.
16. Watanabe, S. *Surf. Sci.* **1995**, *341*(3), 304.
17. Fukidome, H.; Pluchery, O.; Queeney, K. T.; Caudano, Y.; Raghavachari, K.; Weldon, M. K.; Chaban, E. E.; Christman, S. B.; Kobayashi, H.; Chabal, Y. J. *Surf. Sci.* **2002**, *502*, 498.

18. Shimojima, A.; Sugahara, Y.; Kuroda, K. *Bull. Chem. Soc. Jpn.* **1997**, *70*(11), 2847.
19. Tada, H.; Nagayama, H. *Langmuir* **1995**, *11*(1), 136.
20. Morita, M.; Ohmi, T.; Hasegawa, E.; Kawakami, M.; Ohwada, M. *J. Appl. Phys.* **1990**, *68* (3), 1272.
21. Weldon, M. K.; Queeney, K. T.; Chabal, Y. J.; Stefanov, B. B.; Raghavachari, K. *J. Vac. Sci. Technol. B* **1999**, *17*(4), 1795.
22. Uno, T.; Tabata, H.; Kawai, T. *Anal. Chem.* **2007**, *79*(1), 52.
23. Netzer, L.; Iscovici, R.; Sagiv, J. *Thin Solid Films* **1983**, *100*(1), 67.
24. Netzer, L.; Iscovici, R.; Sagiv, J. *Thin Solid Films* **1983**, *99*(1–3), 235.
25. Ulman, A. *Adv. Mater.* **1990**, *2*(12), 573.
26. Brzoska, J. B.; Benazouz, I.; Rondelez, F. *Langmuir* **1994**, *10*(11), 4367.
27. Fadeev, A. Y.; McCarthy, T. J. *Langmuir* **2000**, *16*(18), 7268.
28. Faucheux, N.; Schweiss, R.; Lutzow, K.; Werner, C.; Groth, T. *Biomaterials 2004 25*(14), **2721**.
29. Halliwell, C. M.; Cass, A. E. G. *Anal. Chem.* **2001**, *73*(11), 2476.
30. Kallury, K. M. R.; Krull, U. J.; Thompson, M. *Anal. Chem.* **1988**, *60*(2), 169.
31. Chrisey, L. A.; Lee, G. U.; O'Ferrall, E. O. *Nucleic Acids Res.* **1996**, *24*, 3031.
32. Ulman, A. *Chem. Rev.* **1996**, *96*(4), 1533.
33. Angst, D. L.; Simmons, G. W. *Langmuir* **1991**, *7*(10), 2236.
34. Wasserman, S. R.; Tao, Y. T.; Whitesides, G. M. *Langmuir 1989 5*(4), **1074**.
35. Dugas, V.; Chevalier, Y. *J. Colloid Interface Sci.* **2003**, *264*(2), 354.
36. Steinem, C.; Janshoff, A.; Lin, V. S. Y.; Volcker, N. H.; Ghadiri, M. R. *Tetrahedron* **2004**, *60*(49), 11259.
37. Olmos, D.; Gonzalez-Benito, J.; Aznar, A. J.; Baselga, J. *J. Mater. Process. Technol.* **2003**, *143*, 82.
38. Cypryk, M.; Apeloig, Y. *Organometallics* **2002**, *21*(11), 2165.
39. Adadi, R.; Zorn, G.; Brener, R.; Gotman, I.; Gutmanas, E. Y.; Sukenik, C. N. *Thin Solid Films 2009 518*(8), **1966**.
40. Danahy, M. P.; Avaltroni, M. J.; Midwood, K. S.; Schwarzbauer, J. E.; Schwartz, J. *Langmuir* **2004**, *20*(13), 5333.
41. Gawalt, E. S.; Avaltroni, M. J.; Koch, N.; Schwartz, J. *Langmuir* **2001**, *17*(19), 5736.
42. Helmy, R.; Fadeev, A. Y. *Langmuir* **2002**, *18*(23), 8924.
43. Hanson, E. L.; Schwartz, J.; Nickel, B.; Koch, N.; Danisman, M. F. *J. Am. Chem. Soc.* **2003**, *125*(51), 16074.
44. Gawalt, E. S.; Avaltroni, M. J.; Danahy, M. P.; Silverman, B. M.; Hanson, E. L.; Midwood, K. S.; Schwarzbauer, J. E.; Schwartz, J. *Langmuir* **2003**, *19*(1), 200.
45. Gouzman, I.; Dubey, M.; Carolus, M. D.; Schwartz, J.; Bernasek, S. L. *Surf. Sci.* **2006**, *600*(4), 773.
46. Blattler, T. M.; Binkert, A.; Zimmermann, M.; Textor, M.; Voros, J.; Reimhult, E. *Nanotechnology* **2008**, *19(7)*
47. Crandall, J. K.; Morelfourrier, C. *J. Organomet. Chem.* **1995**, *489*(1–2), 5.
48. Dumas, P.; Chabal, Y. J.; Jakob, P. *Surf. Sci.* **1992**, *269*, 867.
49. Jakob, P.; Chabal, Y. J. *J. Chem. Phys.* **1991**, *95*(4), 2897.

50. Linford, M. R.; Fenter, P.; Eisenberger, P. M.; Chidsey, C. E. D. *J. Am. Chem. Soc.* **1995**, *117*, 3145.

51. Linford, M. R.; Chidsey, C. E. D. *J. Am. Chem. Soc.* **1993**, *115*, 12631.

52. Cicero, R. L.; Linford, M. R.; Chidsey, C. E. D. *Langmuir* **2000**, *16*, 5688.

53. Strother, T.; Hamers, R. J.; Smith, L. M. *Nucleic Acids Res.* **2000**, *28*(18), 3535.

54. Strother, T.; Cai, W.; Zhao, X.; Hamers, R. J.; Smith, L. M. *J. Am. Chem. Soc.* **2000**, *122*(6), 1205.

55. Stewart, M. P.; Buriak, J. M. *J. Am. Chem. Soc.* **2001**, *123*(32), 7821.

56. Stewart, M. P.; Buriak, J. M. *Angew. Chem., Int. Ed.* **1998**, *37*(23), 3257.

57. Colavita, P. E.; Sun, B.; Tse, K. Y.; Hamers, R. J. *J. Am. Chem. Soc.* **2007**, *129*(44), 13554.

58. Franking, R. A.; Landis, E. C.; Hamers, R. J. *Langmuir* **2009**, *25*(18), 10676.

59. Sieval, A. B.; Vleeming, V.; Zuilhof, H.; Sudholter, E. J. R. *Langmuir* **1999**, *15*(23), 8288.

60. Sieval, A. B.; Linke, R.; Heij, G.; Meijer, G.; Zuilhof, H.; Sudholter, E. J. R. *Langmuir* **2001**, *17*(24), 7554.

61. Sieval, A. B.; Demirel, A. L.; Nissink, J. W. M.; Linford, M. R.; van der Maas, J. H.; de Jeu, W. H.; Zuilhof, H.; Sudholter, E. J. R. *Langmuir 1998 14*(7), **1759**.

62. Boukherroub, R.; Wojtyk, J. T. C.; Wayner, D. D. M.; Lockwood, D. J. *J. Electrochem. Soc.* **2002**, *149*(2), H59.

63. Boukherroub, R.; Wayner, D. D. M. *J. Am. Chem. Soc.* **1999**, *121*(49), 11513.

64. Boukherroub, R.; Morin, S.; Bensebaa, F.; Wayner, D. D. M. *Langmuir* **1999**, *15*(11), 3831.

65. Pike, A. R.; Lie, L. H.; Eagling, R. A.; Ryder, L. C.; Patole, S. N.; Connolly, B. A.; Horrocks, B. R.; Houlton, A. *Angew. Chem., Int. Ed.* **2002**, *41*(4), 615.

66. Lasseter, T. L.; Clare, B. H.; Abbott, N. L.; Hamers, R. J. *J. Am. Chem. Soc.* **2004**, *126*(33), 10220.

67. Clare, T. L.; Clare, B. H.; Nichols, B. M.; Abbott, N. L.; Hamers, R. J. *Langmuir* **2005**, *21*(14), 6344.

68. Hamers, R. J.; Chambers, S. A.; Evans, P. E.; Franking, R.; Gerbec, Z.; Gopalan, P.; Kim, H.; Landis, E. C.; Li, B.; McCoy, M. W.; Ohsawa, T.; Ruther, R. *Physica Status Solidi c* **2009**, *7*, 200.

69. Li, B.; Franking, R.; Landis, E. C.; Kim, H.; Hamers, R. J. *ACS Appl. Mater. Interfaces 2009*, *1*(5), **1013**.

70. Böcking, T.; Kilian, K. A.; Gaus, K.; Gooding, J. J. *Adv. Funct. Mater.* **2008**, *18*(23), 3827.

71. Lin, Z.; Strother, T.; Cai, W.; Cao, X.; Smith, L. M.; Hamers, R. J. *Langmuir* **2002**, *18*, 788.

72. Sachidanandam, R.; Weissman, D.; Schmidt, S. C.; Kakol, J. M.; Stein, L. D.; Marth, G.; Sherry, S.; Mullikin, J. C.; Mortimore, B. J.; Willey, D. L.; Hunt, S. E.; Cole, C. G.; Coggill, P. C.; Rice, C. M.; Ning, Z. M.; Rogers, J.; Bentley, D. R.; Kwok, P. Y.; Mardis, E. R.; Yeh, R. T.; Schultz, B.; Cook, L.; Davenport, R.; Dante, M.; Fulton, L.; Hillier, L.; Waterston, R. H.; McPherson, J. D.; Gilman, B.; Schaffner, S.; Van Etten, W. J.; Reich, D.; Higgins, J.; Daly, M. J.; Blumenstiel, B.; Baldwin, J.; Stange-Thomann, N. S.; Zody, M. C.; Linton, L.; Lander, E. S.; Altshuler, D.; Int, S. N. P. M. W. G. *Nature* **2001**, *409*(6822), 928.

73. Venter, J. C.; Adams, M. D.; Myers, E. W.; Li, P. W.; Mural, R. J.; Sutton, G. G.; Smith, H. O.; Yandell, M.; Evans, C. A.; Holt, R. A.; Gocayne, J. D.; Amanatides, P.; Ballew, R. M.; Huson, D. H.; Wortman, J. R.; Zhang, Q.; Kodira, C. D.; Zheng, X. Q. H.; Chen, L.; Skupski, M.; Subramanian, G.; Thomas, P. D.; Zhang, J. H.; Miklos, G. L. G.; Nelson, C.; Broder, S.; Clark, A. G.; Nadeau, C.; McKusick, V. A.; Zinder, N.; Levine, A. J.; Roberts, R. J.; Simon, M.; Slayman, C.; Hunkapiller, M.; Bolanos, R.; Delcher, A.; Dew, I.; Fasulo, D.; Flanigan, M.; Florea, L.; Halpern, A.; Hannenhalli, S.; Kravitz, S.; Levy, S.; Mobarry, C.; Reinert, K.; Remington, K.; Abu-Threideh, J.; Beasley, E.; Biddick, K.; Bonazzi, V.; Brandon, R.; Cargill, M.; Chandramouliswaran, I.; Charlab, R.; Chaturvedi, K.; Deng, Z. M.; Di Francesco, V.; Dunn, P.; Eilbeck, K.; Evangelista, C.; Gabrielian, A. E.; Gan, W.; Ge, W. M.; Gong, F. C.; Gu, Z. P.; Guan, P.; Heiman, T. J.; Higgins, M. E.; Ji, R. R.; Ke, Z. X.; Ketchum, K. A.; Lai, Z. W.; Lei, Y. D.; Li, Z. Y.; Li, J. Y.; Liang, Y.; Lin, X. Y.; Lu, F.; Merkulov, G. V.; Milshina, N.; Moore, H. M.; Naik, A. K.; Narayan, V. A.; Neelam, B.; Nusskern, D.; Rusch, D. B.; Salzberg, S.; Shao, W.; Shue, B. X.; Sun, J. T.; Wang, Z. Y.; Wang, A. H.; Wang, X.; Wang, J.; Wei, M. H.; Wides, R.; Xiao, C. L.; Yan, C. H.; Yao, A.; Ye, J.; Zhan, M.; Zhang, W. Q.; Zhang, H. Y.; Zhao, Q.; Zheng, L. S.; Zhong, F.; Zhong, W. Y.; Zhu, S. P. C.; Zhao, S. Y.; Gilbert, D.; Baumhueter, S.; Spier, G.; Carter, C.; Cravchik, A.; Woodage, T.; Ali, F.; An, H. J.; Awe, A.; Baldwin, D.; Baden, H.; Barnstead, M.; Barrow, I.; Beeson, K.; Busam, D.; Carver, A.; Center, A.; Cheng, M. L.; Curry, L.; Danaher, S.; Davenport, L.; Desilets, R.; Dietz, S.; Dodson, K.; Doup, L.; Ferriera, S.; Garg, N.; Gluecksmann, A.; Hart, B.; Haynes, J.; Haynes, C.; Heiner, C.; Hladun, S.; Hostin, D.; Houck, J.; Howland, T.; Ibegwam, C.; Johnson, J.; Kalush, F.; Kline, L.; Koduru, S.; Love, A.; Mann, F.; May, D.; McCawley, S.; McIntosh, T.; McMullen, I.; Moy, M.; Moy, L.; Murphy, B.; Nelson, K.; Pfannkoch, C.; Pratts, E.; Puri, V.; Qureshi, H.; Reardon, M.; Rodriguez, R.; Rogers, Y. H.; Romblad, D.; Ruhfel, B.; Scott, R.; Sitter, C.; Smallwood, M.; Stewart, E.; Strong, R.; Suh, E.; Thomas, R.; Tint, N. N.; Tse, S.; Vech, C.; Wang, G.; Wetter, J.; Williams, S.; Williams, M.; Windsor, S.; Winn-Deen, E.; Wolfe, K.; Zaveri, J.; Zaveri, K.; Abril, J. F.; Guigo, R.; Campbell, M. J.; Sjolander, K. V.; Karlak, B.; Kejariwal, A.; Mi, H. Y.; Lazareva, B.; Hatton, T.; Narechania, A.; Diemer, K.; Muruganujan, A.; Guo, N.; Sato, S.; Bafna, V.; Istrail, S.; Lippert, R.; Schwartz, R.; Walenz, B.; Yooseph, S.; Allen, D.; Basu, A.; Baxendale, J.; Blick, L.; Caminha, M.; Carnes-Stine, J.; Caulk, P.; Chiang, Y. H.; Coyne, M.; Dahlke, C.; Mays, A. D.; Dombroski, M.; Donnelly, M.; Ely, D.; Esparham, S.; Fosler, C.; Gire, H.; Glanowski, S.; Glasser, K.; Glodek, A.; Gorokhov, M.; Graham, K.; Gropman, B.; Harris, M.; Heil, J.; Henderson, S.; Hoover, J.; Jennings, D.; Jordan, C.; Jordan, J.; Kasha, J.; Kagan, L.; Kraft, C.; Levitsky, A.; Lewis, M.; Liu, X. J.; Lopez, J.; Ma, D.; Majoros, W.; McDaniel, J.; Murphy, S.; Newman, M.; Nguyen, T.; Nguyen, N.; Nodell, M.; Pan, S.; Peck, J.; Peterson, M.; Rowe, W.; Sanders, R.; Scott, J.; Simpson, M.; Smith, T.; Sprague, A.; Stockwell, T.; Turner, R.; Venter, E.; Wang, M.; Wen, M. Y.; Wu, D.; Wu, M.; Xia, A.; Zandieh, A.; Zhu, X. H. *Science* **2001**, *291*(5507), 1304.
74. Fritz, J.; Cooper, E. B.; Gaudet, S.; Sorger, P. K.; Manalis, S. R. *Proc. Natl. Acad. Sci. USA* **2002**, *99*(22), 14142.
75. Robertson, J. *Mater. Sci. Eng. R* **2002**, *37*(4–6), 129.
76. Yang, W.; Hamers, R. J. *Appl. Phys. Lett.* **2004**, *85*, 3626.

77. Sun, B.; Colavita, P. E.; Kim, H.; Lockett, M.; Marcus, M. S.; Smith, L. M.; Hamers, R. J. *Langmuir* **2006**, *22*(23), 9598.
78. Lockett, M. R.; Shortreed, M. R.; Smith, L. M. *Langmuir* **2008**, *24*(17), 9198.
79. Thoms, B. D.; Owens, M. S.; Butler, J. E.; Spiro, C. *Appl. Phys. Lett.* **1994**, *65*, 2957.
80. Allongue, P.; Delamar, M.; Desbat, B.; Fagebaume, O.; Hitmi, R.; Pinson, J.; Savéant, J. M. *J. Am. Chem. Soc.* **1997**, *119*(1), 201.
81. Kuo, T. C.; McCreery, R. L.; Swain, G. M. *Electrochem. Solid State Lett.* **1999**, *2*, 288.
82. Jian, W.; Firestone, M. A.; Auciello, O.; Carlisle, J. A. *Langmuir* **2004**, *20*(26), 11450.
83. Yang, W. S.; Baker, S. E.; Butler, J. E.; Lee, C. S.; Russell, J. N.; Shang, L.; Sun, B. *Chem. Mater.* **2005**, *17*(5), 938.
84. Strother, T.; Knickerbocker, T.; Russell, J. N.; Butler, J. E.; Smith, L. M.; Hamers, R. J. *Langmuir* **2002**, *18*(4), 968.
85. Yang, W. S.; Auciello, O.; Butler, J. E.; Cai, W.; Carlisle, J. A.; Gerbi, J. E.; Gruen, D. M.; Knickerbocker, T.; Lasseter, T. L.; Russell, J. N., Jr., ; Smith, L. M.; Hamers, R. J. *Nat. Mater.* **2002**, *1*(4), 253–257.
86. Prime, K. L.; Whitesides, G. M. *J. Am. Chem. Soc.* **1993**, *115*(23), 10714.
87. Ostuni, E.; Chapman, R. G.; Holmlin, R. E.; Takayama, S.; Whitesides, G. M. *Langmuir* **2001**, *17*(18), 5605.
88. Pale-Grosdemange, C.; Simon, E. S.; Prime, K. L.; Whitesides, G. M. *J. Am. Chem. Soc.* **1991**, *113*(1), 12.
89. Herrwerth, S.; Eck, W.; Reinhardt, S.; Grunze, M. *J. Am. Chem. Soc.* **2003**, *125*(31), 9359.
90. Harder, P.; Grunze, M.; Dahint, R.; Whitesides, G. M.; Laibinis, P. E. *J. Phys. Chem. B* **1998**, *102*(2), 426.
91. Stavis, C.; Lasseter Clare, C.; Butler, J. E.; Radadia, A. D.; Carr, R.; Zeng, H.; King, W.; Carlisle, J. A.; Aksimentiev, A.; Bashir, R.; Hamers, R. J. *Proc. Natl. Acad. Sci. USA* **2011**, *108*(3), 983.
92. Hartl, A.; Schmich, E.; Garrido, J. A.; Hernando, J.; Catharino, S. C. R.; Walter, S.; Feulner, P.; Kromka, A.; Steinmuller, D.; Stutzmann, M. *Nat. Mater.* **2004**, *3*(10), 736.
93. Schasfoort, R. B. M.; Kooyman, R. P. H.; Bergveld, P.; Greve, J. *Biosens. Bioelectron.* **1990**, *5*(2), 103.
94. Kawarada, H.; Araki, Y.; Sakai, T.; Ogawa, T.; Umezawa, H. *Physica Status Solidi a* **2001**, *185*(1), 79.
95. Kharitonov, A. B.; Wasserman, J.; Katz, E.; Willner, I. *J. Phys. Chem. B* **2001**, *105*(19), 4205.
96. Kim, D. S.; Jeong, Y. T.; Lyu, H. K.; Park, H. J.; Kim, H. S.; Shin, J. K.; Choi, P.; Lee, J. H.; Lim, G.; Ishida, M. *Jpn. J. Appl. Phys. Part 1* **2003**, *42*(6B), 4111.
97. Gonzalez-Moreno, R.; Cook, P. L.; Zegkinoglou, L., Liu, X.; Johnson, P. S.; Yang, W. L.; Ruther, R. E.; Hamers, R. J.; Tena-Zaera, R.; Himpsel, F. J.; Ortega, J. E.; Rogero, C. *J. Phys. Chem. C.* **2011**, *115*(37), 18195.
98. Lasseter, T. L.; Cai, W.; Hamers, R. J. *Analyst* **2004**, *129*(1), 3.
99. Yang, W. S.; Butler, J. E.; Russell, J. N., Jr., ; Hamers, R. J. *Langmuir* **2004**, *20*(16), 6778.

100. Yang, W. S.; Butler, J. E.; Russell, J. N.; Hamers, R. J. *Analyst* **2007**, *132*(4), 296.

101. Klein, D. L.; Roth, R.; Lim, A. K. L.; Alivisatos, A. P.; McEuen, P. L. *Nature* **1997**, *389* (6652), 699.

102. Kim, D. S.; Jeong, Y. T.; Park, H. J.; Shin, J. K.; Choi, P.; Lee, J. H.; Lim, G. *Biosens. Bioelectron.* **2004**, *20*(1), 69.

103. Song, K.-S.; Sakai, T.; Kanazawa, H.; Araki, Y.; Umezawa, H.; Tachiki, M.; Kawarada, H. *Biosens. Bioelectron.* **2003**, *19*, 137.

104. Schasfoort, R. B. M.; Bergveld, P.; Kooyman, R. P. H.; Greve, J. *Anal. Chim. Acta* **1990**, *238*(2), 323.

105. Bergveld, P. *Sens. Actuators A* **1996**, *56*(1–2), 65.

106. Gao, Z. Q.; Agarwal, A.; Trigg, A. D.; Singh, N.; Fang, C.; Tung, C. H.; Fan, Y.; Buddharaju, K. D.; Kong, J. M. *Anal. Chem.* **2007**, *79*(9), 3291.

107. Li, Z.; Chen, Y.; Li, X.; Kamins, T. I.; Nauka, K.; Williams, R. S. *Nano Lett.* **2004**, *4*(2), 245.

CHAPTER 14

Perspective and Challenge

FRANKLIN (FENG) TAO AND STEVEN L. BERNASEK

In the past two decades, research in the area of functionalization of semiconductor surfaces has elucidated reaction mechanisms of organic molecules on semiconductor surfaces in many examples. These studies have also led to designed semiconductor surfaces tethered with functional organic materials and biospecies, and have demonstrated molecule-based semiconductor devices and biosensing techniques. However, there remain many open questions.

In order to grow multilayer, three-dimensional, organic architectures, or to immobilize biospecies on semiconductor surfaces, a functionalized surface with outward facing functional groups is necessary. A chemically homogeneous organic monolayer is crucial for significant signal response. To form such a surface with a single functional group facing outward, a bifunctional molecule is typically used. In fact, obtaining methods that exhibit a high selectivity for attachment of only one functionality of the bifunctional molecule to the semiconductor surface is crucial. A protection group to passivate one of the two functional groups can be used in the direct reaction with the semiconductor surfaces. This would be followed by activation of the protected group in the attachment of the second layer of organic molecules. This approach can enhance surface selectivity for some bifunctional molecules. However, the selection of a specific protection group is challenging, and this extra step can often result in damage to the desired functionality. Thus, actually this strategy largely limits the growth of semiconductor surface with pure upward functional group for further applications. More efforts in developing new strategies are necessary.

Many previous studies, including mechanistic studies of organic molecules and the immobilization of organic materials and biospecies, have been performed at room temperature or low temperature. In practice, operational conditions of molecular electronics devices and biosensors at temperatures higher than room temperature are unavoidable. Thus, studies of the thermal evolution of these systems are a piece of

Functionalization of Semiconductor Surfaces, First Edition.
Edited by Franklin (Feng) Tao and Steven L. Bernasek.

information that is important but absent so far for many real applications. In addition, how ambient environments (air, humidity) impact chemical stability of the functionalized surfaces is another important issue to be addressed. Many of the modification bonds (Si–C, Si–O, or Si–N linkages) are actually sensitive to reactive gases such as water vapor and O_2. How such a "slow" but "accumulating" chemical reaction degrades the function of molecule-based devices and sensors should be addressed for real applications in the near future. *In situ* studies of these functionalized semiconductor surfaces under reaction conditions were almost impossible in the past decades. However, the recent advance in *in situ* techniques capable of working under ambient conditions or reactive conditions has made such studies possible. For example, ambient pressure XPS allows the possibility to measure surface composition and study oxidation state and identify chemical environment of surfaces under a pressure of gaseous reactant of 10 Torr or under a liquid layer with a thickness of a few nanometers.

In addition, most of the previous studies were performed on flat homogeneous surfaces. With the advance of nanolithography, functionalization of semiconductor surfaces at micro- or even nanoscale can be realized by using nanopatterned substrates. Through selective passivation of a semiconductor surface, organic molecules can be selectively attached to unpassivated surfaces to form organic or bio-nanopatterns. Such nanostructures could provide spatial selectivity in biosensing technology. This could be an exciting area in the development of biosensing techniques and biomimetic devices.

Another exciting future area in the functionalization of semiconductor surfaces is the incorporation of synthesis of nanoparticles through colloidal chemistry into functions of semiconductor surfaces. Metal nanoparticles are typically coated with a layer of surfactant such as cationic acetyl trimethyl ammonium bromide (C-TAB). The outward functional group of surfactant molecules on metal nanoparticles could be used as a linkage to covalently bind to the outward functional groups tethered on semiconductor surfaces, modifying the semiconductor surface and bringing size-dependent physical, optoelectronic, and chemical properties to the functionalized semiconductor surfaces. This also provides another platform for functionalization of the semiconductor surface since many unique chemical and physical properties of metal or semiconductor nanoparticles could be introduced to the semiconductor surface system.

In any event, there are a number of areas for further research in this exciting field of materials chemistry. This book has provided a survey and review of much of the work that has been carried out in the field of semiconductor surface modification. Applications in organic and eventually molecular electronics, surface modification of the biological and mechanical properties of the semiconductor surface, and the development of chemical and biological sensor devices are apparent in the work described here. There remains much to be done, however, as we have indicated only briefly in these concluding statements. The field of semiconductor surface functionalization and modification will continue to hold excitement for researchers for many years to come.

INDEX

Functionalization of Semiconductor Surfaces, First Edition.
Edited by Franklin (Feng) Tao and Steven L. Bernasek.